PHARMACEUTICAL AMORPHOUS SOLID DISPERSIONS

PHARMACEUTICAL AMORPHOUS SOLID DISPERSIONS

Edited by

Ann Newman
Seventh Street Development Group
Lafayette, Indiana, USA

WILEY

Published by John Wiley & Sons, Inc., Hoboken, New Jersey.
Published simultaneously in Canada.

***Library of Congress Cataloging-in-Publication Data*:**

Pharmaceutical amorphous solid dispersions/edited by Ann Newman.
p.; cm.
Includes index.
ISBN 978-1-118-45520-3 (cloth)
I. Newman, Ann (Pharmaceutical scientist), editor.
[DNLM: 1. Polymers–pharmacokinetics. 2. Technology, Pharmaceutical. 3. Absorption. 4. Drug Carriers–pharmacokinetics. 5. Polymers–chemistry. 6. Solubility. QV 778]
RM301.5
615.7–dc23

2014023090

Printed in the United States of America.

10 9 8 7 6 5 4 3 2 1

This book is dedicated to George Zografi who has been a great source of inspiration to me and all scientists who have worked in this field. This book is also in memory of Marcus Brewster who was a gifted scientist and a gentle giant in this field.

CONTENTS

PREFACE

As the field of pharmaceutical amorphous solid dispersions expands and provides a development avenue for poorly soluble compounds, it seemed to be the right time to collect detailed information on these materials for scientists working in the field. The field is growing so rapidly and the number of papers is so numerous that it is hard to keep up with advances in the field. The goal was to pull together chapters that hit the major areas involved in developing pharmaceutical amorphous solid dispersions, and provide a roadmap from early to late development for those new to the field or for more experienced scientists who may be looking for another approach. Authors from pharmaceutical companies, academia, contract laboratories, and consulting are all included to provide a wide range of views on the development of dispersions.

This book attempts to follow the key development areas from early to late stages. The introduction to amorphous solids and theory (Chapter 1) provides the evolution of dispersions from amorphous drug substance and sets the stage for later chapters. Polymers and surfactants are covered next (Chapter 2) to remind readers that the excipient properties are critical in producing a viable dispersion. The methods and techniques for screening and selection (Chapter 3) are included since every compound will need a tailored plan to find the best dispersion and there are numerous ways to conduct these studies. Characterization (Chapter 4) is an important tool not only to understand what has been made but also to determine how your dispersion may change over time (Chapter 5). The large amount of polymer in the dispersions can influence key properties, such as dissolution and solubility (Chapter 6), and hopefully the excipient will increase supersaturation and help prevent crystallization from solution. Formulating dispersions for early animal and clinical testing (Chapter 7) can be simple or complex depending on the properties of your material. Early animal and bioavailability studies (Chapter 8) are key in determining whether the solubility challenges have been overcome using a dispersion and whether it is a viable option for later development. If a dispersion is moved forward, it has to be produced at large scale, with spray drying (Chapter 9) and melt extrusion (Chapter 10) being the most common choices to date. Once material is available, formulating the dispersion for late clinical trial supplies and as a marketed product (Chapter 11) has its challenges to maintain the amorphous nature and solubility advantages in the long term. Regulatory (Chapter 12) and intellectual properties (Chapter 13) are also considerations from early to late development of amorphous solid dispersions. A compilation of common polymers and surfactants and their properties is provided for easy reference (Chapter 14). Lists of polymers sorted by various properties (such as glass transition temperature, solubility parameter, or molecular weight) are available (Appendix A) to readily find the best polymers for your system.

I hope that the book also shows the interdisciplinary approach that is needed for the research and development of an amorphous dispersion product. Throughout the process, a group of scientists from various departments and fields is needed to help guide the project and address the inevitable challenges that will be encountered along the way.

I would also like to thank all the authors for contributing wonderful chapters in the various areas. It was a pleasure to work with all of them and to learn from their experiences. Finally, I would like to thank George Zografi for acting as a sounding board while the book was being shaped in the early stages and for providing invaluable advice along the way. This book is dedicated to George who has inspired me and many scientists in the field.

Enjoy!

Lafayette, IN
January 2015

Ann Newman

CONTRIBUTORS

Patrick Augustijns, Laboratory for Drug Delivery and Disposition, KU Leuven, Leuven, Belgium

Annette Bak, Discovery Pharmaceutical Sciences, Merck & Co., Kenilworth, NJ, USA

John M. Baumann, Bend Research, Inc., Bend, OR, USA

Jan Bevernage, Drug Evaluation—Pharmaceutical Sciences, Johnson & Johnson Pharmaceutical Research and Development, Janssen Pharmaceutica, Beerse, Belgium

Meinolf Brackhagen, The Dow Chemical Company, Midland, MI, USA

Philip Bransford, Materials Discovery and Characterization, Vertex Pharmaceuticals Incorporated, Boston, MA, USA

Marcus E. Brewster, Drug Evaluation—Pharmaceutical Sciences, Johnson & Johnson Pharmaceutical Research and Development, Janssen Pharmaceutica, Beerse, Belgium

Joachim Brouwers, Laboratory for Drug Delivery and Disposition, KU Leuven, Leuven, Belgium

Patrick R. Connelly, Materials Discovery and Characterization, Vertex Pharmaceuticals Incorporated, Boston, MA, USA

Kieran Crowley, Quotient Clinical Ltd., Nottingham, U.K.

Daniel E. Dobry, Bend Research, Inc., Bend, OR, USA

Majed Fawaz, Materials Discovery and Characterization, Vertex Pharmaceuticals Incorporated, Boston, MA, USA

Andreas Gryczke, Global Development and Technical Marketing Solubilisation, BASF SE, Ludwigshafen, Germany

Abhay Gupta, Division of Product Quality Research, Office of Pharmaceutical Science, U.S. Food and Drug Administration, Silver Spring, MD, USA

Patricia Hurter, Global Pharmaceutical Development and Regulatory Affairs, Vertex Pharmaceuticals Incorporated, Boston, MA, USA

Grace A. Ilevbare, Discovery Pharmaceutical Sciences, Merck & Co., Rahway, NJ, USA

Christopher T. John, Discovery Pharmaceutical Sciences, Merck & Co., West Point, PA, USA

Steve Johnston, Materials Discovery and Characterization, Vertex Pharmaceuticals Incorporated, Boston, MA, USA

Jeff Katstra, Formulation Development, Vertex Pharmaceuticals Incorporated, Cambridge, MA, USA

Mansoor A. Khan, Division of Product Quality Research, Office of Pharmaceutical Science, U.S. Food and Drug Administration, Silver Spring, MD, USA

Jesse L. Kuiper, Analytical Sciences, Merck & Co., West Point, PA, USA

Anuj Kuldipkumar, Materials Discovery and Characterization, Vertex Pharmaceuticals Incorporated, Boston, MA, USA

Jeffrey A. Lindeman, J.A. Lindeman & Co. PLLC, Falls Church, VA, USA

Xia Lu, Crystal Pharmatech, Suzhou, China

Praveen Mudunuri, Materials Discovery and Characterization, Vertex Pharmaceuticals Incorporated, Boston, MA, USA

Padma Narayan, The Dow Chemical Company, Midland, MI, USA

Ann Newman, Seventh Street Development Group, Lafayette, IN, USA

James D. Ormes, Discovery Pharmaceutical Sciences, Merck & Co., Rahway, NJ, USA

Andrey Peresypkin, Materials Discovery and Characterization, Vertex Pharmaceuticals Incorporated, Boston, MA, USA

William W. Porter III, The Dow Chemical Company, Midland, MI, USA

Brian Patrick Quinn, Parthenon - Ernst & Young, New York, NY, USA

Ziyaur Rahman, Division of Product Quality Research, Office of Pharmaceutical Science, U.S. Food and Drug Administration, Silver Spring, MD, USA

Setu Roday, Materials Discovery and Characterization, Vertex Pharmaceuticals Incorporated, Boston, MA, USA

Bill Rowe, Formulation Development, Vertex Pharmaceuticals Incorporated, Boston, MA, USA

Tapan Sanghvi, Formulation Development, Vertex Pharmaceuticals Incorporated, Boston, MA, USA

Dana M. Settell, Bend Research, Inc., Bend, OR, USA

Phillip Snyder, Materials Discovery and Characterization, Vertex Pharmaceuticals Incorporated, Boston, MA, USA

Lynne S. Taylor, Purdue University, West Lafayette, IN, USA

Allen C. Templeton, Analytical Sciences, Merck & Co., Kenilworth, NJ, USA

Hayden Thomas, Formulation Development, Vertex Pharmaceuticals Incorporated, Boston, MA, USA

Christopher Tucker, The Dow Chemical Company, Midland, MI, USA

Guy Van den Mooter, Laboratory for Drug Delivery and Disposition, KU Leuven, Leuven, Belgium

Geert Verreck, Drug Evaluation—Pharmaceutical Sciences, Johnson & Johnson Pharmaceutical Research and Development, Janssen Pharmaceutica, Beerse, Belgium

Frederick G. Vogt, Morgan, Lewis & Bockius LLP, Philadelphia, PA, USA

Hong-Ren Wang, Materials Discovery and Characterization, Vertex Pharmaceuticals Incorporated, Boston, MA, USA

Robert Wenslow, Crystal Pharmatech, North Brunswick, NJ, USA

Wei Xu, Formulation Sciences, Merck & Co., West Point, PA, USA

George Zografi, University of Wisconsin-Madison, Madison, WI, USA

1

INTRODUCTION TO AMORPHOUS SOLID DISPERSIONS

George Zografi[1] and Ann Newman[2]

[1]*University of Wisconsin-Madison, Madison, WI, USA*
[2]*Seventh Street Development Group, Lafayette, IN, USA*

1.1 INTRODUCTION

Over the years one of the major goals of synthetic chemists has been to provide the crystalline form of any active pharmaceutical ingredient (API) being introduced into pharmaceutical development. This is primarily because the symmetrical three-dimensional long-range order and the relatively tight packing of molecules in a crystal lattice most often ensure a high level of chemical purity and solid-state stability. At the same time, an API being developed for oral administration in a solid dosage form generally requires sufficient aqueous solubility upon contact with *in vitro* and *in vivo* dissolution media in order to obtain optimal rates of dissolution and acceptable oral bioavailability. The importance of aqueous solubility in affecting dissolution rates can be shown with the classical Noyes–Whitney equation [1]:

$$\mathrm{d}C/\mathrm{d}t = k_D A(C_s - C_t), \tag{1.1}$$

where $\mathrm{d}C/\mathrm{d}t$ is the dissolution rate, k_D is the dissolution rate constant (dependent on the stirring rate and the diffusion constant), A is the total surface area of the drug particles, C_s

Pharmaceutical Amorphous Solid Dispersions, First Edition. Edited by Ann Newman.

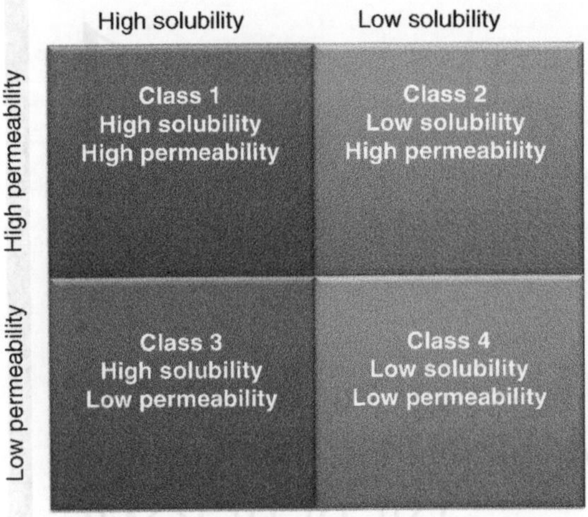

Figure 1.1. Biopharmaceutical classification systems (adapted from Ref. 2).

is the aqueous saturation solubility of the drug, and C_t is the concentration dissolved at time t. Based on this equation, it can be seen that all other factors being constant, the rate of dissolution is proportional to the surface area of the solute particle and to the solubility of the drug. Consequently, drugs with low aqueous solubility would be expected to exhibit low dissolution rates and, likely, poor oral bioavailability. The importance of the rate of dissolution and hence aqueous solubility in acting as a determinant of oral absorption was formally recognized with the establishment of the Biopharmaceutics Classification System (BCS) [2], where, as illustrated in Figure 1.1, the API is classified into four categories: classes 1 and 3 containing molecules with high aqueous solubility, and classes 2 and 4 containing molecules with low solubility; molecules in classes 3 and 4 also exhibit poor biological membrane permeability, another deterrent to drug absorption. Interestingly, over the past few decades there has been a significant increase in the number of APIs under development that have fallen into BSC classes 2 and 4 because of solubility problems. This decrease in dissolution of crystalline API appears to correlate with a corresponding increase in the number of API molecules in the development process that have larger average molecular weights, higher melting temperatures, and a higher degree of hydrophobicity than that observed in previous years. As a consequence, during the past few years, there has been a significantly increased effort to develop strategies that might serve to enhance the rate of dissolution of an API by means of formulation, chemical modification, or processing.

Based on Equation 1.1, we can conclude that there are two major factors that can be used as a basis for enhancing dissolution rates of poorly water-soluble crystalline APIs sufficiently to have some controllable influence on increasing oral bioavailability. These are the surface area of the solid exposed to the aqueous medium and the solubility of the solid in aqueous media. Strategies for enhancing dissolution can be divided further into (i) formulation and processing, (ii) chemical modification, and (iii) use of "high-energy" structurally disordered physical forms of the solid. Starting with the crystalline API, the formulator can simply reduce the particle size of crystalline materials to increase their specific surface area (area per unit mass). Very significant increases in dissolution rate, for example, have been attained by producing particles with diameters on the order of 100–300 nm. One also can increase dissolution rates by adding solubilizers to the formulation, such as surfactants, or complexing agents, such as cyclodextrins, which help to produce a supersaturated solution when the API encounters an aqueous medium. Surfactants can also act as wetting agents to improve access of the aqueous medium to hydrophobic API, thus effectively increasing the available surface area. High levels of supersaturation, upon contact with water, can also be obtained by dissolving the API in liquid lipid-based formulations and administering the product in hard or soft capsule form. Such an approach tends to produce a supersaturated solution upon exposure to aqueous dissolution media. Alteration of the API chemically by forming more highly water-soluble crystalline salts or cocrystals, when possible, can be a very efficient way of increasing dissolution rates as long as the dissolved form of the API can be maintained in a supersaturated state relative to that of the crystalline "free form" of the API itself. Finally, since the high lattice energy of an API crystal, as often reflected at high melting temperatures, can serve as an impediment to attaining adequate thermodynamic solubility, any approach that can change, reduce, or eliminate the crystal lattice energy should be able to enhance the apparent solubility. For example, liquid forms of molecules will generally exhibit greater solubility than their crystalline counterparts (supersaturation), all other factors being equal. Indeed, it is well known that higher energy "less-stable" polymorphic crystal forms of an API generally exhibit greater solubility than the most stable form. It has also been shown that disorder in the crystal lattice introduced as crystal defects can serve to increase dissolution from the defect sites relative to that from the less defective crystal. Consequently, it is not surprising that complete elimination of long-range three-dimensional order in the crystal by forming the amorphous form of an API can greatly enhance apparent solubility and rates of dissolution. Of course, since the amorphous state represents a high-energy form relative to the crystal, this approach can be useful only as long as a supersaturated solution of API can be maintained in the aqueous medium over the time period required for gastrointestinal absorption. Since the overall theme of this book deals with amorphous API-polymer solid dispersions designed to provide enhanced oral bioavailability by creating such supersaturation, it will be useful in this introductory chapter to review some of the important physico-chemical characteristics of amorphous solids as single components and as mixtures of API with other formulation components that might be used to enhance oral bioavailability in drug products. A brief discussion of API-polymer amorphous dispersions, in particular, will serve as an introductory overview of various principles that will be applied in more detail throughout the rest of the book.

1.2 FORMATION OF THE AMORPHOUS STATE AND THE GLASS TRANSITION TEMPERATURE

Let us first consider a single-component system such as an API in its most stable crystalline form. From a classical free energy–temperature diagram [3], as illustrated in Figure 1.2, we can observe a significant reduction in the free energy per mole of the

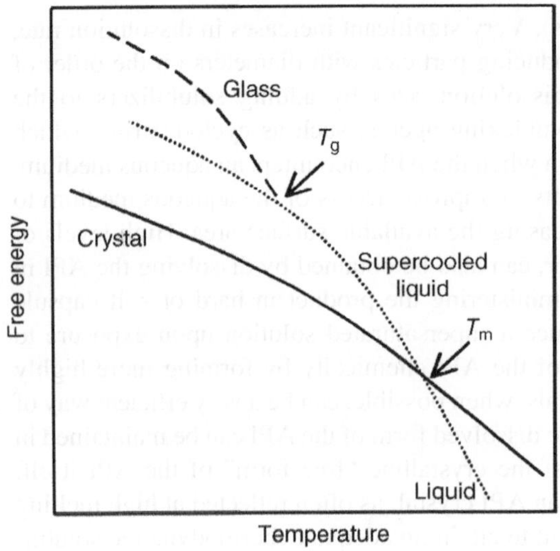

Figure 1.2. Free energy–temperature diagram for a single-component system (reproduced with permission from Ref. 3. Copyright 2001, Elsevier).

crystal as the temperature of the sample is increased until we reach the melting temperature T_m where the crystal undergoes a spontaneous first-order conversion to the liquid form, with the liquid now in a lower free energy state. If the liquid is slowly cooled to below T_m, and there is sufficient time for nucleation and crystal growth to occur, the system will revert to the equilibrium state of the crystal. If, however, as seen in Figure 1.2, the liquid sample is cooled rapidly through T_m so as to kinetically avoid crystallization, the system will show no discontinuities at T_m and maintain the equilibrium properties of the liquid as a supercooled liquid that is metastable relative to the crystal. Upon further cooling and as the viscosity of the supercooled liquid increases and diffusive motions of the molecules decrease, equilibrium can no longer be maintained and a distinct discontinuity in the free energy–temperature diagram occurs with the formation of the unstable glassy state. This occurs at a distinct temperature, designated the glass transition temperature T_g, the value of which for a particular molecule under the same processing conditions is determined by the molecular weight, degree of polarity, and the effect of molecular shape on the closeness of molecular packing. For example, the more polar the solid or the higher the molecular weight, the greater the value of T_g, while the bulkier the shape of the molecule and poorer the packing, the lower the T_g. The value of T_g is experimentally determined most conveniently by using differential scanning calorimetry, where the heat capacity can be measured as the sample temperature is continuously changed at a constant rate from low temperatures to the melting temperature. Because of structural changes that bring about changes in the rate of molecular motions, the heat capacity generally undergoes a distinctly abrupt change at T_g, as illustrated in Figure 1.3. In general, it has been shown that the viscosity of an organic liquid at T_m is on the order of 10^{-2} Pas, while at T_g this value has increased to about 10^{12} Pas, a 14 order of magnitude change! Since this point of discontinuity is associated with such a significant change in viscosity as cooling occurs, experimental values of T_g will depend to a small extent on the rate of cooling: the faster the rate of

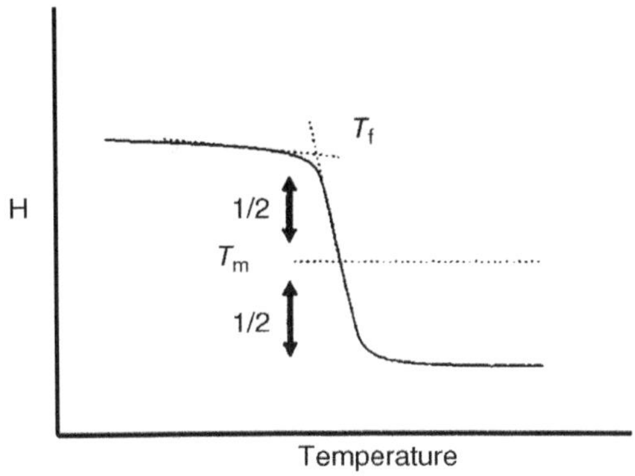

Figure 1.3. Heat capacity change at the glass transition indicating the onset glass transition temperature T_{gonset} and the midpoint glass transition temperature T_{gmid}.

cooling, the greater the T_g. Thus, in reporting the T_g for a particular material, it is important to indicate the conditions used to measure T_g and to form the glassy state. However, despite small differences in T_g that arise with different methods of preparation, it has been observed for small organic molecules and organic polymers exhibiting a crystalline state that when temperature is expressed in Kelvin, the T_g can be approximated empirically to be equal to a value of about $0.67T_m$ [4,5]. Such an empirical equation can be very helpful in determining the likely region of temperature where the T_g of a newly formed amorphous API may be located. Since molecules are kinetically "trapped" in the glassy state, it is not surprising that different rates of cooling generally lead to glasses with slightly different structural features. Because of this, and because the glass is unstable relative to the supercooled liquid, when held at temperatures close to, but below, T_g, the solid generally will exhibit an ability to slowly "age" or "anneal" with accompanying thermodynamic changes, such as a loss of free energy, enthalpy, and entropy and an increase in density, closer to values expected for the supercooled liquid, as illustrated in Figure 1.4 [6]. Thus, we can conclude that determining T_g is central to any characterization of amorphous solids, and that the method

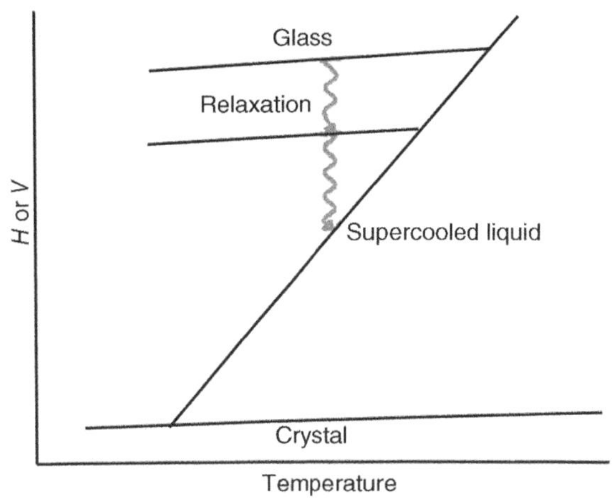

Figure 1.4. Relaxation of a glass toward the equilibrium liquid state due to physical aging.

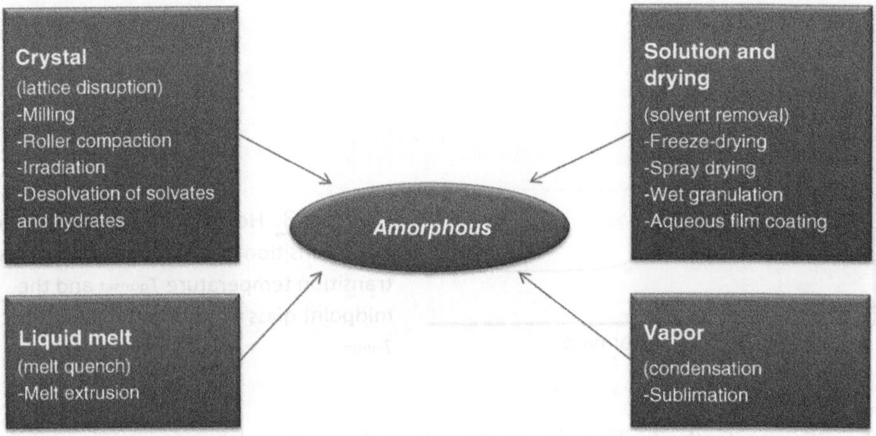

Figure 1.5. Various methods of producing the amorphous state.

of preparation of amorphous solids must be outlined in detail when reporting any value of T_g.

In this regard, so far we have focused only on the preparation of an amorphous solid by melting the crystalline form and then rapidly supercooling the melt to well below the melting temperature so as to avoid crystallization. Indeed, this is the basis for using the hot melt extrusion method to produce amorphous solid dispersions (ASDs), a topic that will be discussed more fully in later chapters. However, as described in Figure 1.5, it is also possible to produce amorphous forms by rapidly condensing molecules directly from the vapor state at low temperatures, or by causing molecules to rapidly precipitate from solution, where in both cases crystallization is kinetically avoided. Although preparation from the vapor state is not currently used as a process to form amorphous pharmaceutical products, there is evidence that such a method can lead to unusually stable glasses [7]. Precipitation from solution to form an amorphous solid is the basis for the widely used processes of lyophilization and spray drying (SD), where lyophilization has proved particularly useful in forming sterile amorphous protein products for parenteral use, and spray drying for the development of solid dispersions for oral and pulmonary use. As seen in Figure 1.5, it is also possible to form the amorphous state directly from a crystal by introducing mechanical stress that is sufficient to create crystal defects that eventually coalesce into a completely amorphous form [8]. Likewise, it has been shown that amorphous forms can be created by the dehydration of crystal hydrates [9] or by the desolvation of crystal solvates [10], where in both cases the desolvated crystal lattice collapses because of the free volume left by removing the solvent from the crystal lattice. Although such methods that disrupt the crystal lattice have not yet been found practical for the preparation of pharmaceutical amorphous systems on a large scale, the importance of such phenomena has been demonstrated in situations where crystalline solids are inadvertently rendered partially amorphous by processes such as milling and drying, leading to small amounts of disorder and unanticipated solid-state instabilities [11]. In conclusion, given that different methods used to form amorphous solids can lead to glasses with somewhat different properties, it is important to recognize that the various pharmaceutical processes used to produce robust amorphous drug products of high quality and performance must be under very careful control with regard to time, temperature, and other process conditions.

1.3 STRUCTURE OF AMORPHOUS SOLIDS

The structural arrangement of molecules in crystals, as determined by intermolecular interactions and molecular size and shape, can be described in terms of a specific local structure reflected by the geometric arrangement of molecules within the unit cell, and the long-range symmetrical three-dimensional extension of the repeating unit cells. The same molecule in the crystalline state may be able to form in different unit cells, and, therefore, to exist in distinctly different polymorphic forms. Single-crystal X-ray diffraction techniques are used to determine the arrangement of molecules in the unit cell, while, as shown in Figure 1.6, powder X- ray diffraction measurements (PXRD) reveal distinct diffraction peaks at characteristic scattering angles that represent the various planes of long-range symmetry within the crystal and can be used to identify the crystal form. Liquids and supercooled liquids, on the other hand, having lost the long-range three-dimensional order of the crystal will exhibit PXRD patterns that are devoid of these distinct peaks, rather than exhibiting a broad halo of X-ray intensity, as seen in Figure 1.6. From extensive studies of liquids and supercooled liquids, it has been

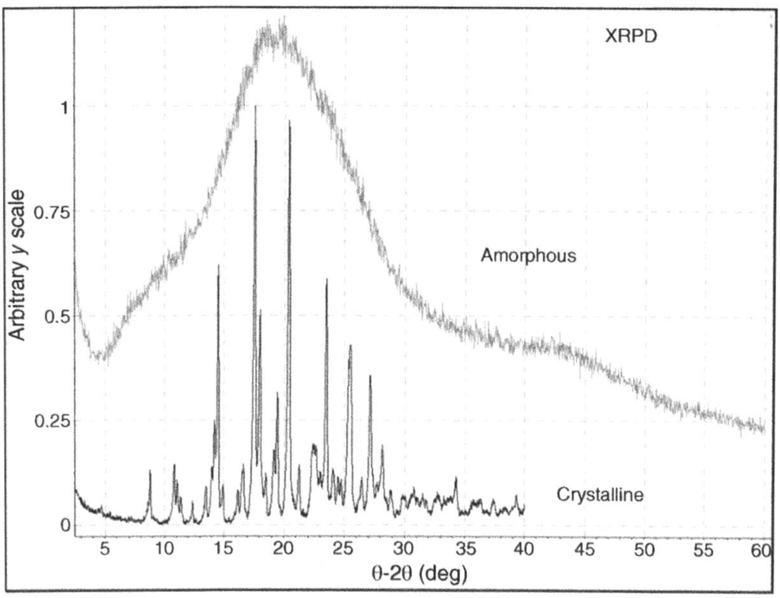

Figure 1.6. Typical powder X-ray diffraction patterns for crystalline and amorphous forms.

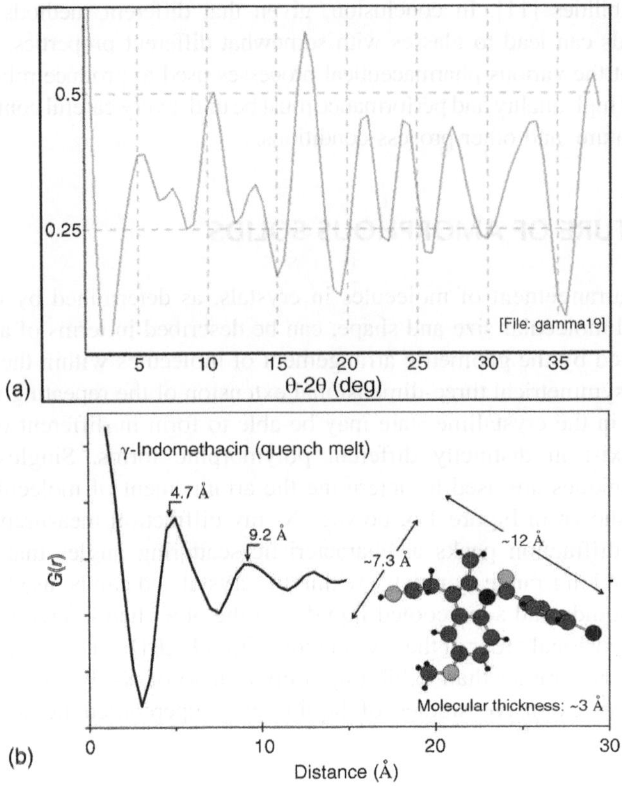

Figure 1.7. Pairwise distribution function. (a) Crystalline indomethacin. (b) Amorphous indomethacin (reproduced with permission from Ref. 12. Copyright 2006, Springer).

established that they maintain a distinct local arrangement of molecules over at least nearest-neighbor (NN) and next nearest-neighbor (NNN) distances, wherein the arrangement is similar to, and sometimes the same as, that in the corresponding crystal unit cell. To more quantitatively describe the local structure of an amorphous solid, it is possible to use PXRD data to determine the pair distribution function (PDF), which is a parameter that describes the probability $G(r)$ of finding the relative location of two atoms within a given volume when they are separated by a radial distance r, as shown in Equation 1.2 [12,13]:

$$G(r) = 4\pi r[\rho(r) - \rho_0], \tag{1.2}$$

where $\rho(r)$ and ρ_0 are the local and average atomic densities, respectively. As shown in Figure 1.7, for the amorphous and crystalline forms of the drug indomethacin, distinct peaks in a plot of $G(r)$ versus distance occur for amorphous indomethacin at distances that correspond very closely to the NN and NNN distances expected for the indomethacin molecule, but not out to greater distances [12]. On the other hand, similar repeating PDF peaks extend out to much greater distances for the crystal, reflecting the greater long-range order in the crystal. From this and many other studies using estimates of the PDF profile of amorphous solids, it has been possible to conclude that local structures of amorphous solids, closely related to the unit cell of the corresponding crystal, are maintained in the amorphous state under all conditions, despite the lack of long-range order.

Although amorphous solids, like liquids, do not exhibit long-range order, it is of interest to have some understanding of the manner in which molecules are organized beyond NN and NNN distances to form the bulk solid structure. Structural features of amorphous solids in the supercooled state at temperatures above T_g can best be understood by what is generally known about the structure of simple liquids, where it is assumed that molecules are packed randomly as polyhedral clusters that minimize the overall free energy of the system without crystallizing [14]. Typically, the densest possible packing of spheres of the same size, as in a face-centered cubic crystal, would have the spheres occupying a maximum of 0.74 of the total volume occupied by the material, while the remainder would be taken up by the volume fraction of void space equal to 0.26. The random close packing (RCP) model is an empirical statistical model that considers the packing of an object that has almost no period packing structure, as when pouring spheres into a container. Mathematical modeling of such a system reveals that at closest packing the spheres must occupy a volume fraction of <0.64 and that such a model can describe the structure of simple liquids quite well. In general, it appears that molecules in the supercooled liquid state contain fairly homogeneously sized polyhedral structures down to temperatures roughly on the order of $1.5T_g$, at which point the molecular structures then become distorted by "jamming up" into a more spatially heterogeneous system with a distribution of cluster sizes [15]. Such a temperature is generally termed the crossover temperature T_c. As will be discussed more fully subsequently, because of such structural changes, at temperatures between T_g and T_c, the mobility of molecules in the system will undergo decrease by orders of magnitude as the system becomes more "solid-like" and approaches the glassy state. When the system goes below T_g into the glassy state, we would expect there to be further "jamming" and a tendency for the molecules to readjust into a distinctly different structure, while retaining essentially the same local structure. Indeed, application of a modified form of the RCP to organic glasses reveals that its structure is best described by small highly dense local clusters of molecules having a size of roughly 2.0–2.5 nm, surrounded by interfacial regions of less densely packed molecules in a higher state of energy [12]. Such a structure might be considered analogous to a polycrystalline mass, containing many small crystallites surrounded by a higher energy region of grain boundaries. An analysis of amorphous indomethacin in the glassy state, for example, led to a structure consistent with this picture, as illustrated in Figure 1.8 [12]. Further analysis suggested that the higher energy region, termed the microstructure, represents about 10% of the total mass, and that it was very likely the region of the glass that spontaneously anneals or ages when held at temperature just below T_g, as illustrated earlier in Figure 1.4. It is also believed to be the likely region that acts to retard the rate at which the local domains nucleate and undergo crystallization.

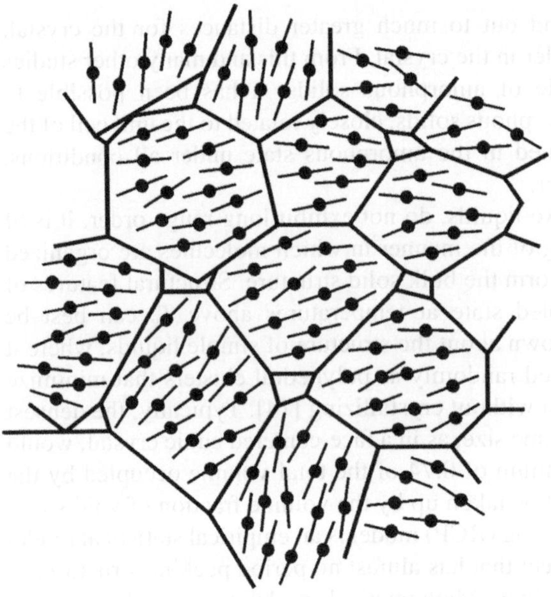

Figure 1.8. Schematic representation of the structure of an amorphous solid in the glassy state (reproduced with permission from Ref. 12. Copyright 2006, Springer).

1.4 MOLECULAR MOBILITY IN AMORPHOUS SOLIDS

As inferred from the discussion so far, during the process of forming amorphous solids, the thermodynamic properties shown in Figure 1.2 are strongly influenced by the kinetic properties of the system, as reflected in the levels of allowed molecular motion, often referred to as molecular mobility. For the present discussion, molecular motions can be placed into three broad categories: (i) high-frequency intramolecular motions, including harmonic bond vibrational and spinning modes; (ii) whole molecule or polymer segmental secondary "caged" or hindered motions, generally referred to as Johari–Goldstein β-relaxations; and (iii) primary whole molecule or polymer segmental translational and rotational diffusive α-relaxations. Central to the properties of molecules in the amorphous state are the highly cooperative diffusive translational and rotational motions that occur at various temperatures above and below T_g. The rate of such translational and rotational motions can be expressed in terms of diffusion constant D_{trans} and D_{rot}, respectively, where for a sphere of radius r in a liquid with viscosity η and at temperature T in the Stokes–Einstein equation,

$$D_{trans} = (k_B T)/(6\pi r \eta), \tag{1.3}$$

and in the Debye equation,

$$D_{rot} = (k_B T)/8\pi r^3 \eta), \tag{1.4}$$

where k_B is the Boltzmann constant. From these equations we can see that at constant molecular size, the rate of such diffusive motions will be critically controlled by both temperature and viscosity.

Another more general way to express molecular mobility quantitatively is to define the relaxation time τ, a parameter that directly indicates the timescale over which a single rotation takes place or the time over which a molecule undergoes translation across a given reference distance; the greater the viscosity and hence the smaller the diffusion coefficient, the greater the translational or rotational relaxation time. To determine the value of τ under a given set of conditions, one can experimentally perturb the system out of equilibrium mechanically, electrically, or magnetically and observe the rate $\phi(t)$ at which the property returns toward the equilibrium state. Dynamic mechanical analysis (DMA) is generally used to measure the rate of relaxation and viscosity; however, primary relaxations can also be measured by electrical perturbation of dipoles in the molecule and measurement of the dipole relaxation back toward an equilibrium state using dielectric spectroscopy (DES). For a system exhibiting a single mode of relaxation that follows first-order kinetics, such as a pure liquid above its melting temperature, one can write

$$\phi(t) = \exp[-t/\tau], \tag{1.5}$$

where τ is the reciprocal of the first-order rate constant. In such a case, one would expect the temperature dependence of τ to follow the Arrhenius equation:

$$\tau(T) = \tau_0 \exp[E_a/RT], \tag{1.6}$$

where τ_0 is the relaxation time at the high temperature limit, on the order of 10^{-12} s, and E_a is the activational energy associated with the process. Indeed, equilibrium liquids generally exhibit Arrhenius kinetics. Supercooled liquids at temperatures approaching T_g, however, appear to exhibit more than a single relaxation time generally described by a "stretch" exponential form of Equation 1.5 that takes into account this distribution of relaxation modes. For example, a widely applicable empirical equation in such situations is the Kohlrausch–Williams–Watts (KWW) equation:

$$\phi(t) = \exp[-t/\tau]^{\beta}, \tag{1.7}$$

where τ is the average relaxation time and β represents the distribution of different relaxation times, with values falling between 0 and 1; a value of 1 indicates a single mode of relaxation and smaller values represent an increasing number of relaxation modes. Most amorphous systems of pharmaceutical interest yield values of β in the range of 0.3–0.6 [16]. When such an equation is applicable, the temperature dependence of τ can be expressed by the nonexponential empirical Vogel–Tammen–Fulcher (VTF) equation:

$$\tau(T) = \tau_0 \exp[DT_0/(T - T_0)], \tag{1.8}$$

where D and T_0 are constants. D is called the "strength parameter," an indication of the activation energy of the diffusive relaxation process, and T_0 is the temperature at which

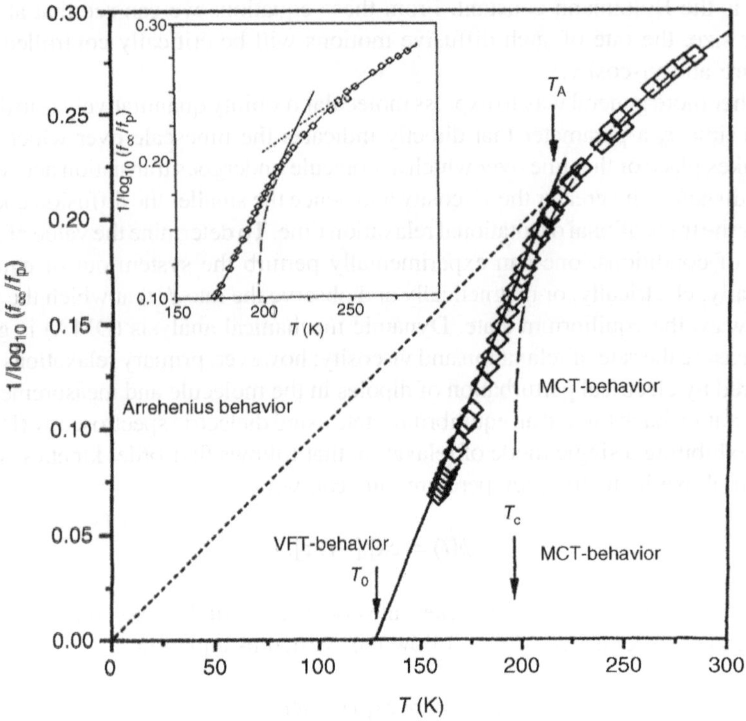

Figure 1.9. A plot representing the reciprocal of relaxation time versus temperature for propylene carbonate (reproduced with permission from Ref. 17. Copyright 1993, American Physical Society).

$\tau(T)$ eventually would reach infinity or zero molecular mobility. For small organic molecules, T_0 generally falls between 40 and 70 K below T_g. Figure 1.9 provides an example of a plot of $1/\tau$ versus T for amorphous propylene carbonate, in the supercooled equilibrium state, having a T_g of 150 K; decreases in the parameter represented on the y-axis reflect an increase in relaxation time or increase in viscosity [17]. Here, we can see that at temperatures above 225 K, or $\sim 1.5T_g$, relaxation times appear to follow Arrhenius kinetics (Equation 1.6) as expected for an equilibrium liquid with a single mode of relaxation. At a temperature labeled T_A, equal to the "crossover temperature" T_c, previously discussed, however, we can observe in Figure 1.9 that there is a marked discontinuity, where the increase in relaxation time now follows the VTF equation (Equation 1.8) with much higher values of relaxation time than predicted by Equation 1.6. Note also in Figure 1.9 the extrapolation of relaxation times to a value of T_0, as presented in Equation 1.8. The data presented in Figure 1.9 show that at temperatures above T_A (T_c), the supercooled liquid is structurally similar to the equilibrium liquid. However, as the system is cooled to below T_c, but above T_g, the supercooled liquid appears to exhibit structural changes and spatial heterogeneity that give rise to multiple modes of relaxation and a very rapid increase in the average τ as the temperature approaches T_g [15]. An

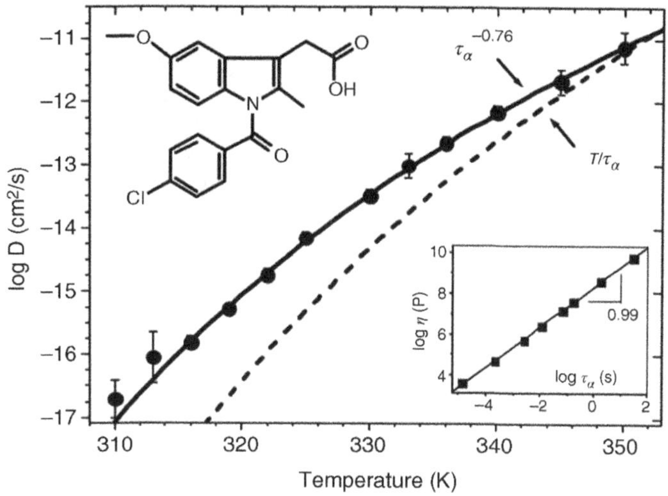

Figure 1.10. Diffusion Coefficients and Rate of Crystal Growth vs. Temperature for Indomethacin Above Its Glass Transition Temperature (experimental points); Expected Diffusion coefficients From Measured Values of Viscosity Assuming the Application of the Stokes-Einstein Equation (see Equation 1.3 in text) (Reproduced with permission from Ref. 18. Copyright © 2011 Royal Society of Chemistry).

indication of the significant structural changes taking place at T_c is provided by the observation, shown in Fig 1.10, that at temperatures below T_c, the relationship between the translational diffusion coefficient and viscosity, as expressed in Equation 1.3, the Stokes–Einstein equation, no longer follows as the temperature dependence of viscosity is uncoupled from that for the diffusion coefficient [18].

From the previous discussion it is clear that the rate at which viscosity and hence diffusive relaxation times increase as the supercooled liquid is cooled toward T_g is an indication of the structural changes caused by changes in molecular packing and intermolecular interactions. Consequently, we would expect such changes to be related to the chemical structure and molecular size and shape of the material and to reflect the activational energy associated with such changes. To put this on a quantitative basis, Angell [19] coined the term fragility and defined it in terms of a fragility index m:

$$m = d\log\tau/d(T_g/T)^{T=T_g}, \tag{1.9}$$

where m is the initial slope of a plot of log τ versus T_g/T taken at the limit of $T = T_g$; the greater the value of m, the greater the fragility, and hence the greater the change in molecular mobility with temperature. Using T_g/T instead of $1/T$ normalizes the data with respect to the value of T_g and allows direct comparison of materials with different values of T_g. For example, as shown in Figure 1.11, plots of log τ versus T_g/T reveal very different initial slopes for three materials with very different structures: silicon dioxide,

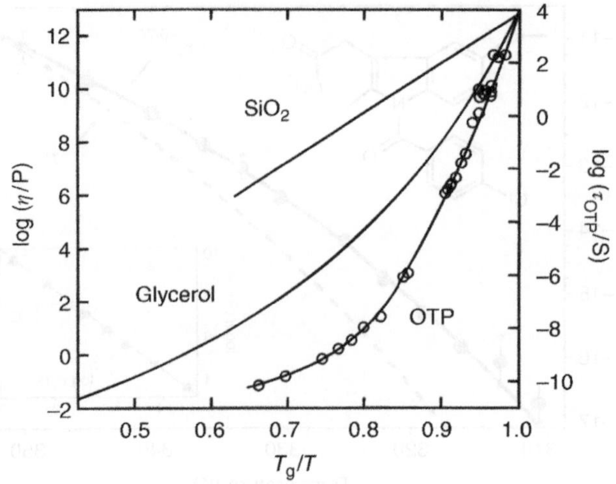

Figure 1.11. Angell Plot for Three Amorphous Materials (Reproduced with permission from Ref. 19. Copyright ©1996 American Chemical Society).

SiO_2, a highly polar inorganic polymeric dense material with $T_g = 1600$ K; *o*-terphenyl, a small molecular weight organic nonpolar molecule with $T_g = 246$ K; and glycerol, a very small molecular weight polar compound with a $T_g = 193$ K. As can be seen, SiO_2 exhibits the smallest value of m, *o*-terphenyl the greatest, and glycerol the intermediate behavior. Generally speaking, organic molecules tend to be fairly fragile and most susceptible to structural changes with changing temperature in the vicinity of T_g, while highly polar inorganic polymers, like SiO_2, tend to be more resistant to structural change; materials exhibiting low fragility are generally said to be "strong" supercooled liquids. By combining Equations 1.8 and 1.9, one can express the fragility index in terms of D, T_g, and T_0:

$$m = \{D(T_0/T_g)\}/\{[1-(T/T_0/T_g)]^2 \ln(10)\}, \tag{1.10}$$

where strong liquids exhibit values of D in the range of 30 or greater, and fragile liquids have values in the range of 7–15. Indeed, it has been shown that a large group of API molecules have values of D that fall in this latter range and thus we can assume that most amorphous API molecules exist as fragile systems with significant sensitivity to changes in temperature above T_g and below T_c [20], the point at which upon cooling molecules tend to "jam" into a spatially heterogeneous state. From previous discussions we have learned that amorphous solids at temperatures lower than T_g form the unstable glassy state that has something like the heterogeneous structure illustrated in Figure 1.8. We also learned that at T_g, the diffusive α-relaxation times on average are on the order of 10^2 s (viscosity of about 10^{12} Pas), indicating that there is still a significant degree of molecular mobility at the initiation of the glassy state. From Equation 1.8 (VTF) we would expect that at temperatures below the crossover temperature T_c, in the supercooled liquid, relaxation times will decrease significantly, the extent of which depends on the level of

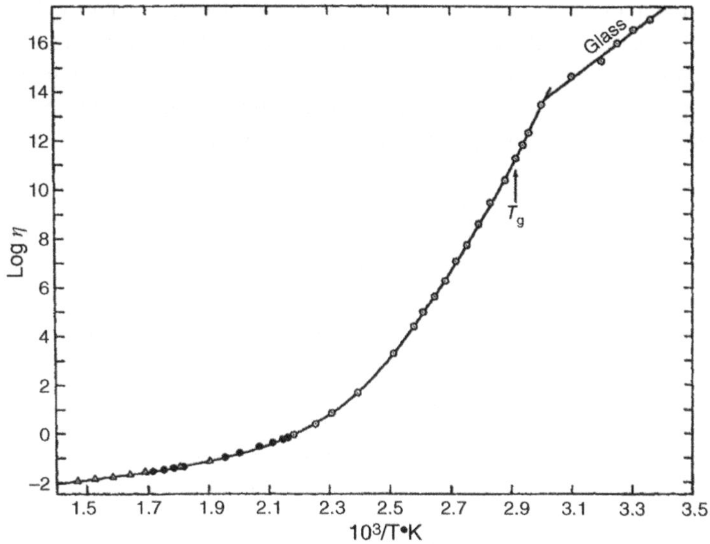

Figure 1.12. Viscosity of amorphous tris-α-naphthylbenzene as a function of the reciprocal of temperature (reproduced with permission from Ref. 21. Copyright 1968, AIP Publishing LLC).

fragility. Since extension of the VTF equation to temperatures below T_g assumes that the supercooled liquid state continues down to T_0, one might expect that the VTF equation would not be able to predict molecular mobility much below T_g. That this is so can be observed in Figure 1.12 where a plot of log η versus $1/T$ for amorphous tris-α-naphthylbenzene indicates that although a discontinuity does not occur at T_g, there is a distinct discontinuity at a temperature that is roughly 15 K below T_g [21]. Below this temperature, viscosity dependence on temperature appears to follow Arrhenius kinetics and exhibit values that are significantly lower than those predicted from the VTF equation. Apparently, closer to T_g there is very rapid aging of the glass toward the supercooled liquid over the time period required to carry out the viscosity measurements (see Figure 1.4 and the earlier discussion of aging of glasses) and viscosity values are those expected for the supercooled liquid. The lower than expected viscosity in the glass would be consistent with the general structure of glasses, as illustrated in Figure 1.8, that contains a microstructure region in which molecules should have higher energy, less density, and higher molecular mobility. Such behavior, that is, greater molecular mobility than predicted from the VTF equation, has also been reported with measurements of relaxation time for amorphous indomethacin at temperatures below T_g using DMA, DES, and thermal analysis [22]. It has been shown further that the extrapolation of relaxation times obtained in the glassy state of indomethacin leads to a value of 3 years (10^8 s) at $T_g - T$ equal to 40 K, 3 years being the desired time of storage generally required for establishing expiration dates of many solid drug products. Thus, long-term stability in this case would require a storage temperature that is about 40 K below T_g. (there will be more discussion of stability subsequently).

An additional factor to consider in discussing molecular mobility in amorphous solids is the fact that molecules at a surface, that is, the solid–vapor interface, which exist at a higher energy than those molecules in the bulk, generally can exhibit greater diffusive molecular motion by means of surface diffusion; the greater degrees of freedom exhibited by a molecule at a surface allows greater lateral translational and rotational motions. This generally is not considered a critical factor with amorphous solids when bulk properties are dominant. However, surface molecular mobility has been shown to be of importance when surface-to-bulk volume ratios become quite great as with thin polymer films or nanosized particles. Indeed, it has been possible to measure bulk and surface diffusion for amorphous indomethacin as a function of temperature, and show that surface diffusion is orders of magnitude greater than bulk diffusion at temperatures well below T_g [23].

1.4.1 Secondary Johari–Goldstein β-Relaxation

So far we have discussed the molecular mobility of amorphous solids in terms of cooperative diffusive translational and rotational motions, generally referred to as primary α-relaxations. Clearly, it is these motions, directly related to viscosity, that determine many of the properties that can affect the pharmaceutical functions and instabilities of amorphous solids. These are also the motions that are responsible for the appearance of the discontinuity in properties that occurs at the glass transition temperature. As already mentioned, molecules in the solid state exhibit other types of motions that occur over very short timescales relative to diffusion. These include intramolecular harmonic bond vibrations and rotations and single molecule or polymer noncooperative segmental hindered or "caged" motions, referred to as Johari–Goldstein (JG) secondary β-motions [24]. Such high-frequency motions, on pico- and nanosecond timescales, are best detected by DES, solid-state nuclear magnetic resonance (SSNMR) spectroscopy, and neutron scattering. The intramolecular harmonic motions do not appear to play an important role in the dynamics of amorphous solids, but the β-motions appear to be important as precursors to events that lead to diffusional motion. Although often difficult to observe in the presence of α-relaxation profiles in DMA or DES measurements, it appears that most organic small molecules undergo JG β-relaxations and that such relaxations involve hindered noncooperative rotational motions of single molecules within the amorphous structure, often envisioned as "molecules rattling in a cage," and located in "islands of mobility" where greater free volume is available. Combined studies using DES and SSNMR indicate that the JG motions occur throughout the amorphous solid, but that roughly 80–90% of the molecules appear to have low amplitudes that are relatively independent of temperature, while 10–20% of the molecules have higher amplitudes and significant temperature dependence [25]. Furthermore, the amplitudes of such motions in the glassy state appear to decrease when the glass undergoes physical aging to more dense structures, as described earlier in Figure 1.4. This suggests that the "islands of mobility" associated with these motions occur primarily in the microstructure region of the glass. The importance of the JG β-motions seems to be related to the fact that these secondary motions provide sufficient critical free volume within the amorphous structure to allow the initiation of the cooperative α-motions that impact many physical properties of the amorphous state [26]. As seen in Figure 1.13, plots of log τ

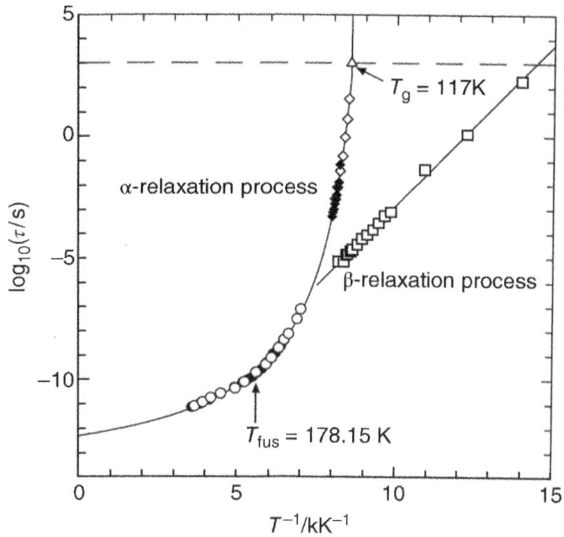

Figure 1.13. Log of α- and β-relaxation times versus reciprocal of temperature for amorphous *ortho*-terphenyl (reproduced with permission from Ref. 27. Copyright 2002, Elsevier).

versus $1/T$ for *o*-terphenyl measurements of α- and β-relaxation times indicate that the two processes decouple at the crossover temperature T_c, discussed previously, and that the β-motions exhibit lower relaxation times than those of α-motions [27]. A term $T_{g\beta}$, obtained by extrapolating values of τ_β to a temperature where these secondary motions cease to be measurable, has been shown to occur at temperatures on the order of 100 K below T_g. This then might be considered the temperature to which a glass should be cooled to eliminate any effects of the JG β-relaxations on the properties of the glass.

1.5 SOLID-STATE CRYSTALLIZATION FROM THE AMORPHOUS STATE

From a thermodynamic perspective, as illustrated in Figure 1.2, molecules in the amorphous state are in a higher free energy state than those in the corresponding crystal at all temperatures. Therefore, we would expect that over some time period the molecules eventually would spontaneously crystallize unless the addition of a crystallization inhibitor and/or a lowering of molecular mobility could act to reduce the rate of crystallization. Such a tendency to crystallize, of course, would negate the use of amorphous forms of API in pharmaceutical products for enhancing aqueous solubility, and hence the dissolution of the API. As we have seen in the previous section, diffusive molecular mobility in the amorphous state can vary by over many orders of magnitude as the temperature is changed; the relaxation times in the supercooled liquid from T_m to T_g can range from 10^{-12} to 10^2 s, respectively. Based on these values, samples stored at temperatures near or above T_g, clearly, would not have a sufficiently lowered molecular mobility to provide long-term storage, for example, 3 years, which represents a relaxation time on the order of 10^8 s. The goal, therefore, is to establish conditions where either temperature T is lowered sufficiently to increase τ to something close to 10^8 s or T_g is raised relative to T with the addition of other amorphous solids having greater T_g values, as with API-polymer amorphous dispersions (to be discussed subsequently). It would also be beneficial if these other amorphous solids, such as polymers, could act as specific crystallization inhibitors.

Studies dealing with crystallization of organic molecules from the amorphous state in the absence of any solvent have shown that the classical picture of homogeneous nucleation and crystal growth from the liquid state can serve as a useful conceptual model [28]. Here, it is assumed that molecules in the liquid state under certain conditions must first undergo spontaneous nucleation, the formation of aggregates or nuclei consisting of a few hundred molecules, followed by the growth of macroscopic-size crystallites. The major thermodynamic factor driving nucleation and crystal growth is the free energy difference per unit volume between molecules in the amorphous state and those of the crystal, ΔG_v, as depicted in Figure 1.14. However, the formation of nuclei requires phase separation to occur with the creation of new surfaces between the nuclei and amorphous matrix, a process that is thermodynamically unfavorable. Because of this, when nuclei form there must be an increase in free energy ΔG_s, which for a spherical nucleus of radius r can be described as

$$\Delta G_s = 4\pi r^2 \sigma, \quad (1.11)$$

where σ is the surface free energy per unit area of surface (the surface tension in liquids) and $4\pi r^2$ is the surface area. Thus, the overall free energy of homogeneous nucleation,

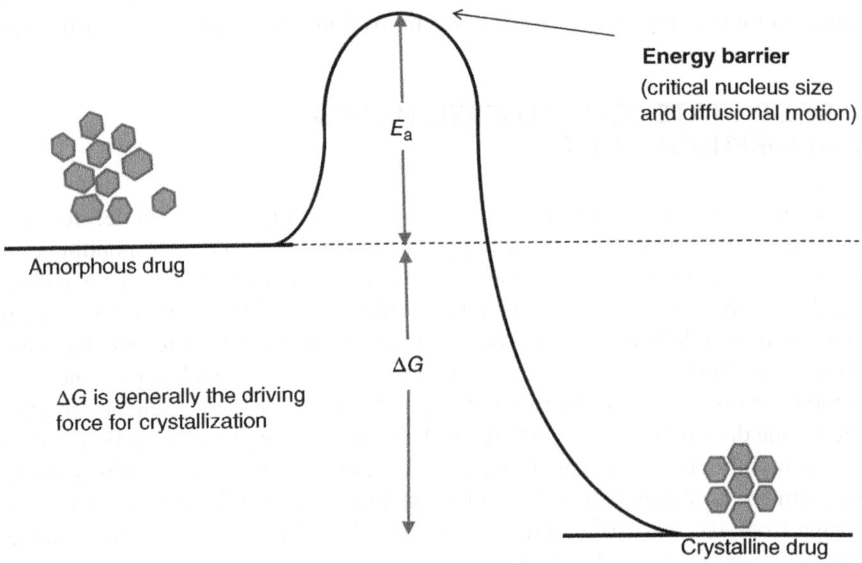

Figure 1.14. Schematic representation of the energetics associated with crystallization from the amorphous state.

ΔG^*, can be described by combining the two free energy terms, one favoring nucleation (ΔG_v) and the other opposing nucleation (ΔG_s):

$$\Delta G^* = \Delta G_v + \Delta G_s = (4/3)\pi r^3 \Delta G_v + 4\pi r^2 \sigma, \quad (1.12)$$

where $(4/3)\pi r^3$ is the volume of a sphere. It can be shown further, as illustrated in Figure 1.14, that the net change in ΔG^* will be positive until a critical radius r of nuclei is reached, above which the loss of free energy due to ΔG_V overcomes ΔG_s and spontaneous nucleation occurs. Thus, if specific nucleation inhibitors could be used to prevent the system from reaching the critical nucleus radius, further crystal growth could be avoided.

As has been implied throughout the discussion so far, an additional important free energy barrier to nucleation arises because of the decrease in diffusional molecular mobility that occurs as temperature is decreased. Expressing this kinetic energy barrier as $\Delta G'$, we can express the overall rate of nucleation, I, as

$$I \sim \exp(-\Delta G^*/kT) \exp(-\Delta G'/kT) \quad (1.13)$$

and can conclude, therefore, that as temperature is decreased (greater supercooling), the thermodynamic barriers to nucleation will decrease, while such cooling will increase the barriers to nucleation as molecular mobility is decreased. Similar analysis of thermodynamic and kinetic factors that affect the rate of crystal growth, u, once nucleation occurs, can be expressed as

$$u = D/\lambda[(1 - \exp(-\Delta G_v/kT)], \quad (1.14)$$

where D is the diffusion constant and λ is the jump distance across the growth interface. Thus, again we see the opposite effects of thermodynamic and kinetic energy requirements.

With this general conceptual picture of nucleation and crystal growth in mind, we can conclude that there are two major strategies that can be used to inhibit crystallization from the amorphous state: the addition of specific nucleation inhibitors, and/or the addition of molecules that will reduce the molecular mobility of the system at expected storage temperatures. In the first case, it should be possible to add molecules to the system that can accumulate at the site of nucleation and directly interact with the API thus interfering with nucleation. In the second case, it is important to first understand the temperature dependence of nucleation and crystal growth as illustrated by the temperature dependence for overall crystallization in Figure 1.15, where we see that the rates of both nucleation and crystal growth first increase with increasing temperature and then reach a maximum followed by a decrease [29]. It has also been suggested, as shown in Figure 1.15, that nucleation begins only when the temperature $T_{g\beta}$ associated with Johari–Goldstein β-relaxations is exceeded, thus providing a basis for the earlier suggestion that the β-relaxations are precursors to diffusive primary α-relaxations [26]. Earlier, it was also shown that $T_{g\beta}$ generally occurs at about 100 K below T_g and that α-motions become significant above 50 K below T_g. This means that strategies involving

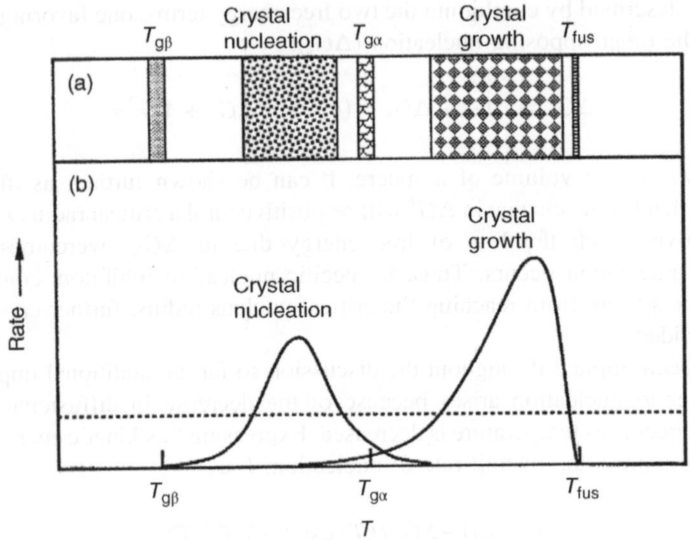

Figure 1.15. Rates of nucleation and crystal growth as a function of temperature in relationship to $T_{g\alpha}$ and $T_{g\beta}$ (reproduced with permission from Ref. 29. Copyright 1997, Elsevier).

a reduction in molecular mobility for inhibiting crystallization over long storage times should require $T_g - T$ to be somewhere in the range of 100–50 K.

1.6 SUPERSATURATION OF API IN AQUEOUS MEDIA FROM THE AMORPHOUS STATE

The major purpose of using the amorphous form of an API in solid dosage forms is to take advantage of the greater apparent solubility from the amorphous state relative to that from the crystal because of the lack of a well-ordered crystal lattice with attractive intermolecular energy that ordinarily tends to reduce solubility in the crystal. Three questions have to be addressed when taking this approach for enhancing API solubility: (i) Can the amorphous solid be maintained in the amorphous state over the time of storage before use? (ii) What level of improvement in solubility in aqueous media can be expected for the amorphous form relative to the crystal? (iii) Can the desired supersaturated concentration of dissolved API from the amorphous form be maintained over the time period required to ensure acceptable oral bioavailability? In the previous sections of this chapter, we have outlined a number of principles that allow the first question to be addressed. Here, we wish to examine the underlying principles that can be applied to address questions 2 and 3.

From a thermodynamic perspective, we would expect the relative solubility of a molecule in the amorphous and crystalline states at any temperature T to be directly determined by the free energy difference between the amorphous and crystalline forms, as illustrated in Figure 1.2. This free energy difference, designated as $\Delta G_T^{A/C}$, can be expressed in terms of the activities of the molecule in the amorphous and crystalline states, a^{A} and a^{C}, respectively:

$$\Delta G_T^{\mathrm{A/C}} = RT\ln(a^A/a^C). \tag{1.15}$$

Assuming activity coefficients equal to 1, $\Delta G_T^{\mathrm{A/C}}$ can be expressed as

$$\Delta G_T^{\mathrm{A/C}} = RT\ln(c^A/c^C), \tag{1.16}$$

where c^{A} and c^{C} are the equilibrium solubilities of the API from the two solid forms. Consequently, knowing the equilibrium solubility of the API in water at temperature T, c^{C}, and the free energy change $\Delta G_T^{\mathrm{A/C}}$, one can estimate the expected value of c^{A}, the supersaturation concentration. One can experimentally determine $\Delta G_T^{\mathrm{A/C}}$ by recognizing that

$$\Delta G_T^{\mathrm{A/C}} = \Delta H_T^{A/C} - T\Delta S_T^{A/C}, \tag{1.17}$$

where $\Delta H_T^{A/C}$ and $\Delta S_T^{A/C}$ are the enthalpy and entropy differences between the amorphous and crystalline forms, respectively. Furthermore, it can be shown that

$$\Delta H_T^{A/C} = \Delta H_T^C - (C_p^A - C_p^C)(T^C - T), \tag{1.18}$$

$$\Delta S_T^{A/C} = \Delta S_T^C - (C_p^A - C_p^C)(\ln[T^C/T]), \tag{1.19}$$

and

$$\Delta S_T^C = \Delta H_T^C/T^C, \tag{1.20}$$

where T^{C} is the melting temperature of the crystal, ΔH_T^C is the heat of fusion of the crystal, ΔS_T^C is the entropy of fusion of the crystal, and C_p^A and C_p^C are the heat capacities of the amorphous and crystal forms, respectively. By experimentally measuring C_p^A and C_p^C using thermal analysis and T^{C} and c^{C} for the crystal at T, it is possible to use Equation 1.16 to predict the value of c^{A}, the apparent equilibrium solubility of the amorphous form that might be expected [30]. Recognizing that the activity of the amorphous solid can be reduced by the presence of any residual absorbed water and that many acidic and basic APIs can undergo some ionization in aqueous solution, a more rigorous thermodynamic expression to predict solubility of amorphous solids under such conditions has been reported [31]. As shown in Figure 1.16, the experimental maximum supersaturation level $c^{\mathrm{A}}/c^{\mathrm{C}}$ for indomethacin at 25 °C before the initiation of crystallization is about 4.9. Using Equation 1.16 the value predicted at 25 °C for indomethacin is on average about 29.0, while correcting for water absorption and ionization the value is reduced to 7.0, much closer to the experimental value of 4.9. Some uncertainty in the agreement between predicted and experimental values of $c^{\mathrm{A}}/c^{\mathrm{C}}$ comes from the fact that crystallization can be triggered by nucleation events that are not easily predicted and

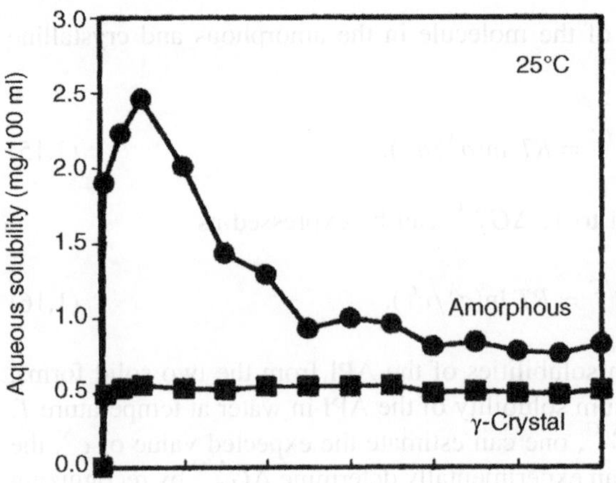

Figure 1.16. Solubility versus time for crystalline and amorphous indomethacin in water (reproduced with permission from Ref. 30. Copyright 2000, Springer).

reproduced. In a recent study, for example, the role of nucleation rates in determining levels of supersaturation of amorphous APIs upon dissolution has been modeled and analyzed by estimating the induction period for the initiation of nucleation and the appearance of crystals (time period for maintaining supersaturation) under various conditions [32].

1.7 MIXTURES OF AMORPHOUS SOLIDS

Mixtures of amorphous solids are of pharmaceutical interest when solid formulation ingredients undergo various types of processing that purposefully or inadvertently can lead to amorphous forms of two or more materials, for example, milling, compaction, drying, hot melt extrusion, and spray drying (see Figure 1.5). As we expect with simple liquids, mixtures of molecules in the amorphous state can exist as completely or partially separated phases, or as a single miscible phase. In the case of amorphous solid dispersions, which often consist of an amorphous mixture of API and polymer, a miscible amorphous system is generally required to obtain appropriate levels of stability, enhanced dissolution, and oral absorption. Consequently, it is important to review some of the underlying principles that determine the extent to which mixtures of different amorphous materials are miscible or immiscible. Of course it must be recognized that although amorphous components may be miscible, the system is still metastable relative to the crystalline forms of these components and eventually can undergo crystallization. We can first address this issue by applying some general thermodynamic principles related to the mixing of liquids. We begin by recognizing that for intermolecular mixing of liquids to occur, there must be an overall loss in the Gibbs free energy of mixing, ΔG_{mix}, at temperature T expressed as

$$\Delta G_{mix} = \Delta H_{mix} - T\Delta S_{mix}, \tag{1.21}$$

where ΔH_{mix} and ΔS_{mix} are the enthalpy and entropy of mixing, respectively. Generally, the mixing of different molecules leads to an increase in entropy, so that the positive entropic term in Equation 1.21 leads to a loss of free energy and the favoring of miscibility. On the other hand, if we consider the molecular mixing of two components A and B, ΔH_{mix} can be either negative or positive depending on the relative strength of the interactions between A:B and between A:A and B:B. Stronger A:B interactions lead to a loss in enthalpy, while stronger A:A and B:B interactions lead to an increase in enthalpy. Thus, the former will contribute to a loss in free energy, while the latter will contribute to an increase in free energy. We can generalize further by saying that two organic molecules having similar levels of polarity and, therefore, capable of undergoing significant intermolecular interaction, for example, hydrogen bonding, would be expected to produce miscible amorphous mixtures, while chemical incompatibility could lead to phase separation of the two or more amorphous forms if not offset by positive entropy changes. An example of such a system is that of amorphous citric acid (quite polar) and indomethacin (relatively hydrophobic) that phase separate beyond a certain concentration [33]. To gain a more quantitative understanding of those factors that control molecular mixing in the amorphous state, it would be desirable to have a theoretical framework that describes the free energy of mixing of liquids at a fundamental level in terms of measurable parameters. The Flory–Huggins lattice theory [34], presented in the Flory–Huggins equation, and originally developed for polymer solutions, describes the Gibbs free energy of mixing per mole for a polymer and a "solvent" as

$$\Delta G_{\mathrm{mix}} = RT(n_A \ln \phi_{\mathrm{A}} + n_B \ln \phi_{\mathrm{B}} + n_A \phi_{\mathrm{B}} \chi_{\mathrm{AB}}), \tag{1.22}$$

where R is the gas constant, ϕ is the volume fraction of each component, χ is the interaction parameter that describes the tendency for interaction between the components A and B, and n is the number of moles of each component. The first two terms in the parenthesis represent the entropic contributions, while the third term represents the enthalpic contributions from intermolecular interactions. For the purposes of this discussion, component A will refer to the "solvent" and component B to the polymer. The smaller or more negative the interaction parameter, the stronger the intermolecular interaction between A and B. From a conceptual perspective, the Flory–Huggins equation is very useful in demonstrating the importance of molecular size in affecting miscibility through a reduction in the entropy of mixing. For example, in Equation 1.23, the entropy of mixing per mole can be described as

$$S_{\mathrm{mix}} = -R(n_A \ln \phi_{\mathrm{A}} + n_B \ln \phi_{\mathrm{B}}). \tag{1.23}$$

If we now express the number of moles of each component in terms of the mass m and molecular weight M, we can rewrite Equation 1.23 as

$$\Delta S_{\mathrm{mix}} = -R[(m_A/M_A) \ln \phi_{\mathrm{A}}) + (m_B/M_B) \ln \phi_{\mathrm{B}}]. \tag{1.24}$$

The important conclusion from this equation is that the greater the molecular weight of any component, the less the contribution of entropy to the free energy of mixing, and thus the less the tendency for miscibility of the system. Essentially, by mixing two macromolecules that individually have many degrees of conformational freedom, any configurational entropy associated with this freedom will be reduced upon mixing and interaction of the two components. Consequently, we would expect that all other factors being equal the miscibility of amorphous binary systems involving at least one polymer would be reduced relative to those for two small molecules and that a mixture of two polymers would have a strong tendency to lose less free energy and possibly phase separate. Indeed, many amorphous polymer blends exhibit significant immiscibility for this reason.

To further examine the thermodynamics of mixing in amorphous solids and the possible relationships with the structural features of the amorphous state, for example, molar volume or density, it will be useful to consider the volumetric changes that can occur with the mixing of components having molar volumes of V_A and V_B. First, recall that the volume per molecule or mole of any component X at any temperature is determined by the volume occupied by the mass of a molecule, V_{occup}, plus any free volume V_{free} that exists between the molecules, as determined by the degree and nature of the molecular packing, as expressed in Equation 1.25:

$$V_X = V_{occup} + V_{free}. \quad (1.25)$$

Consequently, when two molecules A and B are mixed, the net volume change will have contributions from both their occupied and free volumes in proportion to the volume fraction of each of the individual components; as temperature is changed, these volumetric changes will be strongly affected by changes in V_{free}. For an ideal solution, where ΔH_{mix} is equal to zero, we would expect that the volume of the mixture, V_{AB}, would be the sum of the weighted average of the molar volume of each component:

$$V_{AB} = n_A V_A + n_B V_B, \quad (1.26)$$

where V_A and V_B are the molar volumes and n_A and n_B are the mole fractions of components A and B, respectively. This means that for ideal mixing, the occupied and free volumes of each component will be additive in a manner weighted exactly by each component's mole fraction. From the thermodynamic analysis described above, it would be expected that miscible mixtures of amorphous solids, where ΔH_{mix} is not equal to zero, will exhibit nonideal mixing so that there would be an excess molar volume change, V_{excess}, that can be either positive or negative depending on the nature of the intermolecular interactions giving rise to an enthalpy change. In such a case V_{AB} can be written as

$$V_{AB} + n_A V_A + n_B V_B + V_{excess}, \quad (1.27)$$

where very strong interactions between A and B lead to a negative excess volume, that is, a reduction in overall volume, as in the well-known case of the mixing of ethanol and water, while stronger interactions between A and A and B and B than between A and B lead to a positive excess molar volume or an increase in overall molar volume.

From Equation 1.27 and our previous discussion of factors that might influence the glass transition temperature of any amorphous solid, it would appear that the various

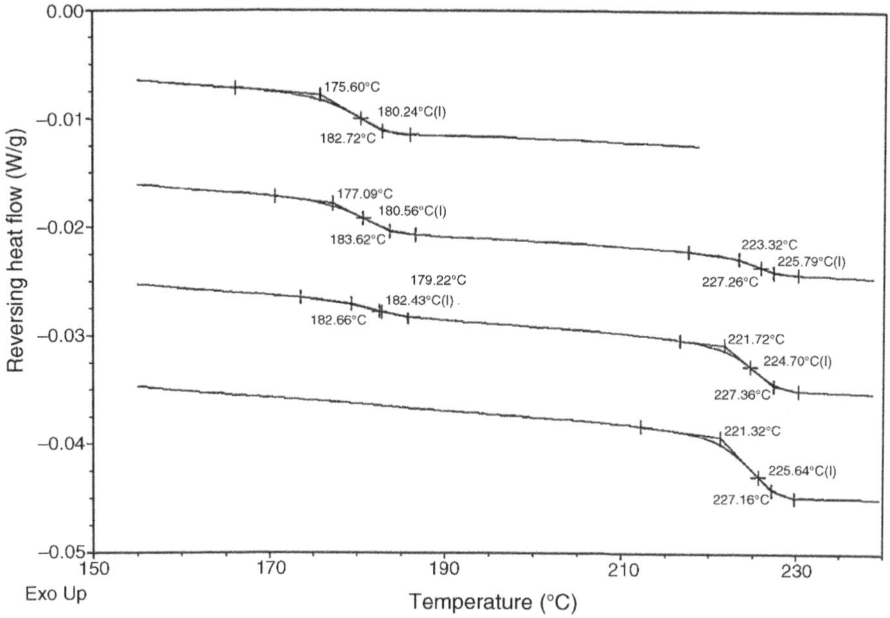

Figure 1.17. Overlay of modulated DSC, reversing heat flow, from top to bottom: PVP, 70 and 30 wt% dextran in PVP dispersions, and dextran (Mn: 64–76 kDa) prepared by freeze-drying (reproduced with permission from Ref. 35. Copyright 2008, Wiley-Liss, Inc).

volumetric changes taking place in the formation of a miscible amorphous mixture, as discussed above, would be reflected in a single value of T_{gmix} intermediate to the values of the individual components. On the other hand, if the two components are immiscible, we would expect there to be two values of T_g equal to the values of the pure individual components. For example, in Figure 1.17, we can observe two values of T_g for a mixture of PVP and dextran, prepared by freeze-drying from an aqueous solution of both components [35]. Note that the values are equal to the T_g of each pure component at all concentrations, indicating complete phase separation. In contrast, consider Figure 1.18 that indicates the values of T_g for pure amorphous indomethacin and PVP, as well as for two mixtures of indomethacin and PVP with different concentrations. Here, we can see that at each concentration there is only one T_g (T_{gmix}) that tends to increase as the concentration of the component having the greater T_g is increased, thus indicating a miscible mixture. Recognizing that it is important to establish the absence or presence of miscibility in amorphous solid dispersions, it would be important to examine some issues related to determining miscibility in amorphous mixtures. As has already been inferred, completely miscible amorphous systems should exhibit a single T_g intermediate to the individual values and dependent on the concentration of components. Typically, DSC measurements, which reflect thermal changes associated with changes in molecular mobility, are not able to produce distinctly separate values of T_g if the T_g values of the individual components are closer than about 10 K, and if the phases that separate have dimensions on the order of <100 nm. Thus, conditions that might lead to nanosized clusters of separated components

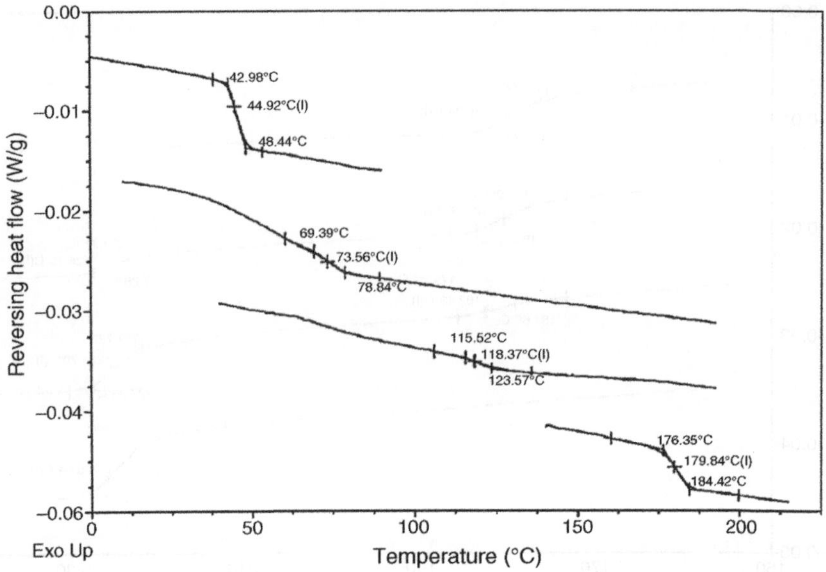

Figure 1.18. Overlay of modulated DSC, reversing heat flow, from top to bottom: indomethacin prepared by melt quench, 70 and 30 wt% indomethacin in PVP dispersions and PVP prepared by flash evaporation (Reproduced with permission from Ref. 35. Copyright 2008, Wiley-Liss, Inc).

would produce a DSC thermal response that indicates only one T_{gmix}. To further establish whether amorphous mixtures truly exist in a single phase, it is possible to use PXRD measurements, as well as SSNMR spectroscopy, since both techniques probe molecular interactions at a very local level. For example, in the case of PXRD, one can compare diffraction patterns of the amorphous mixture with those of the individual components, and if the sum of the weighted average of the patterns of the individual components is the same as that of the mixture, the system is completely phase separated [35]. If the calculated values for the mixture do not coincide with those experimentally determined, the system can be considered to exist as a single phase. One also can determine the PDF of the individual components and the mixture, and similarly compare the sum of the weighted average of the PDFs of the components with that of the mixture. Again, if the weighted average of the PDFs of the individual components is the same as that of the mixture, the system is immiscible, whereas differences indicate miscibility [35]. It is also possible to use solid-state NMR to detect phase separation down to domain sizes less than 100 nm using two-dimensional NMR [36] and by measuring T_1 and $T_{1\beta}$ relaxation times that respond to NN and NNN distance interactions between components. For example, it was recently shown that a lyophilized amorphous mixture of trehalose and dextran produced a single T_g that changed with the concentration of the components. However, PDF analysis of these mixtures, as described above, revealed complete phase separation [35]. SSNMR analysis of this system revealed, indeed, that separate molecular clusters with dimensions on the order of 55–80 nm could be detected [37].

It would be of further interest to know *a priori* how T_{gmix} might change with concentration so that rough predictions might be made by only knowing the individual values of T_g. To see how this might be done, we can start with ideal mixing, as described in Equation 1.26, to show that

$$T_{\mathrm{gmix}} = \phi_{\mathrm{A}} T_{\mathrm{g\,A}} + \phi_{\mathrm{B}} T_{\mathrm{g\,B}}, \tag{1.28}$$

where ϕ_{A} and ϕ_{B} are the volume fractions of components A and B, respectively. For such ideal mixing, we assume that the free volume of the mixture is simply the sum of the free volumes of the individual components. Typically, when working with mixtures of amorphous solids, the concentration of the various ingredients is expressed as weight fraction w rather than volume fraction. Consequently, as shown by Gordon and Taylor [38], Equation 1.28 can be expressed in terms of weight fraction as

$$T_{\mathrm{gmix}} = [(w_A T_{\mathrm{g\,A}} + w_B T_{\mathrm{g\,B}})/(w_A + K w_{\mathrm{B}})], \tag{1.29}$$

where for ideal mixing

$$K = [(\rho_{\mathrm{A}} T_{\mathrm{g\,A}})/(\rho_{\mathrm{B}} T_{\mathrm{g\,B}})], \tag{1.30}$$

where ρ is the density of each component at its respective T_g. The same general equation for ideal mixing, as described in Equation 1.29, has been derived on a thermodynamic basis by Couchman and Karasz [39], with K now defined in terms of the heat capacity changes taking place at T_g for each individual component:

$$K = \Delta C_{p\mathrm{A}}/\Delta C_{p\mathrm{B}}. \tag{1.31}$$

It can be further shown that when the densities of the two components are very similar, as in the case of many polymers, Equations 1.29 and 1.30 lead to the Fox equation [40], where

$$1/T_{\mathrm{gmix}} = [(w_A/T_{\mathrm{g\,A}}) + (w_B/T_{\mathrm{g\,B}})]. \tag{1.32}$$

Further estimations of T_{gmix} for ideal mixing in systems containing more than two amorphous components, for example, API, polymer, and water, although more complex and uncertain, can be carried out as extensions of Equation 1.29 or Equation 1.32 [41,42]. The simplest approach for a system containing n components using an extension of Equation 1.32 would be

$$1/T_{\mathrm{gmix}} = [(w_1/T_{\mathrm{g}1}) + (w_2/T_{\mathrm{g}2}) \cdots (w_n/T_{\mathrm{g}\,n})]. \tag{1.33}$$

Despite the strong possibility of nonideal mixing in most practical systems encountered, these equations for ideal mixing can be useful for estimating the glass transition temperature of a miscible mixture of amorphous components for a number of reasons. Initially, one may want to get a rough estimate of what composition of polymer and API might be required to provide a certain value of T_{gmix}, given only the individual T_g values. This can be done very simply by using Equation 1.32 or Equation 1.33 or by estimating K in Equation 1.30 or Equation 1.31 and using Equation 1.29. To test for ideality or nonideality in the amorphous mixture, one can use these estimates of T_{gmix} and experimentally determine T_{gmix}. Deviations of experimental results from the expected plot would indicate that the mixing of components is nonideal; the deviations can be either negative or positive, meaning that the experimentally determined values of T_{gmix} are either greater or less than predicted. Values greater than those predicted for ideality indicate that the net excess free volume is less than that expected for the weighted sum of individual component free volumes, or that A:B interactions are greater than the A:A and B:B interactions. If the deviations are negative, this means that there is greater excess free volume than expected or stronger A:A and B:B interactions than those between A and B. Examples of T_g versus concentration profiles for ideal and nonideal mixtures with positive and negative deviations from ideality in miscible amorphous mixtures have been reported for pharmaceutical systems [43–45].

1.8 FORMATION AND PROPERTIES OF AMORPHOUS SOLID DISPERSIONS

The focus of the remaining parts of this chapter, and indeed, the rest of this book, will be on amorphous solid dispersions, which we will define as single-phase amorphous mixtures of an API and water-soluble polymer intended to produce enhanced aqueous dissolution and oral bioavailability [46]. The polymer is used (i) to provide long-term storage stability of the amorphous API by inhibiting solid-state crystallization and (ii) to maintain a desirable level of supersaturation in the dissolution medium by preventing solvent-mediated crystallization over the time period needed for the required bioavailability. In some cases a surfactant is included in an ASD to further promote dissolution and/or to facilitate the manufacturing process. Various details of the processing of amorphous dispersions and the ingredients used on a practical scale will be discussed throughout this book. Here, we wish to introduce some general concepts and issues that build on the principles discussed above. Depending on the physical and chemical characteristics of the API and polymer, amorphous solid dispersions are primarily prepared by either hot melt extrusion (HME) of a mixture of powdered API and polymer, or by the spray drying of a solution of API and polymer from a suitable volatile solvent. In HME the dry powder mixture is placed into an extruder at elevated temperatures that melts or softens the components to facilitate mixing. The material is continuously extruded, cooled, and chopped into small fragments, often called lentils, which are further milled with other formulation ingredients into powdered form. Ideally, to facilitate the process of molecular mixing, the crystalline API should be melted, if it is not prone to chemical decomposition at these elevated temperatures, and the polymer should have as low a T_g as possible to promote softening at these temperatures. Surfactants, such as those tabulated in Appendix A, are often added to lower melting and glass transition temperatures. Advantages of the HME process are that it can be carried out continuously, is solvent-free, and can be scaled up to large-scale manufacturing. Spray drying generally involves the spraying of a suitable solution of API and polymer into a heated chamber under tight control of droplet size, spray rate, and other process variables, followed by the rapid evaporation of the solvent and precipitation of the remaining solids to form the amorphous dispersion. Special consideration in the drying process is given to the solubility characteristics of the API and polymer so that miscibility is ensured and maintained as drying occurs. A wide range of solvents and solvent mixtures with varying polarity provide flexibility in the control of the drying process. Spray drying has the additional advantages of being easily scaled up for manufacturing and of providing a means of producing well-controlled particle sizes.

The major polymers used to prepare amorphous solid dispersions are either water soluble under all pH conditions or enteric coating polymers that contain acidic groups that ionize at higher pH values to become water soluble. The most widely used polymers include vinyl-pyrrolidone-based polymers, such as poly(vinyl pyrrolidone) (PVP) and the poly(vinyl pyrrolidone)-poly (vinyl acetate) copolymer (PVP-VA); cellulose-based polymers such as hydroxypropyl methylcellulose (HPMC) and hydroxyl propyl methylcellulose acetate-succinate; and methacrylate-based enteric coating polymer systems such as the Eudragits. The monographs in Chapter 14 and the Appendix A tables list various grades of these and other useful polymers with their corresponding T_g values. Generally speaking, polymer grades that have lower T_g values, for example, in the range of 100 °C, appear to be most useful in the HME process that requires that the ingredients be melted and/or softened to facilitate mixing. It is important to keep in mind that an ASD itself is only an extension of the API and that it must be further formulated and processed to produce a stable and effective oral solid dosage form. Consequently, an important initial goal is to produce an ASD that contains a minimum amount of polymer so that the required "dose" of the API–polymer combination does not become so excessive that it impacts the size of the dosage unit and influences the types and amounts of the other excipients required to ensure proper manufacturing and performance. This minimization of the amount of polymer must be measured, of course, against the primary role of the polymer as an inhibitor of solid-state and solvent-mediated crystallization.

Central to the functioning of a polymer in an ASD is the formation of a single-phase "miscible" system. As already described, the mixing tendencies of amorphous materials are governed by the same general thermodynamic principles that operate with the mixing of liquids. Thus, the concentration of components, their molecular size and shape, and their degree of polarity and ability to interact are all critical characteristics that must be evaluated in choosing an appropriate API–polymer combination. Typically, for simple liquids, miscibility is taken as the mutual *equilibrium solubility* of the components, determined experimentally in terms of concentration and temperature by the establishment of characteristic phase diagrams, which represent well-defined relationships between regions of miscibility and immiscibility. In dealing with mixtures of amorphous materials, prepared by processes that tend to produce metastable states, that is, melting or drying, we can expect most often that the API concentration will be greater than that representing the true equilibrium solubility of the crystalline API in the polymer, and hence still thermodynamically prone to eventual crystallization. If the amount of API is low enough, it is possible that a dispersion can be prepared at a level of API below the equilibrium solubility of the crystal, thus producing both kinetic and thermodynamic solubilities of the system. Recent studies have described a thermal method of measuring the equilibrium solubility of crystal drugs in various polymers, and have shown that these equilibrium solubilities are generally quite low, suggesting that most practical amorphous solid dispersions with the usual therapeutic dose ranges represent metastable solutions of API in polymer [47]. Thus, it is critical that consideration be given to how one can maintain the physical stability of amorphous API for periods required during the storage of the solid dosage form.

1.9 SOLID-STATE CRYSTALLIZATION FROM AMORPHOUS DISPERSIONS

We have already mentioned that the critical steps in preventing solid-state crystallization of an amorphous API are (i) inhibition of nucleation and crystal growth rates by reducing molecular mobility of the API, and (ii) direct interaction of a nucleation inhibitor with the API. The term direct interaction is used here since the formation of any miscible amorphous dispersion generally requires interaction between the API and polymer, most often through hydrogen bonding, whereas inhibition of nucleation requires interaction of the polymer with the specific functional groups on the API that are critical for nucleation. From the previous discussion of the amorphous state and amorphous mixtures, and Equation 1.32, we can now see, in practical terms, how an amorphous excipient with a high T_g and sufficient miscibility with an API would be able to produce relatively high values of T_{gmix}, compared with the T_g of the API alone, and hence reduce the molecular mobility of the API when stored at a particular temperature. Such a decrease in molecular mobility of the API in turn might be enough to reduce tendencies for crystallization of the API under the normally required storage conditions of 2–3 years. For example, if an amorphous API has a T_g value of 320 K (47 °C), alone, it would have to be stored near 0 °C to be at $T_g - T$ equal to at least about 50 K, roughly the temperature range where diffusive molecular mobility is reduced to time periods suitable for preventing crystallization during storage over a few years. To store a sample of this API at 25 °C for a few years without crystallization, for example, one would need to raise the T_g of the system to about 75 °C (348 K) or higher, which according to Equation 1.32, and the individual T_g values would require a minimum of about 30% w/w of a miscible polymer having a T_g of 150 °C (423 K). To store the sample at 40 °C (313 K) for a few years, one would have to raise the T_g to roughly 90 °C (363 K) by using about 50% w/w of this polymer. Figure 1.19 presents experimental studies of the percent of crystallization at 30 °C of amorphous indomethacin with a T_g of 42 °C (315 K) alone and mixed with PVPK90, having an average molecular weight of about 1.5 million and a T_g of 180 °C (453 K) [43]. Here, it can be observed that physical mixtures of the amorphous indomethacin and PVP containing 5% w/w PVP exhibit identical rates of crystallization with indomethacin alone. Such behavior was also observed at all concentrations up to 90% w/w PVP. Note in Figure 1.19 that only 5% w/w PVP in the form of a miscible mixture prevents any crystallization up to a period of 6 months, indicating a significant reduction in the rate of nucleation and crystal growth. It was shown further that increasing the amount of polymer in the dispersion to 30% w/w PVP prevented crystallization for over 2 years.

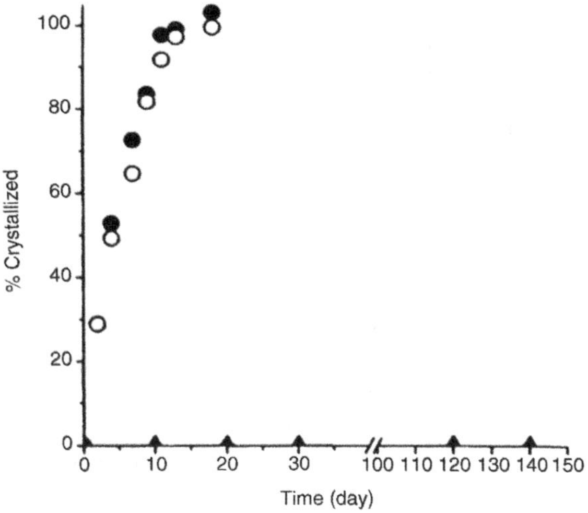

Figure 1.19. Percent crystallization of indomethacin from the amorphous state at 30 °C: alone (dark circles); physically mixed with 5% PVP (open circles); miscible amorphous mixture with 5% PVP (dark triangles) (reproduced with permission from Ref. 43. Copyright 1999, Springer).

Interestingly, whereas the 30% w/w PVP dispersion has a T_g of 61 °C (334 K) so that at 30 °C (303 K) $T_g - T$ is equal to 31 K, 5% w/w PVP with a T_g of 47 °C (320 K) has a $T_g - T$ of 17 K not very different from indomethacin alone with a $T_g - T$ of 12 K. Thus, it appears that the hypothetical analysis and predictions concerning $T_g - T$ and crystallization, presented above using Equation 1.32, are quite reasonable for the 30% w/w dispersion, while PVP at the 5% w/w level is much more effective than might be predicted from effects on molecular mobility alone. Indeed, others have shown that even at levels of 1–2% polymer in a miscible dispersion, there can be significant inhibition of crystallization [48,49] even though there is a negligible increase in T_g over that of the API alone. In the case of 5% w/w dispersions of indomethacin–PVP, evidence from FTIR measurements indicates that PVP interferes through hydrogen bonding with the dimer formation in amorphous indomethacin that is required to initiate nucleation, whereas no such indomethacin hydrogen bonding is observed in the physical mixture [48]. That the 30% w/w PVP dispersion was so effective with a $T_g - T$ of only 31 K, rather than 50 K, further suggests a more specific direct interaction at nucleation sites that contribute to the inhibition of crystallization. We can conclude, therefore, that the inhibition of solid-state crystallization of amorphous API by polymers in an ASD is generally governed by direct interactions that prevent nucleation and by decreasing molecular mobility, that is, increasing diffusive α-relaxation times, and likely increasing secondary β-relaxation times that serve as precursors to the onset of diffusional motions.

Up to this point the ASD systems discussed have been assumed to be essentially free of any water taken up from the atmosphere. Consequently, all T_g values, relaxation times, and quantitative estimations or predictions were made using parameters established for the "dry" API and polymer. Generally, however, when an amorphous solid is exposed to elevated relative humidity (RH), the solid tends to absorb the water vapor into its bulk structure and form a miscible amorphous mixture [50]. The amount of water taken up at a particular RH and temperature will be greater with an increase in degree of polarity of the

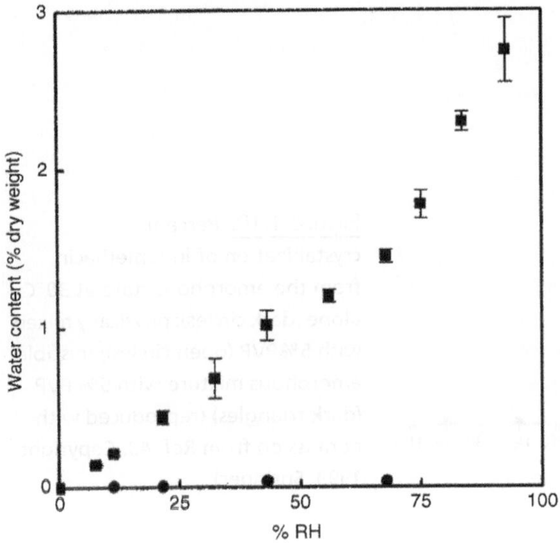

Figure 1.20. Water vapor absorption by amorphous (squares) and crystalline (circles) indomethacin at 30 °C as a function of relative humidity (reproduced with permission from Ref. 52. Copyright 1997, John Wiley & Sons, Inc.).

solid. This is in contrast to water uptake by crystalline solids where water molecules are generally confined to the surface, rarely reaching more than 0.1% w/w adsorbed, or the equivalent of 3–4 molecular layers, even for very polar solids such as NaCl [51]. The relationship between the amount of water vapor absorbed and RH at a particular temperature can be analyzed by determining a water sorption isotherm, as shown for amorphous indomethacin in Figure 1.20 [52], and for three widely used polymers PVP, HPMC, and HPMCAS in Figure 1.21 [53]. Note that the amount of water absorbed at 30 °C by the relatively hydrophobic indomethacin molecule is quite small, for example,

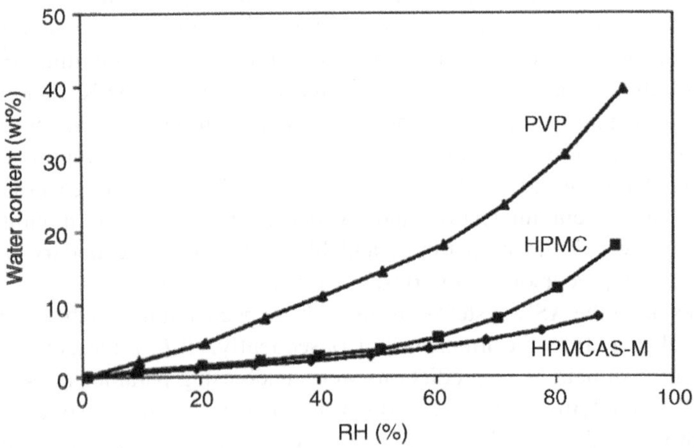

Figure 1.21. Water vapor absorption by three amorphous polymers varying in polarity (reproduced with permission from Ref. 53. Copyright 2008, American Chemical Society).

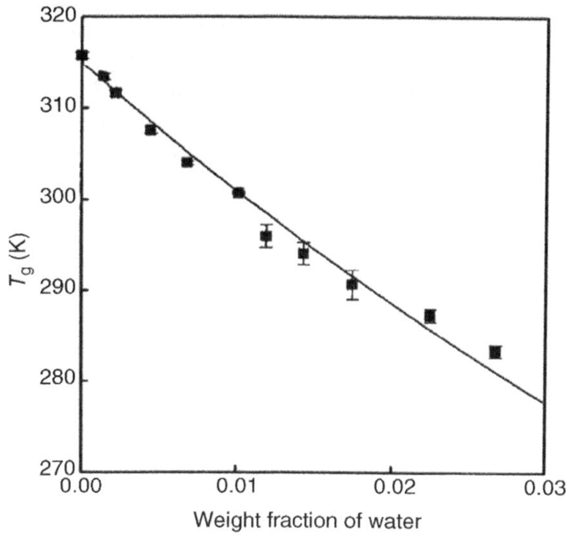

Figure 1.22. Glass transition temperature of amorphous indomethacin as a function of water content (reproduced with permission from Ref. 52. Copyright 1997, John Wiley & Sons, Inc.).

about 1% w/w at 50% RH and 2% w/w at 83% RH, whereas water vapor absorption at room temperature by PVP, HPMC, and HPMCAS at 75% RH is about 26% w/w, 10% w/w, and 7% w/w, respectively. These results are consistent with the general order of polarity and hydrogen bonding ability of these polymers. What they also indicate is that most of the water that is absorbed by a typical API–polymer dispersion would be expected to be associated with the polymer. This is a reasonable generalization since amorphous dispersions are prepared because the API is usually somewhat hydrophobic. Water absorbed into an amorphous solid would be expected to have an effect on the bulk properties of the solid, for example, T_g and molecular mobility, as can be seen in Figure 1.22, where T_g is plotted versus water content for amorphous indomethacin. Here, we can see that in both cases water acts to reduce T_g very significantly, acting as an excellent plasticizer. Such plasticizing effects can be roughly predicted for any water content by applying Equation 1.32 and using the values of 136, 315, and 453 K for the T_g of water, indomethacin, and PVP, respectively. Consider, first, a situation where indomethacin alone is exposed to an environment at 40 °C and 75% RH, resulting in a concentration of water in the API of about 1.7% w/w. From the application of Equation 1.32, it can be shown that T_g would be reduced from 315 to about 308 K, whereas storage of PVP at 40 °C and 75% RH, with 26% w/w water being absorbed, leads to a reduction of the T_g of PVP from 453 to 282 K. Such a significant reduction in T_g close to 0 °C (273 K) would clearly create significant mobility at 40 °C where the sample is stored at 31 K above the new T_g of PVP in the supercooled liquid state. One might mitigate this result to some extent by choosing to use a less polar polymer such as HPMC that has a dry T_g of about 175 °C (448 K) and absorbs 10% w/w water at 75% RH. From Equation 1.31 we can calculate that the T_g of HPMC in the presence of 10% w/w water would be reduced to 364 K, and therefore when stored at 40 °C it would be 52 K below T_g in the glassy state, a much better solution for preventing crystallization in the hydrated

API dispersion. To observe the trade-off between the dry T_g of the polymer and its ability to absorb water, let us finally consider HPMCAS, which has a "dry" T_g of about 115 °C (388 K), and, as shown above, the ability to absorb about 7% w/w water at 75% RH. From Equation 1.32 we would estimate that T_g for HPMCAS with 7% w/w water content would be reduced from 448 K to 344 K, so that at 40 °C, the sample of HPMCAS would have a T_g that is 30 K above the new T_g, a better result than when using PVP that takes up considerably more water. Because of the lower T_g of HPMCAS compared with HPMC, however, this system is still in the supercooled liquid, which indicates the importance of reducing water content to very low values. The fact that HPMCAS is very widely used to form API amorphous dispersions, as will be seen throughout this book, indicates that HPMCAS most likely must also function as a stabilizer by being able to more directly interact with the API and interfere with nucleation. In conclusion, this discussion generally suggests a strategy for storing amorphous solids for long periods without crystallization that includes (i) using polymers with relatively high T_g; (ii) using as high a concentration of polymer in the dispersion as is possible without introducing other negative factors; (iii) using polymers that tend to absorb less water than PVP (see Figure 1.21); (iv) limiting exposure of the dispersion and solid formulation to humid atmospheres by proper processing conditions and packaging; and (v) if possible, including specific nucleation inhibitors that would not adversely increase molecular mobility. Clearly, removal of as much absorbed water as possible is critical.

1.10 DISSOLUTION AND SUPERSATURATION OF API FROM AMORPHOUS SOLID DISPERSIONS

Clearly, the main purpose of using amorphous forms of an API in solid dosage forms is to enhance its dissolution rate and oral bioavailability when the crystalline form cannot be used for such purposes. As shown in Figure 1.16, upon contacting the aqueous dissolution media, the API generally dissolves rapidly and reaches some level of supersaturation primarily determined by the free energy difference between the amorphous and crystalline forms. Upon attaining a certain level of supersaturation in the aqueous phase, the API molecules will generally undergo solvent-mediated nucleation and crystal growth, thus depleting the solution of dissolved API and reducing the solubility advantage sought with such systems; the greater the level of supersaturation attained, the greater the rate of nucleation and the shorter the time period over which supersaturation levels can be maintained. If the rate of dissolution of the added amorphous API, upon contact with the aqueous medium, is slow relative to solid-state crystallization rates, it is also possible that solid API would undergo crystallization due to the plasticizing effects of the water and not attain significant supersaturation because of reduced amounts of amorphous API. As we have already seen, many polymers can form amorphous dispersions with an API and significantly inhibit solid-state crystallization by reducing molecular mobility and by direct interaction at nucleation sites. Consequently, if there is no significant slowing down of the dissolution of the API and polymer when they make contact with the dissolution medium, there will likely be sufficient inhibition of solid-state crystallization to allow the API and polymer to readily dissolve and reach high levels of supersaturation. Having seen that the

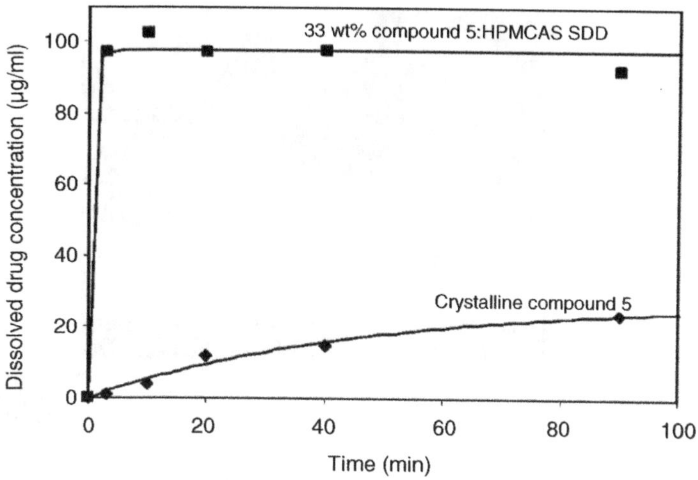

Figure 1.23. Comparison of dissolution rates and levels of supersaturation for a crystalline compound and its amorphous dispersion with HPMCAS (reproduced with permission from Ref. 53. Copyright 2008, American Chemical Society).

formation of API–polymer amorphous dispersions can greatly inhibit solid-state crystallization during storage even at low polymer concentrations, and that the best polymers in this regard tend to directly interact with the API through hydrogen bonding [48], it seems likely that such polymers could also inhibit solvent-mediated crystallization after dissolution of the API and polymer by such interactions in solution. A number of studies, indeed, have shown that polymers, such as PVP, HPMC, and HPMCAS, added to an aqueous solution at solution concentrations that might be expected from a typical API–polymer dispersion significantly inhibit nucleation, decrease the rate of crystal growth, and maintain high levels of supersaturation over extended periods of time [54,55]. That this actually occurs with the dissolution of API–polymer dispersion is illustrated in Figure 1.23 [53]. Recently, increasing evidence has been accumulated to suggest that supersaturation levels of API administered as API–polymer amorphous dispersions can be maintained at high levels by additional possible mechanisms. For example, as illustrated in Figure 1.24, it has been suggested that API–HPMCAS systems upon contact with a dissolution medium initially precipitate as complex high-energy colloidal noncrystalline phases that can retard crystallization and produce a high degree of supersaturation [53]. This behavior has been attributed to the relatively hydrophobic and surfactant-like nature of HPMCAS, and in this regard it has also been suggested that it is likely that such colloidal systems form during *in vivo* dissolution by interacting with colloid forming gastrointestinal materials such as fatty acid derivatives and bile salts. It has also been recently shown that a relatively hydrophobic API, ritonavir, in PVP dispersion at high levels of supersaturation can phase separate as a colloidal API-rich liquid phase that resists crystallization and maintains high levels of supersaturation [56]. Greater detailed discussion of all of these principles related to the inhibition of solid-state

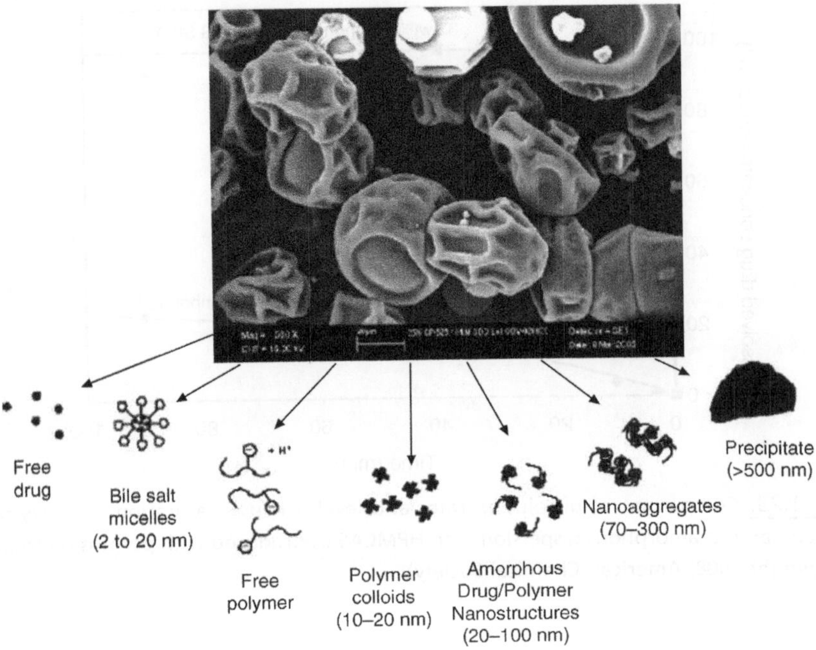

Figure 1.24. Species that can form when HPMCAS solid dispersions are added to aqueous solutions simulating duodenal and intestinal contents (reproduced with permission from Ref. 53. Copyright 2008, American Chemical Society).

and solvent-mediated crystallization during dissolution, including choice of polymers, surfactants, and processes, will be presented in later chapters.

1.11 PHARMACEUTICAL DEVELOPMENT OF AMORPHOUS SOLID DISPERSIONS

Once it has been decided that an ASD will be moved forward, a number of factors need to be considered during the development process. Many scientists consider an ASD as a "formulation" that is then processed into a usable dosage form. Another way to view an ASD is as another solid form. When considered as another solid form, the characterization and property assessment is similar to what has been performed previously for amorphous or crystalline APIs. Factors such as solubility, stability, excipient compatibility, handling, storage, performance, and developability all need to be considered. It is important to outline the critical properties that the dispersion needs to demonstrate in order to produce a successful drug product, and these properties can be used as selection criteria to find the best ASD.

Amorphous solid dispersions have their challenges, such as miscibility, stability, and hygroscopicity, but all of these can be overcome with careful consideration of manufacturing, handling, and downstream processing. Understanding the limits of the ASD and incorporating these limits into the development plan are key decision points during development of these materials.

For early development projects, an ASD may provide the initial solubility boost needed to show proof of concept in a very simple formulation. It may be easy to overcome the challenges of a nonideal ASD for small-scale clinical trials, but these same issues may be insurmountable when moving to large-scale clinical batches. At that point, it must be decided if there is another dispersion available with better properties that can be moved forward or if there is a crystalline alternative that may now produce the needed performance based on the initial pharmacokinetic or metabolism studies. A change in solid form during the development process is a common occurrence as procedures are streamlined in early development to reduce costs and time. Once a drug has proven its worth, a more extensive look at possible forms and dosage forms takes place to find viable alternatives for late stage development. As with any solid form selection, the ASD with all the ideal properties may not be available, but a solid with most of the critical properties can be found and the noncritical properties can be dealt with via other methods, such as processing, formulation, packaging, and the like.

The goal of this book is to outline the various stages of producing, choosing, testing, formulating, and developing ASDs. Polymers and surfactants are necessary components of the dispersion and it is important to have an understanding of polymer properties when dealing with ASD projects. This will also help facilitate which polymers to include in a dispersion screen and the screening methods employed to produce possible dispersions. Chapters on characterization, stability, dissolution, and solubility will provide guidelines on dispersion properties that can be used to select a dispersion for early or late development. Choosing a manufacturing method, usually spray drying or melt extrusion, requires a knowledge of the properties needed for formulating the drug product as well as a comparison with material produced from small-scale processes used in initial studies to ensure that comparable properties are obtained for all materials.

The next phase involves using the ASD in various development activities such as formulation development (early or late) and preclinical/clinical studies. Early formulation development may employ a very simple or streamlined drug product to get an initial assessment of performance. Late-stage formulation development needs to cover a wide range of issues related to large-scale processing, packaging, storage, and use. Early preclinical studies are centered around proof of concept and bioavailability, while clinical studies need to deal with safety, efficacy, and performance of the API. At any stage, the ASD may need to be modified or replaced to move development forward.

It is also important to understand the regulatory components and requirements for dispersions. These include both manufacturing and testing of the dispersion and the final drug product. Information included in early versus late filings will be different and dependent on a number of factors related to the dispersion, API, and formulation. All dispersions found during development can be patented and patent strategies for dispersions are critical for development and life cycle management over the life of the product.

This book will provide information and general strategies that can be used to develop drug products containing ASDs. However, every compound is unique and scientific rigor and understanding will be needed to put together the best development plan to successfully incorporate amorphous solid dispersions in drug products.

REFERENCES

1. Noyes, A.A. and Whitney, W.R. (1897) The rate of solution of solid substances in their own solutions. *J. Am. Chem. Soc.*, 54:930–934.
2. Amidon, G.L., Lennernas, H., Shah, V.P., and Crison, J.R. (1995) A theoretical basis for a biopharmaceutical drug classification: the correlation of *in vitro* drug product dissolution and *in vivo* bioavailability. *Pharm. Res.*, 12:413–420.
3. Hancock, B.C. and Shamblin, S.L. (2001) Molecular mobility of amorphous pharmaceuticals using differential scanning calorimetry. *Thermochim. Acta*, 380:95–107.
4. Lee, W.A. and Knight, G.J. (1970) Ratio of the glass transition temperature to the melting point in polymers. *Br. Polym. J.*, 2:73–80.
5. Fukuoka, E., Makita, M., and Yamamura, S. (1989) Glassy state of pharmaceuticals. III. Thermal properties and stability of glassy pharmaceuticals. *Chem. Pharm. Bull.*, 37(4):1047–1050.
6. Hodge, I.M. (1995) Physical aging in polymer glasses. *Science*, 267:1945–1947.
7. Zhu, L. and Yu, L. (2010) Generality of forming stable organic glasses by vapor deposition. *Chem. Phys. Lett.*, 499:62–65.
8. Vincent, C., Willart, J.F., Lefort, R., Derollez, P., Danede, F., and Decamps, M. (2011) Solid-state amorphization kinetics of alpha lactose upon mechanical milling. *Carbohydr. Res.*, 346:2622–2628.
9. Bates, S., Kelly, R.C., Ivanisevic, I., Schields, P., Zografi, G., and Newman, A.W. (2007) Assessment of defects and amorphous structure produced in raffinose pentahydrate upon dehydration. *J. Pharm. Sci.*, 9(5):1418–1433.
10. Guo, Y., Byrn, S.R., and Zografi, G. (2000) Physical characteristics and chemical degradation of amorphous quinapril hydrochloride. *J. Pharm. Sci.*, 89(1):128–143.
11. Ahlneck, C. and Zografi, G. (1990) The molecular basis of moisture effects on the physical and chemical stability of drugs in the solid state. *Int. J. Pharm.*, 62:87–95.
12. Bates, S., Zografi, G., Engers, D., Morris, K., Crowley, K., and Newman, A.W. (2006) Analysis of amorphous and nanocrystalline solids from their X-ray diffraction patterns. *Pharm. Res.*, 23(10):2333–2348.
13. Dykhne, T., Taylor, R., Florence, A., and Billinge, S.J.L. (2011) Data requirements for the reliable use of atomic pair distribution functions in amorphous pharmaceutical fingerprinting. *Pharm. Res.*, 28:1041–1048.
14. Finney, J.L. (1970) Random packings and the structure of simple liquids: II. The molecular geometry of simple liquids. *Proc. R. Soc. Lond. A*, 319:495–507.
15. Silescu, H. (1999) Heterogeneity at the glass transition: a review. *J. Non-Cryst. Solids*, 243:81–108.

16. Williams, G. and Watts, D.C. (1970) Non-symmetrical dielectric relaxation behavior arising from a simple empirical decay function. *Trans. Faraday Soc.*, 66:80–85.
17. Schonhals, A., Kremer, F., Hofmann, E., Fischer, E.W., and Schlosser, E. (1993) Anomalies in the scaling of the dielectric α-relaxation. *Phys. Rev. Lett.*, 70:3459–3462.
18. Swallen, S.F. and Ediger, M.D. (2011) Self-diffusion of the amorphous pharmaceutical indomethacin near T_g. *Soft Matter*, 7:10339–10344.
19. Angell, C.A. (1988) Perspectives on the glass transition. *J. Phys. Chem. Solids*, 49(8):863–871.
20. Crowley, K. and Zografi, G. (2001) The use of thermal methods for predicting glass-former fragility. *Thermochim. Acta*, 380:79–83.
21. Plazek, D.J. and Magill, J.H. (1968) Physical properties of aromatic hydrocarbons: IV. Analysis of the temperature dependence of the viscosity and the compliance of 1,3,5,-tri-α-naphthylbenzene. *J. Chem. Phys.*, 49:3678–3883.
22. Andronis, V. and Zografi, G. (1998) The molecular mobility of supercooled indomethacin as a function of temperature and relative humidity. *Pharm. Res.*, 15(6):835–842.
23. Zhu, L., Brian, C.W., Swallen, S.F., Straus, P.T., Ediger, M.D., and Yu, L. (2011) Surface self -diffusion of an organic glass. *Phys. Rev. Lett.*, 106:256103.
24. Johari, G.P. and Goldstein, M. (1970) Viscous liquids and the glass transition: II. Secondary relaxations in glasses of rigid molecules. *J. Chem. Phys.*, 53(6):2372–2388.
25. Vogel, M., Tschirwitz, C., Schneider, G., Koplin, C., Medick, P., and Rrossler, E. (2002) A 2H NMR and dielectric spectroscopy study of the slow β-process in organic glass formers. *J. Non-Cryst. Solids*, 307–310:326–335.
26. Ngai, K. and Paluch, M. (2004) Classification of secondary relaxation in glass-formers based on dynamic properties. *J. Chem. Phys.*, 120:857–873.
27. Hatase, M., Hanaya, M., Hikima, T., and Oguni, M. (2002) Discovery of homogeneous nucleation-based crystallization in simple glass-forming liquid of toluene below the glass transition temperature. *J. Non-Cryst. Solids*, 307–310:257–263.
28. Andronis, V. and Zografi, G. (2000) Crystal nucleation and growth of indomethacin polymorphs from the amorphous state. *J. Non-Cryst. Solids*, 271:236–248.
29. Oguni, M. (1997) Intra-cluster rearrangement model for the α-process in supercooled liquids, as opposed to cooperative rearrangement of whole molecules within a cluster. *J. Non-Cryst. Solids*, 210:171–177.
30. Hancock, B.C. and Parks, M. (2000) What is the true solubility advantage for amorphous pharmaceuticals? *Pharm. Res.*, 17:397–404.
31. Murande, S.B., Pikal, M.J., Shanker, R.M., and Bogner, R.H. (2009) Solubility advantage of amorphous pharmaceuticals I. *J. Pharm. Sci.*, 99:1254–1264.
32. Ozaki, S., Minamisono, T., Yamashita, T., Kato, T., and Kushida, I. (2012) Supersaturation–nucleation behavior of poorly soluble drugs and its impact on the oral absorption of drugs in thermodynamically high-energy forms. *J. Pharm. Sci.*, 101:214–222.
33. Lu, Q. and Zografi, G. (1998) Phase behavior of binary and ternary amorphous mixtures containing indomethacin, citric acid and PVP. *Pharm. Res.*, 15:1202–1206.
34. Flory, P.J. (1953) *Principles of Polymer Chemistry*, Cornell University Press, Ithaca, NY.
35. Newman, A., Engers, D., Bates, S., Ivanisevic, I., Kelly, R.C., and Zografi, G. (2008) Characterization of amorphous API:polymer mixtures using X-ray diffraction. *J. Pharm. Sci.*, 97(11):4840–4856.

36. Cheung, M.K., Wang, J., Zheng, S., and Mi, Y. (2000) Miscibility of poly(epichlorohydrin)/poly(vinyl acetate) blends investigated with high resolution solid-state ^{13}C NMR. *Polymer*, 41:1469–1474.
37. Pham, T.N., Watson, S.A., Edwards, A.J., Chavda, M., Clawson, J.S., Stroheim, M., and Vogt, F.G. (2010) Analysis of amorphous solid dispersions using 2D solid-state NMR and ^{1}H T_1 relaxation measurements. *Mol. Pharm.*, 7:2194–2206.
38. Gordon, M. and Taylor, J.S. (1952) Ideal copolymers and the second-order transition of synthetic rubbers: I. Noncrystalline co-polymers. *J. Appl. Chem.*, 2:493–500.
39. Couchman, P.R. and Karasz, F.E. (1978) A classical thermodynamic discussion on the effect of composition on glass transition temperature. *Macromolecules*, 11:117–119.
40. Fox, T.G. (1956) Influence of diluent and copolymer composition on the glass transition temperature of a polymer system. *Bull. Am. Phys. Soc.*, 1:123.
41. Crowley, K.J. and Zografi, G. (2002) Water vapor absorption into amorphous hydrophobic drug/poly(vinylpyrrolidone) dispersions. *J. Pharm. Sci.*, 91:2150–2165.
42. Zhang, J. and Zografi, G. (2001) Water vapor absorption into amorphous sucrose-poly(vinyl-pyrrolidone) and trehalose-poly(vinylpyrrolidone) mixtures. *J. Pharm. Sci.*, 90:1375–1385.
43. Matsumoto, T. and Zografi, G. (1999) Physical properties of solid molecular dispersions of indomethacin with poly(vinylpyrrolidone) and poly(vinylpyrrolidone-*co*-acetate) in relation to indomethacin crystallization. *Pharm. Res.*, 16(11):1722–1728.
44. Shamblin, S.L., Taylor, L.S., and Zografi, G. (1998) Mixing behavior of colyophilized binary systems. *J. Pharm. Sci.*, 87:694–701.
45. Kougaz, K. and Clas, S.D. (2000) Crystallization inhibition in solid dispersions of MK-0591 and poly(vinylpyrrolidone) polymers. *J. Pharm. Sci.*, 89:1325–1334.
46. Newman, A.W., Knipp, G., and Zografi, G. (2012) Assessing the performance of amorphous solid dispersions. *J. Pharm. Sci.*, 101:1355–1377.
47. Tao, J., Sun, Y., Zhang, G.G.Z., and Yu, L. (2009) Solubility of small molecule crystals in polymers: D-mannitol in PVP, indomethacin in PVP/VA and nifedipine in PVP/VA. *Pharm. Res.*, 26(4):855–864.
48. Taylor, L.S. and Zografi, G. (1997) Spectroscopic characterization of interactions between PVP and indomethacin in amorphous molecular dispersions. *Pharm. Res.*, 14:1691–1698.
49. Cai, T., Zhu, L., and Yu, L. (2011) Crystallization of organic glasses: effects of polymer additives on bulk and surface crystal growth in amorphous nifedipine. *Pharm. Res.*, 28:2458–2466.
50. Newman, A.W., Reutzel-Edens, S.M., and Zografi, G. (2008) Characterization of the "hygroscopic" properties of active pharmaceutical ingredients. *J. Pharm. Sci.*, 97:1047–1059.
51. Ewing, G.E. (2005) H_2O on NaCl: from single molecule to clusters, to monolayer, to thin film, to deliquesce. *Intermolecular Forces and Clusters II*, Structure and Bonding, Vol. 116, Springer, pp. 1–25.
52. Andronis, V., Yoshioka, M., and Zografi, G. (1997) Effects of sorbed water on the crystallization of indomethacin from the amorphous state. *J. Pharm. Sci.*, 86:346–351.
53. Friesen, D.T., Shanker, R., Crew, M., Smithey, D.T., Curatolo, W.J., and Nightingale, J.A.S. (2008) Hydroxypropylmethylcellulose acetate succinate-based spray-dried dispersions: an overview. *Mol. Pharm.*, 5:1003–1019.
54. Alonso, D.E., Zhang, G.G.Z., Zhou, D., Gao, Y., and Taylor, L.S. (2009) Understanding the behavior of amorphous pharmaceutical systems during dissolution. *Pharm. Res.*, 27:608–618.

55. Hevbare, G.A., Liu, H., Edgar, K.J., and Taylor, L.S. (2013) Maintaining supersaturation in aqueous drug solutions: impact of different polymers on induction times. *Cryst. Growth Des.*, 13:740–751.

56. Hevbare, G.A. and Taylor, L.S. (2013) Liquid–liquid phase separation in highly supersaturated aqueous solutions of poorly water-soluble drugs: implications for solubility enhancing formulations. *Cryst. Growth Des.*, 13:1497–1509.

2

POLYMERS AND SURFACTANTS

Padma Narayan, William W. Porter III, Meinolf Brackhagen, and Christopher Tucker

The Dow Chemical Company, Midland, MI, USA

2.1 POLYMERS COMMONLY USED IN AMORPHOUS SOLID DISPERSIONS

2.1.1 Overview

An increasing number of drug candidates have a low aqueous solubility relative to the targeted dose that limits the potential efficacy of the drug and can result in the termination of the drug's development. To address this challenge, several strategies based on polymers have been used to form amorphous solid dispersions (ASDs). Early approaches involved the use of polyvinylpyrrolidone (PVP) [1] and polyethylene glycol (PEG) [2]. The number of polymers used as carriers in solid dispersions reported in the literature has expanded dramatically and comprises the following groups (where the asterisk denotes an enteric polymer) [3]:

a. *Polyglycols*: PEG, polyethylene oxide (PEO), and polyethylene–propylene glycol copolymer.

Pharmaceutical Amorphous Solid Dispersions, First Edition. Edited by Ann Newman.

b. *Polyvinyl polymers*: PVP, polyvinyl alcohol (PVA), crospovidone (PVP-CL), polyvinylpyrrolidone–polyvinyl acetate (PVP–VA) copolymer, polyvinyl alcohol phthalate (PVAP),* and polyvinylcaprolactam–polyvinyl acetate grafted on PEG.
c. *Cellulosic polymers*: hydroxypropylmethylcellulose (HPMC), hydroxypropylcellulose (HPC), carboxymethyl ethylcellulose (CMEC),* cellulose acetate phthalate (CAP),* cellulose acetate trimellitate (CAT),* hydroxypropylmethylcellulose phthalate (HPMCP),* and hydroxypropylmethylcellulose acetate succinate (HPMCAS).*
d. Acrylate and methacrylate (co-)polymers (*methacrylates available as an enteric polymer).

As there are wide varieties of polymers that can be utilized in ASD systems, their properties, in combination with the drug substance, can dictate the needs for different manufacturing routes. Important performance factors of polymers include glass transition temperature (T_g), thermal stability, dissolution profiles in organic solvents, and an optimal solubilization of the active pharmaceutical ingredient (API) in the ASD. Despite the available variety of materials, formulators tend to use polymers that have historically proven safety records, often in other pharmaceutical application areas, combined with capability to (a) stabilize amorphous API in the solid state as well as (b) enable and maintain supersaturation in aqueous environment. Consequently, the most commercially marketed ASD products use cellulosics (HPMC, HPMCAS, and HPC) or polyvinylpyrrolidones (PVP and PVP–VA). PEG stands out from this group as it tends to crystallize itself and may act as a wetting agent and solid carrier for stable forms of the API, for example, stable amorphous forms of API or crystals (Gris-PEG®). Table 2.1 summarizes commercial drug products that involve solid dispersions and the types of polymers used [4].

2.1.2 Polymers

The carriers most commonly used in commercialized ASD dosage forms are detailed in this section. Table 2.2 provides detailed information on typical polymers and their pharmacopeia status.

2.1.2.1 Polyvinylpyrrolidone

STRUCTURE AND SYNTHESIS Polyvinylpyrrolidone, with the chemical structure shown in Figure 2.1, is made by free radical polymerization of *N*-vinylpyrrolidone in water using hydrogen peroxide as initiator [5].

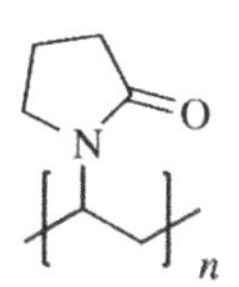

Figure 2.1. Chemical structure of PVP.

TABLE 2.1. Commercial Pharmaceutical Products That Use Solid Dispersion Technology [4]

Product	API	BCS Class	Company	Carrier	Technology	References
Gris-PEG®	Griseofulvin	II	Pedinol Pharmacal Inc.	PEG	Melt extrusion	[4a,c,d]
Nimotop®	Nimodipine	II	Bayer AG	PEG	Spray drying/fluid bed	[4b]
Cesamet®	Nabilone		Valeant Pharmaceuticals	PVP	Spray drying	[4b,c]
Kaletra® tablets	Lopinavir and ritonavir	IV	AbbVie Inc.	PVP–VA	Melt extrusion	[4b,c]
Norvir® tablets	Ritonavir	IV	AbbVie Inc.	PVP–VA	Melt extrusion	[4b,c]
Sporanox®	Itraconazole	II	Janssen Pharmaceuticals	HPMC	Spray layering	[4b,c]
Intelence®	Etravirine	IV	Janssen Pharmaceuticals (Tibotec)	HPMC	Spray drying	[4b,c]
Prograf®	Tacrolimus	II	Astellas Pharma Inc.	HPMC	Spray drying/fluid bed	[4b,c]
Nivadil®	Nilvadipine	II	Fujisawa Pharmaceutical Co., Ltd	HPMC		[4c]
Certican® or Zortress®	Everolimus	III	Novartis Pharmaceuticals	HPMC		[4c]
Isoptin-SRE®	Verapamil	I	AbbVie Inc.	HPC/HPMC	Melt extrusion	[4c]
Incivek®	Telaprevir		Vertex Pharmaceuticals	HPMCAS		[4c]
Zelboraf®	Vemurafenib		Roche	HPMCAS	Coprecipitation	[4c]

Adapted with permission from Ref. 40b.

TABLE 2.2. Polymers Commonly Used in Marketed Solid Dispersions

Polymer Type	Structural Features	Manufacturer	Regulatory Status
PVP	• Neutral	BASF, ISP	USP/Ph. Eur./JPE (povidone)
PVP-VA	• Neutral • Pyrrolidone copolymer	BASF, ISP	USP/Ph. Eur./JPE (copovidone)
MAA-MMA	• Acidic • Polyacrylate	Evonik	USP/NF/Ph. Eur./JPE (polymethacrylates)
HPMC	• Neutral • Cellulosic	Dow, Shin-Etsu, Ashland	USP/Ph. Eur./JPE (hypromellose)
HPMCAS	• Acidic • Cellulosic	Shin-Etsu, Dow (4Q2013), Ashland (announced)	NF/JPE (hypromellose acetate succinate)

The molecular weight can be tailored over a broad range by adjusting the amount of initiator utilized in the synthesis. PVP is usually obtained as a white, hygroscopic powder via spray drying. The available molecular weight (MW) range, determined by light scattering, is 2000–1,500,000 Da. The *K*-value, calculated from the kinematic viscosity of an aqueous solution [6], is used as a substitute to characterize the average degree of polymerization (DP). The *K*-value of commercial products (e.g., Plasdone® from ISP and Kollidon® from BASF) ranges from 12 to 90 [7] and is usually a part of the trade name. Frequently offered grades are K12, K17, K25, K30, and K90. Some of the products with lower degrees of polymerization (K12 and K17) are available as pyrogen-free grades, which are frequently used in injectables. PVP forms complexes with a range of pharmaceutically active ingredients, such as PVP–iodine [8].

Physical Properties PVP is soluble in water over a pH range of 1–10 and is readily soluble in polar organic solvents such as C_1–C_4 alcohols, methylene chloride, and dichloromethane. Its maximum concentration in less polar solvents (acetone, toluene, dioxane, diethyl ether, and ethyl acetate) is less than 2% [5b]. Under strong alkaline conditions at elevated temperatures, cross-linking can occur that leads to a water-insoluble but slightly swellable polymer [5a].

The glass transition temperatures of polyvinylpyrrolidone vary with molecular weight and range from 90 °C (K17 grade) up to 189 °C (K90 grade) [5b,9]. In addition, water uptake of PVP at 50% humidity is approximately 20% and about 40% at 75% humidity [5b].

2.1.2.2 PVP–VA

Structure and Synthesis Vinylpyrrolidone–vinyl acetate copolymer or copovidone, depicted in Figure 2.2, is produced by free radical polymerization of vinyl acetate and vinylpyrrolidone monomers in organic solvents such as ethanol or 2-propanol. The use of organic solvents is dictated by the insolubility of vinyl acetate in water. The weight

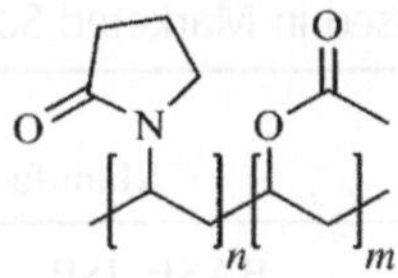

Figure 2.2. Chemical structure of PVP–VA.

ratio of vinylpyrrolidone to vinyl acetate is 3:2 [5b,10]. The *K*-value of commercial grades is about 30 according to the manufacturer's information [5b].

PHYSICAL PROPERTIES The PVP–VA polymer is freely soluble in water and organic solvents such as C_1–C_4 alcohols and dichloromethane. Similar to its parent polymer, PVP, the maximum concentration of PVP–VA in less polar organic solvents is limited (less than 1% in diethyl ether and pentane) [5b]. The glass transition temperature of copovidone is reported to be 103 °C [5b] to 106 °C [9]. In addition, water uptake at 50% humidity is about 20% [9].

2.1.2.3 Methacrylate Copolymers

STRUCTURE AND SYNTHESIS/PHYSICAL PROPERTIES

Poly(methacrylic acid-co-methyl methacrylate) Poly(methacrylic acid-*co*-methyl methacrylate) (p(MAA-*co*-MMA)), an anionic copolymer depicted in Figure 2.3, is a polymerization product of methacrylic acid and methyl methacrylate. The dissolution pH of p(MAA-*co*-MMA) is dictated by the ratio of monomers utilized in the polymerization. For example, a 1:1 molar ratio of methyl methacrylate and methacrylic acid results in a dissolution above pH 6.0. The resulting polymer is designated as *Methacrylic Acid Copolymer Type A* [11]. *Methacrylic Acid Copolymer Type B*, synthesized using a 2:1 molar ratio of methyl ester and carboxylic acid monomers, results in a dissolution pH of >7.0. Glass transition temperatures of these polymers are typically greater than 150 °C [12].

Poly(methacrylic acid-co-ethyl acrylate) Using ethyl acrylate instead of methyl methacrylate in a 1:1 molar ratio with methacrylic acid yields a polymer that is soluble above pH 5.5, as shown in Figure 2.4. This polymer, currently titled *Methacrylic Acid Copolymer Type C*, will be officially titled *Methacrylic Acid and Ethyl Acrylate Copolymer* after December 1, 2015 [11]. The glass transition temperature is about 110 °C [13].

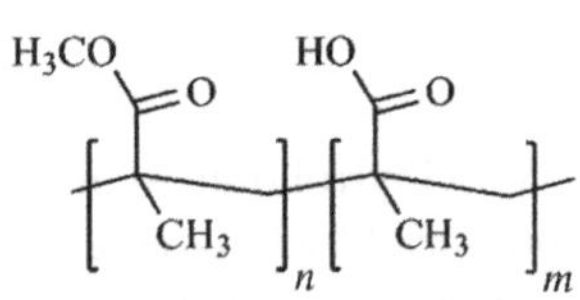

Figure 2.3. Chemical structure of p(MAA-co-MMA): n:m = 1:1 (type A); n:m = 2:1 (type B).

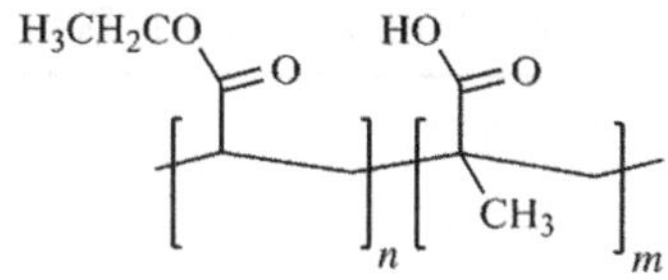

Figure 2.4. Chemical representation of p(MAA-co-EA).

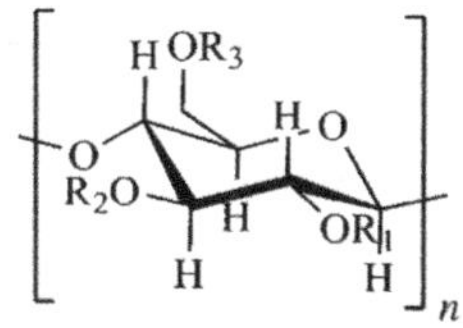

R_1, R_2, R_3 = H, CH_3, $CH_2CH(OH)CH_3$, $CH_2CH(OCH_3)CH_3$

Figure 2.5. Representative structure of HPMC.

2.1.2.4 Hydroxypropylmethylcellulose (Hypromellose)

Structure and Synthesis/Physical Properties The base material for HPMC is cellulose, a natural polymer that is abundant in plants as the primary component of plant cell walls. Cellulose is an extremely long ($MW > 1 \times 10^6$ Da), highly stereoregular polysaccharide composed of anhydroglucose units (AGUs) [14].

HPMC, depicted in Figure 2.5, is made by reaction of cellulose with methyl chloride, propylene oxide, and caustic soda. There are three available OH groups of cellulose and the resulting substituents R_1, R_2, and R_3 are predominantly CH_3, H, $CH_2CH(OH)CH_3$, and $CH_2CH(OCH_3)CH_3$ [15]. The last substituent is formed by further methylation of a hydroxypropyl moiety.

The substitution degree of HPMC is characterized by weight percent of methoxy and hydroxypropoxy groups obtained by Zeisel cleavage with hydrogen iodide [16]. The USP defines four substitution types according to the content of methoxy and hydroxypropoxy groups.

HPMC is available in a wide range of viscosities (measured 2 wt% in water; for viscosity–molecular weight relationship, see Ref. 17) The high-viscosity grades (about 4000–100,000 mPa s) are often used in hydrophilic matrix tablets, while lower viscosity grades (3–15 mPa s), produced by acid-catalyzed cleavage of the cellulose backbone, are excellent film formers and frequently used in tablet coating and capsule manufacture [18]. Table 2.3 shows the types of substitution for HPMC. Glass transition temperatures (T_g) of HPMC are reported in the literature ranging from about 155 to 180 °C [19]. The equilibrium moisture content at 75% relative humidity (RH) was determined to be about 10% for a 2910 substitution-type HPMC [20], which is at the lower end of the range found across HPMC substitution grades. The maximum reported water uptake is in the range of 15–18% at 75% RH for 2208 substitution type [21]. The substitution type most commonly used for preparing ASDs by a solvent evaporation technology is 2910, as it is sufficiently soluble in a range of organic solvents. In addition, this grade is available in

TABLE 2.3. List of Hypromellose Substitution Types as Defined in USP 35 [16a]

Substitution Type	Methoxy (%)	Hydroxypropoxy (%)	Use	Typical Viscosity Range
1828	16.5–20.0	23.0–32.0	Not commonly used	–
2208	19.0–24.0	4.0–12.0	Hydrophilic matrix tablets	4000–10,000
2906	27.0–30.0	4.0–7.5	Film formation (capsules, tablet coating)	3–10
2910	28.0–30.0	7.0–12.0		–

low-viscosity grades suitable for spray drying. An overview of studies performed using HPMC has been composed by Warren et al. [3c]. While HPMC is a polymer that has been successfully used in pharmaceutical applications for decades, recent innovations have demonstrated that there is growing capability to design properties, for example, thermal transition temperatures [22], that might have the potential to bring additional benefits to solve the bioavailability challenge.

2.1.2.5 Hydroxypropylmethylcellulose Acetate Succinate

Structure and Synthesis/Physical Properties HPMCAS is made from HPMC by esterification with acetic acid anhydride and succinic acid anhydride in acetic acid using a basic catalyst such as sodium acetate [23]. The resulting product, depicted in Figure 2.6, is precipitated by addition of water and subsequently purified by washing with additional water. This reaction sequence leads to a plurality of hydrophobic sites and hydrogen bond acceptor and donor capability. Substituents R_1, R_2, and R_3 are predominantly (a) methyl (CH_3), (b) acetyl ($COCH_3$), (c) succinoyl (COC_2H_4COOH), (d) hydroxypropyl ($CH_2CH(OH)CH_3$), and (e) H (Table 2.4). Hydroxypropyl groups may be further substituted at the OH group (not shown).

HPMCAS was first introduced by Shin-Etsu Chemical Co., Ltd., Japan, as an enteric coating agent [24] with three substitution levels designated according to the content of acetyl substituents as L, M, or H. The dissolution pH of HPMCAS ranges from 5.5 (L) to 6.5 (H) [25] depending on the buffer type used for dissolution [9]. Dow Chemical and Ashland Chemical have announced plans to offer HPMCAS [26].

In contrast to HPMC, where substitution levels are specified by the monographs, the range for HPMCAS is not limited to the three commercially available subranges [27].

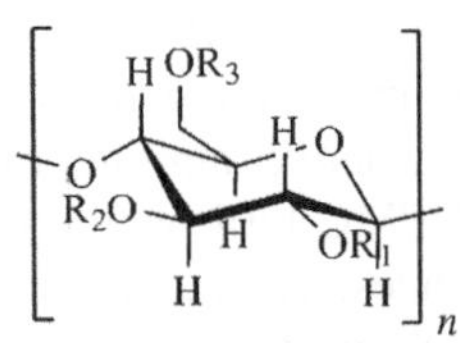

R_1, R_2, R_3 = CH_3, $COCH_3$, COC_2H_4COOH, $CH_2CH(OH)CH_3$, H

Figure 2.6. Chemical structure of HPMCAS.

TABLE 2.4. Range of Substitution of Various HPMCAS Polymers

	Acetyl (%)	Succinoyl (%)	Methoxy (%)	Hydroxypropoxy (%)
USP-NF[a]	2.0–16.0	4.0–28.0	12.0–28.0	4.0–23.0
Aqoat™ AS L[b] AffiniSol™ 716[b]	5.0–9.0	14.0–18.0	20.0–24.0	5.0–9.0
Aqoat™ AS M[b] AffiniSol™ 912[c]	7.0–11.0	10.0–14.0	21.0–25.0	5.0–9.0
Aqoat™ AS H[b] AffiniSol™ 126[c]	10.0–14.0	4.0–8.0	22.0–26.0	6.0–10.0

[a]USP-NF [27].
[b]Technical information, Shin-Etsu Chemical Co., Ltd. [28].
[c]Technical information, Dow Pharma & Food Solutions, The Dow Chemical Company [29].

Dow Pharma & Food Solutions has partnered with Bend Research to develop and commercialize HPMCAS [28] utilizing the full potential range of compendial acetyl and succinoyl substitution levels to solubilize APIs [29].

The performance of the HPMCAS as a dispersion polymer can be optimized by adjusting the acetate and succinate substitution levels to balance the dissolution and sustainment of the active compound. By dissolution testing of 25% API spray-dried dispersions (SDDs) with model compounds, optimum ester substitution levels have been found (Figure 2.7) that allow to maximize the area under the curve (AUC) for several model APIs. The optimum substitution levels vary depending on the physical properties of the active compound [29a,30]. Grasman et al. concluded that having polymers available that span as full a range as possible in ester substitution allows the optimization of ASD performance for a particular drug of interest [30b].

Water absorption and T_g of HPMCAS compared with those of PVP and HPMC at different relative humidities were investigated by Friesen et al. [20]. At 75% RH, water uptake of HPMCAS was found to be 6%, which, according to the authors, leads to a T_g above 70 °C even when equilibrated at 75% RH. The dry T_g has been found to be on the order of 120 °C, which is in agreement with 113 °C reported elsewhere [9].

As HPMCAS forms aggregates in organic solvents, the molecular weight depends on the solvent mixture and the pH. When using saponified HPMCAS (investigating the backbone HPMC), the degree of polymerization is found to be slightly lower than that of the starting material by size-exclusion chromatography with multi-angle laser light scattering (SEC–MALS) [31], while it was significantly higher in organic solvents. Chen [32] found an 8–18 times higher MW for HPMCAS in mixed acetonitrile/aqueous buffer solvent compared with the hydrolyzed sample. From this SEC triple detector study, it was deduced that the higher MW components of HPMCAS form more compact structures (aggregates). The results were confirmed in a recent study using SEC–MALS in mixed acetonitrile/aqueous buffer solvent [33]. The investigated samples of HPMCAS-LF and HPMCAS-MF were found to have a MW of 120,000–140,000 g/mol, while HPMC-HF appeared to be more aggregated (250,000–350,000 g/mol). This compares to a backbone MW of about 20,000 g/mol [31].

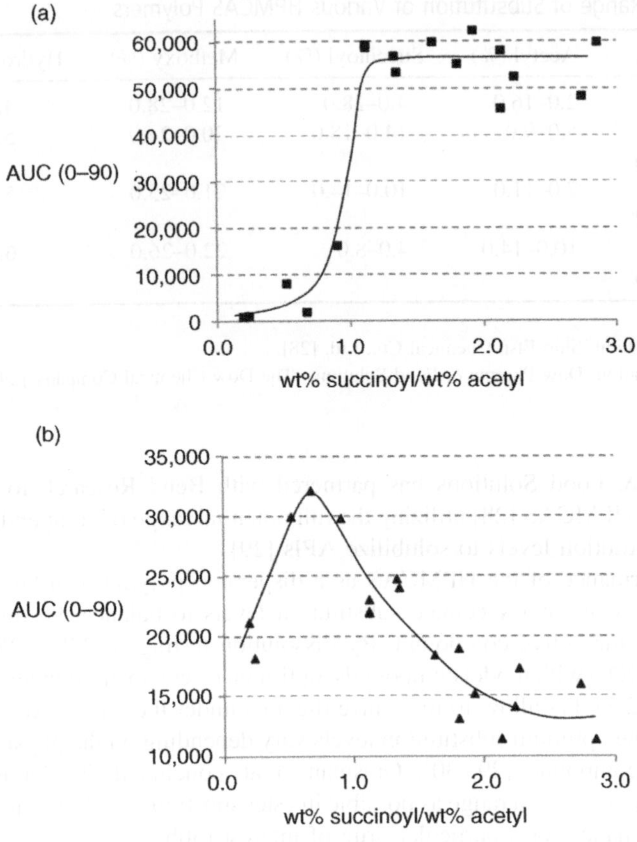

Figure 2.7. AUC(0–90 min) as determined by a microcentrifuge dissolution test of SDD made with 25% API using a range of HPMCAS with varying ester substitution. Dissolution medium = 0.5% SIF in PBS, pH 6.5; dose = 1000 μg/ml. (a) Itraconazole; (b) phenytoin. (Adapted from Ref. 29a.)

2.2 SURFACTANTS COMMONLY USED IN SOLID DISPERSIONS

Surfactants or surface-active agents are defined as substances that adsorb at interfaces, typically having a structure consisting of a hydrophilic group and a hydrophobic group. The hydrophilic portion of the surfactant is water loving and allows the surfactant to be dispersed in water. A representative surfactant structure is shown in Figure 2.8. This is the simplest type of surfactant structure with only one hydrophile and one hydrophobe per molecule. Other surfactant structures include diblock or triblock copolymers and surface-active polymers that contain a random distribution of hydrophobic and hydrophilic structures.

Surfactants are generally classified based on the hydrophilic structure of the molecule, such as anionic, cationic, zwitterionic (also known as amphoteric), and cationic. Within

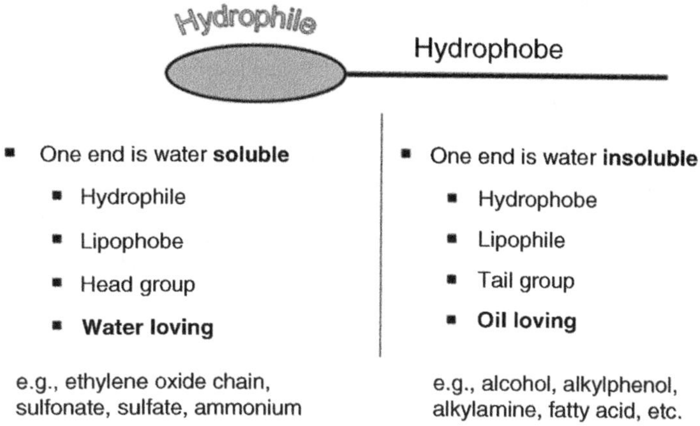

Figure 2.8. Typical surfactant structure with common hydrophile and hydrophobe structures.

each of these classifications, there are subclasses based on the specific structure on the hydrophile (e.g., sulfonate and carboxylate) or other structural features (e.g., block copolymers) [34]. The hydrophobic part of the surfactant, frequently called the tail, consists of linear or branched alkyl groups, substituted alkyl aryl groups, hydrophobic oligomers of alkylene oxides such as polypropylene oxide (PPO), silicone-containing oligomers, or perfluorinated alkyls. Surfactants may be produced synthetically, derived from natural sources (e.g., fatty acid salts), or be present in biological systems naturally (e.g., bile acid salts). The structures of some common surfactants are shown in Figure 2.9.

2.2.1 Interfacial Properties

One of the characteristic properties of surfactants is their tendency to adsorb at interfaces such as liquid–air, liquid–liquid, and solid–liquid interfaces. This results in a number of useful

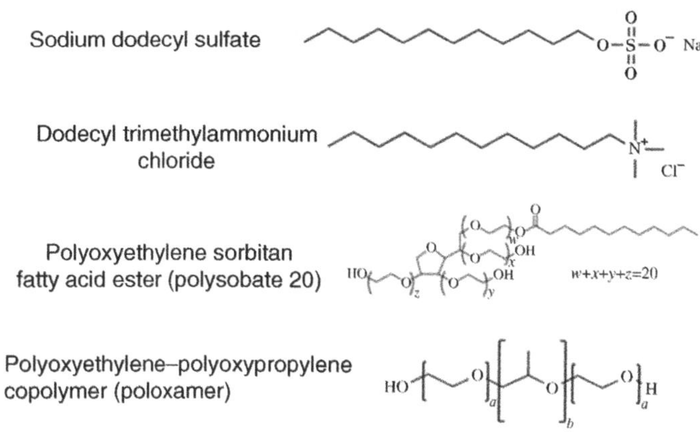

Figure 2.9. Common surfactant chemical structures.

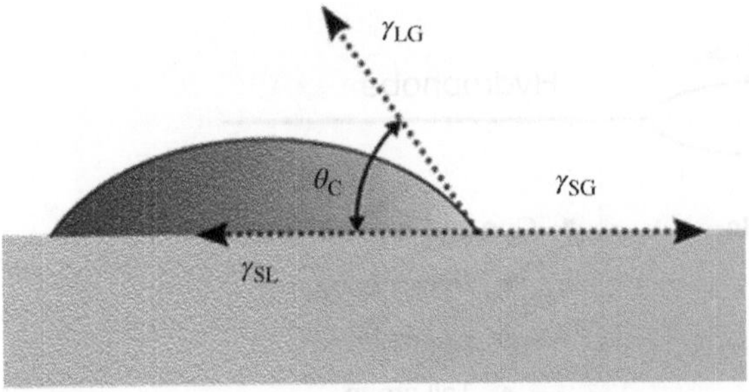

Figure 2.10. Contact angle of liquid on solid substrate.

properties for surfactant-containing systems, such as reduced liquid–gas surface tensions, reduced liquid–liquid interfacial tensions, improved wetting of liquids on solid substrates, and the formation of stable colloidal structures such as emulsions and foams [35]. These useful properties lead to the use of surfactants in a wide range of applications ranging from cleaners (detergents, shampoos soaps, lotions, etc.) to food and pharmaceutical products.

When a surfactant adsorbs at an interface, it tends to change the properties of that interface. The molecules adsorbed at the air–water interface reduce the surface energy (γ_{LG}) of the liquid. Similarly, they can also adsorb at the solid surface, which can affect the surface energy at the solid–liquid interface (γ_{SL}) and the solid–gas interface (γ_{SG}). The combinations of these interfacial energies affect the wetting of aqueous solutions on hydrophobic surfaces and can be determined by measuring the contact angle of the liquid on the solid as shown in Figure 2.10.

The contact angle of the liquid on the solid (θ_C) can be determined by solving Young's equation as shown in Equation 2.1:

$$\cos\theta_C = \frac{\gamma_{SG} - \gamma_{LS}}{\gamma_{LG}}. \tag{2.1}$$

The overall impact of surfactant adsorption at these interfaces is to improve the wetting of solid surfaces by aqueous solutions. This is one of the primary functions of surfactants used in pharmaceutical formulations as wetting of a tablet surface or powder is usually the first step in the dissolution process. An example of this surfactant effect can be seen in Ref. 36, where Lutrol® F127 (an ethylene oxide–propylene oxide block copolymer nonionic surfactant) was added to glipizide–HPMC solid matrices. Amorphous solid dispersions were obtained in both systems, HPMC by itself and an HPMC/Lutrol® F127 blend. However, the surfactant-containing system showed improved solubility that was attributed by the authors to the improved wetting properties of the amorphous dispersion containing surfactant [36].

2.2.2 Self-Assembly

Another property of surfactants is their tendency to form self-assembled structures in solution. When added to water, the "monomeric" molecules first adsorb on the available surfaces (liquid–gas and liquid–solid) to reduce the interaction of the hydrophobic tails

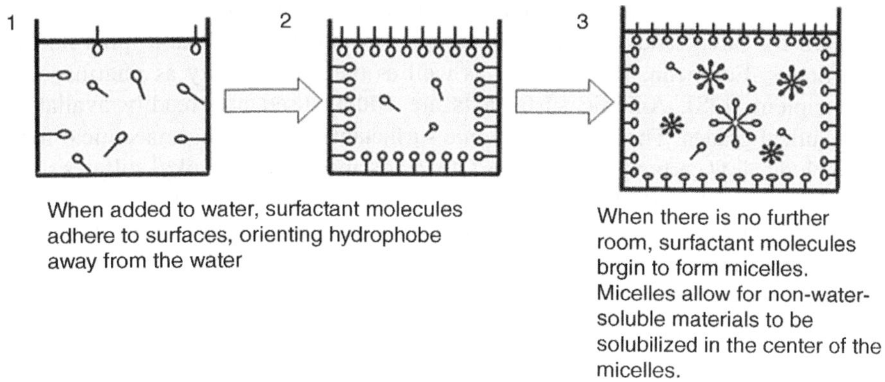

Figure 2.11. Basic behavior of surfactants: surface adsorption, micelle formation, and critical micelle concentration.

with water. Above the critical micelle concentration (CMC), the available interfaces are saturated with the surfactant monomers causing the remaining monomers to self-assemble into macromolecular structures to reduce the interaction of the hydrophobes with water. This process is illustrated in Figure 2.11. These structures include micelles (which may be spherical or cylindrical) and vesicles (which are basically spherical bilayers). Micelles can generally be thought of as having a hydrophilic shell surrounding a hydrophobic core. The CMC is a function of the surfactant structure and is commonly used as a measure of its efficiency. A lower CMC indicates a more efficient surfactant.

In many cases, it is possible for hydrophobic materials to be incorporated or "solubilized" in the interior of micelles. This key property of surfactants in solution is another reason for their incorporation in pharmaceutical formulations. Solubilization occurs when one forms thermodynamically stable isotropic solutions of normally insoluble materials in a solvent (usually water) through the addition of surfactants. The structures that form may be thought of as swollen micelles, although it is somewhat ambiguous where the transition between swollen micelles or micro/macroemulsions occurs. Generally, solubilization occurs at a high ratio of surfactant to hydrophobic solute and is smaller than what is seen with amorphous forms. The ability of a particular surfactant to solubilize a given solute is highly dependent upon the structural characteristics of both moieties and the micelles formed. The apparent increase in solubility has been used by itself to improve the bioavailability of poorly soluble drugs and is also involved in increasing the dissolution rate of poorly soluble drugs by increasing transport rates from the solid surface. An example of a surfactant-enhanced bioavailability is shown by El Eini et al., where the solubility of a model drug, dexamethasone, increased linearly with added nonionic surfactant [37].

2.2.3 Applications of Surfactants

In pharmaceutical applications, surfactants are used as emulsifiers, wetting agents, dispersants, foaming agents, and penetrating agents. The types of surfactants used in pharmaceutical products range across the different classes of surfactants but are constrained by safety and toxicity and acceptance into the applicable pharmacopeia (United States, European, British, etc.) as well as their availability as pharmaceutical grade excipients [38]. Anionic surfactants are widely used and readily available in pharmaceutical grades. The types of anionic surfactants used in pharmaceutical formulations include alkyl carboxylates such as fatty acids and their salts, alkyl sulfates such as sodium dodecyl sulfate (SDS), and sulfonates such as sodium dioctyl sulfosuccinate. Nonionic surfactants are also widely used in pharmaceutical formulations due to their wide structural diversity and generally perceived lower toxicity. Examples of nonionic surfactants used in pharmaceutical formulations include glycol and glycerol esters, ethoxylated alcohols, ethoxylated fatty acids, sorbitan fatty acid esters, and ethoxylated sorbitan esters. There are also a number of examples where naturally occurring materials such as Vitamin E are ethoxylated to generate surfactants of use in pharmaceutical formulations. There are also a number of naturally occurring surfactants such as bile salts, phospholipids, and lecithins that are used in pharmaceutical formulations. Cationic and zwitterionic surfactants are rarely used in pharmaceutical applications. Most of the above surfactants, when used in solid dispersions, are used as additives to improve the wetting, dispersion, or breakup. A list of typical surfactants used in pharmaceutical applications is shown in Table 2.5 [9].

TABLE 2.5. Common Surfactants Found in Pharmaceutical Products

Surfactant	Structural Features	Applications	Regulatory Status
Sodium lauryl sulfate	Anionic Sulfate	Wetting aid, emulsifier, solubilizing agent	USP, Ph. Eur., BP, JP
Dioctyl sulfosuccinate	Anionic Sulfonate	Wetting, dissolution aid	USP, BP
Stearic acid salts	Anionic Carboxylate	Wetting aid, emulsifier, solubilizing agent	Ph. Eur., USP, JPE
Oleic acid salts	Anionic Carboxylate	Wetting aid, emulsifier, solubilizing agent	Ph. Eur., USP, BP
Benzalkonium chloride	Cationic Quaternary ammonium	Wetting aid, solubilizing agent	USP, Ph. Eur., BP, JP
Glycerol monostearate	Nonionic Glycerol ester	Emulsifier, solubilizing agent	USP, Ph. Eur., BP, JP
Polyoxyethylene alkyl ethers	Nonionic PEG ethers	Wetting aid, emulsifier, solubilizing agent	USP, Ph. Eur., BP, JP
Polyoxyethylene caster oil derivatives	Nonionic PEG ethers	Emulsifier, solubilizing agent	Ph. Eur., USP, BP
Polyoxyethylene sorbitan fatty acid esters	Nonionic PEG ethers	Wetting aid, emulsifier, solubilizing agent	USP, Ph. Eur., BP, JP
Polyoxyethylene–polyoxypropylene copolymers	Nonionic Block copolymer	Emulsifier, solubilizing agent	USP, BP

A category of nonionic surfactants with good solubilizing properties includes block copolymers of ethylene oxide and propylene oxide. These are widely used in pharmaceutical and personal care applications and have also been utilized for solid dispersions of pharmaceutical agents. These block copolymers, referred to as poloxamers, are available under a variety of trade names such as Pluronic® from BASF. Poloxamers are available in a variety of molecular weights and ethoxylated contents and have the benefit of being both effective surfactants and polymeric. Their polymeric nature makes poloxamers amenable to many of the standard processes used to prepare ASDs. In addition, their structure results in improved miscibility compared with nonpolymeric surfactants with many pharmaceutical actives. In a number of cases, poloxamers have been used by themselves to prepare ASDs. For example, Moneghini et al. prepared an ASD of nimesulide in poloxamer 188 via dry blending followed by microwave irradiation to melt the surfactant and generate a solution of nimesulide in poloxamer [39].

The broad utility of surfactants has resulted in their inclusion in ASDs for a number of reasons. First, they are added to ASDs for the same reasons they are included in other pharmaceutical products, primarily to improve wetting and dispersion of powders in liquids. Surfactants have also been used in combination with lipids or other oleophilic materials to form emulsions or nanosuspensions of poorly water-soluble drugs [40]. In this approach, the drug is maintained as a solution in a lipid carrier. The primary role of the surfactant is to act as an emulsifier to form stable emulsions or microemulsions of the drug containing oil in water.

Surfactants can also increase the dissolution rate of poorly water-soluble drugs through a combination of wetting and solubilization. The mechanism for the enhanced dissolution of poorly soluble drugs in surfactant solutions proceeds through a number of steps, including [41] (1) diffusion of surfactant to the solid–liquid interface, (2) formation of a surfactant hemimicelle at the solid surface, (3) mixing of drug with the hemimicelle aggregates, (4) formation and desorption of drug-loaded micelles from the solid surface, and (5) diffusion of the loaded micelles away from the solid surface into the fluid phase.

2.3 SYNERGIES BETWEEN SURFACTANTS AND POLYMERS IN SOLID DISPERSION SYSTEMS

Surfactants have been combined with polymers in many delivery systems that are used in both solid and liquid forms for both oral and parenteral administration [4c,42]. Surfactants have also been used in many lipid-based formulations to improve the solubility and bioavailability of hydrophobic drugs. The mechanisms involved in liquid-based systems, such as interactions of micelles and hydrated polymers with drugs in solution, can also be applied to the design of solid dispersions and solid dosage forms. Synergistic effects between these functional materials can also occur within the solid state. A thorough review of surfactant and solubilization systems in relation to drug delivery systems can be found in the literature [4c,43]. Mechanistic interactions between various types of polymers and surfactants (ionic and nonionic) have been investigated by Goddard [44]. This section focuses on the criteria for designing amorphous solid dispersions with combinations of various surfactant-type molecules and polymers for stabilization of the supersaturated or high-energy form, and to improve the bioavailability of poorly soluble drugs. Food effects, common to many lipid-soluble drugs, can be overcome by maximizing the available drug in solution with the best choice of excipients.

ASDs can be classified into three generations: low molecular weight, highly water-soluble carriers such as urea, short-chain carboxylic acids (citric/succinic acid), and sugars (sucrose, mannitol, and trehalose); polymeric carriers such as PVP, PEG, and cellulose derivatives; and self-emulsifying and surfactant/polymer-based systems [4c]. The small molecule- and/or polymer-based carriers in the first two types function to stabilize the drug in an amorphous state, eutectic mixture, or molecular dispersion. Polymeric carriers are often preferable due to their ability to be processed at lower temperatures, such as melt blending/extrusion, in addition to spray drying or congealing to form solid matrices. With the addition of surfactants in the third type, improved properties such as wetting, dispersion, and solubilization in aqueous media can result and are important especially for very poorly soluble and hydrophobic drugs. Polyethoxylated surfactants such as Gelucire®, Tween®, and Labrasol® have shown improved dissolution and oral bioavailability when incorporated into ASDs [4c]. Other surfactants such as ionic surfactants (i.e., sodium lauryl sulfate) and nonionic surfactants (i.e., PEO–PPO block copolymers) have been used to improve dissolution and solubility of the active ingredient; however, many of these materials tend to be waxy at room temperature complicating their incorporation into solid dosage forms such as tablets. Alternatives such as direct filling of melts into hard gelatin capsules have been explored, to ensure effective containment and physical stability of the matrix. However, higher molecular weight blends of PEG (MW up to 8000) can be combined with surfactants to enable them to process into solids that can be tableted and easily disintegrated in aqueous media. Several examples have been cited by Leuner and Dressman [3b] that show synergies of amphiphilic or surfactant-like materials with PVP, cellulosic polymer derivatives, polyacrylates, and other classes as solid dispersion systems. Favorable interactions have been reported with PEG 4000, 6000, and 20,000 and Tween® 80 or sodium lauryl sulfate with drugs such as naproxen and griseofulvin. Other synergistic interactions to form solid solutions have been found with alkali dodecyl sulfates, bile salts, and cholesterol esters. Increased wettability of the dosage forms and improved solubilization have been attributed to bioavailability enhancement *in vivo* [3b].

The drug, polymer, and surfactant interactions, where homogeneous solid dispersions form, may be predicted by constructing a phase diagram. In certain cases, solid eutectic mixtures may occur at a unique composition, where very fine particles crystallize out from each component and comprise a physical mixture. This system does not involve the molecular dispersion of the drug within a homogeneous polymer–surfactant matrix. On the other hand, solid solutions, where all components are molecularly dispersed within a common matrix, can form based on controlling specific compositional and temperature ranges for that system [3b,4c]. Figure 2.12 shows a typical example of a binary phase diagram for predicting regions forming solid solutions.

In a eutectic mixture, the components may be insoluble in the solid state but miscible in the molten liquid state. Solubilization enhancement and increased dissolution may occur due to the high surface area to volume ratio of the fine particles and enhanced wettability with the presence of surfactant and polymer. In the solid solution, defined

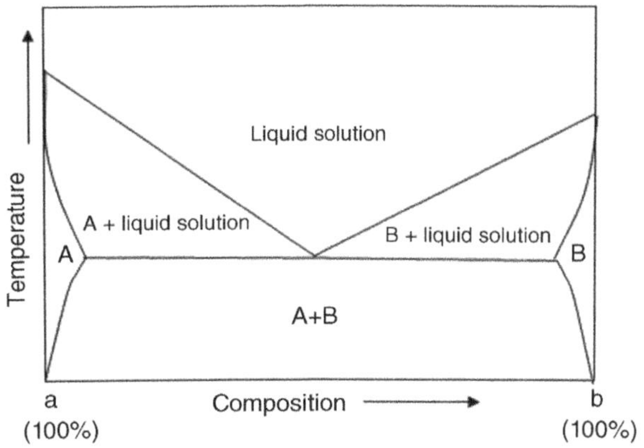

Figure 2.12. Phase diagram of a two-component discontinuous solid solution. (Adapted from Ref. 3b.)

typically as 5% or greater miscibility of one component in another [3b], supersaturation of the drug may be achieved to a greater extent due to the more intimate arrangement between drug, polymer, and surfactant. Before conducting extensive studies for the best solid dispersion components, a simple physical mixture can be quickly screened to gain a qualitative understanding of the functionality by blending or granulating components together and filling into a capsule or compressing into tablet form. The prototype dosage form can be studied in representative dissolution media and also simulated fed and fasted conditions. The effectiveness of drug dissolution, solubilization, and precipitation inhibition can be estimated by monitoring the concentration of drug in solution as a function of time.

2.3.1 The "Spring and Parachute" Model

Drug bioavailability is dependent upon the propensity to remain soluble or supersaturated in the gastrointestinal (GI) tract. In the supersaturated state, the drug maintains a high thermodynamic activity and flux through the intestinal tract, increasing its absorption. Agents such as polymers or surfactants used alone in a solid dispersion may either improve the aqueous solubility of a crystalline drug or improve the longevity of the dissolved amorphous form. However, a combination of ingredients can be beneficial in cases where kinetic and thermodynamic approaches to stabilization require refinement to achieve a certain drug concentration, and where the solubilized drug is labile enough to precipitate into more stable crystalline forms. For example, the oral bioavailability of tacrolimus and itraconazole was found to increase with the use of solubilizers and precipitation inhibitors [45]. To overcome this phenomenon, a combined "spring and parachute" model to improve drug solubility and stabilization over a longer time window has been proposed by Guzman and several other researchers [46] to define a drug salt form, polymer, or surfactant system to improve the solubility/supersaturation (or "spring") and curtail the onset of precipitation or crystallization by adding an inhibitor (or "parachute") such as a polymer (cellulosic, PVP, or PEG). The "spring" is typically a higher energy material that facilitates dissolution and supersaturation of the drug in the GI tract, and the "parachute" is the excipient that delays nucleation of the soluble high-energy drug form. Figure 2.13 explains

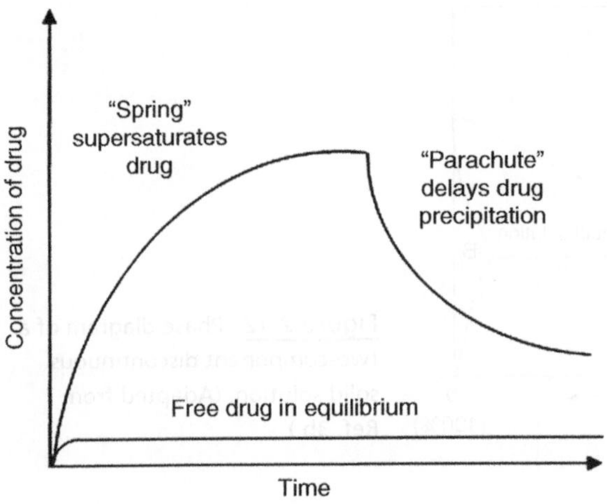

Figure 2.13. Spring and parachute mechanism. (Adapted from Ref. 46a.)

the spring and parachute concept. The mechanism of precipitation inhibition involves additives that attempt to reduce the collision between monomeric drug molecules in solution and prevent nucleation and crystallization. This is not merely by increasing the local viscosity of the solution, but possibly by adsorbing to newly formed drug nuclei and disrupting lattice formation [4c,46a].

For additives that control the rate of crystallization, the rate of nucleation (I) can be defined by the differential between the activation free energy for molecular diffusion from the nucleus to the crystal lattice (ΔG_a) and free energy associated with formation of a spherical nucleus of a critical size (ΔG^*):

$$I = A \exp\left(\frac{-(\Delta G_a + \Delta G^*)}{kT}\right). \tag{2.2}$$

The free energy for nucleus formation also depends on the balance of free energy between the amorphous and crystalline phases (ΔG_v) and the energy to form a new solid surface (ΔG_s). Nucleation is favored by ΔG_v, dependent upon volume, as formation of the crystalline phase is more energetically favorable. However, ΔG_s is dependent upon surface area and interfacial tension, counteracts nucleation as formation of a new surface requires energy, and this value is generally more positive than ΔG_v [4c]. The magnitude of change between the two free energies as a function of nucleus diameter can determine the critical nucleus size to induce crystallization. Polymers can inhibit the diffusion rate of nuclei, reduce drug diffusivity, and adsorb to interfaces preventing crystallization. In addition, polymers and surfactants that can lower the interfacial tension can stabilize smaller nuclei and prevent growth to a "critical size" for crystal lattice formation.

Surfactants in formulations can be limited in their use for certain poorly soluble drugs as solubilization capacity has been reported to be less than 20 mg of drug in micelles per gram of surfactant used [46b]. Although large doses of surfactants may be less than practical for commercial use, smaller quantities can improve the wettability of solid dispersions and also enhance polymer interactions. Parameters such as CMC, cloud point, and interaction energies with the drug molecule can also influence the effectiveness of the surfactant in the dispersion formulation [46b]. Compatibility between the drug, polymer, and other additives in the solid dispersion is critical in the prevention of phase separation within the matrix and optimal dissolution during contact with aqueous media in the GI tract. The free energy of mixing between two components can be described as follows (from Flory–Huggins theory for polymer systems) [4c]:

$$\Delta G_{\mathrm{mix}} = RT\left[\left(\frac{\varphi_{\mathrm{A}}}{r_{\mathrm{A}}}\right)\ln\varphi_{\mathrm{A}} + \left(\frac{\varphi_{\mathrm{B}}}{r_{\mathrm{B}}}\right)\ln\varphi_{\mathrm{B}} + \chi\varphi_{\mathrm{A}}\varphi_{\mathrm{B}}\right]. \tag{2.3}$$

This relationship depends on the entropy and enthalpy of mixing the drug (A) and polymer material (B). r is the number of monomer units (for small-molecule drugs, $r=1$), φ is the volume fraction of the components, and χ is the interaction parameter. The greater the value of χ, the higher the enthalpy value and the system is prone to immiscibility. An increase in the entropic terms (first two) favors mixing. Phase diagrams to predict phase separation boundaries can be determined by plotting the free energy of mixing and temperature versus volume fractions of mixture components.

As an example, the diffusion of the drug dexamethasone within systems of a triblock PEO–PPO–PEO copolymer in conjunction with surfactants such as SDS, dodecyltrimethylammonium bromide (DTAB), and polyoxyethylene(10)isooctyl phenyl ether (TX-100) was investigated [47]. The interaction parameters between the surfactants and block copolymer systems were calculated by iteration between the following equations (assuming component 1 is the surfactant and component 2 is the polymer):

$$\beta = \frac{\ln(\alpha_1 \mathrm{CMC}_{\mathrm{mix}}/X_1 \mathrm{CMC}_1)}{(1-X_1)^2}, \tag{2.4}$$

$$\frac{X_1^2 \ln(\alpha_1 \mathrm{CMC}_{\mathrm{mix}}/X_1 \mathrm{CMC}_1)}{(1-X_1)^2 \ln[\{(1-\alpha_1)\mathrm{CMC}_{\mathrm{mix}}\}/(1-X_1)\mathrm{CMC}_2]} = 1, \tag{2.5}$$

$$\frac{1}{\mathrm{CMC}_{\mathrm{mix}}} = \frac{\alpha_1}{\mathrm{CMC}_1} + \frac{1-\alpha_1}{\mathrm{CMC}_2}, \tag{2.6}$$

where β is the interaction parameter between the surfactant and the polymer, α_1 is the mole fraction of component 1 in solution, X_1 is the mole fraction of component 1 in the mixed micelles, and CMC is the critical micelle concentration of respective components and their mixtures. DTAB and SDS both had negative values of β ranging between −3.0 and −4.0, demonstrating favorable synergistic behavior with the triblock copolymer. When the copolymer alone was mixed with dexamethasone at low concentrations (i.e., below 0.2 mM), its solubilization in water doubled. However, at concentrations of 1 mM or greater, the drug release profile in aqueous media showed a decline in the time to reach its maximum concentration, showing stronger associations between the polymer and the drug. In the presence of surfactants, the diffusivity of drug increased with increasing SDS, DTAB, or TX-100 concentration. This behavior was shown when all components were mixed in an agar controlled-release matrix. From understanding the interaction parameters between surfactants and polymers, ASDs for both immediate- and controlled-release profiles may be tailored, while maintaining supersaturation of the drug.

In addition to understanding the thermodynamic interactions between the API solute and excipient systems, which affect both physical stability in the ASD and API solubility, determining the kinetics of API dissolution is also important, which ultimately influences the bioavailability of the drug in the GI tract. The dissolution rate of a drug (dM/dt) from a solid matrix as a function of time can be described by the Noyes–Whitney equation [48]:

$$\frac{dM}{dt} = \frac{DA}{h}(C_0 - C_t). \tag{2.7}$$

The dissolution rate is influenced by various factors such as its diffusivity from a solid matrix into the solution phase (D), surface area (A), boundary layer thickness (h), equilibrium solubility (C_0), and concentration in solution at time t (C_t). As a drug is finely dispersed within the excipients down to its molecular size, the resistance for dissolution and diffusion is reduced and its surface area to volume ratio increases.

The expression to determine the rate of the "spring" and "parachute" phases during dissolution can be generally derived by comparing the dissolution profiles of the theoretical amorphous and crystalline phases with the "real" amorphous solid system in the presence of crystallizing drug. Figure 2.14 shows the typical shape of dissolution profiles of the various drug forms.

The initial rise in drug concentration x can be related to time t by the "spring" rate constant K_s (where $x = K_s t$). Assuming that X_a (pure amorphous drug in solution) is $\gg X_c$ (pure crystalline drug in solution), the rate constant K_s can be defined as [46b]

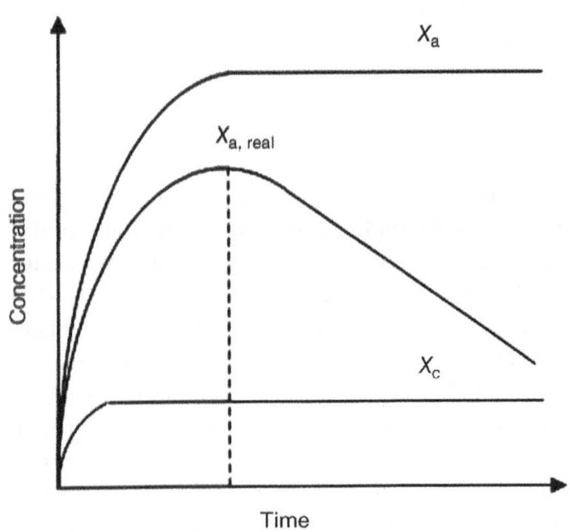

Figure 2.14. Dissolution profiles of three types of drug species: ideal amorphous (X_a), "real" amorphous/supersaturated ($X_{a,ideal}$), and ideal crystalline (X_c). (Adapted from Ref. 46b.)

$$K_s = \frac{X_a - X_c}{t_s} = \frac{X_c}{t_s} \exp\left(\frac{\Delta G_{a-c}}{RT}\right), \tag{2.8}$$

where t_s is the characteristic time for the solid dispersion system affected by factors such as wetting, surface area, and/or particle size, and ΔG_{a-c} is the free energy of the drug from the amorphous to the most stable crystalline state. Based on this expression, the "spring" period (where t is related to d, the time for onset of crystallization) can be expressed as

$$\text{“Spring”period } (t \leq d): \quad x = \frac{X_c}{t_s} \exp\left(\frac{\Delta G_{a-c}}{RT}\right) t. \tag{2.9}$$

After t approaches d, x approaches a maximum value (X_{max}), crystals begin to form, and the "parachute" phase begins where the drug concentration in solution diminishes in an exponential fashion as a function of time:

$$\text{“Parachute”period } (t \geq d): \quad x = (X_{max} - X_c)\exp\left\{-K_p \ln\left(\frac{X_{max}}{X_c}\right)(t-d)\right\} + X_c, \tag{2.10}$$

where K_p is the rate constant for crystallization of drug. In the above cases, the "spring" period is enhanced by the difference in free energy favoring the amorphous form and the "parachute" period can be extended by minimizing K_p or extending d. As an example, the addition of PVP to indomethacin increased the dissolution rate (K_s) by up to 7 times that of baseline and decreased K_p by approximately 0.2 times that of baseline. HPMC was also shown to improve both phases of the curve [46b].

2.3.2 Mechanistic Interactions Between Polymers and Surfactants

Early "spring and parachute" dissolution studies were focused on celecoxib, a poorly soluble NSAID drug, where the "spring" was thought to be mediated by more soluble salt forms of the drug, such as the sodium, sodium propylene glycol, and *N*-methylpyrrolidone solvates [46a]. When the pure soluble salt form was dissolved in gastric media, the kinetic stability of the supersaturated form in solution was limited, and the crystalline form of the drug evolved, limiting the bioavailability. Hence, in order to improve the longevity of the soluble drug form to retard precipitation, various excipients were investigated as "parachutes." These comprised of surfactants such as the Pluronic-type PEO–PPO–PEO triblock copolymer, sodium dodecyl sulfate, Cremophor® (presently known as Kolliphor®), D-α-tocopheryl polyethylene glycol 1000 succinate (Vitamin E TPGS), and several polysorbates. In addition, polymers such as PEG and HPC were studied for the application as a "parachute." These excipients could be combined at various levels in a solid dosage form or dispersion to enable intimate contact with an API. In studying the extent of supersaturation, the Pluronic with higher PPO block units and lower values of CMC tended to inhibit precipitation of celecoxib and provide the best performance compared with other classes of surfactants. The surfactants were most effective in improving the longevity of the supersaturated form at concentrations above their CMC. Overall, Pluronics P103, 104, 105, and 123 were found to be most promising candidates. In addition to providing a "parachute" for the drug, these increased the drug solubility and therefore the "spring" effect.

Guzman et al. also studied the performance of the polymers HPC and PEG alone and in combination with Pluronic F-127 and Vitamin E TPGS [46a]. A significant delay in precipitation time of the neutral drug form was found by adding HPC with both surfactants and enhanced effects were reported on combining the surfactant and polymer systems with PEG 400. Based on the *in vitro* dissolution results, various formulations were studied in the canine model to establish an *in vitro–in vivo* correlation. A linear relationship was noted with a range of excipients studied. The authors do not claim a synergistic mechanism between polymers and surfactants for providing a "spring and parachute" phenomenon for celecoxib, since the spring was intended to be drug salt form. However, other researchers have indicated that association of both types of materials, especially at their hydrophobic moieties, can improve the solubilization effect of hydrophobic molecules, by increasing the capacity of the micelles to incorporate more "greasy" solute.

To better understand the synergistic mechanisms between excipients and drugs, the interactions between surfactant and polymers in aqueous systems with lyophilic solutes were studied. Concentrations of the combined polymer–surfactant system to solubilize an active can often be achieved below the CMC of the surfactant [34]. The polymer and surfactant can associate together such that the hydrophobic regions of the surfactant complex or strongly associate with the nonpolar regions of the polymer. This then causes the polymer and surfactant to have their hydrophilic groups readily available to associate with the solute and incorporate it into the modified micelles [34]. For example, in polymers such as polyoxyethylenated glycols with a broad molecular weight distribution, the lower molecular weight glycols tend to associate with the polar head groups of the surfactant micelle and the longer chains form random coils complexing more strongly with the surfactant hydrophobes. This behavior can occur with nonionic, anionic, and cationic surfactants, depending upon the functional groups of the polymer and surfactant moieties. Hydrophobic polymers exhibit a higher degree of adsorption onto surfactant hydrophobes. The order of adsorption for anionic surfactants onto nonionic polymers appears to follow this order: PVP = polypropylene glycol (PPG) > polyvinyl acetate > methylcellulose (MC) > PEG > PVA. Cationic polymers may interact weakly with polymers that can be protonated, such as PVP, and nonionic polymers in general have weaker interactions with the above types of macromolecules [34].

In addition to increasing the thermodynamic solubility or supersaturation of a solute, the rate of dissolution may also be enhanced if the cloud point (nonionic species only) or CMC of the surfactant is affected by the addition of a polymer. The dissolution rate of solutes has shown to increase proportionally with the concentration of surfactant above the CMC, and is highly dependent upon temperature in the region of the cloud point [34]. This may be explained by the diffusion of micelles to the oil/water interface, dissociation of the assembly, and readsorption to the surface as monomers that can further interact with the polymer and dispersed drug solute and reform micelles. This kinetic process is affected by temperature and concentration. Solutes that are nonpolar such as hydrocarbons and saturated compounds tend to increase the cloud point of the system as they are entrapped in the interior of the micelle, in an oil-in-water system. The radius of the micelle may increase due to the entrapment of the solute, maximizing the surface of the micelle to interact favorably with the aqueous continuous phase and promote hydration of polar groups. In contrast, polar solutes such as fatty acids, highly ionizable species, and small-chain alcohols become solubilized in the outer regions of the micelle and can disrupt the surfactant monomer–water associations [34]. Polymers can be used in the latter case to stabilize the disruptions to the micelles.

The solubilization power between surfactant and polymer systems has been studied by various researchers using fluorescence techniques, with model hydrophobic solutes. Fundamentally understanding these interactions is beneficial for many fields other than the pharmaceutical industry where solubilization and delivery of hydrophobic actives are important. Sivadasan and Somasundaran investigated the interactions of cellulosic polymers such as hydroxyethylcellulose (HEC) with anionic SDS and a 12-carbon (C_{12}) nonionic ethoxylated surfactant, dodecyloxyheptaethoxyethyl alcohol [49]. Some cellulosic polymers can improve wettability and lower the interfacial tension in the presence of oil-soluble species. Pyrene was used as a model fluorescent hydrophobic probe and fluorescence spectroscopy was used to trace the pyrene molecule and determine degree of entrapment using the ratio of the third and first vibrational bands of monomer emission, which are measures of the effective polarity of the medium. Hydrophobically modified HEC (HMHEC) (Natrosol Plus, Aqualon, Inc.) was compared with standard HEC (Natrosol 250GR, Aqualon, Inc.) to contrast the interactions between the solute, surfactant, and functionality of the polymer. Without the presence of surfactant, HEC did not appear to entrap any pyrene. However, uptake was seen with HMHEC, most likely due to the hydrophobic associations at a critical concentration of about 1000 ppm in solution. Pyrene associated closely with the C_{12} and SDS surfactants alone, attributed to entrapment in micelles above the CMC. A dramatic increase in the entrapment of pyrene was seen in the presence of both HMHEC and C_{12}, and about 1.5 times more with HMHEC and SDS. The SDS–HMHEC interaction provided a more hydrophobic environment than C_{12}–HMHEC, as the former possibly clustered into "pseudomicelles" and stabilized by the charged head groups. The HMHEC probably associated with the C_{12} surfactant by adsorption onto its hydrophobic regions, rather than forming micelle-like structures. Other systems with synergies could include polyethylene oxide–SDS, polyvinyl alcohol–SDS, and polyvinyl acetate–dodecyl benzenesulfonate. A more thorough analysis of uncharged polymer–charged surfactant interactions has been reported [44a]. Selection and screening of synergistic materials may be possible using fluorimetric techniques especially when intending to design an ASD with optimal "spring" and "parachute" characteristics for improving the kinetics and thermodynamics for drug dissolution and physical stability in the GI tract.

As these materials are common in pharmaceutical and other industrial applications, various cellulosic derivatives with SDS have been characterized demonstrating that changes in the dynamic surface tension and fluorescence spectra occur as a function of polymer hydrophobicity [50]. A linear relationship was found between the critical polymer–surfactant aggregation concentration (CAC) and the surface activity of the polymer. An extensive review on characterizing polymer surfactant interactions with fluorescence techniques can be found in Ref. 51. Molecular modeling, using Monte Carlo simulations, has also been conducted to predict self-assembly of nonionic surfactants with amphiphilic polymers [52].

Additionally, a unique example is presented, where the ability to entrap hydrophobic molecules in a surfactant–polyelectolyte complex that precipitates to form nanoparticles has been demonstrated [53]. The system included an anionic surfactant, SDS, with a cationic polyelectrolyte, polydiallyldimethylammonium chloride, where particles could be formed to about 15–30 nm. Molecules such as Nile Red and pyrene were studied as hydrophobic solutes, using absorbance and fluorescence, respectively, to estimate the degree of association with the nanoparticles. SDS was able to solubilize Nile Red within a micellar structure, but solubilized pyrene at much lower levels. In combination with the polymer, Nile Red maintained its solubility within the nanoparticle system, toward the hydrophobic domains of the complex based on data collected from absorbance spectra. In contrast, pyrene was found to solubilize at the more polar regions of the polymer–SDS complex due to an increase in the first vibrational band of monomer emission compared with SDS–pyrene. Analytical techniques such as fluorescence and absorbance can be used to predict whether a solute (such as an API) can have favorable interactions with a surfactant/polymer-based system in providing enhanced solubility or stability of the dissolved form. These considerations are interesting in fabricating functionalized polyelectolyte micron- to submicron-range solid particles for the entrapment and stabilization of active molecules.

2.3.3 Examples of "Third-Generation" ASDs

The previous sections focused mainly on the mechanisms governing the interactions between surfactants, polymers, and small-molecule solutes. Specific examples of various systems are cited here regarding "third-generation" ASDs, where the polymer, surfactant, and API can be combined together for enhanced bioavailability of the API. Polymers such as PEGs, PVPs, and cellulosic derivatives have been frequently studied with surfactant systems such as polysorbates and SDS in ASDs.

2.3.3.1 PEG-Based Systems A three-component ASD system was studied using PEGs of molecular weight from 4000 to 20,000, polysorbate 80, and ofloxacin, an antibiotic [54]. The ternary system was beneficial over the binary components; however, the lower molecular weight PEG was superior in solubilizing the ofloxacin. Improved wettability was also noted with addition of surfactant, in contrast to the pure PEG systems. The interaction of various PEGs with nonionic surfactants, such as polyoxyethylene oleyl and cetyl ethers, can decrease the cloud point of these species and is sensitive to PEG concentration and molecular weight [55]. Phase separation into more polymer-rich and surfactant-rich phases can occur, when these molecules are mixed together in critical concentrations. The drug can therefore migrate between the two phases, which can affect its stability and solubility. In an earlier article by Schwuger [56], the interactions of PEG/PPG with ionic surfactants are explained to form complexes, and the tendency increases with the molecular weight of the polymer. However, there is a contrast between the behavior of anionic and cationic surfactants (model systems investigated were SDS and *n*-dodecyltrimethylammonium chloride, respectively). Anionic surfactants tend to be attracted to the positively charged ether linkages in PEGs or PPGs, as the lone electron pair on the ether oxygen tends to leave a slightly positive tendency. This promotes complexation of the surfactant head groups and adsorption of the hydrophobic tail to the carbon-containing region of the polymer chains. In contrast, cationic surfactants tend to repel from the polyglycols due to repulsion from the ether oxygen, hence showing weak associations in aqueous solution. The effects of pH on associations for anionic surfactants are greater due to their electrostatic attractions; however, they are less significant for cationic surfactants where their associations are poor with polyglycols. Especially with pharmaceutical formulations where pH effects are significant throughout the GI tract, these considerations must be made when choosing a polymer–surfactant system in conjunction with the drug molecule.

Efforts to improve the bioavailability of some naturally derived compounds have been studied using PEG–surfactant-based systems. In an example by Hu et al., ASDs were designed for daidzein, a cardiovascular drug extracted from Chinese herbal medicine [57]. The components studied were PEG 10,000 and Tween 80, coprocessed with the drug to make a solid dispersion solidified from ethanol. PEG alone at high PEG to drug ratios (8:1) had good solubilizing power for daidzein. The addition of Tween 80 slightly improved the solubility of the drug at lower PEG:drug ratios (2:1 and 5:1); however, it seemed to have a negative effect at 8:1. The authors eluded that the solidity of the matrix and processing of the dispersion may be factors that affect the ternary interactions. Another naturally derived compound studied was oleanolic acid (OA), a triterpenoid compound that exists widely in natural plants, for anti-inflammatory and other protective functions [58]. In this case, polysorbate 80 was found to increase the solubilization and release profile for OA when it was formulated in a PVP K30 solid dispersion matrix. The solid dispersion provided dissolution enhancement of up to twofold over the simple physical mixture of components.

A PEG-incorporated amphiphilic graft copolymer that consists of polyvinylcaprolactam–polyvinyl acetate–polyethylene glycol, with a trade name of Soluplus®, has also been used in many ASD applications [59]. This polymer may have both solubilizing and delaying of precipitation characteristics to enable better drug absorption. Three drugs, danazol, itraconazole, and fenofibrate, were studied using this polymer to make solid solutions by hot melt extrusion (HME) techniques. These drugs are known to have food effects, typical of many BCS class II, low-solubility/high-permeability compounds. The absorption studies in dogs showed significant increases in bioavailability over the crystalline drug and fairly good agreement between the *in vitro* dissolution studies. Due to the improvement in bioavailability, food effects may be less likely with these formulations as solid solutions. It is possible that the graft copolymer and drugs may form mixed micelles in the gastrointestinal tract with components such as bile salts and phospholipids present in gastric fluids.

2.3.3.2 PVP-Based Systems In addition to PEG-based systems, PVP derivatives have often been used in second-generation ASDs. However, combinations of PVP derivatives with surface-active molecules have been beneficial to many compounds. The dissolution and precipitation behavior of itraconazole, an antifungal, was studied from a spray-dried dispersion of polyvinylpyrrolidone/vinyl acetate 64 (PVP/VA 64) and doped with D-α-tocopheryl polyethylene glycol 1000 succinate (TPGS 1000) [60]. Binary mixtures of itraconazole and PVP/VA 64 were able to form stable dispersions, with calculated theoretical glass transition temperatures (T_g) ranging above 90 °C; however, addition of TPGS 1000 at around 10% caused an increase in the dissolution (initial supersaturation) but more rapid precipitation occurred. The theoretical T_g of the ternary mixture was calculated to be about 20 °C less than that of the binary drug–polymer dispersion. Experimentally, it was found that one truly amorphous phase was difficult to form with the ternary mixture; hence, theoretical and experimental results for T_g had discrepancies in their values. In another example, the mechanism of micelle formation and partitioning of the drug ABT-102, a TRPV1 antagonist for pain, was studied within an amorphous solid dispersion of PVP/VA 64 and various surfactants [61]. The drug is poorly soluble with a log P of ~5.2 and an aqueous solubility of ~0.05 μg/ml in buffer. A melt extrusion process was used to produce the solid matrix consisting of the excipients polysorbate 80, sucrose palmitate, PVP/VA 64, and poloxamer 188. Amorphous ABT-102 within this matrix was found to increase its apparent solubility up to 200 times in comparison with the crystalline drug. The contents of the matrix were analyzed after dissolution in aqueous media to understand the physical state and interactions of the drug with matrix components. Methods such as centrifugation and field flow fractionation enabled the separation of the dissolved matrix into three phases: polymeric, micellar, and microparticulate. Various analytical techniques such as powder X-ray diffraction (PXRD), UV absorbance, Fourier transform infrared (FTIR) spectroscopy, and microscopy were used to identify the components. The polymer (poloxamer 188) and the mixed micelles formed from polysorbate 80 and sucrose palmitate were clearly identified as solubilizing agents due to the presence of nonprecipitated ABT-102. However, the microparticulate phase also contained amorphous ABT-102, which appeared to be associated with and stabilized by the surfactants. All parts of the matrix contributed to crystallization inhibition of the drug and marked improvement in its solubility.

A newer type of inulin derivative surfactant (Inutec® SP1) was studied in comparison with inulin alone and PVP K30 for various actives such as diazepam, fenofibrate, ritonavir, and efavirenz [62]. Inutec contains a polar inulin head group with two alkyl hydrophobes covalently linked to it. Unmodified inulin and PVP both do not have traditional surfactant-like structural features and show minimal reduction of surface tension at various concentrations. Inutec reduces the surface tension in aqueous solutions to 54 mN/m at similar concentration ranges and until its CMC value. Solid dispersions were made by a spray freeze drying method, where solutions were sprayed into liquid nitrogen and then lyophilized. The Inutec dispersion was able to produce stable dispersions for all model drugs and with relatively higher T_g values compared with PVP and inulin dispersions. Accelerated aging at 40 °C/75% RH at 3 months showed insignificant changes in dissolution from tablets made with the dispersions.

2.3.3.3 Carbohydrate-Based Systems Carbohydrate oligomers such as cyclodextrins (CDs) and surfactants have been combined together using spray drying to produce dispersions of tacrolimus, an immunosuppression agent, using hydroxypropyl-β-cyclodextrin (HPBCD) and dioctyl sulfosuccinate (DOSS) [63]. The drug:

HPBCD:DOSS ratio (1:8:0.1) was found to optimally improve solubilization. The processing method was found to influence drug stabilization in the matrix, where complete solubilization of the system prior to spray drying formed an amorphous dispersion of tacrolimus, as opposed to an inclusion complex, which is typical of most interactions with cyclodextrins. The conformation within the dispersion was corroborated with differential scanning calorimetry (DSC) and PXRD data. Other polymer and surfactant combinations have been studied for their amphiphilic properties, where solid dispersions were made from alkyl polyglucosides (APGs), SLS, and the nonsteroidal anti-inflammatory agent aceclofenac [64]. The APGs consisted of fatty acid chains of C_{10}, C_{12}, and mixed C_{12}/C_{14} moieties. SLS and C_{12}/C_{14} APGs formed the best solid mixture to provide high drug release and prevention of crystallization. It was calculated that the CMC of the mixture was lower than that of the individual components, improving the solubility of the hydrophobic drug within the mixed micelle system [64]. Interestingly, sucrose laurate, not as commonly used in oral pharmaceuticals, was studied in combination with PEG 6000 in a melt-solidified dispersion for improving the delivery of gemfibrozil (GEM), a lipid reducing agent. This drug has an aqueous solubility of only 0.0291 mg/ml at a pH of ~6.2 [65]. At a level of 10% in the solid formulation, the release of GEM in dissolution media was increased by almost fourfold [65]. Sugar-based surfactants are typically biodegradable and the hydrophilic group consists of a polar carbohydrate, which is coupled to a fatty acid hydrophobe. One must take caution in adding surfactants that might be difficult to process as a solid form, because they degrade, do not fully melt (as sucrose palmitate or sucrose stearate) to produce a homogeneous dispersion, or may overly plasticize or soften the matrix [65].

The synergies between surfactants and polymers in solid dispersions and solid dosage forms have shown to be important in designing systems to improve the solubility and bioavailability of hydrophobic drug molecules. Understanding the mechanisms of polymer–surfactant interactions and phase behavior is important in choosing the best combination of materials for a solid matrix. Processing of the materials (such as extrusion, spray drying, or simple blending) can affect the dissolution performance of the drug. In addition, the properties of surfactants and polymers such as T_g, molecular weight, and chemical functionality can influence the processability and long-term stability of the dispersion. There are many excipients that are suitable to commercializing solid formulations of poorly soluble drugs; however, opportunities to develop novel materials tailored to better processing and stability are certainly warranted based on the continued introduction of unique hydrophobic/BCS class II type compounds to the market for various therapeutic areas.

2.4 PHYSICAL PROPERTIES OF MATERIALS AND CONSIDERATIONS IN DESIGNING SOLID DISPERSIONS

The successful formulation of an ASD requires the generation of basic physical properties of both the drug substance and the carrier agent. For the API, these properties are typically generated throughout the drug development process regardless of the intended formulation technique. For the drug substance, physical properties to be considered include melting point of the crystalline solid, glass transition temperature (T_g) of the amorphous solid, solubility in organic solvents, stability and solubility over a range of pH values, solubility parameters, and hydrogen bond acceptors and donors. In the case of the polymer or surfactant utilized as the carrier for the ASD, physical properties to be considered include molecular weight, T_g, solubility in organic solvents, thermal stability, melt viscosity, and the ability to process the polymer by various methods. The ability to utilize these important API and polymer properties combined with the critical parameters for the process used for creating the amorphous solid dispersion can guide a formulator to create a stable ASD.

2.4.1 Important Attributes

2.4.1.1 Solubility Parameters A significant attribute of any physically stable ASD is the miscibility of the drug in the carrier matrix [66]. The cohesive energy and the solubility parameters are the key components to determine the extent to which a drug will be miscible with the carrier matrix. The cohesive energy represents the total energy of all intermolecular interactions including those arising from electrostatic and van der Waals interactions and hydrogen bonding [67]. Typically, cohesive energy is referred to in terms of the cohesive energy density, or the amount of energy required to vaporize a mole of liquid per unit volume, and detailed discussions of the calculation of the cohesive energy density have been previously published [67,68]. The solubility parameter is defined as the square root of the cohesive energy density [69]. More practically, the solubility parameter is made up of three main factors: dispersion, polar, and hydrogen bonding. Experimental methods for determining solubility factors include direct methods such as measuring the solubility in solvents with known parameters using the heat of vaporization through calorimetry [68] or indirect methods such as inverse gas chromatography [70]. Despite a variety of methods to determine the solubility parameters of a compound, the most common approach is to use group contribution theory. In this method, each compound is broken into fragments with known parameters and the sum of the parts gives the estimate of solubility parameters. Group contribution methods fail to account for long-range interactions and compounds that have highly directional interactions [71].

Despite limitations in solubility parameters, they are frequently utilized to predict the miscibility of a drug in a carrier matrix [71,72a]. Although commonly used, there is no clear-cut rule for how similar the solubility parameter for a drug and carrier matrix needs to be to ensure full miscibility. Nevertheless, Greenhalgh et al. demonstrated for ibuprofen with various polymers that full miscibility of the drug and carrier was observed when the difference in solubility parameters between the two was less than $7\,MPa^{0.5}$ [72b]. However, larger variations in the difference of carrier polymers should not serve as an exclusion for assessing a specific polymer–drug combination due to other factors such as intermolecular forces that could stabilize the ASD.

2.4.1.2 Glass Transition Temperature Amorphous solids and ASDs are frequently characterized by their glass transition temperature (T_g), the temperature at which the amorphous material begins to undergo physical changes. The mechanism of a T_g transformation is complex and there are several reviews on the issue that can be consulted for an in-depth consideration of the topic [73]. In brief, the T_g is a second-order transition that is associated with continuous changes in the material. As such, the measurement of a material's T_g is sensitive to factors such as sample history, impurities, and moisture.

In general, materials with a high T_g are preferred as a high T_g will be stable at room temperature and under typical product storage conditions. While a high T_g is preferred, Chokshi et al. found that the low-T_g compound indomethacin formed a stable ASD with Eudragit E PO due to strong molecular interactions between the polymer and the drug [74]. The T_g of a solid dispersion can be predicted using the Gordon–Taylor equation [75] (Equation 2.11) in which T_g represents the T_g of the ASD, w_1 and w_2 are the weight fractions of each component in the blend, T_{g1} and T_{g2} are the T_g values of the pure components, and k is a constant derived from the true density (ρ) and the difference between expansion coefficients of the melt and the glass ($\Delta\alpha$). The constant is given by Equation 2.12. For simplicity, k can be approximated by the Simha–Boyer rule (Equation 2.13) [76].

$$T_g = \frac{w_1 T_{g1} + k w_2 T_{g2}}{w_1 + k w_2}, \qquad (2.11)$$

$$k = \frac{\rho_1 \Delta\alpha_1}{\rho_2 \Delta\alpha_2}, \qquad (2.12)$$

$$k = \frac{\rho_1 T_{g1}}{\rho_2 T_{g2}}. \qquad (2.13)$$

The comparison of an ASD's experimental T_g with the theoretical T_g from the Gordon–Taylor equation can provide significant details regarding the ASD. If the experimental and theoretical T_g values are equivalent, it indicates the formation of dispersed ASD with ideal mixing. Deviations between experimental and theoretical values typically indicate nonideal mixing [77]; however, when the experimental T_g is greater than the predicted T_g, it can be indicative of molecular interactions occurring between the drug and the carrier agent [78].

2.4.1.3 Specific Interactions Intermolecular interactions between drugs and the matrix can impart stability to the highly unstable amorphous drug form. These interactions exist as ionic interactions, hydrogen bonding, dipole–dipole interactions, and van der Waals interactions, and a relative ranking of each type of interaction to prioritize the role each type of interaction plays is shown in Table 2.6 [79].

As Table 2.6 indicates, ASDs in which the carrier and drug can form strong ionic and hydrogen bonding have the highest likelihood of the interactions stabilizing the amorphous state of the drug and in fact there are several examples in the literature [80]. In addition, there are several recent in-depth reviews of thermodynamic solubility and kinetic miscibility available to provide a theoretical understanding of these topics [81].

TABLE 2.6. Typical Bond Energy and Relative Strength of Intermolecular Forces

Type of Interaction	Bond Energy (kJ/mol)	Approximate Relative Strength
Ionic interactions	850–1700	1000
Hydrogen bonding	50–170	100
Dipole–dipole interaction	2–8	10
van der Waals interaction	~1	1

Adapted with permission from Ref. 79.

More recent approaches have utilized interactions measured by DSC in combination with Flory–Huggins solution theory to construct drug–polymer phase diagrams [82].

2.4.1.4 Hygroscopicity The performance and stability of ASDs are drastically impacted by moisture that is always present even in systems that are considered dry. As such, the hygroscopicity of the polymer is a key parameter in the long-term chemical and physical stability of any ASD. The chemical stability can be impacted through increased hydrolytic degradation caused by the increased presence of water [83]. In addition, water can act as a plasticizer in an ASD causing the overall T_g to lower and thus increase molecular mobility [84]. Another mechanism by which water can destabilize amorphous dispersions is by disrupting the interactions between a drug and a polymer resulting in phase separation and eventual crystallization of the drug [85]. The myriad of problems caused by the absorption of water into an ASD have resulted in the suggestion that polymers with low or no water solubility such as methacrylates and HPMCAS will help stabilize ASDs by minimizing water interactions [74].

2.4.2 Preparation of Amorphous Solid Dispersions

In addition to determining how well a drug is stabilized in the amorphous dispersion, the physical properties of the polymers, surfactants, and drug also influence the method utilized to create the solid dispersion. While spray drying and HME are the most industry-relevant techniques to produce ASDs, there are many techniques that can be utilized. Such techniques, which include supercritical fluid processing, co-grinding, melt quenching, and fluid bed layering, are limited in use for reasons such as poor or partial conversion and slow rates of conversion.

Spray drying is a robust industrial process utilized to transform a solution, slurry, suspension, or emulsion into a dry powder. For pharmaceutical applications, the process has been reviewed in the literature [86] and is covered in detail in Chapter 9. The spray drying procedure entails the atomization of the feed solution that contains the drug and the carrier agent through a nozzle. The atomized droplets are mixed with heated drying gas that rapidly evaporates the solvent from the droplets. The resulting particles are then isolated via a filter or cyclone and further dried as necessary. Through the control of process parameters such as feed solution flow rate and pressure, drying gas inlet and outlet temperatures, and flow rates, the size, shape, and bulk density of the resulting particles can be tailored. Among the advantages of spray drying are the rapid removal of solvent to form an amorphous state and the limited exposure of an API to heat that may cause degradation. In contrast, spray drying is limited by feed solution viscosity that must be low enough to be pumped through the spray nozzle. If a high-viscosity polymer is required to obtain the desired drug release profile or supersaturation enhancement, it will not be viable for spray drying. In addition, the solvent selection should be such that the vaporization temperature is at least 10 °C below the T_g to ensure the stable formation of an ASD [87].

The other industrial process for producing amorphous dispersions is HME, a technique that is covered in detail in Chapter 10 and has been extensively covered in the literature [86]. Amorphous solid dispersions are created via a process of injecting a mixture of polymer and drug into one or two rotating screws under elevated temperatures. The configuration and rotation of the screws combined with the elevated temperatures create a fluid state that homogeneously mixes the starting materials and passes the resulting mixture through a die to create a uniform extrudate. The final extrudate can be molded or cut into precise dosage forms or sized for encapsulation or pressed into a pellet. HME is advantageous over spray drying in that it is a solvent-free process and does not require additional processing of the extrudate. However, HME is limited by the melt viscosity of the polymer used and the high temperatures and sheer rates that the process utilizes, which can cause degradation of both the polymer and the drug [86]. Recent advances in the HME technology have resulted in KinetiSol technology in which rotating blades produce enough kinetic and thermal energy that additional heat sources are not utilized [88], thus mitigating the thermal degradation issues traditionally associated with HME.

2.4.3 Polymer Selection

Despite the robust understanding of the physical properties of drug compounds and carrier agents used for amorphous solid dispersions and the processing parameters of the method used to create amorphous solid dispersions, the choice of carrier agent is serendipitous. One advantage of using a carrier agent with a drug compound is that the carrier agent can provide stabilization to a supersaturated solution and prevent crystallization of the drug upon dissolution of the ASD. Recently, the precipitation inhibition of polymers as a key performance marker to identify trends in performance with different classes of drug compounds was reported. In the study, the precipitation inhibition from supersaturated solutions for 42 different classes of polymers covering a range of grades and molecular weights for 78 total polymers was tested with nine different poorly soluble drugs [89]. The drugs were selected to represent three major drug categories: nonelectrolytes, weak acids, and weak bases. To analyze the vast array of polymers and drugs tested, principal component analysis was applied to the experimental results. In general, cellulose-based polymers (particularly HPMC and its derivatives) provided robust precipitation inhibition across the range of drugs tested [89]. In addition, it was found that polymers with primary amine functional groups enhanced precipitation rates and polymer molecular weight has only a minor influence on the precipitation inhibition of the polymer. Furthermore, polymers with the opposite charge of the drug provided additional stabilization through ionic interactions [89].

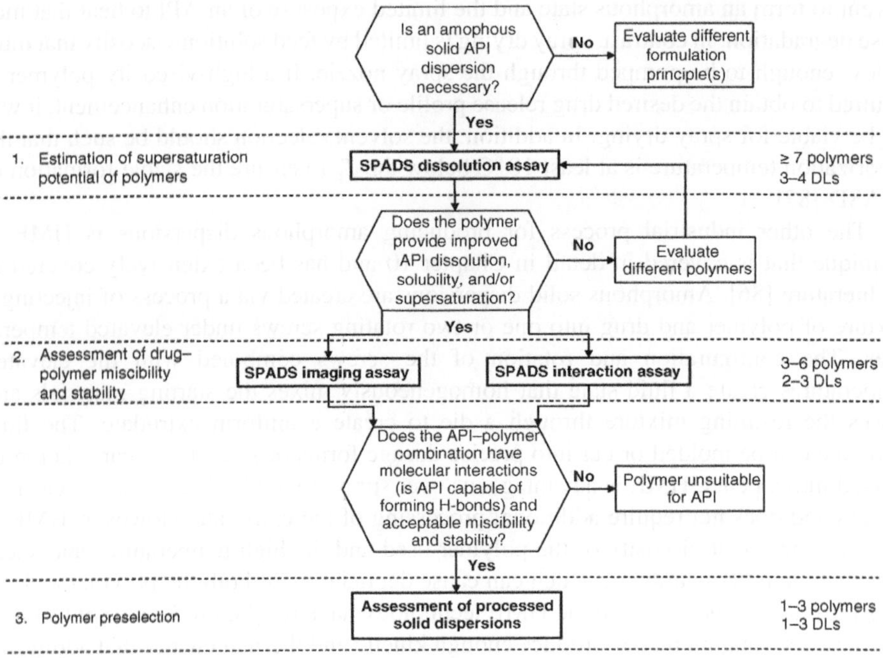

Figure 2.15. Flow diagram of the SPADS screening process. Only promising polymers and drug loadings advance to each subsequent stage. (Adapted from Ref. 90.)

In an effort to streamline the polymer selection process for ASDs and minimize time and resources, Hoffmann-La Roche introduced a miniaturized method that is referred to as screening of polymers for amorphous drug stabilization (SPADS) [25b,90]. The SPADS method is a three-stage process, outlined in Figure 2.15, in which amorphous films consisting of at least seven polymers and multiple drug loadings are prepared. In the first step, the dissolution of the films is assessed in 96-well plates. Polymers and drug loadings that demonstrate improved API solubility and supersaturation are then subjected to an imaging assay utilizing atomic force microscopy to determine molecular homogeneity of the ASD and an interaction assay via FTIR to investigate intermolecular interactions. ASDs of polymers and drug loadings that have favorable interactions and molecular homogeneity are then prepared via spray drying and tested with traditional dissolutions studies and 6-month stability studies [90].

2.4.4 Stability of Amorphous Solid Dispersions

Due to the nature of ASDs in which the API is molecularly dispersed within the carrier surfactant or polymer matrix, there is an increased chance of physical interaction between the two species. Careful evaluation of an ASD must occur to ensure that it is both chemically and physically stable to ensure that the desired performance of the ASD is achieved and maintained. A general method of screening for chemical stability and compatibility between a drug entity and an excipient has been described by Serajuddin et al. [91]. In this method, physical mixtures of the drug and excipient are placed at elevated temperatures with added moisture and evaluated for chemical transformations. As this is a generalized method, it can be used to screen all components of a drug formulation of chemical incompatibility and applicable to selecting a polymer or surfactant carrier for the formation of an ASD. The chemical stability of an ASD can be detected through common chromatographic and vibrational spectroscopy methods.

The chemical stability of ASDs can be influenced by an array of parameters. As an example, derivatives of cyclodextrin can have profound impact on the stability. Inclusion complexes of CD derivatives with ziprasidone suggest that when the CD contains electron-donating side chains it will catalyze the oxidative drug degradation in solution [92]. In addition, the inclusion of any side chains on the CD increased molecular mobility in the solid state creating a physically unstable ASD. Another example of chemical instability is the formation of reactive intermediates that in turn can cause drug degradation. In one instance, the formation of formaldehyde from oxidation of PEG 400 in solution resulted in the degradation of O^6-benzylguanine and was the predominant form of O^6-benzylguanine degradation [93]. Similarly, polyoxyethylene surfactants readily form formaldehydes and peroxides via air oxidation [94], which in turn has resulted in cross-linking in gelatin capsules that affects the dissolution profile of the drug formulation [94b,c,e].

An additional cause of chemical instability in ASDs is water that is in the dispersion as residual from processing or from the environment. Drug molecules that are molecularly dispersed in the matrix are more susceptible to chemical reactions such as hydrolysis than their crystalline form. This phenomenon has recently been investigated in the literature where relationships between water content and stability were reviewed [83].

2.4.4.1 Physical Stability The most challenging aspect of formulating an ASD is stabilizing the high-energy amorphous state and preventing crystallization of the drug in the solid state. Failure to stabilize the amorphous drug can result in crystallization of the drug, slower dissolution kinetics, and lower bioavailability. Despite the vast wealth of information on the physical properties of drug compounds, polymers, and the methodology for making ASDs, the stability of ASDs remains unpredictable. In fact, the use of accelerated conditions to extrapolate the long-term stability of a drug formulation (ICH Q1E guidelines) cannot be applied to ASDs [95]. Typically, long-term stability studies must be performed requiring longer development time and increased resources. Recently, there have been several reported methods to predict and describe the stability of ASDs [95,96].

As a method to predict whether an ASD will be stable under long-term conditions, a general suggestion of the storage temperature being less than $T_g - 50\,°C$ has been provided [97]. This $T_g - 50\,°C$ "rule of thumb" stems from a body of work in which indomethacin was allowed to crystallize in its native state and in dispersions of PVP. It was found that the amorphous indomethacin, with a T_g 20 °C above storage temperature, recrystallized in less than 6 weeks, whereas ASDs of indomethacin in PVP, with a T_g approximately 50 °C above storage temperature, inhibited crystallization. This general

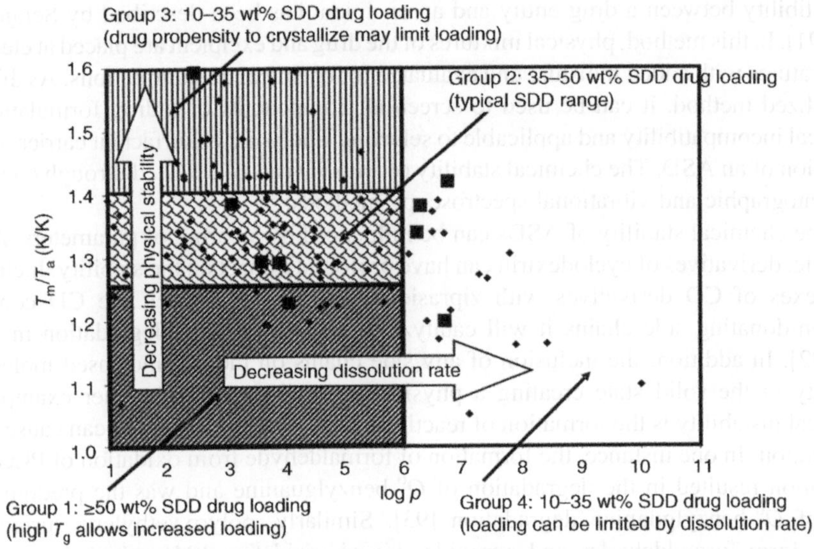

Figure 2.16. Map of T_m/T_g versus log *P* for 139 compounds formulated as spray-dried dispersions with HPMCAS. As T_m/T_g ratio increases, the physical stability and drug loading capacity decrease. As log *P* increases, the dissolution rate of the spray-dried dispersion decreases. (Adapted from Ref. 20.)

rule assumes that if the T_g is greater than 50 °C above storage temperatures the molecular mobility in the glassy state will be suppressed to the point to provide long-term stability.

Another significant attribute that dictates the long-term stability of an ASD is the drug loading level. Systems in which the drug loading is high may result in high local concentrations of drugs, which results in high molecular mobility and faster nucleation. As such, ASD stability is lowest when the ASD is not homogeneous or when the drug loading is greater than the drug–polymer miscibility [81b,98]. Thus, to formulate the most stable ASD, the drug–polymer miscibility regime must be studied to optimize the drug loading level.

Friesen et al. used 139 poorly soluble compounds formulated as spray-dried dispersions with HPMCAS to identify trends to predict overall stability and performance [20]. This work, summarized in Figure 2.16, reveals that a map of the ratio of melting temperature (T_m) and T_g versus log *P* for the compounds provided insight into formulation strategies for low-solubility drugs. Overall, compounds with low T_m/T_g ratios (<1.25) and a log *P* value less than 6, group 1 compounds, were excellent glass formers and could successfully sustain high drug loadings in the ASD. As the T_m/T_g ratio increases above 1.25 for compounds with log *P* values less than 6, group 2 and group 3 compounds, the overall drug loading that is sustainable as an ASD decreases. For compounds with log *P* values greater than 6, group 4 compounds, it was reported that lower drug loadings must be used in the ASD to have an acceptable dissolution rate [20]. In addition to providing a template for mapping the performance for the formulation of ASDs, this work emphasizes the importance of drug loading levels in the formulation.

2.4.5 Performance of ASDs

Solubility enhancement of a poorly soluble drug is frequently accepted as a route to improved bioavailability of that drug and thus a number of techniques for solubility enhancement have been developed. While these methods have proven improved dissolution compared with the native species, the permeability of the drug through the intestinal membrane is often overlooked [99]. Interplay between solubility and permeability in which the increase in solubility corresponds with a decrease in permeability has been demonstrated for cyclodextrin complex solubilization [100], micellar solubilization [101], and cosolvent solubilization [102]. The implication of the trade-off between solubility and permeability is that an increase in solubility may not translate to increased bioavailability of the drug. Miller et al. found that amorphous solid dispersions of progesterone with HPMCAS showed an increase in apparent solubility without a change in permeability [99]. The increase in apparent solubility thus resulted in an increased flux across the membrane and improved bioavailability [99], a distinct advantage for ASDs over other solubility-enhancing technologies as demonstrated in Figure 2.17.

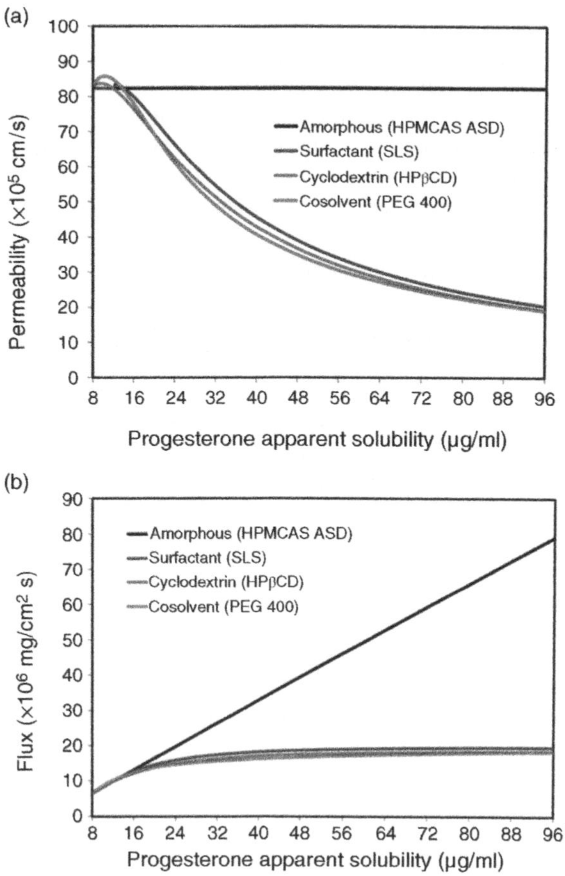

Figure 2.17. Dispersions of HPMCAS with progesterone compared with solubilization of progesterone with surfactant, cyclodextrin, and cosolvent demonstrating that (a) the permeability of the ASD is constant over an increasing apparent solubility and (b) the flux across the membrane increases with apparent solubility. (Adapted from Ref. 99.)

The advantage of ASD over other solubility-enhancing technology was more broadly demonstrated using nifedipine with HPMCAS, copovidone, and PVP [103]. In this study, it was found that ASD of nifedipine with all three polymers resulted in increased apparent solubility without a loss in permeability. It was, however, noted that the increase in apparent solubility and the flux across the membrane were found to be highly polymer dependent [103]. Despite the polymer dependence on overall performance, amorphous solid dispersions were demonstrated to be advantageous over other solubility-enhancing delivery mechanisms.

The design of solid amorphous dispersions involves the selection of key excipients, such as polymers and surfactants, which enhance the bioavailability of drug compounds by improving their solubility and physical stability in the GI tract. The selection of the best functional materials tailored to suit a drug comprises many complex factors; hence, understanding the thermodynamics and kinetics of the multicomponent systems is important, as well as the functionality of the solid dispersion matrix.

REFERENCES

1. (a) Chiou, W.L. and Riegelman, S. (1969) Preparation and dissolution characteristics of several fast-release solid dispersions of griseofulvin. *J. Pharm. Sci.*, 58:1505–1510. (b) Chiou, W.L. and Riegelman, S. (1970) Oral absorption of griseofulvin in dogs: increased absorption via solid dispersion in polyethylene glycol. *J. Pharm. Sci.*, 59:937–942. (c) Simonelli, A.P., Mehta, S.C., and Higuchi, W.I. (1976) Dissolution rates of high energy sulfathiazole–povidone coprecipitates II: characterization of form of drug controlling its dissolution rate via solubility studies. *J. Pharm. Sci.*, 65:355–361.
2. Chiou, W.L. and Riegelman, S. (1971) Pharmaceutical applications of solid dispersion systems. *J. Pharm. Sci.*, 60:1281–1302.
3. (a) Curatolo, W., Nightingale, J.A., and Herbig, S.M. (2009) Utility of hydroxypropylmethylcellulose acetate succinate (HPMCAS) for initiation and maintenance of drug supersaturation in the GI milieu. *Pharm. Res.*, 26(6):1419–1431. (b) Leuner, C. and Dressman, J. (2000) Improving drug solubility for oral delivery using solid dispersions. *Eur. J. Pharm. Biopharm.*, 50:47–60. (c) Warren, D.B., Benameur, H., Porter, C.J.H., and Pouton, C.W. (2012) Using polymeric precipitation inhibitors to improve the absorption of poorly water-soluble drugs: a mechanistic basis for utility. *J. Drug Target.*, 18(10):704–731. (d) Paudel, A., Worku, Z.A., Meeus, J., Guns, S., and Van den Mooter, G. (2013) Manufacturing of solid dispersions of poorly water soluble drugs by spray drying: formulation and process considerations. *Int. J. Pharm.*, 453(1):253–284.
4. (a) DiNunzio, J.C., Zhang, F., Martin, C., and McGinity, J.W. (2012) Melt extrusion. In: Williams, R.O., III, Watts, A.B., and Miller, D.A. (Eds), *Formulating Poorly Water Soluble Drugs*, Springer, New York, pp. 311–362 . (b) Duarte, I., Temtem, M., Gil, M., and Gaspar, K. (2011) Overcoming poor bioavailability through amorphous solid dispersions. *Ind. Pharm.*, 30: 4–6. (c) Williams, H.D., Trevaskis, N.L., Charman, S., Shanker, R.M.A., Charman, W.N., Pouton, C.W., and Porter, C.J.H. (2013) Strategies to address low drug solubility in discovery and development. *Pharmacol. Rev.*, 65(1): 315–499. (d) Riegelman, S. and Chiou, W.L. (1979) Increasing the absorption rate of insoluble drugs. US Patent 4,151,273A.

5. (a) Hallensleben, M.L. (2000) Polyvinyl compounds, others. *Ullmann's Encyclopedia Ind. Chem.*, doi: 10.1002/14356007.a21_743. (b) Bühler, V. (2008) Kollidon®: Polyvinylpyrrolidone Excipients for the Pharmaceutical Industry, 9th revised edition, BASF SE, Pharma Ingredients & Services, Ludwigshafen, Germany.
6. Anonymous (2013) *United States Pharmacopeia and National Formulary*, Vol. USP 36-NF 31, pp. 6037–6039.
7. Folttmann, H. and Quadir, A. (2008) Polyvinylpyrrolidone (PVP)—one of the most widely used excipients in pharmaceuticals: an overview. *Drug Deliv. Technol.*, 8(6):22–27.
8. (2013) *United States Pharmacopeia and National Formulary*, Vol. USP 36-NF 31, p. 4861.
9. (2010) *Handbook of Pharmaceutical Excipients*, 6th edition, Pharmaceutical Press, Washington, DC.
10. (2013) United States Pharmacopeia and National Formulary, Vol. USP 36-NF 31, pp. 1968–1970.
11. (2013) United States Pharmacopeia and National Formulary, Vol. USP 36-NF 31, pp. 2087–2089.
12. Evonik Industries. Eudragit® L 100. Available at http://eudragit.evonik.com/product/eudragit/en/products-services/eudragit-products/enteric-formulations/l-100/Pages/default.aspx (accessed Nov. 13, 2014).
13. Evonik Industries. Eudragit® L 100-55. Available at http://eudragit.evonik.com/product/eudragit/en/products-services/eudragit-products/enteric-formulations/l-100-55/pages/default.aspx (accessed Nov. 13, 2014).
14. Klemm, D., Heublein, B., Fink, H.P., and Bohn, A. (2005) Cellulose: fascinating biopolymer and sustainable raw material. *Angew. Chem., Int. Ed.*, 44(22):3358–3393.
15. (a) Adden, R., Müller, R., and Mischnick, P. (2006) Analysis of the substituent distribution in the glycosyl units and along the polymer chain of hydroxypropylmethyl celluloses and statistical evaluation. *Cellulose*, 13:459–476. (b) Engelskirchen, K. (1987) Umwandlung von Cellulose. In: *Methoden der organischen Chemie (Houben-Weyl)*, Bd. E20, Makromolekulare Stoffe (4. Aufl.), Thieme, pp. 2051–2065.
16. (a) (2013) United States Pharmacopeia and National Formulary, Vol. USP 36-NF 31, pp. 5993–5995. (b) Hodges, K.L., Kester, W.E., Wiederrich, D.L., and Grover, J.A. (1979) Determination of alkoxyl substitution in cellulose ethers by Zeisel-gas chromatography. *Anal. Chem.*, 51(13):2172–2176.
17. Keary, C.M. (2001) Characterization of METHOCEL cellulose ethers by aqueous SEC with multiple detectors. *Carbohydr. Polym.*, 45:293–303.
18. Al-Tabakha, M. (2010) HPMC capsules: current status and future prospects. *J. Pharm. Pharm. Sci.*, 13(3):428–442.
19. Doelker, E. (1993) Cellulose derivatives. *Adv. Polym. Sci.*, 107:199–265.
20. Friesen, D.T., Shanker, R., Crew, M., Smithey, D.T., Curatolo, W.J., and Nightingale, J.A. (2008) Hydroxypropyl methylcellulose acetate succinate-based spray-dried dispersions: an overview. *Mol. Pharm.*, 5(6):1003–1019.
21. The Dow Chemical Company (2002) METHOCEL Cellulose Ethers: Technical Handbook. Available at http://www.dow.com/dowwolff/en/pdf/192-01062.pdf (accessed Nov. 13, 2014).
22. Adden, R., Brackhagen, M., Knarr, M., and Zhao, J. (2010) Preparation of novel cellulose ethers for use in capsules or in coatings for dosage forms. WO2012051035.

23. Onda, Y., Muto, H., and Maruyama, K. (1978) Ether–ester derivatives of cellulose and their applications. U.S. Patent 4,226,981.
24. Shin-Etsu (2010) Shin-Etsu Aqoat hypromellose acetate succinate NF: enteric coating agent. http://www.metolose.jp/e/pharmaceutical/aqoat.shtml (accessed Nov. 13, 2014).
25. (a) Shah, N., Sandhu, H.K., Choi, D.S., Kalb, O., Page, S., and Wyttenbach, N. (2012) Structured development approach for amorphous systems. In: Williams, R.O., III, Watts, A.B., and Miller, D.A. (Eds.), *Formulating Poorly Water Soluble Drugs*, Springer, New York, pp. 267–310. (b) Kokubo, H., Obara, S., Minemura, K., and Tanaka, T. (1997) Development of cellulose derivatives as novel enteric coating agents soluble at pH 3.5–4.5 and higher. *Chem. Pharm. Bull.*, 45(8):1350–1353. (c) Tanno, F., Nishiyama, Y., Kokubo, H., and Obara, S. (2004) Evaluation of hypromellose acetate succinate (HPMCAS) as a carrier in solid dispersions. *Drug Dev. Ind. Pharm.*, 30(1):9–17.
26. Harrison, S. and Knabe, N. Dow Chemical and Cambrex Collaborate to Manufacture Drug-Solubility Solution. Available at http://www.dow.com/news/press-releases/article/?id=6166 (accessed Nov. 13, 2014).
27. (2013) United States Pharmacopeia and National Formulary, Vol. USP 36-NF 31, pp. 2045–2048.
28. Ray, R. (2012) Addressing challenges with low-solubility compounds: the importance of functional excipients in the formulation of amorphous dispersions. *Drug Dev. Deliv.*, 12(8):26–34.
29. (a) Morgen, M., Lyon, D., Schmitt, R., Brackhagen, M., and Petermann, O. (2013) New excipients for solubilizing APIs: expansion of the HPMCAS chemistry space. *Tablets Capsules*, 11(5):10–15. (b) The Dow Chemical Company (2013) AffiniSol Product Information (Solving the Insoluble with Dow). Available at http://www.dow.com/dowwolff/en/pdf/affinisol.pdf (accessed Nov. 13, 2014).
30. (a) Porter, W., III, Grasman, N., Petermann, O., Brackhagen, M., Guillaudeu, S., Murri, B., Miller, W., Morgan, R., and Morgen, M. (2012) *AIChE National Meeting*, Pittsburgh, PA. (b) Grasman, N., Porter, W., III, Petermann, O., Brackhagen, M., Guillaudeu, S., Murri, B., Miller, W., Morgan, R., and Morgen, M. (2012) Impact of Substitution Level on Spray Dried Dispersions of Hypromellose Acetate Succinate: A Quality by Design Approach, Chicago, IL.
31. Fukusawa, M. and Obara, S. Molecular weight determination of hypromellose acetate succinate (HPMCAS) using size exclusion chromatography with a multi-angle laser light scattering detector. *Chem. Pharm. Bull.*, 52(11):1391–1393.
32. Chen, R. (2009) Characterization of hypromellose acetate succinate by size exclusion chromatography (SEC) using Viscotec triple detector. *Int. J. Polym. Anal. Charact.*, 14:617–630.
33. Chen, R., Ilasi, N., and Sekulic, S.S. (2011) Absolute molecular weight determination of hypromellose acetate succinate by size exclusion chromatography: use of a multi angle laser light scattering detector and a mixed solvent. *J. Pharm. Biomed. Anal.*, 56(4):743–748.
34. Rosen, M.J. (2004) *Surfactants and Interfacial Phenomena*, 3rd edition, Wiley.
35. Holmberg, K., Lindman, B., Joensoon, B., and Kronberg, B. (2002) *Surfactants and Polymers in Aqueous Solution*, 2nd edition, Wiley.
36. Saindane, N. and Vavia, P. (2012) Osmotic pellet system comprising osmotic core and in-process amorphized drug in polymer–surfactant layer for controlled delivery of poorly water-soluble drug. *J. Pharm. Sci.*, 101(9):3169–3179.
37. El Eini, D.I.D., Barry, B.W., and Rhodes, C.T. (1976) Micellar size, shape, and hydration of long-chain polyoxyethylene nonionic surfactants. *J. Colloid Interface Sci.*, 54(3): 348–351.

38. Nielloud, F. and Marti-Mestres, G. (2000) *Pharmaceutical Emulsions and Suspensions*, Marcel Dekker.

39. Moneghini, M., Zingone, G., and De Zordi, N. (2009) Influence of the microwave technology on the physical–chemical properties of solid dispersion with nimesulide. *Powder Technol.*, 195(3):259–263.

40. Muller, R.H., Benita, S., and Bohm, B.H.L. (1998) *Emulsions and Nanosuspensions for the Formulation of Poorly Soluble Drugs* (Proceedings of Invited Lectures and Communications Presented at the Colloidal Drug Carriers 3rd Expert Meeting, May 29–31, 1997, Berlin, Germany), MedPharm Scientific Publishers.

41. (a) Shaeiwitz, J.A., Chan, A.F.C., Cussler, E.L., and Evans, D.F. (1981) The mechanism of solubilization in detergent solutions. *J. Colloid Interface Sci.*, 84(1):47–56. (b) Sun, W., Larive, C.K., and Southard, M.Z. (2003) A mechanistic study of danazol dissolution in ionic surfactant solutions. *J. Pharm. Sci.*, 92(2):424–435.

42. Strickley, R.G. (2004) Solubilizing excipients in oral and injectable formulations. *Pharm. Res.*, 21(2):201–230.

43. Torchilin, V.P. (2001) Structure and design of polymeric surfactant-based drug delivery systems. *J. Control. Release*, 73(2–3):137–172.

44. (a) Goddard, E.D. (1986) Polymer surfactant interaction. 1. Uncharged water-soluble polymers and charged surfactants. *Colloids Surf.*, 19(2–3):255–300. (b) Goddard, E.D. and Hannan, R.B. (1976) Cationic polymer–anionic surfactant interactions. *J. Colloid Interface Sci.*, 55(1):73–79.

45. Craig, D.Q. (2002) The mechanisms of drug release from solid dispersions in water-soluble polymers. *Int. J. Pharm.*, 231(2):131–144.

46. (a) Guzman, H.R., Tawa, M., Zhang, Z., Ratanabanangkoon, P., Shaw, P., Gardner, C.R., Chen, H., Moreau, J.P., Almarsson, O., and Remenar, J.F. (2007) Combined use of crystalline salt forms and precipitation inhibitors to improve oral absorption of celecoxib from solid oral formulations. *J. Pharm. Sci.*, 96(10):2686–2702. (b) Kawakami, K. (2012) Modification of physicochemical characteristics of active pharmaceutical ingredients and application of supersaturatable dosage forms for improving bioavailability of poorly absorbed drugs. *Adv. Drug Deliv. Rev.*, 64(6):480–495.

47. Thummar, A.D., Sastry, N.V., Verma, G., and Hassan, P.A. (2011) Aqueous block copolymer–surfactant mixtures: surface tension, DLS and viscosity measurements and their utility in solubilization of hydrophobic drug and its controlled release. *Colloids Surf. A*, 386 (1–3):54–64.

48. Shah, S., Maddineni, S., Lu, J., and Repka, M.A. (2013) Melt extrusion with poorly soluble drugs. *Int. J. Pharm.*, 453(1):233–252.

49. Sivadasan, K. and Somasundaran, P. (1990) Polymer–surfactant interactions and the association behavior of hydrophobically modified hydroxyethylcellulose. *Colloids Surf.*, 49(3–4):229–239.

50. Persson, B., Nilsson, S., and Sundelof, L.O. (1996) On the characterization principles of some technically important water-soluble nonionic cellulose derivatives. 2. Surface tension and interaction with a surfactant. *Carbohydr. Polym.*, 29(2):119–127.

51. Winnik, F.M. and Regismond, S.T.A. (1996) Fluorescence methods in the study of the interactions of surfactants with polymers. *Colloids Surf. A*, 118(1–2):1–39.

52. Balazs, A.C., Huang, K.L., and Pan, T. (1993) Modeling of amphiphilic polymers and their interactions with nonionic surfactants. *Colloids Surf. A*, 75:1–20.

53. Nizri, G. and Magdassi, S. (2005) Solubilization of hydrophobic molecules in nanoparticles formed by polymer–surfactant interactions. *J. Colloid Interface Sci.*, 291(1):169–174.
54. Okonogi, S. and Puttipipatkhachorn, S. (2006) Dissolution improvement of high drug-loaded solid dispersion. *AAPS PharmSciTech*, 7(2).
55. Pandit, N.K., Kanjia, J., Patel, K., and Pontikes, D.G. (1995) Phase behavior of aqueous solutions containing nonionic surfactant–polyethylene glycol mixtures. *Int. J. Pharm.*, 122(1–2):27–33.
56. Schwuger, M.J. (1973) Mechanism of interaction between ionic surfactants and polyglycol ethers in water. *J. Colloid Interface Sci.*, 43(2):491–498.
57. Hu, L., Zhang, N., Yang, G., and Zhang, J. (2011) Effects of Tween-80 on the dissolution properties of daidzein solid dispersion *in vitro*. *Int. J. Chem.*, 3(1):68–73.
58. Liu, L.X. and Wang, X.C. (2007) Improved dissolution of oleanolic acid with ternary solid dispersions. *AAPS PharmSciTech*, 8(4):5.
59. Linn, M., Collnot, E.M., Djuric, D., Hempel, K., Fabian, E., Kolter, K., and Lehr, C.M. (2012) Soluplus® as an effective absorption enhancer of poorly soluble drugs *in vitro* and *in vivo*. *Eur. J. Pharm. Sci.*, 45(3):336–343.
60. Janssens, S., Nagels, S., de Armas, H.N., D'Autry, W., Van Schepdael, A., and Van den Mooter, G. (2008) Formulation and characterization of ternary solid dispersions made up of itraconazole and two excipients, TPGS 1000 and PVPVA 64, that were selected based on a supersaturation screening study. *Eur. J. Pharm. Biopharm.*, 69(1):158–166.
61. Frank, K.J., Westedt, U., Rosenblatt, K.M., Holig, P., Rosenberg, J., Magerlein, M., Fricker, G., and Brandl, M. (2012) The amorphous solid dispersion of the poorly soluble ABT-102 forms nano/microparticulate structures in aqueous medium: impact on solubility. *Int. J. Nanomed.*, 7:5757–5768.
62. Srinarong, P., Hamalainen, S., Visser, M.R., Hinrichs, W.L.J., Ketolainen, J., and Frijlink, H.W. (2011) Surface-active derivative of inulin (Inutec® SP1) is a superior carrier for solid dispersions with a high drug load. *J. Pharm. Sci.*, 100(6):2333–2342.
63. Joe, J.H., Lee, W.M., Park, Y.J., Joe, K.H., Oh, D.H., Seo, Y.G., Woo, J.S., Yong, C.S., and Choi, H.G. (2010) Effect of the solid-dispersion method on the solubility and crystalline property of tacrolimus. *Int. J. Pharm.*, 395(1–2):161–166.
64. Patel, A.R. and Joshi, V.Y. (2008) Evaluation of SLS:APG mixed surfactant systems as carrier for solid dispersion. *AAPS PharmSciTech*, 9(2):583–590.
65. Szuts, A., Lang, P., Ambrus, R., Kiss, L., Deli, M.A., and Szabo-Revesz, P. (2011) Applicability of sucrose laurate as surfactant in solid dispersions prepared by melt technology. *Int. J. Pharm.*, 410(1–2):107–110.
66. Crowley, K.J. and Zografi, G. (2003) The effect of low concentrations of molecularly dispersed poly(vinylpyrrolidone) on indomethacin crystallization from the amorphous state. *Pharm. Res.*, 20(9):1417–1422.
67. Hancock, B.C., York, P., and Rowe, R.C. (1997) The use of solubility parameters in pharmaceutical dosage form design. *Int. J. Pharm.*, 148(1):1–21.
68. Gupta, J., Nunes, C., Vyas, S., and Jonnalagadda, S. (2011) Prediction of solubility parameters and miscibility of pharmaceutical compounds by molecular dynamics simulations. *J. Phys. Chem. B*, 115(9):2014–2023.
69. Hildebrand, J.H. and Scott, R.L. (1950) *The Solubility of Nonelectrolytes*, 3rd edition, Reinhold Publishing Corporation.

70. Egawa, H., Maeda, S., Yonemochi, E., Oguchi, T., Yamamoto, K., and Nakai, Y. (1992) Solubility parameter and dissolution behavior of cefalexin powders with different crystallinity. *Chem. Pharm. Bull.*, 40(3):819–820.

71. Langer, M., Holtje, M., Urbanetz, N.A., Brandt, B., Holtje, H.D., and Lippold, B.C. (2003) Investigations on the predictability of the formation of glassy solid solutions of drugs in sugar alcohols. *Int. J. Pharm.*, 252(1–2):167–179.

72. (a) Albers, J., Alles, R., Matthee, K., Knop, K., Nahrup, J.S., and Kleinebudde, P. (2009) Mechanism of drug release from polymethacrylate-based extrudates and milled strands prepared by hot-melt extrusion. *Eur. J. Pharm. Biopharm.*, 71(2):387–394. (b) Greenhalgh, D.J., Williams, A.C., Timmins, P., and York, P. (1999) Solubility parameters as predictors of miscibility in solid dispersions. *J. Pharm. Sci.*, 88(11):1182–1190.

73. (a) Bhattacharya, S. and Suryanarayanan, R. (2009) Local mobility in amorphous pharmaceuticals: characterization and implications on stability. *J. Pharm. Sci.*, 98(9):2935–2953. (b) Johari, G.P. and Shanker, R.M. (2010) On determining the relaxation time of glass and amorphous pharmaceuticals' stability from thermodynamic data. *Thermochim. Acta*, 511(1–2):89–95.

74. Chokshi, R.J., Shah, N.H., Sandhu, H.K., Malick, A.W., and Zia, H. (2008) Stabilization of low glass transition temperature indomethacin formulations: impact of polymer-type and its concentration. *J. Pharm. Sci.*, 97(6):2286–2298.

75. Gordon, M. and Taylor, J.S. (1952) Ideal copolymers and the second-order transitions of synthetic rubbers. I. Noncrystalline copolymers. *J. Appl. Chem.*, 2:493–500.

76. Simha, R. and Bayer, R.F. (1962) General relation involving the glass temperature and coefficients of expansion of polymers. *J. Chem. Phys.*, 37:1003–1007.

77. Six, K., Berghmans, H., Leuner, C., Dressman, J., Van Werde, K., Mullens, J., Benoist, L., Thimon, M., Meublat, L., Verreck, G., Peeters, J., Brewster, M., and Van den Mooter, G. (2003) Characterization of solid dispersions of itraconazole and hydroxypropylmethylcellulose prepared by melt extrusion, part II. *Pharm. Res.*, 20(7):1047–1054.

78. Khougaz, K. and Clas, S.D. (2000) Crystallization inhibition in solid dispersions of MK-0591 and poly(vinylpyrrolidone) polymers. *J. Pharm. Sci.*, 89(10):1325–1334.

79. Yang, Z. and Han, C.D. (2008) Rheology of miscible polymer blends with hydrogen bonding. *Macromolecules*, 41(6):2104–2118.

80. (a) Forster, A., Hempenstall, J., Tucker, I., and Rades, T. (2001) The potential of small-scale fusion experiments and the Gordon–Taylor equation to predict the suitability of drug/polymer blends for melt extrusion. *Drug Dev. Ind. Pharm.*, 27(6):549–560. (b) Forster, A., Hempenstall, J., Tucker, I., and Rades, T. (2001) Selection of excipients for melt extrusion with two poorly water-soluble drugs by solubility parameter calculation and thermal analysis. *Int. J. Pharm.*, 226(1–2):147–161. (c) Yoo, S.U., Krill, S.L., Wang, Z., and Telang, C. (2009) Miscibility/stability considerations in binary solid dispersion systems composed of functional excipients towards the design of multi-component amorphous systems. *J. Pharm. Sci.*, 98(12):4711–4723.

81. (a) Janssens, S., De Zeure, A., Paudel, A., Van Humbeeck, J., Rombaut, P., and Van den Mooter, G. (2010) Influence of preparation methods on solid state supersaturation of amorphous solid dispersions: a case study with itraconazole and Eudragit E100. *Pharm. Res.*, 27(5):775–785. (b) Paudel, A., Van Humbeeck, J., and Van den Mooter, G. (2010) Theoretical and experimental investigation on the solid solubility and miscibility of naproxen in poly(vinylpyrrolidone). *Mol. Pharm.*, 7(4):1133–1148. (c) Zhao, Y.Y., Inbar,

P., Chokshi, H.P., Malick, A.W., and Choi, D.S. (2011) Prediction of the thermal phase diagram of amorphous solid dispersions by Flory–Huggins theory. *J. Pharm. Sci.*, 100(8):3196–3207.

82. Tian, Y.W., Booth, J., Meehan, E., Jones, D.S., Li, S., and Andrews, G.P. (2013) Construction of drug–polymer thermodynamic phase diagrams using Flory–Huggins interaction theory: identifying the relevance of temperature and drug weight fraction to phase separation within solid dispersions. *Mol. Pharm.*, 10(1):236–248.

83. Ohtake, S. and Shalaev, E. (2013) Effect of water on the chemical stability of amorphous pharmaceuticals. I. Small molecules. *J. Pharm. Sci.*, 102(4):1139–1154.

84. (a) Hancock, B.C. and Zografi, G. (1994) The relationship between the glass transition temperature and the water content of amorphous pharmaceutical solids. *Pharm. Res.*, 11(4):471–477. (b) Roos, Y.H. and Karel, M. (1991) Plasticizing effect of water on thermal behavior and crystallization of amorphous food models. *J. Food Sci.*, 56(1):38–43. (c) Shalaev, E.Y. and Zografi, G. (1996) How does residual water affect the solid-state degradation of drugs in the amorphous state? *J. Pharm. Sci.*, 85(11):1137–1141.

85. Marsac, P.J., Konno, H., Rumondor, A.C.F., and Taylor, L.S. (2008) Recrystallization of nifedipine and felodipine from amorphous molecular level solid dispersions containing poly (vinylpyrrolidone) and sorbed water. *Pharm. Res.*, 25(3):647–656.

86. Ghebre-Sellassie, I. and Martin, C. (2003) *Pharmaceutical Extrusion Technology*, Marcel Dekker.

87. Sollohub, K. and Cal, K. (2010) Spray drying technique. II. Current applications in pharmaceutical technology. *J. Pharm. Sci.*, 99(2):587–597.

88. (a) DiNunzio, J.C., Brough, C., Miller, D.A., Williams, R.O., and McGinity, J.W. (2010) Applications of KinetiSol® Dispersing for the production of plasticizer free amorphous solid dispersions. *Eur. J. Pharm. Sci.*, 40(3):179–187. (b) DiNunzio, J.C., Hughey, J.R., Brough, C., Miller, D.A., Williams, R.O., and McGinity, J.W. (2010) Production of advanced solid dispersions for enhanced bioavailability of itraconazole using KinetiSol® Dispersing. *Drug Dev. Ind. Pharm.*, 36(9):1064–1078.

89. Warren, D.B., Bergström, C.A.S., Benameur, H., Porter, C.J.H., and Pouton, C.W. (2013) Evaluation of the structural determinants of polymeric precipitation inhibitors using solvent shift methods and principal component analysis. *Mol. Pharm.*, 10(8):2823–2848.

90. Wyttenbach, N., Janas, C., Siam, M., Lauer, M.E., Jacob, L., Scheubel, E., and Page, S. (2013) Miniaturized screening of polymers for amorphous drug stabilization (SPADS): rapid assessment of solid dispersion systems. *Eur. J. Pharm. Biopharm.*, 84(3):583–598.

91. Serajuddin, A.T.M., Thakur, A.B., Ghoshal, R.N., Fakes, M.G., Ranadive, S.A., Morris, K.R., and Varia, S.A. (1999) Selection of solid dosage form composition through drug–excipient compatibility testing. *J. Pharm. Sci.*, 88(7):696–704.

92. Hong, J.Y., Shah, J.C., and McGonagle, M.D. (2011) Effect of cyclodextrin derivation and amorphous state of complex on accelerated degradation of ziprasidone. *J. Pharm. Sci.*, 100(7):2703–2716.

93. Bindra, D.S., Williams, T.D., and Stella, V.J. (1994) Degradation of *O*-6-benzylguanine in aqueous polyethylene glycol 400 (PEG 400) solutions: concerns with formaldehyde in PEG 400. *Pharm. Res.*, 11(7):1060–1064.

94. (a) Bergh, M., Magnusson, K., Nilsson, J.L.G., and Karlberg, A.T. (1997) Contact allergenic activity of Tween® 80 before and after air exposure. *Contact Dermatitis*, 37(1):9–18. (b) Bergh, M., Magnusson, K., Nilsson, J.L.G., and Karlberg, A.T. (1998) Formation of

formaldehyde and peroxides by air oxidation of high purity polyoxyethylene surfactants. *Contact Dermatitis*, 39(1):14–20. (c) Bergh, M., Shao, L.P., Hagelthorn, G., Gafvert, E., Nilsson, J.L.G., and Karlberg, A.T. (1998) Contact allergens from surfactants. Atmospheric oxidation of polyoxyethylene alcohols, formation of ethoxylated aldehydes, and their allergenic activity. *J. Pharm. Sci.*, 87(3):276–282. Donbrow, M. (1987) Stability of the polyoxyethylene chain. In: Schick, M.J. (Ed.), *Nonionic Surfactants: Physical Chemistry*, Surfactant Science Series, Vol. 23, pp. 1011–1072. (e) Frontini, R. and Mielck, J.B. (1995) Formation of formaldehyde in polyethyleneglycol and in poloxamer under stress conditions. *Int. J. Pharm.*, 114(1):121–123.

95. Greco, S., Authelin, J.-R., Leveder, C., and Segalini, A. (2012) A practical method to predict physical stability of amorphous solid dispersions. *Pharm. Res.*, 29(10):2792–2805.

96. (a) Lauer, M.E., Siam, M., Tardio, J., Page, S., Kindt, J.H., and Grassmann, O. (2013) Rapid assessment of homogeneity and stability of amorphous solid dispersions by atomic force microscopy: from bench to batch. *Pharm. Res.*, 30(8):2010–2022. (b) Yang, J., Grey, K., and Doney, J. (2010) An improved kinetics approach to describe the physical stability of amorphous solid dispersions. *Int. J. Pharm.*, 384(1–2):24–31.

97. (a) Hancock, B.C., Shamblin, S.L., and Zografi, G. (1995) Molecular mobility of amorphous pharmaceutical solids below their glass transition temperatures. *Pharm. Res.*, 12(6):799–806. (b) Hancock, B.C. and Zograf, G. (1997) Characteristics and significance of the amorphous state in pharmaceutical systems. *J. Pharm. Sci.*, 86(1):1–12.

98. (a) Marsac, P., Li, T., and Taylor, L. (2009) Estimation of drug–polymer miscibility and solubility in amorphous solid dispersions using experimentally determined interaction parameters. *Pharm. Res.*, 26(1):139–151. (b) Qian, F., Huang, J., and Hussain, M.A. (2010) Drug–polymer solubility and miscibility: stability consideration and practical challenges in amorphous solid dispersion development. *J. Pharm. Sci.*, 99(7):2941–2947. (c) van Drooge, D.J., Hinrichs, W.L.J., Visser, M.R., and Frijlink, H.W. (2006) Characterization of the molecular distribution of drugs in glassy solid dispersions at the nano-meter scale, using differential scanning calorimetry and gravimetric water vapour sorption techniques. *Int. J. Pharm.*, 310(1–2):220–229.

99. Miller, J.M., Beig, A., Carr, R.A., Spence, J.K., and Dahan, A. (2012) A win–win solution in oral delivery of lipophilic drugs: supersaturation via amorphous solid dispersions increases apparent solubility without sacrifice of intestinal membrane permeability. *Mol. Pharm.*, 9(7):2009–2016.

100. (a) Dahan, A., Miller, J.M., Hoffman, A., Amidon, G.E., and Amidon, G.L. (2010) The solubility–permeability interplay in using cyclodextrins as pharmaceutical solubilizers: mechanistic modeling and application to progesterone. *J. Pharm. Sci.*, 99(6):2739–2749. (b) Miller, J.M. and Dahan, A. (2012) Predicting the solubility–permeability interplay when using cyclodextrins in solubility-enabling formulations: model validation. *Int. J. Pharm.*, 430 (1–2):388–391.

101. Miller, J.M., Beig, A., Krieg, B.J., Carr, R.A., Borchardt, T.B., Amidon, G.E., Amidon, G.L., and Dahan, A. (2011) The solubility–permeability interplay: mechanistic modeling and predictive application of the impact of micellar solubilization on intestinal permeation. *Mol. Pharm.*, 8(5):1848–1856.

102. (a) Beig, A., Miller, J.M., and Dahan, A. (2012) Accounting for the solubility–permeability interplay in oral formulation development for poor water solubility drugs: the effect of PEG-400 on carbamazepine absorption. *Eur. J. Pharm. Biopharm.*, 81(2):386–391. (b) Miller, J.M., Beig, A., Carr, R.A., Webster, G.K., and Dahan, A. (2012) The solubility–permeability

interplay when using cosolvents for solubilization: revising the way we use solubility-enabling formulations. *Mol. Pharm.*, 9(3):581–590.

103. Dahan, A., Beig, A., Ioffe-Dahan, V., Agbaria, R., and Miller, J.M. (2013) The twofold advantage of the amorphous form as an oral drug delivery practice for lipophilic compounds: increased apparent solubility and drug flux through the intestinal membrane. *AAPS J.*, 15(2):347–353.

3

AMORPHOUS SOLID DISPERSION SCREENING

Ann Newman

Seventh Street Development Group, Lafayette, IN, USA

3.1 INTRODUCTION

Finding the optimal form for drug development can involve a number of different screening activities, including a search for polymorphs, salts, cocrystals, and/or amorphous solid dispersions. Solid form screening activities can be performed in early or late development with different strategies at each stage of development (Figure 3.1). Amorphous solid dispersions are often included as part of this screening strategy, but they can also be part of a formulation strategy, especially for preclinical studies, based on solubility risk assessment [1]. High-risk compounds will require special formulations that include amorphous solid dispersions. Whether the focus is on API or formulation, an amorphous solid dispersion may be necessary to find an optimal form for further development.

An amorphous dispersion screen is performed to find a suitable amorphous dispersion that exhibits acceptable properties. The screening performed in early development may focus on parameters such as solubility, supersaturation, short-term physical and chemical stability, and animal bioavailability to get into early-phase clinical trials quickly and establish proof of concept. Characterization is usually limited to essential information, such as amorphous structure, glass transition temperature (T_g), water/solvent content, and

Pharmaceutical Amorphous Solid Dispersions, First Edition. Edited by Ann Newman.

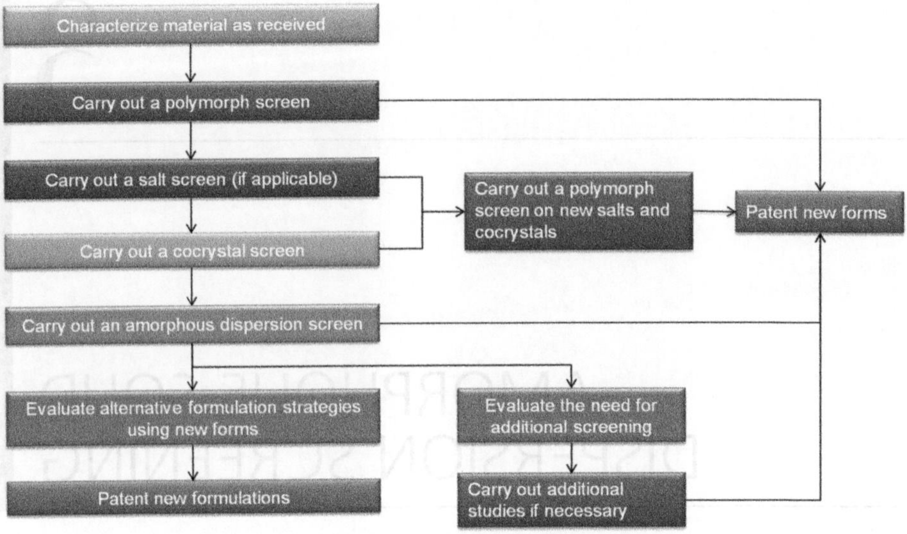

Figure 3.1. Solid form screening strategy including amorphous solid dispersions.

miscibility. Early formulations will be relatively straightforward with drug in a capsule or simplified oral formulations. Later in development, other parameters will be investigated to determine long-term formulatability, shelf life, and performance. So, a different screen may be needed to collect this type of information and assess risk management.

There are a number of steps that need to be considered for an amorphous dispersion screen. These include experimental variables, characterization and property assessment, and finally selection.

3.2 AMORPHOUS DISPERSION SCREENING

3.2.1 API and Polymer Properties

3.2.1.1 API Characterization Before starting an amorphous dispersion screen project, information on the active pharmaceutical ingredient (API) is needed. This may include chemical structure, solid form, physical stability, chemical stability, log *P*, hydrogen bond donors/acceptors, calculated solubility parameters, and known solubilities of the solid in various solvents/biorelevant media. In early development, only some of these properties will be known and it will be necessary to obtain additional data as needed to move forward.

Solid-state characterization of the initial API as received is the first step and can include techniques summarized in Table 3.1 that would determine properties such as crystallinity/form, thermal properties, volatile content, and water uptake. This initial set of data will establish a baseline for later comparison with amorphous solid dispersions prepared during the screen.

TABLE 3.1. Common Characterization Techniques for Initial API and Polymer Components

Technique	Crystalline Materials	Amorphous Materials	Amorphous Dispersions
X-ray powder diffraction	Crystallinity/form	Amorphous structure	Amorphous structure Crystalline "impurity"
Differential scanning calorimetry	Melting point (T_m) Transitions upon heating	Glass transition temperature (T_g) Transitions upon heating	T_g Miscibility Transitions upon heating T_m for crystalline "impurity"
Hot-stage microscopy	Thermal event identification	Thermal event identification	Thermal event identification
Thermogravimetric analysis	Volatile content; if volatiles are present, they need to be identified by another method	Volatile content; if volatiles are present, they should be identified, amorphous material should be dried, and T_g remeasured	Volatile content; if volatiles are present, they should be identified, amorphous material should be dried, and T_g remeasured
Moisture uptake	Water uptake; determine handling conditions	Water uptake; determine handling conditions	Water uptake; determine handling conditions
Spectroscopy (IR, Raman, SSNMR)	Baseline data; compare with dispersions to determine interactions and miscibility	Baseline data; compare with dispersions to determine interactions and miscibility	Baseline data; compare with dispersions to determine interactions and miscibility

The data collected on the API can be used to help compile a list of screening experiments. Determination of the melting point of a crystalline API will be necessary if melt experiments will be used in the screen or if melt extrusion will be the primary method for making larger quantities of material. It should also be determined whether the API melts with or without decomposition so that a suitable set of melt experiments can be devised to prevent decomposition. Estimated solubility values of the as-received API (crystalline or amorphous) in a variety of solvents are also needed for solvent-based screening experiments or if spray drying will be the primary method for scale-up. For spray drying of amorphous solid dispersions, typical solvents include dichloromethane, acetone, methanol, tetrahydrofuran, ethanol, ethyl acetate, water, and various combinations of these [2]. Solvent mixtures can be added as needed for the screen (based on the solubility values of the API and polymers) with the understanding that the solvent system may have to be modified if spray drying is used to make larger quantities. If the API is poorly soluble in many solvents, this may limit the solvents and polymers available for a specific screen. If the API exhibits low solubility in all commonly used organic solvents, melt experiments may be a better option.

It is also useful to prepare the amorphous form of the API for comparison with the dispersions produced in the screen. A variety of methods are available to produce amorphous material and the methods chosen will depend on the properties of the crystalline material. Methods to produce the amorphous form include melt quench, grinding, and solvent methods (crash precipitation, rotary evaporation, lyophilization, and spray drying) [3]. Once amorphous material is produced, characterization can include the techniques shown in Table 3.1. Solubility in water, biorelevant media, and organic solvents would also be helpful for comparison with the crystalline material and to determine supersaturation levels.

Once these data are obtained for the API material, they will be used to help plan the dispersion screen and determine specific parameters for experiments.

3.2.1.2 Polymer Properties Most polymer properties can be obtained from literature references or from the vendor. Considerations when selecting a polymer may include molecular weight, glass transition temperature, hydrophilic/hydrophobic properties, hydrogen bond donors/acceptors, solubility parameter, pH–solubility profile, thermal stability, water uptake, and monomer levels. Some of these properties are compiled in Chapter 14 and Appendix A. Other considerations that will be specific to the dispersion system include solubility and/or miscibility of the drug in the polymer, degree of supersaturation during dissolution, and prevention of API crystallization from the solid or solution. Additional properties may be important for the method of manufacture, as outlined in Chapters 9 and 10 for spray drying and melt extrusion, respectively.

3.2.2 Screening Considerations

A number of factors need to be determined for each project that are dependent on the properties of the molecule, the stage of development (early versus late), and the development/formulation plan. These factors will help define the goals and scope of the dispersion screen. The following sections will outline some of the parameters to consider for the screen, such as polymer selection, API and polymer composition, and use of a surfactant.

3.2.2.1 Polymer Selection There are a number of approaches that can be used when choosing polymers for a screen, and some of these are summarized in Table 3.2. It is obvious that a wide variety of methods have been reported for finding the best combination of API and polymer. Some methods are straightforward for narrowing the list of polymers, such as an empirical approach using only common polymers [4] or using polymers specific to melt extrusion. Other methods attempt to choose polymers based on simple tests of miscibility, such as information obtained from differential scanning calorimetry (DSC) (T_g, T_m, and T_c) [5–7] or calculation of solubility parameters [8]. Interactions between components are also considered from simple hydrogen bonding

TABLE 3.2. Approaches for Choosing Polymers in a Dispersion Screen

Method	Description	Comments	References
Empirical	Choose common polymers	Use a variety of polymers; melting point or solubility may be considered	[4]
Manufacturing	Choose polymers based on the desired large-scale manufacturing (melt extrusion and spray drying)	Need low-melting polymers for melt extrusion, need solubility in solvent for spray drying	[17,18]
Solubility parameter	Choose polymers with a similar solubility parameter as the API	Similar solubility parameter maximizes mixing and promotes miscibility	[8,19]
Miscibility	Choose polymers that are miscible with API	Miscible systems may show better physical stability	[5]
Polymer properties	Choose polymer with high T_g	High T_g of polymer will increase T_g of the dispersion; need T_g of around 70–80 °C to be about 50 °C above room temperature	[4,20]
	Choose polymer with low water sorption	Water will lower T_g and may decrease physical stability	[4,10]
	Review pH where polymer dissolves compared with pH of targeted absorption site in GI tract	Compare pH with that of API properties (acid versus base) and site of absorption (gastric versus intestinal)	[21]
DSC	Determine T_m/T_g ratio	Choose polymers with high T_g values to increase T_g of the dispersion compared with the amorphous API; high ratios may crystallize more easily	[6]
	Reduced crystallization temperature	Crystallization peak in DSC increases in temperature with increased interactions between API and polymer	[7]
	Determine width of T_g	Wide T_g may indicate partial miscibility	[22]
Interactions	Look at common H-bonding motifs, ion–dipole interactions, or acid–base interactions between drug and polymer	Try to maximize interactions in dispersion to disrupt bonding in crystalline materials (e.g., PVP disrupts indomethacin dimers)	[9–11,23]
	Acid–base interactions	Ionic interactions produce strong interactions for stable amorphous solid dispersions	[23,24]

(continued)

TABLE 3.2. *(Continued)*

Method	Description	Comments	References
	Complex interactions	Use GEMANOVA analysis based on XRPD microscopy data	[25]
	Ab initio calculations	Rank order crystallization inhibitors	[11]
	Determine binding constants using affinity capillary electrophoresis	Drug–polymer binding constants based on changes in effective electrophoretic mobility of the drug upon binding to the polymer	[26]
Supersaturation	Determine polymers that improve supersaturation or stability in solution	Stable supersaturated solutions provide API:polymer ratios; spray drying used to produce dispersions	[27–29]
Drug solubility in polymer	Determine level of drug that will dissolve in polymer	Gives upper concentration for unsaturated dispersion	[5,30,31]

motifs [9,10] to *ab initio* calculations [11]. Additional approaches are related to specific properties of the polymer such as solubility parameter, hygroscopicity, or pH (for polymers used as enteric coatings). Statistical analysis using design space and QbD concepts has also been reported [12]. A number of studies assess properties of the dispersions, such as solubility, physical stability, and crystallization inhibition (supersaturation), to choose polymers; in some respects, these parameters may be more applicable as part of the dispersion selection process but when automated may be included as part of the screening process [13–16]. Many of the methods in Table 3.1 are easily incorporated into early screens, while others require more time and expertise, and may be more suitable for screening in later development. A summary of select approaches will be given in this section.

Consideration of Large-Scale Manufacturing At larger scales needed for early animal studies or clinical supplies, amorphous dispersions typically are formed by combining the API and polymer in a melt extruder [17,32,33], by dissolving them in a solvent, followed by solvent removal such as spray drying [18], or, less often, lyophilization. The type of large-scale preparation targeted for the project can focus on the types of polymers, solvents, and conditions used during the screen. Screening experiments can include melt methods, using no solvent to mimic melt extrusion, and rapid evaporation experiments using solvent to mimic spray drying. For some projects, it may be helpful to have a mix of both methods and for some situations a more targeted screen may be applicable. Typical polymers used for amorphous solid dispersions are given in Appendix A; this is not meant to be a comprehensive list but covers commonly used polymers in amorphous solid dispersions.

Solubility Parameter The solubility parameter (δ) originated from the Hildebrand solution theory used to predict the solubility of organic compounds. It represents the attractive forces holding the molecule together and includes contributions from dispersion, polar, and hydrogen bonding forces. A comparison of solubility parameters for the API and various polymers is generally used to select polymers for dispersion studies. It has been suggested that components will be miscible if the difference in solubility parameters is less than 7 $mPa^{0.5}$ and will be immiscible if the difference in solubility parameters is greater than 10 $mPa^{0.5}$ [8]. Calculating the solubility parameter of the API and finding polymers with similar values is one way to narrow down the list of polymers in a screen. This should not be the only criterion used to select polymers because there may be other properties, such as ionic interactions, that may override solubility issues. However, it is useful as an initial starting point for determining a polymer list.

Polymer Properties Polymer properties must be considered to make the dispersion stable and manufacturable. Polymers with a higher T_g have less mobility, leading to better inhibition of drug crystallization in the solid state. The Gordon–Taylor equation or the simplified Fox equation (Equation 3.1) can be used to estimate the T_g of the dispersion based on the T_g values of the individual components and the API:polymer ratio:

$$1/T_{\mathrm{gmix}} = (w_{\mathrm{A}}/T_{\mathrm{gA}}) + (w_{\mathrm{B}}/T_{\mathrm{gB}}). \tag{3.1}$$

where TgA and TgB are the individual Tgs (in K) of the individual components and wA and wB are the weight fractions. Based on this relationship, the polymers with higher T_g values will result in dispersions with higher T_g values when the same API is used. If the amorphous API exhibits a T_g above 70–80 °C, then the polymer T_g may not be as important. When choosing polymers with an elevated T_g, it must be considered that higher temperatures will be required during the extrusion process if this manufacturing method is chosen.

In general, polymers that sorb small amounts of water under typical relative humidity conditions expected for handling and storage are desirable. The hygroscopicity of the polymer can negatively affect physical and chemical stability of the dispersion, by increasing mobility and decreasing the T_g of the dispersion when water is sorbed by the system [34]. Polymers that tend to sorb water extensively, such as PVP, are commonly used during screening and may work for early formulations, but may need special packaging if used in marketed products.

Many of the polymers used in amorphous solid dispersions are enteric coatings, such as the methacrylate/methacrylic acid copolymers (Eudragit®), which are designed to dissolve in the small intestine under more neutral pH conditions rather than the acid pH of the stomach. If the API is expected to dissolve under the acidic conditions of the stomach, then using enteric polymers may not provide the dissolution properties necessary for development of that compound. It is important to choose polymers that coincide with the properties of the API in this regard.

For solvent screening experiments, it is important to know the solubility of the polymers in a variety of solvents, especially those solvents with high volatility that can be removed readily to trap the amorphous API, such as those commonly used in spray drying. Adequate solubility of both the API and polymer is needed in the solvents chosen for the screen. Solvent solubilities of common polymers are given in Chapter 14.

Melt experiments do not use solvent and processing must be performed at temperatures above the T_g of the polymer and high enough for the API to either melt and/or dissolve into the polymer matrix. When using melt experiments, it is important to ascertain the extent of chemical instability of the drug at elevated temperatures. For many large-scale extrusion processes, a surfactant may also be needed and this is considered in a separate section.

Thermal Data The melting point and glass transition ratio, T_m/T_g, can be combined with the log P information to estimate the propensity for crystallization and drug loading in the dispersion (Figure 3.2); this figure is based on the use of HPMCAS only [6]. By plotting log P versus T_m/T_g, it was found that an increase along the y-axis reduced physical stability and an increase along the x-axis reduced dissolution rate. The plot area was then divided into four sections depending on the API properties, such as propensity for drug to crystallize at high T_m/T_g or limited API loading due to dissolution rate at high log P. Based on these properties, suggested polymer loadings are incorporated into the plot for each section. More research is needed to determine whether this approach will work for other polymers, but it can be used to help guide the polymer selection process.

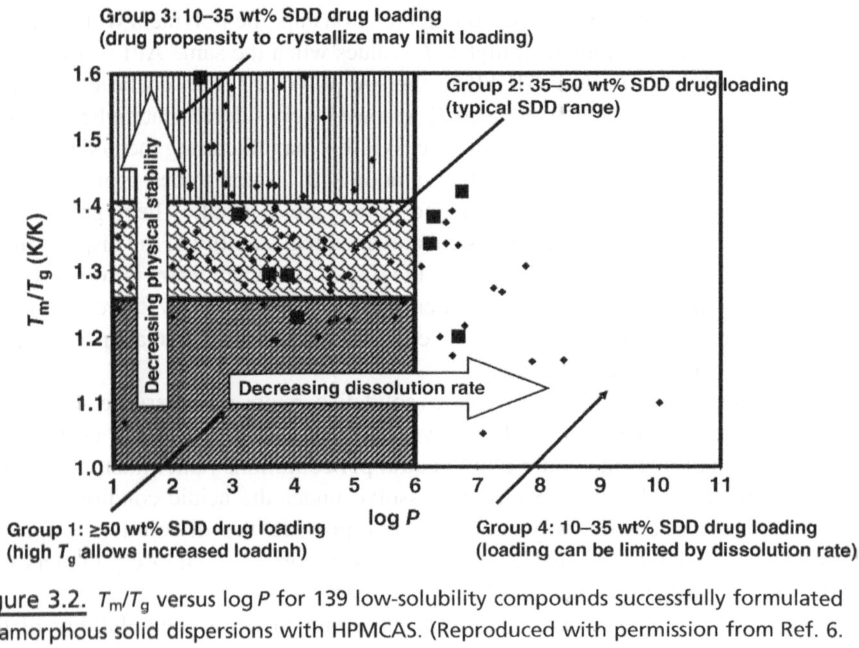

Figure 3.2. T_m/T_g versus log P for 139 low-solubility compounds successfully formulated as amorphous solid dispersions with HPMCAS. (Reproduced with permission from Ref. 6. Copyright 2008, American Chemical Society.)

The crystallization exotherm observed in the DSC can also be used to probe interactions between the API and the polymer using the reduced crystallization temperature (T_{RC}) and the following equation [7]:

$$T_{RC} = (T_c - T_g)/(T_m - T_g), \quad (3.2)$$

where T_c is the crystallization exotherm temperature, T_g is the glass transition temperature, and T_m is the melting endotherm temperature.

The exotherm temperature is measured with three to four different polymer loadings. If the crystallization exotherm shifts to higher temperature with increased polymer, then there are interactions between the components indicating a more stable system. If there is no shift in the exotherm with increased polymer levels, then there is minimal interaction. A plot of the polymer content versus the reduced crystallization temperature results in the crystallization parameter based on the slope of the line. This crystallization parameter can be used to semiquantitatively rank order the polymers with the higher values indicating better stability. Work with nifedipine showed good interactions with PVP K25, PVP/VA 64, and Soluplus®, and it was confirmed with other methods for determining interactions with nifedipine or a related compound felodipine, such as solubility and Flory–Huggins interaction parameters [30], miscibility assessment [35], and stability testing [36]. Studies on a new chemical entity using the reduced crystallization temperature showed good interactions with only one polymer (PAA). Physical stability studies performed on the API:PAA dispersion at 40 °C/75% RH for up to 6 months in a closed container with desiccant showed that the dispersion was physically stable.

3.2.2.2 Binary Versus Ternary Systems For early dispersion systems, binary mixtures of a polymer and the API will be the easiest to investigate. Once acceptable binary mixtures have been identified and characterized, other components can be added to see whether the properties can be improved [37,38]. These components can include polymers, surfactants, lipids, pH modifiers, or even small molecules, such as citric acid or magnesium oxide. The choice of another component will depend on the API and the property that needs to be modified. A sampling of some of these components is given in the following section.

Another approach to selecting ternary systems is to include a variety of polymers and surfactants during the initial screen. Using a plate screen, a larger number of combinations can be tried. An example of a ternary dispersion screen for a development compound used six polymers and eight surfactants [15]. An excipient:compound ratio of 10:1 was used for 14 binary mixtures and a polymer surfactant ratio of 3:1 was used for 48 ternary formulations. Initial dispersions were evaluated for solubility in the plate as represented in Figure 3.3. Those dispersions exhibiting acceptable solubility were moved to the next tiers involving dissolution and *in vivo* testing. An HPMCP:Vitamin E TPGS amorphous solid dispersion was found to give the best *in vivo* performance resulting in an oral bioavailability similar to a solution.

	None	SLS	Poloxamer 407	Crodesta F110	Crodesta F160	Tween 80	Vitamin E TPGS	Volpo 10	Cremophor EL
None	0	0	1	8	31	56	85	87	89
Eudragit RS100	0	5	1	0	0	3	1	3	13
Kollidan VA 64	0	**48**	21	31	**70**	32	**52**	64	18
Plasdone K29/32	0	17	20	14	**59**	72	**56**	66	**79**
HPC SL	4	65	0	22	**36**	4	**54**	44	17
Eudragit L100	17	4	**59**	40	**56**	50	**59**	62	**70**
HPMCP 50	20	5	1	18	39	8	**43**	64	38

Figure 3.3. Example of plate screen for ternary amorphous solid dispersions. The number in each cell is the average value of % dissolved after 1 h of incubation in SIF ($n = 3$ or 6). The color of the cells indicates whether % dissolved was <25% (gray), between 25 and 50% (white), or >50% (black). The top row contains the results for surfactant-only formulations; the left column contains the results for polymer-only formulations; the upper-left corner contains the results for unformulated compound (no excipients) that was processed by solvent casting in an otherwise identical manner to the formulations; and the remaining cells contain the results for polymer/surfactant formulations. The 13 formulations that were scaled up using the melt press method are identified by the use of a bold/underlined font (e.g., **79**) for % dissolved. (Reproduced with permission from Ref. 15. Copyright 2008, Elsevier.)

If a second component is chosen, it is important to understand any possible interactions that can occur between the components. In the case of two polymers, it is helpful to understand if they are miscible and how the API or processing will affect the miscibility or other properties [12,39]. In one study, it was found that the solvent used to spray dry a ternary dispersion of griseofulvin, PHPMA (poly(*N*-(2-hydroxypropylmethacrylate)), and PVP (polyvinylpyrrolidone) resulted in different morphologies, stabilities, and dissolution properties of the solid dispersion. The observed differences were attributed to different polymer conformations leading to different relaxation rates [39].

Surfactants Surfactants are surface-active agents comprised of both a hydrophilic group and a hydrophobic group (see Chapter 2 for more details and Appendix B for a list of common surfactants). These molecules are generally added to amorphous solid dispersions for a number of reasons, for example, to improve wetting and dispersion of the powders in liquids or to increase the rate of dissolution of poorly soluble drugs using a combination of wetting and solubilization. Surfactants also are commonly combined with polymers in amorphous solid dispersions in order to improve properties such as dissolution and bioavailability [37]. They can act as plasticizers when added to the polymers and help facilitate thermal processing during melt extrusion [40]. However, adding a surfactant can also result in a more complicated system with drawbacks such as a reduction of the API melting point, T_g of the polymer, or T_g of the amorphous solid dispersion. In the case of an API–Plasdone S-630–Tween 80 system, a reduction in T_g from 97 to 67 °C was observed with a 10% loading of the surfactant [40]. A reduction in the overall T_g of the dispersion system could result in less stable dispersions depending on the value of the T_g; however, physical stability was not measured for this system. It also important to understand how the polymer and surfactant interact and how this will affect the stability and performance of the amorphous solid dispersion [41]. In this study, interactions were studied with HPMC E5, anionic surfactants (bile salts and SDS), and a nonionic surfactant (Tween 80). The formation of a polymer–anionic surfactant complex resulted in high supersaturation of the poorly soluble drug in solution; however, the nonionic surfactant promoted polymer dissolution and the drug was found to rapidly recrystallize during dissolution.

If a surfactant is needed, different types can be tried and properties assessed to determine the best combination of surfactant and polymer. For solvent-based experiments, similar steps need to be performed to determine adequate solubility of the surfactant, polymer, and API in specific solvents. Common surfactants are listed in Appendix B. The next step is to determine whether the surfactant will improve the solubility of the API in water. This can initially be tested by determining API solubility in acceptable surfactant solutions, such as 5–10% aqueous solutions containing common surfactants. The surfactants that significantly improve solubility would be useful additions to the screen to produce ternary systems with a polymer. Additional studies to understand the mechanism of improved dissolution or supersaturation can also be performed, such as evaluating the role of polymer–surfactant interactions during dissolution [41]. Conversely, it has been shown that dispersions with surfactants can show faster crystallization from solution than those without surfactants [42].

The T_g and water uptake of the ternary system should also be measured to obtain initial information on factors that might influence handling and storage. Further investigation into the miscibility of the three components (API, polymer, and surfactant) may require additional data, such as more thorough characterization of the interactions between the components.

If a surfactant is added, it is important to determine the optimal amount required, which adds to the complexity of the screen. Low amounts of surfactant, such as 2–5%, should be tried initially in the solid dispersion. This value may be higher if the surfactant is needed to help with the melt extrusion process.

OTHER COMPONENTS Additional materials can be added to amorphous solid dispersions such as pH modifiers [38,43]. Telmisartan, for example, is known to have a higher solubility at elevated pH conditions; therefore, pH modifiers were added to affect the microenvironmental pH of the drug to enhance the dissolution rate of the drug. Basic pH modifiers (sodium carbonate, sodium hydroxide, magnesium oxide, and calcium carbonate) were screened with five polymers to produce ternary dispersions with improved dissolution properties. A dispersion containing 2:0.5:3 API:PVP K30:sodium carbonate was chosen for an animal study in rats based on characterization, solubility, and dissolution data. The dispersion exhibited substantially better performance than the marketed dosage form with a C_{max} increase of 6.4-fold and an AUC improvement of 2.1-fold.

Cyclodextrins have been used to complex the API to improve solubility and are being incorporated into amorphous solid dispersions as binary systems with the API [44,45] or ternary systems with other components such as polymers [46]. It should be recognized that these types of systems employ complexation with the cyclodextrins, which is a different method of supersaturation compared with dispersions made with polymers. Improved performance such as dissolution rate and blood levels has been shown with these more complex cyclodextrin systems [46], but additional characterization data may be needed to understand the mechanism of the enhancement. For early studies, these more complicated systems may be a second tier of screening if more simple binary systems have not produced the desired properties.

Lipids have also been added to dispersions to enhance supersaturation by producing micelles [47,48]. Examples of lipids used in amorphous solid dispersion systems include Gelucire®, Vitamin E TPGS (D-α-tocopheryl polyethylene glycol 1000 succinate), and phosphatidylcholine. The ability to supersaturate solutions is due to formation of lipid vesicles that trap a certain amount of drug and enable it to stay in solution. It is also important to understand the nature of any interactions between the polymer and the lipid in these ternary dispersions. In one study, it was found that different lipids were incorporated into the PEG 4000 structure in different ways [49]. Small hydrophilic lipids were found to increase the folding of PEG, while large nonpolar lipids retarded the unfolding during secondary crystallization, resulting in different crystalline regions in the polymer that could ultimately affect the physical stability of an amorphous solid dispersion.

Small molecules, such as amino acids, have also been used in place of polymers in binary systems [50,51] or with polymers to form ternary systems [46,52]. Binary co-amorphous drug–drug or drug–amino acid mixtures have been shown to stabilize the amorphous form and increase the dissolution due to molecular interactions between the components [50,51]. Further work is needed to determine how this concept will translate to a wide range of drug molecules, but it does offer another avenue for dispersion formation.

3.2.2.3 Amorphous Dispersion Composition The desired composition of a dispersion is based on (1) miscibility, (2) the extent of antiplasticization needed (T_g), and (3) impact on dissolution rate and supersaturation levels. In early development and screening, limited information will be available on these aspects. Once the solvents, polymers, and possible surfactants to be used are determined, other factors need to be considered when setting up the screening experiments. One important consideration is the amount of polymer needed to maintain physical stability. It has been shown that

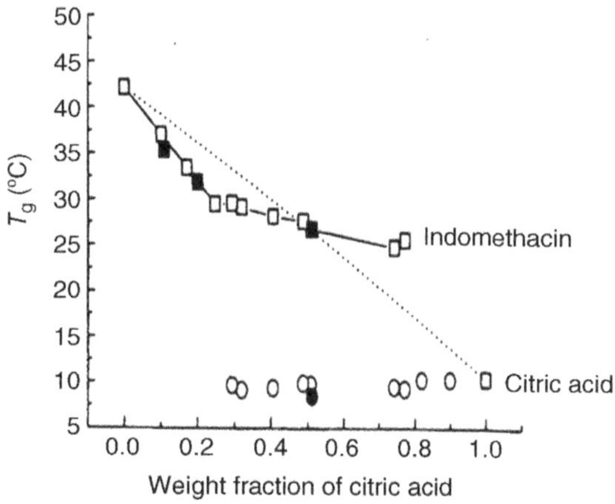

Figure 3.4. Example of miscibility changing with concentration. Values of T_g for binary mixtures of citric acid–indomethacin as a function of weight fraction of citric acid. The square symbols correspond to the values of T_{g2}, while the circles show the values of T_{g1}. (Reproduced with permission from Ref. 53. Copyright 1998, Springer.)

miscibility can change with concentration [53] as shown with indomethacin and citric acid in Figure 3.4; below a citric acid weight fraction of 0.3, only one T_g is observed suggesting a miscible system, but above this weight fraction two T_g values are observed indicating a nonmiscible system. This observation should be applicable to API:polymer systems as well.

Based on miscibility studies, it is important to look at a range of concentrations of API and polymer. A common perception is that more polymer will provide more stability, but it has been shown that even small amounts of polymer (such as 5%) can provide stabilization for years [54]. Many pharma applications target polymer levels above 50% polymer, but the final dose of the API also has to be considered, since a large amount of polymer can significantly increase the size of the capsule or tablet to be ingested. A suggested initial range is 20–50% API, depending on the formulation plans of the compound. Guidelines shown in Figure 3.2 can also be used if applicable.

The minimum amount of polymer needed to maintain physical stability is another aspect to consider. One approach that has been reported is to make dispersions with low polymer concentrations, equilibrate under extreme stress conditions, and quantify the amount of crystalline API that is produced [55]. An improved kinetics approach describes the physical stability of amorphous solid dispersions. A low level of polymer is used to encourage crystallization of the API for quantitation, and a plot of the weight fraction crystallized versus the amount of polymer results in a minimum value of polymer needed to maintain an amorphous material, as shown in Figure 3.5 for an efavirenz:PVP dispersion. In this example, a minimum polymer concentration of 24% PVP was determined, and this value was confirmed using a 75:25 API:PVP sample stored at 72 °C/53% RH that showed no crystallization for 24 months. For early development projects, collecting physical stability data on three to five dispersions at extreme

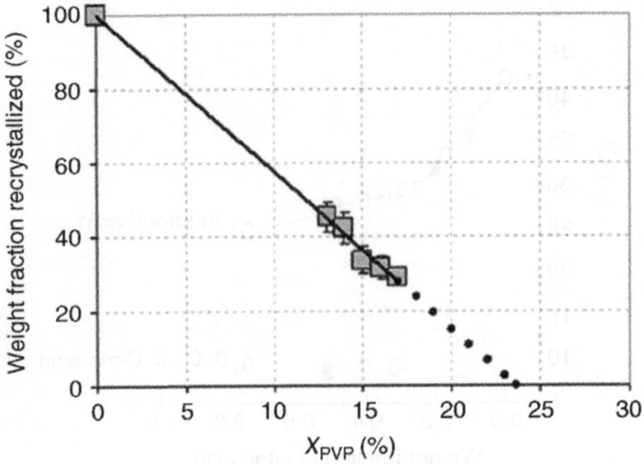

Figure 3.5. Final efavirenz crystallinity as a function of PVP content at 72 °C/52% RH storage (error bar represents one standard deviation). (Reproduced with permission from Ref. 55. Copyright 2010, Elsevier.)

conditions can provide significant information on the minimum polymer concentration needed to maintain physical stability; it is likely that the same amount of polymer would provide protection under more normal handling conditions used in development.

3.2.3 Screening Methods

While it is important to determine the final large-scale preparation method, the goal of the screen is to cover a wide dispersion space using various polymers, solvents, ratios, and conditions to produce stable amorphous solid dispersions. This concept is similar to covering a wide crystallization space during a polymorph screen [56]. The screen should be considered a search for viable amorphous solid dispersions and not necessarily a search for a process. This implies that a variety of techniques be incorporated into the screen to widen the possibility of finding a suitable material for development. Other factors, such as time and material, may limit the number of experiments that are possible in initial early screens. It is advisable to start with the simplest formulation containing a polymer and the API, and a range of API:polymer ratios should be incorporated into the experimental plan. However, additional screens can be performed as the compound moves through development to widen the dispersion screening experiments and to find a material that provides better characteristics and properties for later more complex formulations.

This section will focus on melt and solvent-based methods; it is not meant to be a comprehensive list of experimental methods but a guide to illustrate options using these methods.

3.2.3.1 Melt Techniques For melt methods, it is important to evaluate the melting properties of the components and ensure that there is no decomposition associated with the melt. Initial experiments can be performed with DSC, hot-stage microscopy, and thermogravimetric analysis (TGA) to determine the melting point and investigate weight loss and/or discoloration during melting. Samples can also be analyzed with other methods, such as solution NMR or HPLC, to look for degradation.

The simplest thermal screening method is melt mixing, where a simple mixing apparatus such as a stir bar or impeller is used with a heating source [57]. For these experiments, an excipient, usually a low molecular weight polymer, is liquefied at elevated temperature and combined with the drug and, possibly, other components, such as a surfactant. Viscosity limitations restrict this method to low molecular weight polymers, and this can severely limit the number of polymers used in the screen. Cooling conditions also need to favor producing a miscible dispersion containing amorphous API rather than crystallization of the drug.

DSC has been used to evaluate API and polymer systems in a screen [58]. Pure API and API:polymer blends are melted in a DSC pan using an appropriate ramp rate, such as 10 K/min, cooled rapidly (a liquid nitrogen cooling assembly is helpful), and then rescanned to look for thermal events. Positive DSC results would show one glass transition event and no melting endotherms for the components. Temperature information obtained from this method can be applied to other methods at larger scale.

Melt compression can be used to produce about 100 mg of dispersion [15]. Selected formulation blends of API, polymer, and surfactant can be geometrically mixed in a mortar and pestle and then melt compressed using a heated press to produce films. A typical force of 1500 lb and a dwell time of 30 s can be used while the appropriate temperature is maintained (approximately 20 °C above softening temperature of the polymer). Specific conditions need to be determined using a placebo excipient mixture.

Gram quantities of dispersions have been made using a melt mixer [15]. The components can be processed for 5–10 min at a temperature previously determined from DSC or other melt experiments. Ternary mixtures containing API, polymer, and surfactant can also be produced using this method.

Small-scale melt extruders are also available and can be employed for screening experiments if larger amounts of API are available at the screening stage. Often laboratory melt extruders will be used to produce gram quantities of lead dispersions for additional studies [15,58]. Temperatures and processing conditions will be based on previous small-scale experiments. Conditions can also be determined at a larger scale on the melt extruder if previous data are not available.

A combination of these methods is commonly used to produce different amounts of material throughout the screening and selection process. For example, a combination of DSC (5–10 mg), stainless steel beakers and a hot plate (10 g), and small-scale extruder (>10 g) was used to find amorphous solid dispersions of indomethacin, lacidipine, nifedipine, piroxicam, and tolbutamide with PVP [58]. Initial melting temperatures were obtained from the DSC results and applied to the beaker and extrusion methods. A second example involves initial evaluation using solvent casting with subsequent melt compression, melt mixer, and melt extrusion methods used for screening and scale-up of ternary amorphous solid dispersions using six polymers and eight surfactants [15].

Melt techniques offer a distinct advantage for poorly soluble compounds that do not readily dissolve in organic solvents and because this is a "green" production method that does not involve solvent. API properties, such as melting point and degradation, need to be considered for this method, but may not be an issue if the API dissolves in a melted polymer or excipient mix (polymer and surfactant) below the melting point of the API.

3.2.3.2 Solvent Techniques For solvent-based methods, it is important to measure the API solubility in a variety of solvents and investigate any degradation that may occur over time. Solvents that show acceptable API solubility then need to be cross-referenced with the solubility values of possible polymers/surfactants. It is critical that both the API and polymer show sufficient solubility and chemical stability in the solvents chosen for the screening experiments. While it is helpful to have a wide variety of solvents for screening studies and eventual scale-up, the reality is that solvent choices may be limited for many poorly soluble compounds in development.

The goal of the screening experiments is to remove solvent quickly and trap the amorphous API in a miscible dispersion. There are a number of ways that have been used to remove solvent, including rotary evaporation [3], freeze drying [4], film preparation [15], electrospinning [59,60], and spray drying [18]. This is not meant to be an exhaustive list but does provide examples of possible methods to incorporate into a screen. The experiments planned for a screen will be dependent on the properties of the API, the amount of material available, the applicable solvents to be used, and the equipment on hand. For example, the solvents that can be used for freeze drying are limited, and if the API does not have adequate solubility in those solvents, then this method should not be chosen for the screen.

For small-scale experiments, 96-well plates can be used for dispersion screening [15,61,62]. Stock solutions of polymers, surfactants, and API can be made in appropriate solvents or miscible solvent mixtures and added to the wells in the desired ratios. Compound masses of 60 μg per well can be used to minimize the amount of API used in the screen. Replicates for each experiment can readily be prepared with this method. Once the solutions are dispensed, the plates can be vortexed to help mix the solutions. The plates are placed in a vacuum centrifuge to remove the solvent; rotor speed, temperature, and vacuum can all be controlled for each plate. Solvent evaporation can take approximately 1.5–2.5 h depending on the solvents and conditions used. In some cases, a hold at room temperature overnight will provide time for the unstable samples to crystallize to help narrow down acceptable dispersions in a short period of time.

A common method to produce amorphous solid dispersions is rotary evaporation [4]. Samples as small as 30–50 mg can be easily produced for a screen. The use of volatile solvents will help trap the amorphous API in the miscible dispersion by decreasing the amount of time the API is dissolved in the solvent. Larger sample sizes and solvent volumes will increase the time the API is in solution and may result in crystalline rather than amorphous material or the evaporation rate may result in less stable dispersions [12]. While the rotary evaporation method is not ideal for some systems, the equipment is readily available in most laboratories and will give an idea of the ease of preparation of an amorphous solid dispersion under crude conditions. If a stable amorphous solid dispersion is obtained using this method, it will likely be produced by other fast evaporation methods such as spray drying.

Freeze drying or lyophilization is another method that may be available in laboratories [4,63]. Water is commonly used for this method, which poses problems when dealing with poorly soluble compounds. Organic solvents for this method are limited, so adequate solubility in appropriate solvents (such as dioxane, *t*-butanol, pyridine, DMSO, or aqueous mixtures of these when applicable) is necessary and modifications to the equipment to accommodate organic solvents are needed. Small (milligram) and large (gram) quantities can be produced using this method. Using manifold equipment, the number of samples is limited by the ports available on the equipment. Using a tray dryer allows for a larger number of vials to be processed at the same time. Determining the optimal conditions for the various stages of prefreezing, primary drying, and secondary drying may be needed to produce amorphous material during the screen.

Producing films is commonly reported in the literature for small-scale dispersion preparation. Different methods have been reported including film casting [64] and spin coating [24,65,66]. In film casting, a solution containing the API and polymer is poured onto a glass plate covered with a Teflon film and the solution is left to dry at room temperature. For spin coating, a similar procedure is used with the solution (1–3 drops or ~200 μl) being added to cover slip or chip and the sample being spun at high speeds (such as 2000–8000 rpm) for short periods of time (3–20 s) to disperse the solution and evaporate the solvent. Samples can be analyzed immediately after preparation.

Spray drying has also been used for screening studies [37,61,62]. A variety of organic solvents (such as methanol, ethanol, toluene, hexane, acetone, ethyl acetate, and dichloromethane) can be used, but the spray dryer has to be equipped with certain features to make it compatible for use with these solvents. Mini spray dryers have been reported to produce samples ranging from 1 mg to 1 g [67], which would help to extend the number of experiments in a screen. For typical laboratory-scale spray dryers, sample sizes of 0.5 g to hundreds of grams can be produced. These spray dryers require larger amounts of material than previous methods and the number of experiments may be limited based on the amount of sample available. In addition to being used for larger scale screening, spray drying with laboratory spray dryers is commonly used to scale-up a limited number of dispersions for additional investigation, such as animal bioavailability studies.

Solvent-based screening methods cover a wide range of processing variables and sample sizes. The choice of solvent is critical for many of these processes and quick removal of solvent is usually a key parameter to produce amorphous instead of crystalline material. A combination of techniques can be useful during a screen, for example, using a 96-well plate for initial screening and then spray drying [61,62] or melt extrusion [15] to produce larger amounts of lead dispersions.

3.2.3.3 *Initial Characterization* Initial testing of the amorphous solid dispersion samples can cover physical testing (X-ray powder diffraction (XRPD), DSC, miscibility, hygroscopicity, and solvent content) and/or properties (solubility, dissolution, physical/chemical stability, and bioavailability). The amount of testing will depend on the amount of material available as well as the purpose of the screen and the desired properties. This will vary from project to project, but an overview of approaches will be presented here.

A number of physical tests can be performed on small amounts of material, as shown in Table 3.1. XRPD can be used to determine if an amorphous material has been produced and to help identify minor amounts of crystalline material that may be present. This information will be important in assessing conditions to produce the amorphous dispersion and changing parameters as needed to generate the desired amorphous material. DSC and modulated DSC (mDSC) will identify the glass transition temperature. A higher glass transition temperature (such as greater than about 50 °C above room temperature) could result in improved physical stability. DSC will also help determine miscibility; one T_g usually indicates a miscible dispersion, whereas two T_g events indicate a nonmiscible system [20]. Small amounts of crystalline material can be detected with DSC and crystallization exotherms will help determine thermal stability of the various dispersions. The amount of water/solvent in the materials is important and can be initially assessed with TGA to get a total volatile content. Other techniques such as Karl Fisher, solution NMR, or gas chromatography will be needed to identify other solvents present in the dispersion. Solvents will act as plasticizers to lower the T_g and cause instability even at low levels; therefore, it is important to understand how solvent contents will affect dispersion properties. Spectroscopy can be used to investigate interactions between components and determine miscibility. It can also be used to detect small amounts of crystalline material that may exist. Moisture sorption with an automated system will establish water uptake at various RH conditions, which can provide information on handling and storage. Crystallization is sometimes evident in these curves, which can help rule out certain dispersions. Other characterization techniques that can be used for screening and scale-up are given in Chapter 4.

Other properties of dispersions related to performance are also important to evaluate during the screen and should be related to the properties required in the development plan for the compound. Dispersions are commonly used to improve solubility and this parameter is commonly measured as part of the screen. Miniaturized solubility measurements are available that can be integrated into well plate screens [15,68,69]. Solution stability, or the ability of the dispersion to be dispersed or dissolved in simulated GI fluids without crystallization, provides information on whether the material will stay solubilized *in vivo*. For most systems, this will be attributed to the polymer used in the dispersion and this type of information has been used to perform early screens in order to select polymers [24] or can be performed after the preliminary screen as part of the dispersion selection. These studies can be performed with relatively small amounts of material, but amounts will depend on the analysis needed for the solid or solution. Dissolution rate and physical/chemical stability can be compared for different dispersions, but additional material may be needed for these studies.

For many companies, the ultimate test in performance is bioavailability and dispersions/formulations will be dosed to animals to assess blood levels and other pharmacokinetic parameters. In early development, the dispersion is commonly compared with the crystalline material to determine the advantage gained with the dispersion. For late development compounds, the dispersion may be compared with marketed products or other high-level formulations [21].

A combination of physical testing and property determination will be needed for comparison and assessment of various amorphous solid dispersions. Not all properties will be important for all molecules and it is important to collect relevant data for the system being studied.

3.2.3.4 Scale-Up After initial assessment of the screening samples, certain samples will be chosen for further study and more material will be needed. The amounts can range from a couple of grams to tens or hundreds of grams. Laboratory methods such as rotary evaporation can be used, or a move to larger scale processing such as spray drying or melt extrusion can be used if laboratory equipment is available. These scale-up samples can help in determining parameter ranges for large-scale production. It is important to characterize the materials and compare back with the original properties of the screening samples to determine whether larger batches provide the same properties and advantages as seen previously. As with crystallization, it is not always straightforward to produce larger scale dispersions with the same properties when moving from screening scale to laboratory/processing scale.

Drying is also an important step for solvent-based methods. Larger batches can commonly contain a higher solvent content than smaller batches and different methods may be needed to remove it. For spray drying, this is called secondary drying [15,70]. It is important to use previous data to choose drying conditions that will remove solvent, but will not cause the API to crystallize. For example, TGA data can be used to pick a temperature where solvent will be readily removed and DSC will show whether crystallization will occur in this temperature range

Once materials are scaled up, it is important to store them under conditions that will maintain their amorphous nature. At this point, physical stability may not have been assessed, or only limited information is available. In these cases, it is helpful to have materials with a low solvent content and to store the material at about 50 °C below T_g and at reduced RH conditions to prevent water uptake from the atmosphere. As more information becomes available, the storage requirements will change and hopefully an amorphous solid dispersion with robust handling properties will be available.

3.3 AMORPHOUS SOLID DISPERSION SELECTION

3.3.1 Properties

Selection of a lead amorphous solid dispersion, and possible backup dispersions, is based on properties needed for development and the proposed development plan for the molecule. Questions to help determine the path forward are the following:

- What do I have?
  - What forms/dispersions are available?
  - What are their properties?
- What do I need?
  - What properties are needed to move forward?
  - What dosage form is targeted short (and long) term?

- Where are the holes?
  - What properties are missing in the current materials?
  - Are these properties needed for the development plan?
- How do I fill them?
  - Will a dispersion fit the need?
  - Will I need a different dosage form to accommodate it?
  - Are there other considerations short and long term?

These questions form the basis for the solid form selection. The properties of the compound may need improvement, such as solubility, dissolution, or bioavailability. These issues should be included in the selection criteria to ensure that there is an improvement compared with the original form. Other properties that can play a role are T_g, hygroscopicity, physical/chemical stability, and crystallization tendency in simulated biological fluids.

Early versus late formulations are a factor at this stage. For early formulations, there may be certain properties that are essential in moving your molecule forward, such as an increase in solubility or bioavailability. A simple formulation with the necessary safeguards (such as low temperature and humidity) may be very doable for small clinical supplies but may not be a viable formulation for larger or longer clinical studies. As the required supplies increase and the formulation is modified, other processing, handling, and storage conditions need to be considered. Changing to a different dispersion that covers a different set of selection criteria may also be necessary.

The properties needed for a specific formulation (such as a suspension, tablet, or capsule) will be different and these need to be included in the selection process. Even simple formulations, such as powder in a capsule, need to consider possible transfer of water in a gelatin capsule shell to the dispersion, as well as its effect on the stability of the amorphous solid dispersion. The humidity of the room during capsule filling can also be an issue for some dispersions if it is not controlled and fluctuates above 50–60% RH. Formulating capsules or tablets also involves using other excipients. It is important to understand possible interactions with these components, especially if they contain water that can be transferred to the amorphous solid dispersion causing physical or chemical instability. Many times anhydrous excipients are used in initial formulations because the physical stability of the dispersion with water is not known. As more is learned about the physical stability of the dispersion, hydrated excipients can be added if the dispersion will not easily crystallize in the presence of water.

Formulation processes also need to be considered for the desired formulation. An example of formulation processes used for oral dosage forms is given in Figure 3.6 [71]. It is important to understand which processes to avoid or modify when working with amorphous solid dispersions. In most cases, wet granulation is not going to be the best first option. Other steps such as drying, milling, and even film coating also need to be evaluated based on the properties of the dispersion in order to prevent crystallization or degradation.

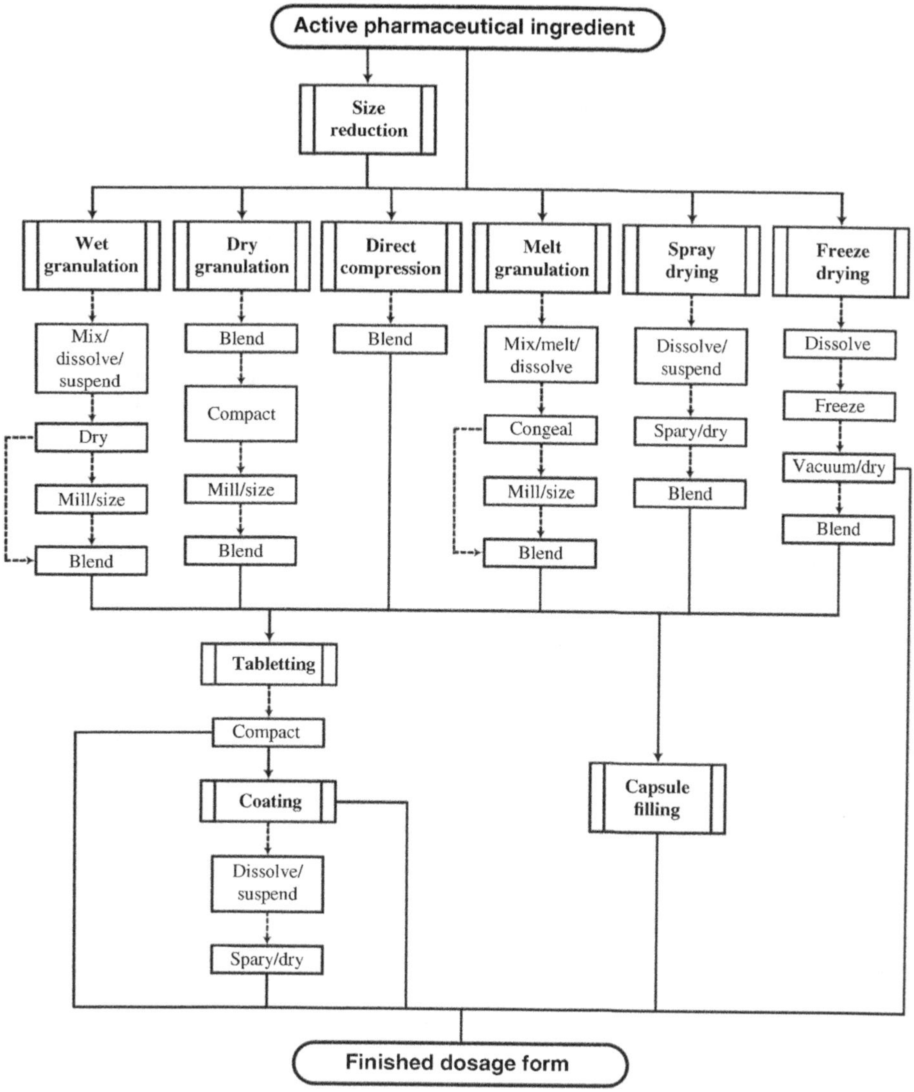

Figure 3.6. Common processes for preparing solid oral dosage forms. (Reproduced with permission from Ref. 71. Copyright 2004, Elsevier.)

3.3.2 Selection Tools

Once a list of desired properties is in hand, there are a number of ways to review and compare data during form selection. Two common approaches are a decision tree and a form matrix.

A decision tree narrows down the options based on a series of decision points. Those dispersions that do not exhibit the desired properties are removed from consideration and

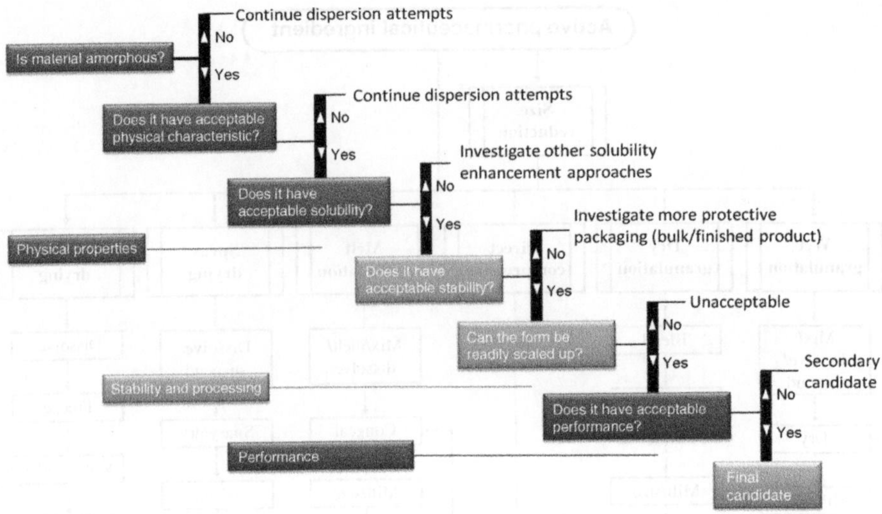

Figure 3.7. Amorphous dispersion decision tree. (Reproduced with permission from Ref. 4. Copyright 2010, Wiley.)

the rest are moved to the next decision point. An example of a decision tree for an amorphous solid dispersion is given in Figure 3.7. The decision points can be grouped based on various properties, such as physical properties, stability/processing, and performance. In this example, the first decision point is obtaining a material that is amorphous; any materials showing crystalline peaks would not move forward. The second decision point is acceptable physical characteristics, such as T_g, solvent content, and hygroscopicity. The next is an estimated solubility that is acceptable for development. Stability and processing are assessed next and finally performance. Performance can be any number of parameters ranging from dissolution rate, lack of crystallization upon dissolution, to bioavailability. Additional decision points can be added depending on the development needs and issues. For some companies, solubility and dissolution may be more important than T_g because the material can easily be stored at reduced temperatures for early studies. Decision points can be moved as needed to fit the development plan of the molecule.

A form matrix can also be used to identify materials with the most acceptable properties for development. The materials and selected properties are put into a table and the acceptable properties are highlighted. The dispersion with the best properties is then selected for development. As with the decision tree, it is important to understand which properties should be included for the best selection and these properties may be different for an early versus late development program. The properties listed in the matrix table will be specific to the system and the development plan. An example is given in Table 3.3 for AMG 517 amorphous solid dispersions [37]. Dispersions containing two polymers at two drug loadings were prepared by spray drying. A variety of properties were evaluated for the dispersions, including amorphous nature by XRPD, total volatile content, T_g,

TABLE 3.3. Form Matrix Used for Selecting Amorphous Solid Dispersions for AMG 517

Parameter	Formulation					
	A (Lot 1)	A (Lot 2)	B	C	D (Lot 1)	D (Lot 2)
Polymer	HPMCAS-MF	**HPMCAS-MF**	HPMCAS-MF	HPMC-E5	HPMC-E5	HPMC-E5
Drug (wt%)	15	**15**	50	15	50	50
XRPD	Amorphous	**Amorphous**	Amorphous	Amorphous	Amorphous	Amorphous
% Weight loss TGA	2.39	**0.32**	2.84	1.94	1.88	1.44
T_g (°C)	106	**101**	98	117	106	107
DSC	No melt	**No melt**	Minor melt	No melt	Minor melt	Minor melt
Physical stability[a]	No change	**No change**	No change	No change	Minor crystallization	Minor crystallization
Dissolution[b] (μg/ml)	NA	**2.5**	0.9	1.8	1.6	NA
Yield	76	**85**	55	61	12	60

Shaded cells are indicative of acceptable properties. Formulation A, lot 2 (bold) showed the best properties and was moved forward in development. (Data from Ref. 37.)
[a]40 °C/75% RH in an open dish for 6 months.
[b]Dissolution in pH 6.8 phosphate buffer at 120 min.

presence of crystalline material by DSC, physical stability, dissolution, and yield. By highlighting the acceptable values in the table, it is readily apparent that the dispersion with the most favorable properties is the HPMCAS-MF dispersion at a 15% loading. Based on these data, further studies were performed to optimize the dissolution behavior by adding a surfactant (5% SDS) and ultimately this ternary dispersion was used in an animal study to show improved bioavailability over a micronized suspension formulation.

When selecting any form, there is always the possibility that one material will not have all the desired characteristics. In this case, choose the best form based on the needs of the project. There may also be other solutions that can be used to help move the dispersion forward. If the API is not soluble in organic solvents, which would make spray drying unattractive, melt extrusion can be explored. If the solid is sensitive to moisture, controlled relative humidity during capsule filling or special packaging may be needed.

3.4 CASE STUDY

3.4.1 Miniaturized Studies

A miniaturized screen called SPADS (screen polymers for amorphous drug stabilization) was developed to rapidly identify amorphous solid dispersions with maximum supersaturation and solid stability using small amounts of material [16]. A model compound, CETP(2), was used to demonstrate the screen; it is a poorly water-soluble BCS class II compound. The screen included seven polymers (Eudragit E PO, Eudragit L 100-55, PVP/VA 64, PVP K30, HPMCAS-MF, HPMCP HP-50, and cellulose acetate phthalate (CAP)) and four drug loads (50, 30, 20, and 10% (w/w)). Initial assays were combined to determine the supersaturation, drug–polymer miscibility, and stability of possible dispersions. The second step investigated interactions using FTIR and molecular homogeneity using AFM (atomic force microscopy). The third step involved spray drying scale-up, dissolution testing, and a 6-month physical stability study of the lead dispersions from the screen. An overview of the SPADS methodology is shown in Figure 3.8.

The initial dissolution assay was performed in a 96-well plate. Amorphous drug–polymer films were prepared by solvent-based film casting in 96-well plates. Solvents used were acetone or acetone/ethanol, depending on the solubility of the polymer. Seven polymers at four drug loads resulted in 28 dispersions and all experiments were performed in triplicate. A stirring bar and FaSSIF were added to each plate maintained at a temperature of 37 °C. Samples were agitated for 60 or 180 min. Solutions were collected and analyzed by UPLC. Values were compared with the equilibrium solubility of the crystalline material to calculate the supersaturation value for each dispersion. It was found that no polymer gave consistently high supersaturation at all ratios. The highest drug concentrations were obtained with Eudragit E PO (30% API), Eudragit L 100-55 (30% API), CAP (20% API), and HPMCP HP-50 (20% API). All of these polymer systems were moved into the second tier of the screen. PVP/VA 64 was included as a negative control.

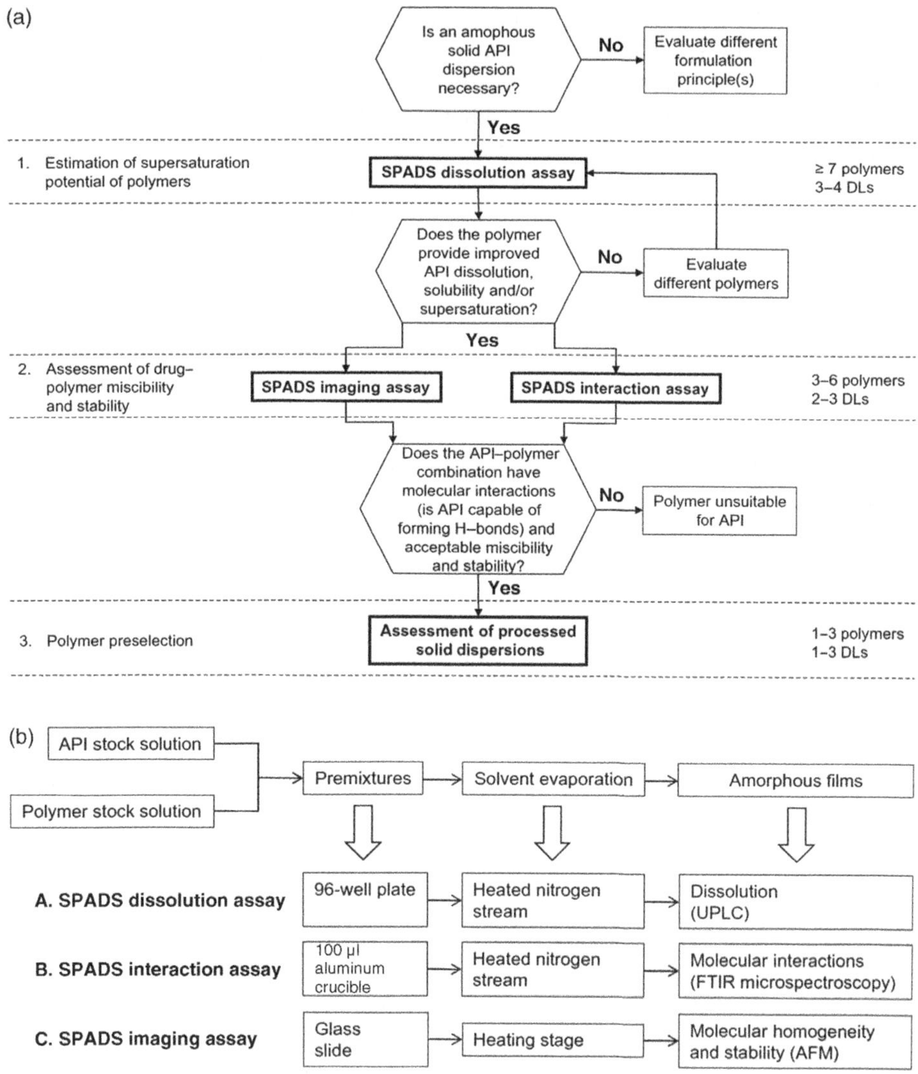

Figure 3.8. Schematic for (a) the SPADS dispersion screening approach and (b) experimental procedure for SAPDS assays. (Reproduced with permission from Ref. 16. Copyright 2013, Elsevier.)

The interaction assay was performed on drug–polymer films prepared in aluminum DSC pans using the same solvents as those in the dissolution assay. The 20–100 nm thick films were analyzed by FTIR microscopy. The amide I carbonyl stretching vibrations (between 1624 and 1640 cm^{-1}) were used to investigate the amide group as a hydrogen bond acceptor. Peak positions were compared with the pure drug and shifts between

−8 and +16 cm^{-1} were used to investigate interactions at a 20% drug loading. Four different loads of the polymers were used to help differentiate the systems. Eudragit E PO and PVP/VA 64 caused a blue (positive) shift and CAP and HPMCP HP-50 resulted in a very weak red (negative) shift. As the amount of API decreased, the amide I shifts generally became larger and the peak intensity decreased.

The imaging assay involved analyzing quench-cooled melts by AFM at nanometer scales. Ten different regions were examined for miscibility. Established criteria were used to identify miscible dispersions. Binary mixtures at two loadings (20 and 50%) with the polymers that moved into the interaction assay were also used for the imaging. Surface roughness of the dispersions was compared with that of the pure polymer; if the roughness was less than that of the pure polymer, then the dispersions were considered to be miscible. Good miscibility was found for Eudragit E PO (≤50% API), Eudragit L 100-55 (≤20% API), PVP/VA 64 (≤50% API), and HPMCP HP-50 (≤50% API). Poor miscibility was found for CAP. Both CAP dispersion and 50% Eudragit L 100-55 dispersion were not moved forward to the stability studies.

Physical stability was assessed by surface demixing on fractured surfaces exposed to 40 °C/75% RH for 2 h. Established stability criteria were used to identify physically stable dispersions. The fractured surfaces of all samples were found to change upon exposure and several samples exhibited segregated material. Segregation was not observed for the Eudragit E PO (20 and 50% API) and PVP/VA 64 (20% API) samples.

Based on the SPADS assays, only the Eudragit E PO showed good drug release, high drug concentrations, acceptable supersaturation, good miscibility, and desired physical stability. To confirm the results, three polymers with different advantages were scaled up to 10 g using spray drying: Eudragit E PO (best-performing polymer), PVP/VA 64 (poor supersaturation but good stability), and CAP (high supersaturation but poor stability). A variety of properties were assessed for the spray-dried samples, such as yield, particle size, drug content, residual solvent content, T_g, amorphous nature by XRPD, dissolution, and physical stability testing. Two T_g values were obtained for the CAP dispersion, indicating a physical mixture rather than a miscible dispersion. Only one T_g value was observed for the other two polymer systems. The dissolution testing using a USP-1 apparatus gave the same results as the SPADS dissolution test with a ranking of Eudragit E PO > CAP > PVP/VA 64. Physical stability testing at 40 °C/75% RH and 40 °C/25% RH for 6 months showed no drug recrystallization in the Eudragit E PO dispersions (20 and 50% API) under both conditions. Crystals were observed by optical microscopy for the PVP/VA 64 and CAP samples after 6 months at 40 °C/75% RH. These results were not expected for PVP/VA 64 based on the SPADS test, but were expected for the CAP samples.

A matrix table was compiled to compare the data (Table 3.4) and it was evident that Eudragit E PO dispersion was the best choice. It was concluded that the miniaturized experiments were a good representation of larger scale experiments and they could be used to identify appropriate polymers in a few days time using minimal amounts of API. This approach can help accelerate the development of suitable dispersions for preclinical and early clinical trials.

TABLE 3.4. Matrix Table for SPADS Imaging and SPADS Interaction Results of Selected API–Polymer Combinations Compared with the Physical Stability Results of Spray-Dried Dispersions

		50% Drug Load							20% Drug Load						
		SPADS			SDD				SPADS			SDD			
Polymer	Storage Conditions	AFM Mix[a,b]	AFM Stab[b,c]	IR-mic[d]	IR-class[e]	PLM[f]	XRPD[f]	DSC[f]	AFM Mix[a,b]	AFM Stab[b,c]	IR-mic[d]	IR-class[e]	PLM[f]	XRPD[f]	DSC[f]
Eudragit E PO	Initial analysis	O	O	$+3\,cm^{-1}$	$+2\,cm^{-1}$	O	O	O	O	O	$+10\,cm^{-1}$	$+11\,cm^{-1}$	O	O	O
	6 months, 40 °C, 25% RH	–[g]	–	–	–	O	O	O	–	–	–	–	O	O	O
	6 months, 40 °C, 75% RH	–	–	–	–	O	O	O	–	–	–	–	O	O	O
PVP/VA 64	Initial analysis	O	O	$+4\,cm^{-1}$	$+3\,cm^{-1}$	O	O	O	O	O	$+16\,cm^{-1}$	$+4\,cm^{-1}$	O	O	O
	6 months, 40 °C, 25% RH	–	–	–	–	O	O	O	–	–	–	–	O	O	O
	6 months, 40 °C, 75% RH	–	–	–		X	O	O	–	–	–		X	O	O
CAP	Initial analysis	X	X	$-2\,cm^{-1}$	$-3\,cm^{-1}$	O	O	O	X	X	$-2\,cm^{-1}$	$-1\,cm^{-1}$	O	O	O
	6 months, 40 °C, 25% RH	–	–	–	–	X	O	X	–	–	–	–	O	O	O
	6 months, 40 °C, 75% RH	–	–	–	–	X	X	X	–	–	–	–	X	O	O

[a]Nanometer homogeneity API/excipient mixtures as determined by AFM in the SPADS imaging assay directly after sample preparation.

[b]The symbol "O" indicates homogeneous (i.e., no separated phases detected by AFM) and the symbol "X" indicates heterogeneous (i.e., separated phases observed by AFM).

[c]Nanometer API/excipient mixtures as determined by AFM in the SPADS imaging assay directly after short stress storage for 2 h at 40 °C, 75% RH.

[d]Amide shift determined by FTIR microspectroscopy in the SPADS interaction assay.

[e]Amide shift determined by classical FTIR measurements.

[f]The symbol "O" indicates that the material was amorphous at that condition and time point. The symbol "X" indicates that the material contained crystalline drug.

[g]Not determined.

3.5 CONCLUSIONS

This chapter has outlined a number of factors to consider when performing an amorphous solid dispersion screen. All the aspects discussed will not be relevant to every compound or every screen, but it is important to look at the properties of the API and the development plan as a starting point. From there, the goal of the screen can be determined and the experimental details can be chosen. One set of screening conditions will not work in all cases; it is important to understand the options available when developing the screen and to tailor the conditions to maximize success in obtaining relevant dispersions that will work for the compound being developed.

REFERENCES

1. Zheng, W., Jain, A., Papoutsakis, D., Dannenfelser, R.-M., Panicucci, R., and Garad, S. (2012) Selection of oral bioavailability enhancing formulations during drug discovery. *Drug Dev. Ind. Pharm.*, 38(2):235–247.
2. Bee, T. and Rahman, M. (2010) Insolubility solved by spray drying. Manufacturing Chemist, March 2010. Available at http://www.manufacturingchemist.com/technical/article_page/Insolubility_solved_by_spray_drying/46618 (accessed January 23, 2014).
3. Guillory, J.K. (1999) Generation of polymorphs, hydrates, solvates, and amorphous solids. In: Brittain, H.G. (Ed.), *Polymorphism in Pharmaceutical Solids*, Marcel Dekker, New York, pp. 183–226.
4. Engers, D., Teng, J., Jimenez-Novoa, J., Gent, P., Hossack, S., Campbell, C., Thomson, J., Ivanesevic, I., Templeton, A., Byrn, S., and Newman, A. (2010) A solid-state approach to enable early development compounds: selection and animal bioavailability studies of an itraconazole amorphous solid dispersion. *J. Pharm. Sci.*, 99(9):3901–3922.
5. Marsac, P.J., Shamblin, S.L., and Taylor, L.S. (2006) Theoretical and practical approaches for prediction of drug–polymer miscibility and solubility. *Pharm. Res.*, 23(10):2417–2426.
6. Friesen, D.T., Shanker, R., Crew, M., Smithey, D.T., Curatolo, W.J., and Nightingale, J.A.S. (2008) Hydroxypropyl methylcellulose acetate succinate-based spray-dried dispersions: an overview. *Mol. Pharm.*, 5(6):1003–1019.
7. Bhugra, C., Schwabe, R., Telang, C., and Li, Z. (2012) Rational selection of polymers for amorphous dispersions: using reduced crystallization temperature to assess stability potential. AAPS National Meeting Poster Presentation, Chicago, IL.
8. Greenhalgh, D.J., Williams, A.C., Timmins, P., and York, P. (1999) Solubility parameters as predictors of miscibility in solid dispersions. *J. Pharm. Sci.*, 88(11):1182–1190.
9. Taylor, L.S. and Zografi, G. (1997) Spectroscopic characterization of interactions between PVP and indomethacin in amorphous molecular dispersions. *Pharm. Res.*, 14(12):1691–1698.
10. Kaushal, A.M., Gupta, P., and Bansal, A.K. (2004) Amorphous drug delivery systems: molecular aspects, design, and performance. *Crit. Rev. Ther. Drug Carrier Syst.*, 21(3):133–193.
11. Van Eerdenbrugh, B. and Taylor, L.S. (2011) An *ab initio* polymer selection methodology to prevent crystallization in amorphous solid dispersions by application of crystal engineering principles. *CrystEngComm*, 13:6171–6178.

12. Wu, J.X., Yang, M., van den Berg, F., Pajander, J., Rades, T., and Rantanen, J. (2011) Influence of solvent evaporation rate and formulation factors on solid physical stability. *Eur. J. Pharm. Sci.*, 44:610–620.
13. Dai, W.-G., Dong, L.C., and Den, Z. (2008) Combination of Pluronic/Vitamin E TPGS as a potential inhibitor of drug precipitation. *Int. J. Pharm.*, 355:31–37.
14. Dai, W.-G. (2010) *In vitro* methods to assess drug precipitation. *Int. J. Pharm.*, 393:1–16.
15. Shanbagh, A., Rabel, S., Nauka, E., Casadevall, G., Shivanand, P., Eichenbaum, G., and Mansky, P. (2008) Method for screening of solid dispersion formulations of low-solubility compounds: miniaturization and automation of solvent casting and dissolution testing. *Int. J. Pharm.*, 351:209–218.
16. Wyttenbach, N., Janas, C., Siam, M., Lauer, M.E., Jacob, L., Scheubel, E., and Page, S. (2013) Miniaturized screening of polymers for amorphous drug stabilization (SPADS): rapid assessment of solid dispersion systems. *Eur. J. Pharm. Biopharm.*, 84(3):583–598.
17. Crowley, M.M., Zhang, F., Repka, M.A., Thumma, S., Upadhye, A.B., Battu, S.K., McGinnity, J.W., and Martin, C. (2007) Pharmaceutical applications of hot-melt extrusion: part I. *Drug Dev. Ind. Pharm.*, 33:909–926.
18. Dobry, D.E., Settell, D.M., Baumann, J.M., Ray, R.J., Graham, L.J., and Beyerinck, R.A. (2009) A model-based methodology for spray-drying process development. *J. Pharm. Innov.*, 4:133–142.
19. Shah, N., Sandhu, H., Choi, D.S., Kalb, O., Page, S., and Wyttenbach, N. (2012) Chapter 8: Structured development approach for amorphous systems. In: Williams, R.O., III, Watts, A.B., and Miler,D.A. (Eds.), *Formulating Poorly Water Soluble Drugs*, Springer, New York, pp. 267–310.
20. Newman, A. and Munson, E. (2012) Characterizing miscibility in amorphous solid dispersions. *Am. Pharm. Rev.*, (April):92–98.
21. Newman, A., Knipp, G., and Zografi, Z. (2012) Assessing the performance of amorphous solid dispersions. *J. Pharm. Sci.*, 101(4):1355–1377.
22. Song, M., Hammiche, A., Pollock, H.M., Hourston, D.J., and Reading, M. (1996) Modulated differential scanning calorimetry. 4. Miscibility and glass transition behaviour in poly(methyl methacrylate) and poly(epichlorohydrin) blends. *Polymer*, 37(25):5661–5665.
23. Yoo, S.-U., Krill, S.L., Wang, Z., and Telang, C. (2009) Miscibility/stability considerations in binary solid dispersion systems composed of functional excipients towards the design of multi-component amorphous systems. *J. Pharm. Sci.*, 98(12):4711–4723.
24. Van Eerdenbrugh, B. and Taylor, L.S. (2010) Small scale screening to determine the ability of different polymers to inhibit drug crystallization upon rapid solvent evaporation. *Mol. Pharm.*, 7(4):1328–1337.
25. Wu, J.X., Van Den Berg, F., Sogaard, S.V., and Rantanen, J. (2013) Fast-track to a solid dispersion formulation using multi-way analysis of complex interactions. *J. Pharm. Sci.*, 102(3):904–914.
26. Jia, J., Choi, D.S., and Chokshi, H. (2013) Determination of drug–polymer binding constants by affinity capillary electrophoresis for aryl propionic acid derivatives and related compounds. *J. Pharm. Sci.*, 102(3):960–966.
27. Janssens, S., Nagels, S., Novoa de Armas, H., D'Autry, W., Schepdael, A., and Van den Mooter, G. (2008) Formulation and characterization of ternary solid dispersions made up of itraconazole and two excipients, TPGS 1000 and PVPVA 64, that were selected based on a supersaturation screening study. *Eur. J. Pharm. Biopharm.*, 69:158–166.

28. Murdande, S.B., Pikal, M.J., Shanker, R.M., and Bogner, R.H. (2011) Solubility advantage of amorphous pharmaceuticals. Part 3. Is maximum solubility advantage experimentally attainable and sustainable? *J. Pharm. Sci.*, 100(10):4349–4356.
29. Vandercruys, R., Peeters, J., Verreck, G., and Brewster, M.E. (2007) Use of a screening method to determine excipients which optimize the extent and stability of supersaturated drug solutions and application of this system to solid formulation design. *Int. J. Pharm. Sci.*, 342:168–175.
30. Sun, Y., Tao, J., Zhang, G.Z., and Yu, L. (2010) Solubilities of crystalline drugs in polymers: an improved analytical method and comparison of solubilities of indomethacin and nifedipine in PVP, PVP/VA, and PVAc. *J. Pharm. Sci.*, 99(9):4023–4031.
31. Tao, J., Sun, Y., Zhang, G.G.Z., and Yu, L. (2009) Solubility of small-molecule crystals in polymers: D-mannitol in PVP, indomethacin in PVP/VA, and nifedipine in PVP/VA. *Pharm. Res.*, 26(4):855–864.
32. Breitenbach, J. (2002) Melt extrusion: from process to drug delivery technology. *Eur. J. Pharm. Biopharm.*, 54:107–117.
33. Repka, M.A., Battu, S.K., Upadhye, S.B., Thumma, S., Crowley, M.M., Shang, F., Martin, C., and McGinitiy, J.W. (2007) Pharmaceutical applications of hot-melt extrusion: part II. *Drug Dev. Ind. Pharm.*, 33:1043–1057.
34. Greco, S., Authelin, J.-R., Leveder, C., and Segalini, A. (2012) A practical method to predict physical stability of amorphous solid dispersions. *Pharm. Res.*, 29:2792–2805.
35. Rumondor, A.C.F., Ivanisevic, I., Bates, S., Alonzo, D.E., and Taylor, L.S. (2009) Evaluation of drug–polymer miscibility in amorphous solid dispersion systems. *Pharm. Res.*, 26(11):2523–2534.
36. Ivanisevic, I. (2010) Physical stability studies of miscible amorphous solid dispersions. *J. Pharm. Sci.*, 99(9):4005–4012.
37. Kennedy, M., Hu, M., Gao, P., Li, L., Ali-Reynolds, A., Chal, B., Gupta, V., Ma, C., Mahajan, N., Akrami, A., and Surapaneni, S. (2008) Enhanced bioavailability of a poorly soluble VR1 antagonist using an amorphous solid dispersion approach: a case study. *Mol. Pharm.*, 5(6):981–993.
38. Tran, P.H.L., Tran, H.T.T., and Lee, B.-J. (2008) Modulation of microenvironmental pH and crystallinity of ionizable telmisartan using alkalizers in solid dispersions for controlled release. *J. Control. Release*, 129:59–65.
39. Al-Obaidi, J., Brocchini, S., and Buckton, G. (2009) Anomalous properties of spray dried dispersions. *J. Pharm. Sci.*, 98(12):4724–4737.
40. Ghebremeskel, A.N., Vemavarapu, C., and Lodaya, M. (2007) Use of surfactants as plasticizers in preparing solid dispersions of poorly soluble API: selection of polymer–surfactant combinations using solubility parameters and testing the processability. *Int. J. Pharm.*, 328:119–129.
41. Qi, S., Roser, S., Edler, K.J., Pigliacelli, C., Rogerson, M., Weuts, I., Van Dycke, F., and Stokbroekx, S. (2013) Insights into the role of polymer–surfactant complexes in drug solubilisation/stabilization during drug release from solid dispersions. *Pharm. Res.*, 30: 290–302.
42. Ilevbare, GA. (2013) Influence of additives on phase transformations in highly supersaturated solutions of poorly water-soluble drug compounds. Ph.D. dissertation, Purdue University, West Lafayette, IN. Available at http://docs.lib.purdue.edu/dissertations/AAI3591252/.

43. Marasini, N., Tran, T.H., Poudel, B.K., Cho, H.J., Choi, Y.K., Chi, S.-C., Choi, H.-G., Yong, C.S., and Kim, J.O. (2013) Fabrication and evaluation of pH-modulated solid dispersion for telmisartan by spray-drying technique. *Int. J. Pharm.*, 441:424–432.
44. Soliman, O.A.E., Kimura, K., Hirayama, F., Uekama, K., El-Sabbagh, H.M., El-Gawad, A.E.H., and Hashim, F.M. (1997) Amorphous spironolactone–hydroxypropylated cyclodextrin complexes with superior dissolution and oral bioavailability. *Int. J. Pharm.*, 149: 73–83.
45. Nagarsenkar, M.S., Meshram, R.N., and Ramprakash, G. (2000) Solid dispersion of hydroxypropyl β-cyclodextrin and ketorolac: enhancement of *in-vitro* dissolution rates, improvement in anti-inflammatory activity and reduction in ulcerogenicity in rats. *J. Pharm. Pharmacol.*, 52:949–956.
46. Mora, P.C., Cirri, M., Allolio, B., Carli, F., and Mura, P. (2003) Enhancement of dehydroepiandrosterone solubility and bioavailability by ternary complexation with α-cyclodextrin and glycine. *J. Pharm. Sci.*, 92:2177–2184.
47. Law, S.L, Lo, W.Y, Lin, F.M., and Chaing, C.H. (1992) Dissolution and absorption of nifedipine in polyethylene glycol solid dispersions containing phosphatidylcholine. *Int. J. Pharm.*, 84:161–166.
48. Khoo, S.-M., Porter, C.J.H., and Charman, W.N. (2000) The formulation of halofantrine as either non-solubilising PEG 6000 or solubilising lipid based solid dispersions: physical stability and absolute bioavailability assessment. *Int. J. Pharm.*, 205:65–78.
49. Unga, J., Matsson, P., and Mahlin, D. (2010) Understanding polymer–lipid solid dispersions: the properties of incorporated lipids govern the crystallization of PEG. *Int. J. Pharm.*, 386:61–70.
50. Lobmann, K., Grohganz, H., Laitinen, R., Strachan, C., and Rades, T. (2013) Amino acids as co-amorphous stabilizers for poorly water soluble drugs—part I: preparation, stability, and dissolution enhancement. *Eur. J. Pharm.*, 85(3):873–881.
51. Lobmann, K., Laitinen, R., Strachan, C., Rades, T., and Grohganz, H. (2013) Amino acids as co-amorphous stabilizers for poorly water soluble drugs—part II: molecular interactions. *Eur. J. Pharm. Biopharm.*, 85(3):882–888.
52. Pan, R.-N., Chen, J.-H., and Chen, R.R. (2000) Enhancement of dissolution and bioavailability of piroxicam in solid dispersion systems. *Drug Dev. Ind. Pharm.*, 26(9):989–994.
53. Lu, Q. and Zografi, G. (1998) Phase behavior of binary and ternary amorphous mixtures containing indomethacin, citric acid, and PVP. *Pharm. Res.*, 15(8):1202–1206.
54. Matsumoto, T. and Zografi, G. (1999) Physical properties of solid molecular dispersions of indomethacin with poly(vinylpyrrolidone) and poly(vinylpyrrolidone-*co*-vinyl-acetate) in relation to indomethacin crystallization. *Pharm. Res.*, 16(11):1722–1728.
55. Yang, J., Grey, K., and Doney, J. (2010) An improved kinetics approach to describe the physical stability of amorphous solid dispersions. *Int. J. Pharm.*, 384:24–31.
56. Newman, A. (2013) Specialized solid form screening techniques. *Org. Process Res. Dev.*, 17:457–471.
57. Keen, J.M., McGinity, J.W., and Williams, R.O., III (2013) Enhancing bioavailability through thermal processing. *Int. J. Pharm.*, 450(1–2):185–196.
58. Forster, A., Hempenstall, J., Tucker, I., and Rades, T. (2001) The potential of small-scale fusion experiments and the Gordon–Taylor equation to predict the suitability of drug/polymer blends for melt extrusion. *Drug Dev. Ind. Pharm.*, 27(6):549–560.

59. Brettman, B.K., Cheng, K., Myerson, A.S., and Trout, B.L. (2013) Electrospun formulations containing crystalline active pharmaceutical ingredients. *Pharm. Res.*, 30:238–246.
60. Nagy, Z.K., Balogh, A., Dravavolgyi, G., Ferguson, J., Pataki, J., Vajna, B., and Marosi, G. (2013) Solvent-free melt electrospinning for preparation of fast dissolving drug delivery system and comparison with solvent-based electrospun and melt extruded systems. *J. Pharm. Sci.*, 102(2):508–517.
61. Moser, J.D., Broyles, J., Liu, L., Miller, E., and Wang, M. (2008) Enhancing bioavailability of poorly soluble drugs using spray dried solid dispersions. Part I. *Am. Pharm. Rev.*, (September–October):68–73.
62. Moser, J.D., Broyles, J., Liu, L., Miller, E., and Wang, M. (2008) Enhancing bioavailability of poorly soluble drugs using spray dried solid dispersions. Part II. *Am. Pharm. Rev.*, (November–December):94–99.
63. Schwegman, J.J., Hardwick, L.M., and Akers, M. (2005) Practical formulation and process development of freeze-dried products. *Pharm. Dev. Technol.*, 10:151–173.
64. Janssens, S., De Zeure, A., Paudel, A., Van Humbeeck, Rombaut, P., and Van den Mooter, G. (2010) Influence of preparation methods on solid state supersaturation of amorphous solid dispersions: a case study with itraconazole and Eudragit E100. *Pharm. Res.*, 27(5):775–785.
65. Ng, Y.C., Yan, Z., McAuley, W.J., and Qi, S. (2013) Stabilisation of amorphous drugs under high humidity using pharmaceutical thin films. *Eur. J. Pharm. Biopharm.*, 84:555–565.
66. Lee, T. and Lee, J. (2003) Drug–carrier screening on a chip. *Pharm. Technol.*, (January):40–48.
67. Baumann, J., Dobry, D., and Ray, R. (2013) Amorphous dispersion formulation development: phase-appropriate integrated approaches to optimizing performance, manufacturability, stability and dosage form. *Drug Dev. Deliv.*, 13(6):30–37.
68. Alsenz, J. and Kansy, M. (2007) High throughput solubility measurement in drug discovery and development. *Adv. Drug Deliv. Rev.*, 59:546–567.
69. Alsenz, J., Meister, E., and Haenel, E. (2007) Development of a partially automated solubility screening (PASS) assay for early drug development. *J. Pharm. Sci.*, 96(7):1748–1762.
70. Hsieh, S., Yue, H., Nicholson, S., Roberts, D., Schild, R., Gamble, J., and Lindrud, M. (2013) Modeling of the secondary drying and the fate of organic solvents for spray dried dispersion (SDD) product. AIChE Annual Meeting. Abstract available at http://www3.aiche.org/proceedings/Abstract.aspx?PaperID=319900 (accessed January 25, 2014).
71. Zhang, G.G.Z., Law, D., Schmitt, E.A., and Qiu, Y. (2004) Phase transformation considerations during process development and manufacture of solid oral dosage forms. *Adv. Drug Deliv. Rev.*, 56:371–390.

4

SOLID-STATE CHARACTERIZATION OF AMORPHOUS DISPERSIONS

Frederick G. Vogt

Morgan, Lewis & Bockius LLP, Philadelphia, PA, USA

4.1 INTRODUCTION

Pharmaceutical amorphous solid dispersions present unique challenges for analytical and physical characterization. These challenges largely arise from the desire to characterize the nature of interactions between the drugs and excipients used to form the dispersion, detect low levels of recrystallization of the drug, and observe phase transformations upon dissolution of the dispersion. The amorphous nature of the dispersion presents a hurdle to analysis, because many of the major solid-state analytical techniques used in pharmaceutical development are most effective in applications to crystalline systems [1]. While some of these techniques can be used to characterize dispersions in the same manner as crystalline systems, they often require adaptation to allow for useful applications to amorphous solid dispersions. Currently, many analytical techniques are available to address the characterization challenges compared with other approaches to drug formulation, and research continues into new approaches.

This chapter reviews established and emerging analytical and physical techniques for the characterization of amorphous solid dispersions. Many of the techniques reviewed here are spectroscopic in nature and probe the interaction of electromagnetic fields of different energies with the sample of interest. Spectroscopic techniques access molecular or atomic energy levels, which in molecular solids are primarily dependent on

Pharmaceutical Amorphous Solid Dispersions, First Edition. Edited by Ann Newman.

short-range structure and thus provide information about local structure in amorphous solids. The energy levels accessed may include electronic transitions, vibrational transitions, and nuclear spin transitions. X-ray diffraction techniques probe the interference of radiation with periodic changes in electron density in materials. These effects are most clearly observed in materials that possess long-range order (such as crystals) and often provide less information in studies of amorphous systems. The use of single-crystal X-ray diffraction (SCXRD), one of the most powerful methods for characterizing crystalline pharmaceutical phases, is generally precluded in amorphous systems. However, advances in powder X-ray diffraction (PXRD) analysis have enabled its use in the characterization of amorphous solid dispersions. The thermal analysis methods used for studies of amorphous dispersions are often focused on observation of second-order events such as the glass transition and its characteristic temperature (T_g). The detection of melting and other first-order events is generally pursued only as a means to detect unwanted crystalline content in amorphous solid dispersions. In many cases, information from a range of characterization techniques is combined to build the most complete picture possible of the dispersion.

In the evaluation of the structure of binary solid dispersions consisting of a drug and a polymer, wherein the material lacks long-range order and is described as amorphous, several major scenarios of interest frequently arise. The drug may be fully miscible in the amorphous polymer, leading to a situation often referred to as a glass solution [2,3]. In a glass solution, the drug and polymer are homogeneously mixed at the molecular level, and consequently a glass solution is also commonly referred to as a molecular dispersion. A glass solution is a type of solid solution, where the latter term also includes crystalline systems in which one component is incorporated into the crystal structure of the other component [4]. If the drug and polymer are not fully miscible, a solid suspension can be formed wherein the drug and polymer are present predominantly in individual amorphous domains. The domains may consist entirely of one component, or may themselves be mixtures of components. In the latter situation, for example, a dispersion may consist of a drug-rich amorphous phase discontinuously dispersed as droplets in a polymer-rich continuous phase, with both phases containing significant amounts of both drug and polymer. The domain sizes can range down to dimensions of nanometers, and consequently the system may be described as an amorphous nanosuspension. Amorphous nanosuspensions can be difficult to distinguish from glass solutions using traditional analytical techniques. The detection of phase-separated amorphous domains and the evaluation of miscibility are the major challenges to which recent analytical developments have been addressed. The structural situation becomes more complex with the addition of other components to the dispersion, such as surfactants, which can disperse differently in a polymer than the drug. More recently developed classes of amorphous dispersion, such as dispersions of drugs within mesoporous silica or on the surface or within nanomaterials, also create potentially complex characterization challenges for analytical methods [5]. The presence and miscibility of adventitious components, such as water or residual solvents, and their effect as potential plasticizers, are also of interest in the characterization of phase miscibility in amorphous solid dispersions. Many of the analytical techniques discussed in this chapter seek to determine whether the components of dispersions are miscible.

Beyond the question of miscibility and phase separation, the individual noncovalent interactions between components are also of interest in amorphous solid dispersions. These interactions can include those observed in molecular crystals or in solution phase, such as hydrogen bonding, π-stacking involving aromatic rings or other π-electron systems, π-cation interactions, halogen bonds, and dispersive (van der Waals) forces [6]. These interactions, which are often readily observed in crystalline materials by SCXRD and spectroscopic techniques, can be a challenge to detect in the inherently disordered environments that occur in amorphous solid dispersions. Sensitive spectroscopic techniques based on vibrational and solid-state nuclear magnetic resonance (SSNMR) spectroscopy are particularly useful in detecting such interactions.

Characterization techniques can be used to study changes in the physical state of amorphous solid dispersions, such as structural relaxation toward an equilibrium amorphous state. Structural relaxation refers to the relaxation of the amorphous state toward a lower energy, higher density state [7–9]. Subtle relaxation effects often occur with aging of amorphous solid dispersions, resulting in an increase in density and a decrease in free volume, and endothermic events corresponding to the enthalpy of relaxation can be observed by calorimetric methods and other methods discussed in this chapter. However, the major temperature-dependent state change that often occurs in amorphous materials, including amorphous solid dispersions, is the glass transition in the vicinity of the temperature T_g, where the mobility, viscosity, and other properties of the amorphous solid dispersion transition from those of a glass to those of a rubber (or supercooled liquid) [7–9]. A number of theoretical models of T_g are available, although none fully explains all of the properties of the glassy state [10,11]. Most importantly, the T_g can vary depending on the glass preparation conditions, and in the case of a melt preparation, T_g typically decreases as the cooling rate of the melt is reduced because the amorphous phase can relax to a more energetically favorable state [11]. Below the T_g, the glass can still exhibit mobility and structural relaxation on a short timescale until it is further cooled to the temperature T_K (known as the Kauzmann temperature), where the entropy curve of the glass intersects that of a reference crystalline phase [11]. T_K values can often range well below the T_g, with values of 40–190 °C lower than T_g determined in some cases [12]. The T_K is often found to be near to the theoretical Vogel–Fulcher temperature, where the relaxation time associated with conformational molecular motions begins to deviate from that associated with vibrational motions [11]. In a polymer, the passage through T_g is often pictured as a cooperative, non-Arrhenius process wherein the polymer chains overcome internal resistance and tangling, become less viscous, and begin to slide past one another more freely. Far above the glass transition temperature, the components of the dispersion move rapidly as in a liquid phase and the material exhibits Arrhenius-type behavior [11]. The glass transition is not a true thermodynamic transition, but is instead referred to as a kinetic transition [11].

The glass transition is often referred to as a primary or α-relaxation process or a series of α-relaxation processes [9]. Other secondary processes are also observed and are commonly referred to as β-processes [9,10,13]. The β-processes are described as being specific molecular interactions that may occur in spatial regions of a dispersion with different densities and exhibit Arrhenius-type behavior, and are sometimes referred to as "noncooperative" motional processes [9]. Conceptually, β-relaxation processes may be thought of as short-range reorganizations within the amorphous solid dispersion. The occurrence of "polyamorphism," or two distinct amorphous states separated by a phase transition, is also a potential consideration [14,15]. However, the occurrence of true polyamorphism is rare, and can be easily confused with structural relaxation [9].

Because amorphous forms of a drug are generally metastable with respect to crystalline forms of a drug, crystallization can occur within dispersions if the drug is sufficiently concentrated to allow for crystal nucleation and growth [2]. Analytical methods are therefore necessary to detect and quantify the presence of crystalline content at low levels in the presence of a large amount of amorphous material. Many of the techniques that are widely employed to detect undesired polymorphs in fully crystalline dosage forms can be adapted for this purpose [1,16]. The most important analytical techniques currently used for detection of crystalline content in an amorphous solid dispersion will be discussed in this chapter.

While the analytical methods discussed here often produce data that can be directly interpreted, for example, by measuring peak maxima in a spectrum or by analyzing the position of a thermal event, in some cases data can be sufficiently complex to warrant the use of multivariate analysis (MVA) methods. MVA is commonly used to analyze analytical data in scenarios where significant overlap occurs (e.g., to extract trends from subtle changes in a poorly resolved spectrum), or in cases where a large amount of data must be comparatively interpreted (e.g., the results of large-scale screening experiments). A full discussion of common MVA methods can be found elsewhere [17]. Several of the key methods used in the analysis of amorphous solid dispersions include direct classical least squares (DCLS), partial least squares (PLS), and principal component analysis (PCA). These techniques are often used to analyze complex spectral data produced in analysis of dispersions.

4.2 THERMAL ANALYSIS METHODS

The thermal analysis methods used in most reported studies of amorphous solid dispersions include differential scanning calorimetry (DSC), thermogravimetric analysis (TGA), and isothermal microcalorimetry (ITC) [18]. Thermal analysis methods are uniquely suited toward the characterization of amorphous solid dispersions because of the temperature-dependent behavior that is commonly exhibited by amorphous solids as well as enthalpic changes that occur upon relaxation of an amorphous solid [7–9]. For example, thermal analysis methods can probe T_g in amorphous solids or observe the occurrence of multiple T_g values. As previously noted, in the temperature range around T_g, the properties of an amorphous solid gradually change from the solid-like glassy state prevalent at lower temperatures to the supercooled liquid-like rubbery state prevalent at higher temperatures. Since T_g is normally a critical parameter in the characterization of amorphous solid dispersions, methods to observe it and other thermal or energetic parameters are important for analytical characterization.

The most common thermal analysis methods used to characterize amorphous solid dispersions are based on DSC. Most modern DSC systems are developed around the heat flux design, in which the difference in heat flow into the sample and a reference material is measured while the sample temperature is linearly increased [18,19]. Conventional DSC methods involving the measurement of total heat flow as a function of sample temperature are widely used to detect the formation of small amounts of crystalline content in amorphous solid dispersions through the observation of sharp melting endotherms. Conventional DSC normally observes T_g via a step change in heat capacity at constant pressure (C_p) as the sample is heated. The onset or midpoint value of the step change, which can cover 10 °C or more, is usually reported as T_g. The T_g measured by a DSC experiment is dynamic, in that it is strongly dependent on the heating rate employed and the extent of annealing of the glass [18,20]. It is possible to obtain "equilibrium" T_g values using an extrapolation approach [20]. In practice, the dynamic T_g values reported in most studies can be highly dependent on the heating rate used and the degree of annealing of the amorphous solid dispersion and thus can vary between batches, instruments, and laboratories [18].

The detection of T_g using conventional DSC techniques can be difficult in some cases because of the small size of the change in C_p and the possibility of enthalpic relaxation and solvent loss in the same temperature region where the T_g of interest occurs [18,21]. To address these situations, temperature-modulated DSC (MDSC or TMDSC) is often employed [18,22,23]. MDSC is a widely used technique that superimposes a waveform (often a sinusoidal modulation) onto the conventional linear heating rate [18,22,23]. The modulation amplitude is typically about 0.1–2 °C with a period of 40–100 s. In a MDSC experiment, the total heat flow ($\mathrm{d}H/\mathrm{d}t$) is given by the sum of a term arising from the temperature-dependent heat capacity of the sample (C_p) and a term f arising from "excess" heat production or absorption by the sample from events such as crystallization or decomposition [18,22,23]:

$$\frac{\mathrm{d}H}{\mathrm{d}t} = C_p \frac{\mathrm{d}T}{\mathrm{d}t} + f(T, t), \tag{4.1}$$

where H, T, and t represent heat, temperature, and time, respectively. The conventional DSC experiment measures these terms as one signal, while MDSC allows for the separation of the "reversing" heat capacity term from the "non-reversing" heat flow f by Fourier transformation or similar approaches. The reversing component contains information about changes in C_p, including the T_g of interest in amorphous solid dispersions as well as information about reversible melting of crystalline materials. In addition to crystallization and decomposition events, the non-reversing heat flow reports on evaporation, enthalpic relaxation in amorphous solids, denaturation, and some types of melting. The use of MDSC to separate the measurement of T_g from other interfering effects is highly useful in studies of amorphous solid dispersions. MDSC also provides superior results when residual solvents are present in a dispersion, such as water and organic solvents from spray drying [18]. MDSC operates under the assumption that the sample only exhibits slow non-reversing phenomena. Alternatives to sinusoidal modulation, such as the use of stochastic temperature modulation, have been developed to address situations where this assumption is not met [24]. Instead of a slow sinusoidal temperature change, in these methods a stochastic (or random) series of step changes is impressed upon a linear temperature ramp.

MDSC measurements of T_g are commonly used to assess miscibility of amorphous solid dispersions. Observation of a single T_g by MDSC is consistent with the presence of a homogeneous, miscible dispersion. In Figure 4.1a, a T_g measurement was performed using MDSC on three amorphous solid dispersions of tenoxicam:L-arginine in polyvinylpyrrolidone (PVP). All three of the dispersions show a single T_g, in agreement with other data showing miscibility at the molecular level [25]. In contrast, the observation of two T_g values by MDSC is indicative of phase separation. In Figure 4.1b, two T_g values are observed by MDSC analysis for dispersions of dextran and PVP that are consistent with the T_g values observed for the individual amorphous components [26]. The observation of T_g by MDSC can be challenging in some polymer systems that do not exhibit a clear T_g even in the pure polymer, such as in hydroxypropylcellulose (HPC) [27]. In dispersions, the potential occurrence of amorphous nanosuspensions also presents a challenge to MDSC analysis. Detection of a T_g event by MDSC for a particular amorphous domain generally requires a domain size of approximately 30 nm or greater [26,28].

The T_g values measured by DSC and MDSC can be empirically interpreted to compare the properties of different amorphous solid dispersions. T_g values are generally increased by the presence of stronger intermolecular forces such as hydrogen bonding and van der Waals forces. This is related to the effect noted earlier, wherein near to T_g molecules and polymers overcome intermolecular forces and begin to move past one another and exhibit liquid-like motion. Beyond this qualitative interpretation, quantitative models are also available and are widely used in the study of amorphous solid dispersions. One of the most common quantitative models for T_g involves the Gordon–Taylor (GT) equation [29]. The GT equation predicts the T_g of an amorphous binary mixture (such as a drug and a polymer) from the properties of the individual amorphous components. The GT equation assumes that the free volumes of both components are additive and that no specific interactions occur between the two components. The free volume is the volume available in the solid to permit motion of nearby atoms [11]. Under these assumptions, the T_g can be predicted as

$$T_{g,\mathrm{mix}} = \frac{w_1 T_{g,1} + K_{\mathrm{GT}} w_2 T_{g,2}}{w_1 + K_{\mathrm{GT}} w_2}, \tag{4.2}$$

where w_1 and w_2 are the weight fractions of each component and $T_{g,1}$ and $T_{g,2}$ are the glass transition temperatures of each component. The constant K_{GT} is given by

$$K_{\mathrm{GT}} = \frac{\rho_1 \Delta\alpha_2}{\rho_2 \Delta\alpha_1}, \tag{4.3}$$

where ρ_1 and ρ_2 are the true densities of each component and $\Delta\alpha_1$ and $\Delta\alpha_2$ are the change in thermal expansivity of T_g of each component. The true densities are commonly

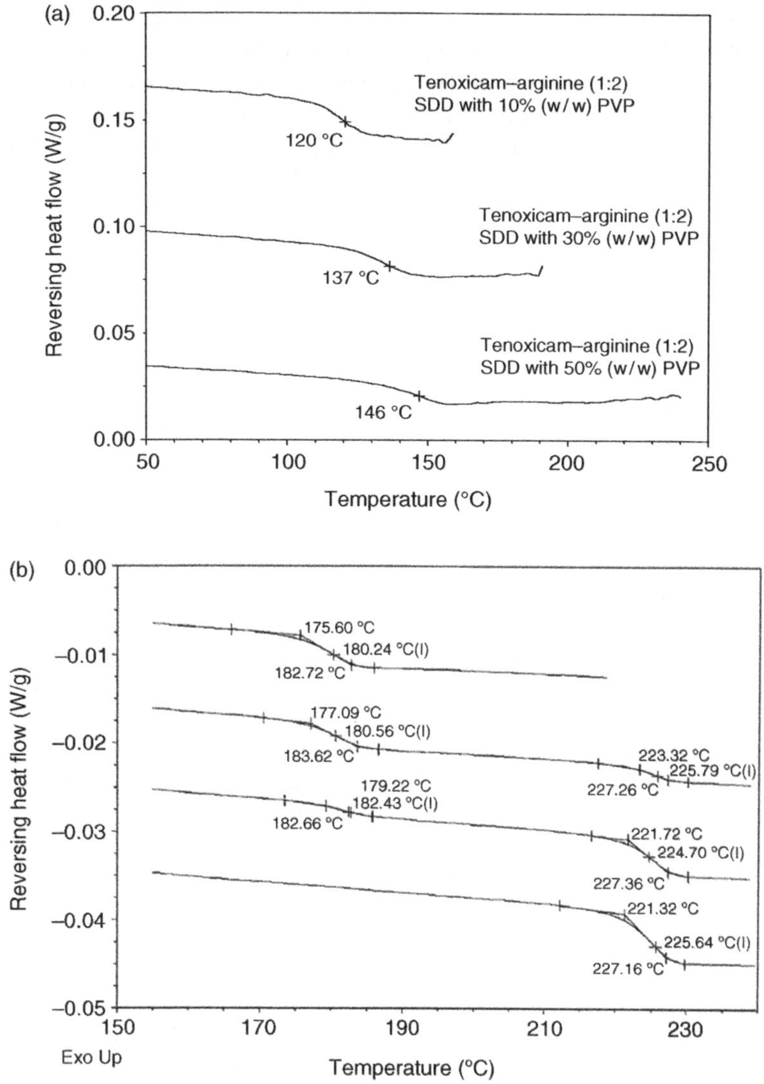

Figure 4.1. (a) MDSC thermograms of amorphous solid dispersions of 1:2 tenoxicam:L-arginine in PVP, prepared as a spray-dried dispersion, showing the single stepwise change in heat capacity at T_g that is characteristic of a miscible system. Reversing heat flows are plotted. (Reproduced with permission from Ref. 25. Copyright 2012, Wiley-Liss, Inc. and the American Pharmacists Association.) (b) MDSC thermograms of an immiscible (or phase-separated) system of PVP and dextran, illustrating the two T_g transitions observed in mixtures. The thermograms from top to bottom were obtained from pure PVP, 70% (w/w) dextran in PVP, 30% (w/w) dextran in PVP, and pure dextran. Reversing heat flows are plotted. (Reproduced with permission from Ref. 26. Copyright 2008, Wiley-Liss, Inc. and the American Pharmacists Association.)

determined by helium displacement pycnometry for the separate pure components. The changes in thermal expansivity for each pure component can be approximated using the values of T_g for each component [30]. The GT equation only applies to binary mixtures in which the components are fully miscible over the entire composition range. The Couchman–Karasz (CK) equation is another popular approach to prediction of T_g and is particularly well-suited to cases where the drug acts as a plasticizer upon the polymer [31]. The T_g of a mixture is again predicted using Equation 4.2 but with K_{GT} replaced by K_{CK} [18]:

$$K_{CK} = \frac{\Delta C_{p,2}}{\Delta C_{p,1}}. \tag{4.4}$$

Here, $\Delta C_{p,1}$ and $\Delta C_{p,2}$, respectively, represent the changes in specific heat capacity for components 1 and 2 measured at the T_g values separately obtained from pure amorphous components. These values are readily measured if it is possible to prepare pure amorphous components, although not all drugs are easily prepared as amorphous phases. Other equations have also been developed to predict glass transitions in binary amorphous systems, but are not as commonly employed, and at present no universal equation capable of modeling all systems has been proposed [18].

Both the GT and CK equations predict an increase in T_g that is proportional to the mass fraction of the higher T_g component, and an experimental finding of this prediction is often used to support the proposition that the two components are miscible [18]. Deviations from the GT and CK equations are normally interpreted to mean that a violation of the assumption of non-interacting components has occurred. For example, T_g values obtained by DSC for dispersions of a compound known as MK-0591 with PVP were higher than those predicted by the Gordon–Taylor equation, indicating the presence of drug–polymer interactions [32]. Negative deviations from the Gordon–Taylor equation have also been observed in other DSC studies [18].

Fast-scanning DSC is a more recently developed technique that also has applications to amorphous solid dispersions [33]. The key benefit of fast DSC is that the heat is applied rapidly, so that changes in temperature during the measurement do not have enough time to affect the miscibility of the drug and the polymer. As a result, fast-scanning DSC can allow for a better understanding of miscibility in amorphous solid dispersions. DSC experiments on polymers can also be performed under variable pressure in different gaseous atmospheres, which affects the heat capacity and measured T_g value of an amorphous material, potentially allowing new observations or more detailed analysis in challenging systems [34].

ITC is a complementary thermal analysis technique that generally requires a larger amount of sample but can measure much smaller heat flows [35]. ITC is more commonly employed in the detection of small amounts of amorphous content in crystalline pharmaceuticals [36]. However, applications of ITC to the study of two-phase amorphous systems such as dispersions have appeared [37]. In particular, ITC can be used to study relaxation of amorphous solid dispersions, as these amorphous states continually evolve toward lower energy arrangements. In these studies, ITC can be used to study detect the rate of heat release during structural relaxation of an amorphous solid, and the relaxation time can be obtained by fitting to theoretical models such as the Kohlrausch–Williams–Watts (KWW) equation [38].

4.3 DIELECTRIC RELAXATION METHODS

Dielectric relaxation spectroscopy (DRS) allows for the investigation of molecular motion in solids [39]. In DRS, electric dipoles within a molecule are perturbed from their equilibrium configuration by an external electric field. When the field is removed, the relaxation back to equilibrium of the electric dipoles within the molecules is monitored via the electric susceptibility. The rate at which the electric dipoles relax depends on the extent to which the local electric dipole-dependent transitions are modulated by the molecular motion. The experiment reports on the mole fraction of dipole reorientation at the measurement frequency, which is often in the MHz to GHz range. Relaxation in DRS has some parallels with nuclear spin relaxation measured by solid-state NMR, mainly in the use of similar spectral density functions to explain relaxation phenomena, although the techniques measure different phenomena [40]. Although DRS has not been widely applied in studies of amorphous solid dispersions, the closely related technique of thermally stimulated current (TSC) spectroscopy, also referred to as thermally stimulated depolarization current (TSDC) spectroscopy, has been applied in several studies [41,42]. TSDC is a method that can be used to study low-frequency molecular motions in the frequency range of 10^{-3} to 10^{-5} Hz [43]. TSDC is able to better resolve lower frequency relaxation mechanisms as a function of temperature than DRS [44]. A TSDC experiment results in a thermogram with peaks corresponding to the depolarization current intensity as a function of temperature [41]. Sharper peaks tend to correspond to α-relaxation, while β-relaxation peaks tend to be broad and cover a wider temperature range [43]. Specialized thermal windowing (or thermal cleaning) experiments can be used to confirm α-relaxation processes [42]. Thermal windowing allows for observation of a distribution of temperature-dependent relaxation times, in contrast to the single average value generally available from other techniques [42]. Kinetic parameters obtained from TSDC experiments can be used to probe molecular mobility below T_g. An extensive study of dispersions containing 10–60% (w/w) of a small-molecule drug with PVP was performed using TSDC [41,42]. The amorphous drug showed two events in the TSDC spectrum, while DSC could only observe one thermal event. Analysis of pure PVP with an average molecular weight of 40 kDa (commonly known as "K30" grade) by TSDC showed two distinct motional modes, a higher temperature mode assigned to α-relaxation processes near to T_g and a lower temperature motional mode (at 132.3 °C) assigned to β-processes that involve localized, noncooperative motional processes with weak interactions between nearby atoms and correspondingly low activation enthalpies [42].

4.4 MOISTURE SORPTION METHODS

Amorphous materials tend to exhibit increased hygroscopicity relative to crystalline solids, as water can be absorbed into their internal structure in addition to being adsorbed on particle surfaces [45]. Many of the commonly used polymers in amorphous solid dispersions are hydrophilic in nature and contain polar functional groups, and thus can absorb moisture from the atmosphere. The total water content in an amorphous solid dispersion can be quantitatively measured using standard approaches such as direct or oven-assisted Karl Fischer titrations for purposes of quality control testing [46]. In pharmaceutical development, the measurement of the level of water with a particular dispersion can be critical because the presence of water can have a plasticizing effect that can lower T_g and increase mobility in the dispersion at a given temperature [47,48]. The increase in mobility can potentially be detrimental to chemical or physical stability.

More fundamental information about the interaction between moisture and a solid can be obtained by measuring water sorption and desorption isotherms. The technique known as gravimetric vapor sorption (GVS) or dynamic vapor sorption (DVS) is the most common approach used for measurement of sorption isotherms [48]. The GVS instrument integrates a relative humidity (RH)- and temperature-controlled environment with a sensitive balance to measure sample mass as a function of RH in a convenient manner. An older, less convenient approach to the same measurement that is still employed, particularly when long RH equilibration periods are needed, involves placing preweighed samples into sealed chambers containing saturated salt solutions with known RH values [49]. This method can be used to obtain the equilibrium moisture uptake for an amorphous solid dispersion at a given humidity (e.g., 60% RH) and temperature, either gravimetrically or by performing a water content measurement such as a Karl Fischer titration after storage of the dispersion for an extended period. Unlike salt chambers, GVS instruments can also perform unique humidity ramping experiments to study processes such as recrystallization from amorphous solid dispersions. The use of GVS is facilitated for amorphous solid dispersion studies because instruments are widely available in many analytical laboratories due to their range of applications; for example, methods based on GVS are often used to quantify amorphous forms at low levels in crystalline drug substances [50].

The adsorption of water vapor in an amorphous solid dispersion can be modeled up to an RH of about 40% using the well-known Brunauer–Emmett–Teller (BET) equation that is also used for modeling of crystalline solids [48]. The full RH range can be modeled for many amorphous substances using an extended form of the BET equation known as the Guggenheim–Anderson–de Boer (GAB) equation (or simply the Anderson equation) [48,51]. Over a range of relative pressures (P/P^0), the GAB equation accounts for adsorption of water through the processes of initial site adsorption, intermediate bound water states, and condensation at higher RH:

$$W = \frac{W_m C_G K(P/P^0)}{[1 - K(P/P^0)]\,[1 - K(P/P^0) + C_G K(P/P^0)]}, \tag{4.5}$$

where W is the mass of water absorbed per gram of solid at a relative pressure (P/P^0), W_m is the capacity of an adsorption monolayer, K is a constant given by

$$K = B \exp\left[\frac{H_L - H_m}{RT}\right], \tag{4.6}$$

and C_G is a constant given by

$$C_G = D\exp\left[\frac{H_L - H_m}{RT}\right]. \tag{4.7}$$

In Equations 4.6 and 4.7, B and D are constants, H_L is the heat of conduction of the bulk adsorbate, and H_m is the heat of adsorption of vapor adsorbed as intermediate bound water [48]. The GAB equation allows for modeling of the GVS isotherm for amorphous solid dispersions of interest and quantitative comparisons of these fitted parameters. In GVS studies of amorphous dispersions, kinetic effects often lead to hysteresis, in that the increasing and decreasing RH isotherms are not reversible [48,52]. This effect is also commonly attributed to changes in polymer conformation resulting from the plasticizing effect of water [52].

A detailed study of the effect of RH on T_g for PVP was performed using variable-temperature GVS and thermal analysis, and is illustrative of the type of characterization data that can be obtained for dispersion-forming polymers [53]. The amount of water absorbed at a given RH increased as temperature decreased, with a corresponding change in the shape of the isotherm. PVP was observed to change from a highly viscous glass to a much less viscous rubber in the same region where absorbed water behaves as if it is in a liquid-like mobile state. The appearance of tightly bound water was found to correlate with the polymer entering the glassy state. GVS instruments also enable a number of practical applications for amorphous solid dispersions. For example, GVS studies of the kinetics of the uptake of moisture by dispersions can be useful in predicting the results of longer term stability studies at elevated RH. Finally, GVS techniques that utilize organic solvent vapors instead of moisture are also available, and may be useful in the design and development of organic spray drying processes and subcritical or supercritical fluid processes for manufacturing amorphous solid dispersions [54].

4.5 VIBRATIONAL SPECTROSCOPY AND MICROSPECTROSCOPY

In the pharmaceutical sciences, vibrational spectroscopic methods are generally used to probe molecular vibrations and phonon modes in crystalline substances. The principal techniques used at present are based on infrared (IR) spectroscopy and Raman spectroscopy [55]. In IR spectroscopy, the absorption, transmission, or reflectance of IR radiation by a sample is measured. The IR spectrum is divided into three regions according to the wavelength or frequency of the radiation: the near-IR (NIR) region between 0.78 and 2.5 μm (12,800 to 4000 cm^{-1}), the mid-IR (mid-IR) region between 2.5 and 50 μm (4000 to 200 cm^{-1}), and the far-IR region between 50 and 1000 μm (200 to 1 cm^{-1}). All three regions have been employed in studies of amorphous solid dispersions. Vibrational bands observed in the mid-IR region typically arise from single-quantum ($\Delta\nu = \pm 1$) vibrational modes, while the NIR region generally measures overtone bands with a higher $\Delta\nu$ value. The far-IR region (200 to 10 cm^{-1}) generally observes low-frequency phonon modes (also known as lattice vibrations) in crystalline materials, which can be useful to detect crystallinity in amorphous solids. Amorphous and disordered solids generally exhibit broader bands in this region because of their lack of long-range order, and may not exhibit any bands at all [56].

A number of sampling modes are available for IR spectroscopic analysis of bulk substances [55]. One of the most popular is attenuated total reflectance (ATR) sampling, which consists of contacting a solid sample surface with a material with a different refractive index and reflecting the IR beam from the interface [55]. An evanescent wave forms and travels along the surface region, which leads to IR absorbance that can be detected in the reflected IR radiation. ATR sampling is a simple and robust method, particularly for routine analysis, but yields reduced absorbance at higher frequencies (corresponding to higher wavenumbers) and can lead to difficulty in observation of stretching vibrations of interest in hydrogen bonding studies such as those involving O–H and N–H groups. Transmission IR spectroscopy using KBr or CsI pellets or thin films of solvent-evaporated dispersions on NaCl plates is also used to analyze solid materials [55]. These methods do not suffer from the ATR signal attenuation issue at higher wavenumbers. However, the preparation of the pellet necessarily involves mixing the sample with the salt and applying pressure to achieve cold flow, which can cause changes to the sample including changes in water and solvent content. Diffuse reflectance IR measurements (commonly referred to as DRIFTS) can be used directly on powders with minimal sample preparation in cases where neither ATR nor transmission methods are applicable, albeit with a loss of sensitivity [55]. DRIFTS can typically be performed on neat samples with modern accessories, but dilution in materials such as KBr and CsI may be necessary in some cases if bands of interest are saturated, again with the potential for undesirable effects on the water content of the sample. Transmission IR and DRIFTS methods also generally expose the sample to dry N_2 purge gas (unlike ATR sampling), again with potential unwanted effects.

IR spectroscopy can be used to identify drug–polymer-specific interactions in amorphous solid dispersions through changes in peak shape or position that are related to molecular interactions. In particular, mid-IR spectroscopy is a useful probe of hydrogen bonding for specific functional groups particularly through analysis of the vibrational frequencies of O–H, N–H, and C=O bonds [57,58]. For example, the stretching frequencies of O–H and N–H groups acting as hydrogen bond donors tend to decrease and their bands tend to broaden as hydrogen bond donor–acceptor distances get shorter [57,58]. The detection of hydrogen bonding between the drug and polymer can be important because such an interaction can inhibit crystallization in an amorphous solid dispersion. For example, in a study of indomethacin in PVP, the inhibition of crystallization was linked to the observation by mid-IR spectroscopy of drug–polymer hydrogen bonds, which prevented formation of a competing dimeric interaction between indomethacin molecules that stabilized the crystalline form [59]. In this study, a range of indomethacin dispersions in PVP were probed with IR spectroscopy using transmission sampling in KBr pellets [60]. The spectra of the dispersions were compared to identify interactions between the drug and the polymer. Model solution-phase systems of indomethacin, methylpyrrolidone, and acetic acid were used to help qualitatively interpret the IR spectra to understand the carbonyl band frequencies arising from free pyrrolidone, hydrogen-bonded dimers of carboxylic acids, and carboxylic acids engaged in hydrogen bonding to pyrrolidone carbonyl acceptors. In Figure 4.2, expanded regions

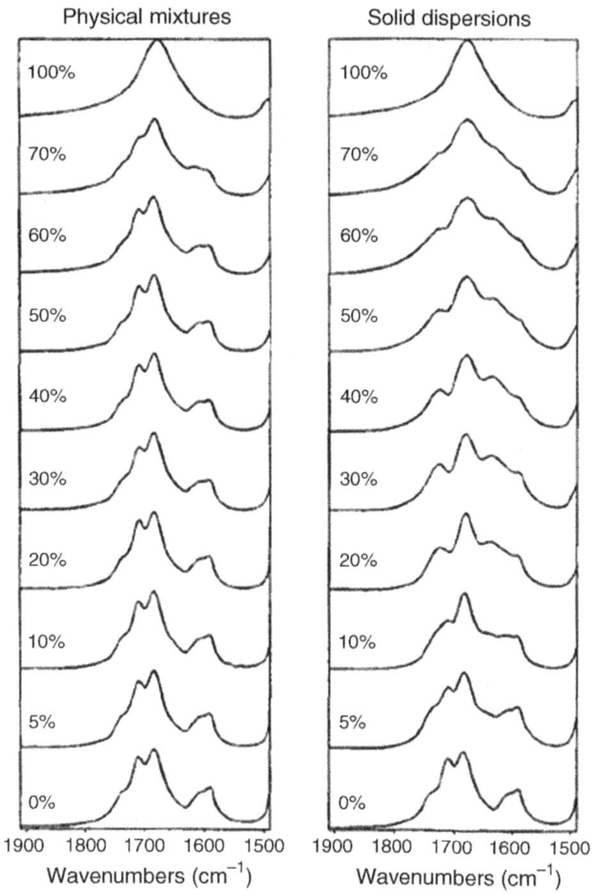

Figure 4.2. Expanded regions of the IR spectra of solid dispersions of indomethacin in PVP displaying the carbonyl band region, shown in comparison with physical mixtures of amorphous indomethacin with PVP. The spectra were obtained by transmission IR spectroscopy using KBr pellets containing approximately 1% of the sample. The percentages correspond to the amount of PVP present in the sample, with 0% corresponding to pure amorphous indomethacin and 100% corresponding to pure PVP. (Reproduced with permission from Ref. 60. Copyright 1997, Plenum Publishing and Springer.)

of the IR spectra obtained for indomethacin dispersions in PVP over a wide concentration range are shown in comparison with physical mixtures of separately prepared amorphous drug and PVP produced by powder blending [60]. The changes in carbonyl band structure in the dispersion can be related to hydrogen bonding between the drug and the polymer. Other studies have employed transmission IR spectroscopy of thin dispersion films to observe drug–polymer interactions in dispersions consisting of nifedipine–PVP, ketoprofen–PVP, droperidol–PVP, and pimozide–PVP [61]. The use of IR spectroscopy in this manner is a powerful and popular method to characterize dispersions, but is limited in some cases by the resolution of the IR spectra and by the inability to prepare stable physical mixtures of amorphous drugs in polymers for some systems. Studies using different IR sampling modes have obtained similar results. For example, an ATR IR spectroscopic study of ibuprofen impregnated into PVP using a supercritical fluid detected interactions between the drug and the C=O group within PVP [62]. A study of a dispersion of ketoconazole and PVP utilized IR as well as ^{13}C SSNMR and DSC to study interactions between the drug and the polymer [63]. IR spectroscopy was also applied to study interactions of indomethacin, lacidipine, nifedipine, and tolbutamide in PVP and polyvinylpyrrolidone-*co*-vinyl acetate (PVP–VA) in dispersions produced by hot melt extrusion [64].

NIR spectroscopy has also been used in studies of amorphous solid dispersions. The NIR spectrum of organic substances is typically less specific than other regions of the IR spectrum, and is made up of sum and difference combination bands and first, second, and third overtone bands arising from the fundamental vibrational modes observed in the mid-IR region. NIR spectroscopy has advantages over mid-IR spectroscopy because of the weak nature of overtone and combination bands, which allows for studies of powdered solid samples without saturation [65]. Many IR spectrometers can be configured to analyze both NIR and mid-IR spectra using an appropriate set of sources, detectors, and optical components (e.g., beam splitters). Applications of NIR spectroscopy have been reported for studies of amorphous solid dispersions. For example, a combination of NIR spectroscopy and multivariate analysis using PLS has been employed as a sensitive, nondestructive measure of crystalline content in tacrolimus dispersions after calibration of the method using dispersions spiked with crystalline material [66]. Drug–polymer interactions in an amorphous solid dispersion of tranilast in the polymethacrylate polymer Eudragit® E PO with enhanced solubility and improved oral bioavailability were studied using a combination of NIR and mid-IR spectroscopic methods [67]. NIR spectroscopy has also been applied to the study of semicrystalline dispersions of cyclosporin A and polyethylene glycol (PEG) with enhanced dissolution, primarily to estimate the quantity of drug present, with mid-IR spectroscopy again applied to probe potential interactions between the drug and the polymer (in this case detecting no such interaction) [68].

Far-field IR microscopy can be used to perform chemical imaging studies of amorphous solid dispersions, although spatial resolution is limited by diffraction to about 25 μm [69]. NIR diffuse reflectance microscopy or imaging is used in a similar manner to assess homogeneity of dispersions, although multivariate methods are generally a requirement for data analysis [68]. Imaging studies of formulations can be facilitated using mid-IR analysis with wide-area ATR sampling and focal planar array (FPA) detectors [70,71]. In this approach, a diamond or ZnSe ATR crystal with dimensions in the ~5 mm range (several times larger than a conventional ATR crystal used in IR analysis) is employed with specialized focusing optics to allow the FPA detector to obtain mid-IR spectra simultaneously from all regions of the sample in good contact with the ATR "window," usually with spatial resolution in the tens of μm. This approach has been applied to image amorphous solid dispersions of small-molecule drugs in polymers as well as their release from the dispersion [72,73].

The coupling of IR spectroscopy with atomic force microscopy (AFM) has opened up new possibilities for the microscopic characterization of dispersions with extremely

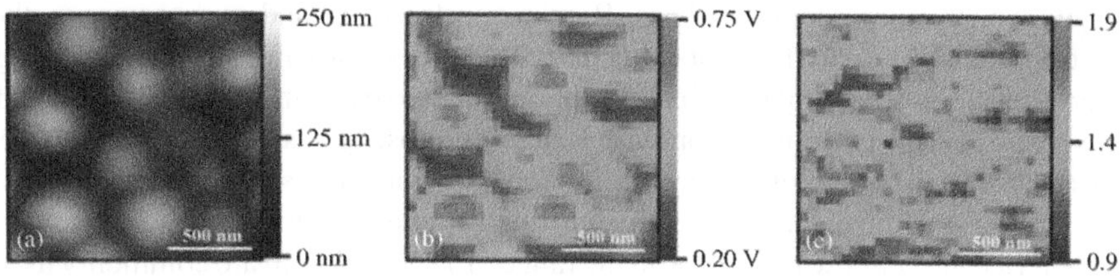

Figure 4.3. (a) AFM topological image of a 50% (w/w) felodipine in PAA. (b) Image showing IR band intensity at 1500 cm^{-1}. This band is specific for amorphous felodipine. (c) Band ratio image showing relative intensity of bands at 1500 and 1700 cm^{-1}, with the latter band specific for PAA. A correlation is observed between the topographical AFM map and the band intensity that indicates that the discontinuous nanophase (with domains on the order of approximately 200–500 nm) is rich in felodipine, while the continuous phase is rich in PAA. (Reproduced with permission from Ref. 75. Copyright 2012, Wiley Periodicals, Inc. and the American Pharmacists Association.)

high spatial resolution [74]. AFM is a near-field microscopy technique that uses an atomically-sharpened tip connected to a cantilever to profile the surface of a solid sample. AFM IR spectroscopy, like other near-field microscopy techniques that place a very small sensor near to the sample, is not limited by the diffraction effects that limit the spatial resolution of far-field IR microscopy [74]. AFM IR has been applied to the study of a dispersion of felodipine in polyacrylic acid (PAA), where a spatial resolution of approximately 100 nm per pixel was obtained [75]. Submicrometer discontinuous domains rich in felodipine were observed, which were surrounded by a continuous phase with a higher concentration of PAA. The results of the AFM IR study on this system are shown in Figure 4.3, where the spatial resolution of this approach can be seen to allow for chemically specific imaging of features on the order of 200–500 nm. A study using AFM IR to characterize dispersions of dextran and maltodextrin with PVP prepared by solvent evaporation has reported similarly promising results [76].

The far-IR region in condensed phases has only become accessible in recent years with the advent of THz band spectroscopic methods [77,78]. The portion of the THz spectral region offering the best access to structural information is usually in the range of 0.1–4 THz (3.3–133 cm^{-1}). Phonon modes in amorphous materials are generally weaker than those in crystalline solids, allowing THz spectroscopy to be used to detect crystalline content [78]. THz radiation is able to penetrate the organic and inorganic materials used in amorphous solid dispersions, and thus may be particularly useful for 2D and 3D imaging crystalline content in large particles of a dispersion produced, for example, by hot melt extrusion. However, this region of the IR spectrum is at present largely unexplored in applications to pharmaceutical amorphous solid dispersions.

Raman spectroscopy is based on the phenomenon of Raman scattering and offers complementary information as well as distinct advantages in sampling and specificity in comparison with IR techniques [55,65]. Both dispersive and FT Raman techniques are in wide use in the pharmaceutical industry. Because of the shorter wavelengths of light used in Raman spectroscopy relative to IR spectroscopy, it is often more amenable to combination with microscopic analysis. Raman spectroscopy and microscopy methods benefit from their ability to detect specimens behind glass, allowing for easy sampling of dynamic effects and studies of unstable amorphous phases sealed in vials or between slides. When spatial control is desired, confocal Raman microscopy is often the technique of choice, as it achieves selectivity in the usual x- and y-dimensions as well as the z-dimension (with the latter enabling depth profiling into the sample, assuming sufficient transparency) [65]. Because laser wavelengths in the range of 785 to 633 nm are commonly used in confocal Raman microscopy applications, the diffraction limit is smaller and effective spatial resolutions of a few μm can be reached. FT Raman spectra are normally obtained with longer wavelength radiation (typically 1064 nm), which is more compatible with the interferometers used to modulate the irradiating beam. FT Raman is commonly used for analysis of bulk amorphous dispersion materials as it provides a rapid means of assessing the identity and drug content. Typical FT Raman spectra for dispersions containing 30% (w/w) of the drug diflunisal in three polymers are shown in Figure 4.4 in comparison with similarly obtained spectra of the polymers alone [79]. Diflunisal contains many aromatic bonds that exhibit a strong change in hyperpolarizability while undergoing molecular

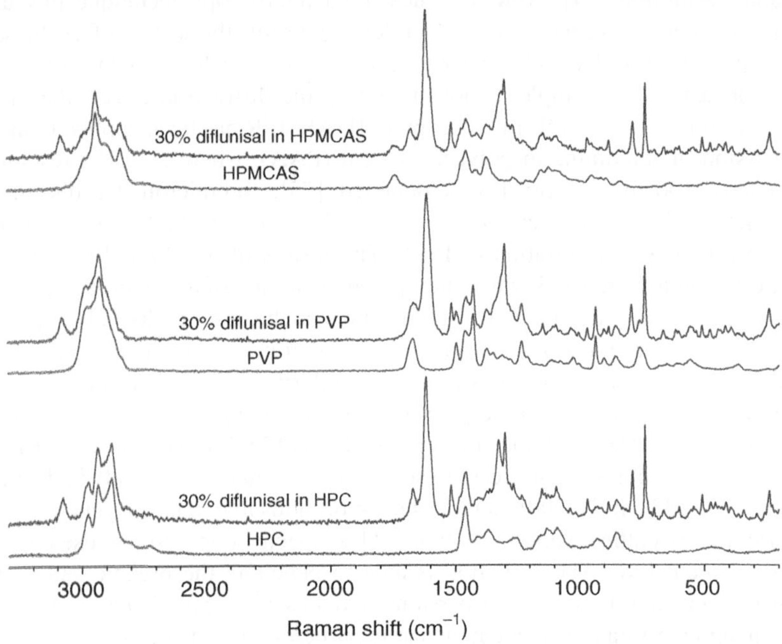

Figure 4.4. Comparison of FT Raman spectra of three amorphous solid dispersions containing 30% (w/w) diflunisal in PVP, HPMCAS, and HPC prepared by solvent evaporation with reference spectra of the polymers. Spectra were obtained on samples of 50 mg of powder in a borosilicate glass tube using a Bruker Optics MultiRAM FT-Raman spectrometer equipped with a 1064 nm laser and using 400 mW of laser power, 4 cm^{-1} resolution, and 512 accumulated scans. The diflunisal dispersion in HPMCAS is known to show evidence of phase separation or domain formation [79].

vibrations, leading to fairly intense Raman scattering and pronounced spectra. The Raman bands of diflunisal seen in Figure 4.4 thus dominate those of all three of the polymers used, despite the drug being the minor component of these dispersions. Although a significant amount of overlap between drug and polymer bands is observed, several specific bands for diflunisal can be found in each of the three dispersions.

Confocal Raman microscopy has been widely employed in studies of pharmaceutical amorphous solid dispersions [80–82]. An example of confocal Raman microscopic analysis used for chemical imaging of an amorphous solid dispersion is illustrated in Figure 4.5. A dispersion containing 30% (w/w) of diflunisal in PVP was prepared by rapid evaporation from acetone solution, and a piece of the thin film obtained was mapped using confocal Raman microscopy as shown in Figure 4.5a. The false-color map imposed on the optical bright-field image in Figure 4.5a is obtained by dividing the band area between 3060 and 3160 cm^{-1} (specific for diflunisal) by the area between 2845 and 3050 cm^{-1} (specific for PVP). This produces a band ratio plot that factors out other influences on signal intensity (such as sample topology and density) and is thus useful for assessing dispersion homogeneity. Raman spectra extracted from the high- and low-intensity points of the map are compared in Figure 4.5b and illustrate the homogeneity of the drug in the PVP at a spatial resolution of 10 μm. Although heterogeneity with sub-μm dimensions in a dispersion cannot be completely resolved because of the diffraction limit, it can be indirectly detected using conventional confocal Raman microscopy systems by carefully comparing concentrations from a mapped region and looking for spectral changes that correspond to concentration differences between the drug and the polymer as in Figure 4.5. A study employing confocal Raman microscopic mapping of nanocrystalline domains of the drug ebselen in PVP–VA illustrates the effects that can be observed in the case of sub-μm heterogeneity [82]. Because it can be focused onto a small spatial area of a sample and obtain highly sensitive results for that region, confocal Raman microscopy can also detect potential miscibility between a drug and a polymer when other techniques cannot. For example, in the aforementioned study of a nanocrystalline ebselen dispersion in PVP–VA, confocal Raman microscopy detected the presence of potential miscible amorphous regions in the dispersion that could not be detected using other characterization techniques [82]. Confocal Raman microscopy using 785 nm laser irradiation has been applied to study heterogeneity in dispersions of dextran and PVP and in biopharmaceutical materials produced by lyophilization [83,84]. Biopharmaceutical formulations are similar in many respects to amorphous solid dispersions of a small-molecule drug and a polymer, except that the larger molecule is the drug (usually a protein) and the small molecules are excipients. More sophisticated multivariate analysis methods such as the band target entropy minimization (BTEM) method have also been applied in conjunction with Raman microscopy studies of dispersions [85]. Confocal Raman microscopy can also be employed to study phase transformations during dissolution of dispersions using suitable sample preparation (e.g., positioning of a dispersion particle between glass slides), yielding similar information to that obtained using the aforementioned FPA ATR IR method.

In some cases, where poorly soluble but highly potent drugs must be delivered, amorphous solid dispersions are prepared with low amounts of drug. The observation of drug using conventional Raman (or IR) spectroscopic methods can then be problematic.

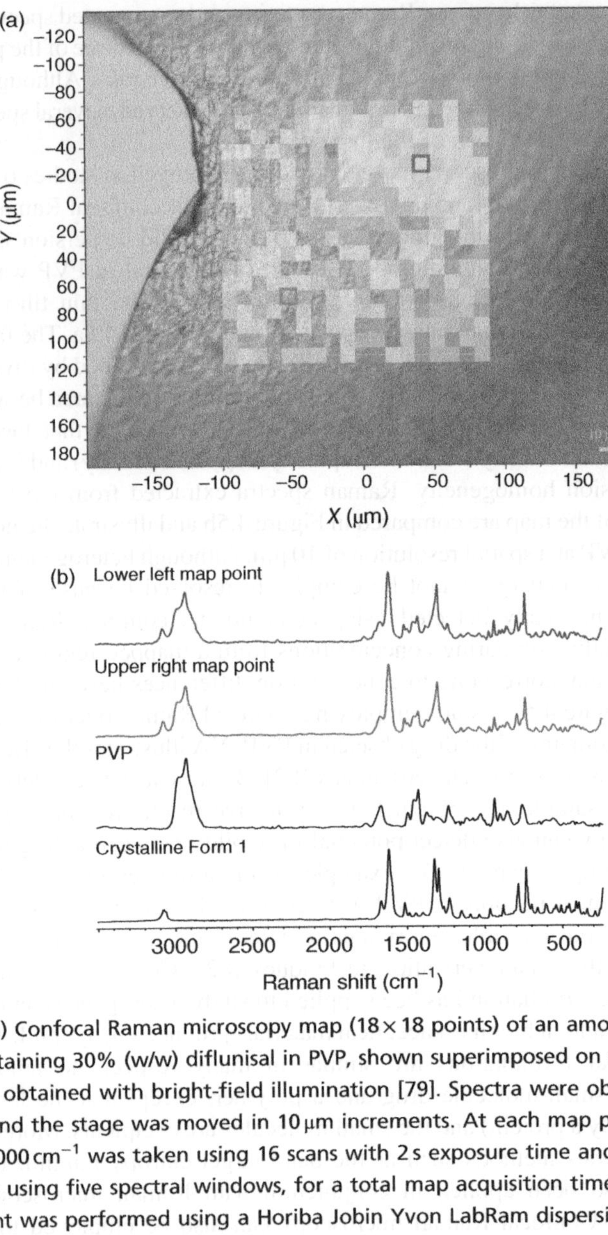

Figure 4.5. (a) Confocal Raman microscopy map (18 × 18 points) of an amorphous dispersion containing 30% (w/w) diflunisal in PVP, shown superimposed on an optical image of the particle obtained with bright-field illumination [79]. Spectra were obtained using a 633 nm laser and the stage was moved in 10 μm increments. At each map point, a spectrum from 200 to 4000 cm^{-1} was taken using 16 scans with 2 s exposure time and with the full spectrum built using five spectral windows, for a total map acquisition time of about 14 h. The experiment was performed using a Horiba Jobin Yvon LabRam dispersive Raman spectrometer equipped with an Olympus BX41 confocal microscope using a 20× objective. In (a), a false-color band ratio plot of the band area between 3060 and 3160 cm^{-1} (diflunisal) divided by the area between 2845 and 3050 cm^{-1} (PVP) is superimposed on a bright-field optical image. More intense map points indicate a higher ratio of diflunisal to PVP. (b) Raman spectra extracted from the map points marked with rectangles in (a) are shown in comparison with reference spectra of crystalline diflunisal Form 1 and PVP.

Several possibilities exist to enhance the sensitivity of Raman spectroscopy. For example, use of ultraviolet (UV) laser irradiation at wavelengths of, for example, 266 nm allows for significant enhancements in sensitivity for typical drug molecules both by resonance Raman enhancement and under normal Raman scattering [86]. The availability of resonance Raman enhancements obtained by use of a 266 nm laser allows for significant signal enhancements for the drug diflunisal when present at 1% (w/w) in an amorphous dispersion in PVP, as shown in Figure 4.6. The corresponding FT Raman spectra obtained using a 1064 nm laser, also shown in Figure 4.6, exhibit the low signal intensity for diflunisal that would normally be associated with drug at this level. This approach allows for more

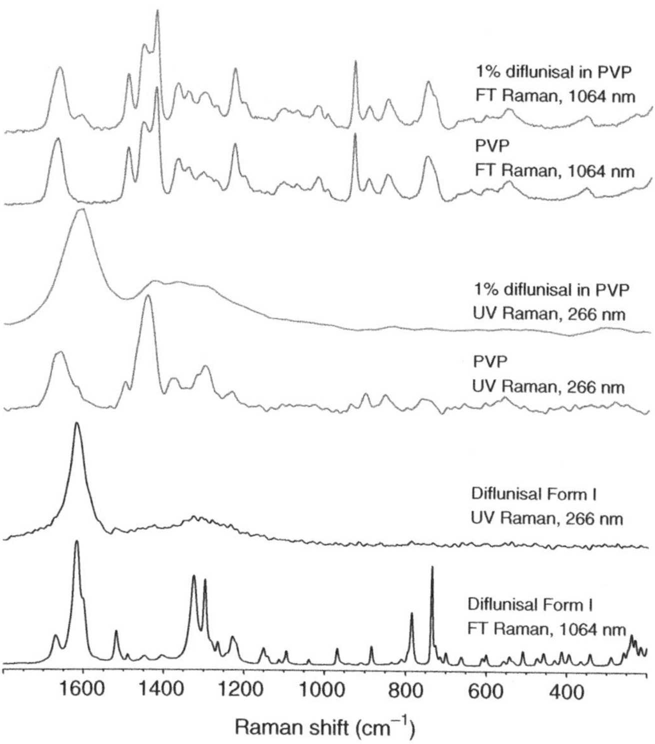

Figure 4.6. Spectra obtained by confocal UV Raman microscopy analysis of a 1% (w/w) dispersion of diflunisal in PVP [86]. The UV Raman spectrum of the dispersion was obtained using 266 nm laser irradiation, 25 mW laser power, a 20x objective, and 32 s acquisition time, and was a single point taken from a 22 × 22 map that required 4 h. The FT Raman spectrum was obtained using conventional backscattering detection on bulk powder, 1064 nm laser irradiation, 400 mW laser power, and 18 min acquisition time. Similarly obtained spectra of crystalline diflunisal Form I are also shown for comparison. Resonance Raman enhancement is observed for the band at 1612 cm^{-1} in diflunisal. FT Raman spectra were obtained using a Bruker Optics MultiRAM spectrometer equipped with a 1064 nm laser and using 400 mW of laser power, 4 cm^{-1} resolution, and 512 accumulated scans.

efficient studies of low-dose, highly potent drugs formulated as dispersions. Although sensitive, resonance Raman spectra typically exhibit a broadened appearance that reduces their specificity, and samples must be checked for potential UV laser damage [86].

Another sensitivity-enhanced Raman spectroscopic technique is coherent anti-Stokes Raman scattering (CARS) microscopy, which has emerged as a highly sensitive tool for chemical imaging in biological applications and has now begun to find applications in pharmaceutical development [87–90]. In CARS, the sample is excited by a probe beam with frequency ω_{pump}, a Stokes beam with frequency ω_{Stokes}, and a probe beam with frequency ω_{probe}. Typically, the frequencies ω_{pump} and ω_{probe} are chosen to be equal. When a sample is irradiated using a tightly focused laser with ω_{pump} and ω_{Stokes} radiation, and the frequency $\omega_{pump} - \omega_{Stokes}$ matches the frequency of a Raman band, a strong signal is generated at the anti-Stokes frequency $\omega_{anti\text{-}Stokes}$. CARS uses tightly focused beams delivered via a microscope to achieve a phase matching condition necessary for the coherent process in a narrow sample area, so that images may be formed by moving the stage as in conventional confocal Raman microscopy. Traditional CARS microscopy experiments produce an image for a single Raman band and are sometimes made specific for a given molecule by isotopic labeling. CARS microscopy has the advantage of very high sensitivity compared with other methods of obtaining Raman spectral information, allowing imaging at video rates in some applications [87]. A CARS study of a semi-solid dispersion of 15% (w/w) paclitaxel in a blend of PEG and poly(lactide-*co*-glycolic acid) (PLGA) detected phase separation between PEG and PLGA and provided evidence of the drug forming a dispersion with the PEG in preference to the PLGA [91]. Newer broadband, multiplex CARS microscopy methods allow for simultaneous imaging of multiple vibrational bands, so that bands arising from a drug and a polymer could be imaged quickly in an amorphous solid dispersion [92]. Studies of drug miscibility in a conventional amorphous solid dispersion employing a drug and a polymer using CARS have not yet been reported but are expected to provide a useful tool in the future.

Other methods capable of accessing molecular vibrational quantum states have been employed in studies of dispersions to a lesser extent, but hold promise for future applications. Neutron vibrational spectroscopy (NVS), also known as inelastic neutron scattering (INS), has been applied to study structure in mesoporous dispersions of ibuprofen [93]. NVS detects the energy transfer to or from neutrons by a scattering molecule, which depends on vibrational transitions, and allows more sensitive detection of vibrational transitions in systems that involve hydrogen because of the high incoherent scattering cross section of this element relative to other elements [57,94]. This effect is useful in organic systems such as amorphous solid dispersions that contain a large molar ratio of hydrogen. NVS produces spectra that are more sensitive to vibrational modes involving hydrogen than IR or Raman spectroscopy and is consequently useful in studies of hydrogen bonding [57]. NVS also offers extremely high resolution from 16 to 4000 cm^{-1} and detects all vibrational transitions, not just those allowed by optical selection rules [93,94]. A major disadvantage of the technique is the requirement for a neutron source and spectrometer, which are currently restricted to several major research facilities because of their size and expense.

4.6 SOLID-STATE NMR SPECTROSCOPY

SSNMR detects nuclear spin energies by probing a sample positioned in a high magnetic field using additional inductive magnetic fields with frequencies in the radiofrequency (RF) range [95,96]. The anisotropic nature of the interactions that dominate the energies of nuclear spin states generally leads to broadened spectra if conventional solution-state NMR methods are applied to solids. The broadening is largely caused by nuclear spin interactions that are normally averaged by molecular tumbling in solutions that does not typically occur in solids. These interactions, which include magnetic dipolar coupling between nuclear spins, anisotropic chemical shift interactions, and coupling between quadrupolar nuclei and electric field gradients (EFGs), dominate the appearance of SSNMR spectra and define the experimental approaches [95,96]. The experimental approaches used in SSNMR are also often highly specific to the nucleus or type of nucleus to be observed. Nuclei with a nuclear spin quantum number of ½ can offer highly resolved and informative spectra in solids. The spin-½ nuclei of importance in studies of amorphous solid dispersions include the ^{1}H, ^{13}C, ^{15}N, ^{19}F, ^{29}Si, and ^{31}P isotopes. These and other NMR-accessible nuclei can be used to selectively detect one or more components in an amorphous solid dispersion, and most studies to date involve only spin-½ nuclei. However, developments over the past 15 years have extended the reach of routine SSNMR techniques to many nuclei with nuclear spin quantum numbers greater than ½, offering access to more chemical environments of interest in amorphous solid dispersions. These nuclei are known as quadrupolar nuclei and can interact with the electric field gradient in a material [95,96]. Quadrupolar nuclei of particular interest in pharmaceutical amorphous solid dispersions include ^{2}H, ^{14}N, ^{17}O, ^{23}Na, $^{35/37}Cl$, and $^{79/81}Br$, many of which are naturally present in drug molecules, counterions, polymers, surfactants, and other dispersion components or can be inserted via labeling methods. Through these nuclei, SSNMR provides a powerful characterization method for the study of amorphous solid dispersions, which are relatively complex amorphous organic substances, because of its demonstrated capability in the study of complex inorganic substances and organic polymers. For example, SSNMR has been extensively used to characterize inorganic glasses and amorphous organic polymers [97–99]. Both 1D and 2D SSNMR methods have been widely employed in these applications. Like pharmaceutical systems, inorganic glasses and polymers contain a number of useful NMR-active nuclei that have been accessed using SSNMR to provide detailed information about chemical structure, including ^{11}B, ^{13}C, ^{19}F, ^{23}Na, ^{27}Al, ^{29}Si, and ^{31}P [97–99].

The dominant experimental technique used in most applications of SSNMR to amorphous dispersions is known as magic-angle spinning (MAS) [95,96]. MAS involves gently packing the sample in a ceramic rotor and rapidly spinning it in a small gas turbine inclined at an angle of 54.74° relative to the z-axis of the static magnetic field (B_0). MAS causes partial averaging of many anisotropic interactions observed in solids by spinning the sample at a rate (ν_r) that is typically of the same order of the size of an internal interaction in the sample. This has the beneficial effect of narrowing the broadened solid-state resonances seen in SSNMR spectra that occur in the absence of rapid, isotropic molecular tumbling as in solution. Under sufficiently fast conditions (e.g., $\nu_r > 25$ kHz), the anisotropic chemical shift interaction, heteronuclear dipolar coupling, and homonuclear dipolar coupling are more completely averaged. In practice, however, homonuclear dipolar coupling between abundant ^{1}H nuclei and heteronuclear dipolar coupling between ^{1}H and X nuclei (such as ^{13}C) are often not fully averaged, and additional decoupling using RF pulse sequences is required to further suppress broadening (see below). Anisotropic *J*-coupling interactions are also averaged by MAS, but the isotropic *J*-coupling familiar from solution-state NMR remains unless decoupled using RF pulse sequences. Quadrupolar interactions are only partially averaged under MAS conditions; more complete averaging requires specialized pulse sequences and/or hardware described below. Other interactions that are not averaged by MAS and commonly contribute to line broadening in amorphous solid dispersions include dipolar-mediated residual quadrupolar coupling and anisotropic bulk magnetic susceptibility [100,101]. The dominant cause of line broadening that remains in most SSNMR spectra of amorphous solid dispersions after experimental optimization arises from the range of chemical shift effects caused by the range of disordered structural environments found in an amorphous solid [98]. Although this "amorphous" broadening can be overcome using 2D SSNMR experiments that correlate local spin pairs (which tend to exhibit a narrower range of chemical shifts), this approach has been primarily applied to inorganic glasses to date and has not yet been reported in studies of amorphous solid dispersions [102].

The maximum ν_r obtainable in SSNMR experiments is primarily limited by rotor diameter and probe design. Decreasing the diameter to increase maximum ν_r must be balanced against sensitivity losses from decreasing sample volume as well as the practical aspects of rotor packing and the cleaning of small-volume rotors. Common rotor sizes range from 7.5 down to 1.0 mm, with corresponding maximum ν_r settings ranging from approximately 8 to 70 kHz [103]. The use of faster MAS rates causes significant frictional heating, with rotor temperature increases of 50 °C common in the absence of cooling, and also causes pressure effects that scale as ν_r^2 and may potentially affect amorphous solid dispersion analyses (e.g., by inducing recrystallization in a less stable system) [104]. No reports appear to have been published where a direct link between MAS and recrystallization or another physical change in an amorphous solid dispersion was established, but the potential does exist and can be monitored by careful experiments. The use of N_2 as a spinning, bearing, and temperature control gas for MAS experiments, which is commonly done to allow for a wide range of temperatures to be probed, also has the effect of slowly purging paramagnetic O_2 gas from the sample (depending on the rotor and probe design), which can lead to changes in relaxation times as discussed below.

SSNMR is often seen as a complex technique because of the wide variety of experiments that can be performed and the potentially complex spectral interpretation required. However, most SSNMR studies of amorphous solid dispersions make use of a limited subset of SSNMR experiments. The most common experimental approaches currently of use in the study of amorphous solid dispersions are summarized in Table 4.1. Basic 1D ^{13}C cross-polarization (CP) experiments, including CP-MAS and CP total sideband suppression (CP-TOSS) experiments, are generally the first experiments applied to a novel amorphous solid dispersion. The natural abundance (1.1%) and relative sensitivity of the spin-½ ^{13}C isotope allows for high-quality data collection for a

TABLE 4.1. Overview of Selected SSNMR Experiments Useful in Studies of Amorphous Solid Dispersions

Experiment Name	Description and Information Obtained	Typical Duration	References
^{1}H DP-MAS or DUMBO	Basic 1D ^{1}H experiment, generally performed at the highest B_0 field available. Spectra can readily detect hydrogen-bonded proton environments at >9 ppm.	10 min	[116,117]
$^{1}H-^{13}C$ CP-MAS or CP-TOSS	Basic 1D ^{13}C experiment, generally performed at moderate B_0 fields. Detects both broadened lineshapes for drug and polymer in dispersions, provides general structural information, and can provide useful hydrogen bonding information via carbonyl resonance effects.	5–10 h	[107]
$^{1}H-^{19}F$ or ^{1}H–X CP-MAS	Basic 1D experiment for observation of ^{19}F or other X heteronuclei (e.g., ^{31}P).	Minutes to hours	[119]
^{13}C- or X-detected ^{1}H T_1 measurement	Determines ^{1}H T_1 by saturation recovery using indirect detection via the spectrum of ^{13}C, ^{19}F, or another heteronucleus. Detects phase heterogeneity in dispersions.	12–24 h	[25,119]
^{13}C- or X-detected ^{1}H $T_{1\rho}$ determination	Determines ^{1}H $T_{1\rho}$ by adding a spin-lock period prior to the CP transfer and performing detection via the spectrum of ^{13}C, ^{19}F, or another heteronucleus. Detects phase heterogeneity in dispersions.	12–24 h	[25,106]
$^{1}H-^{13}C$ or ^{1}H–X CP-HETCOR	Detects dipolar connectivity between ^{1}H and ^{13}C spins, either directly or via spin diffusion when used with longer CP periods (250 μs to 10 ms). Can be performed using a longitudinal mixing period for longer spin diffusion mixing. Detects molecular association in dispersions.	12–24 h	[119,126]
$^{1}H-^{13}C$ or ^{1}H–X LGCP-HETCOR	Detects dipolar connectivity between ^{1}H and ^{13}C spins with suppression of spin diffusion. Detects molecular association in dispersions.	12–24 h	[119,127]
$^{1}H-^{14}N$ HMQC	Detects dipolar connectivity between ^{1}H and ^{14}N spins. Detects molecular association in dispersions.	4–8 h	[121,135]

The typical experimental durations for ^{13}C-based experiments listed are estimated for a dispersion containing typical small-molecule drugs in the 300–800 Da mass range and a drug loading in the range of 25–75% (w/w). Longer times may be needed for materials outside of this range. Experiments that detect sensitive nuclei such as ^{19}F, ^{23}Na, or ^{31}P can be much shorter and/or more sensitive.

typical dispersion to be completed after several hours using these experiments [105]. The CP process provides a significant enhancement of ^{13}C spectral signal relative to single-pulse direct polarization MAS (DP-MAS) experiments [106]. ^{13}C spectra of dispersions often contain many resonances, and the potential for overlap with MAS sidebands is high. Because these can complicate interpretation, a common approach is to remove the sidebands experimentally using TOSS pulse sequences [107]. Finally, the performance of the decoupling of ^{1}H nuclei is critical to the quality of ^{13}C spectra, and decoupling pulse sequences such as SPINAL-64 are commonly performed at RF powers of approximately 100 kHz to ensure good results on dispersions [108].

Several types of information can be obtained from a 1D ^{13}C SSNMR spectrum. First, the resonances are expected to be broad because the amorphous nature of the material leads to a range of ^{13}C sites with different chemical shielding environments. The observation of sharp resonances is normally an indication of crystalline content. In Figure 4.7, 1D ^{13}C

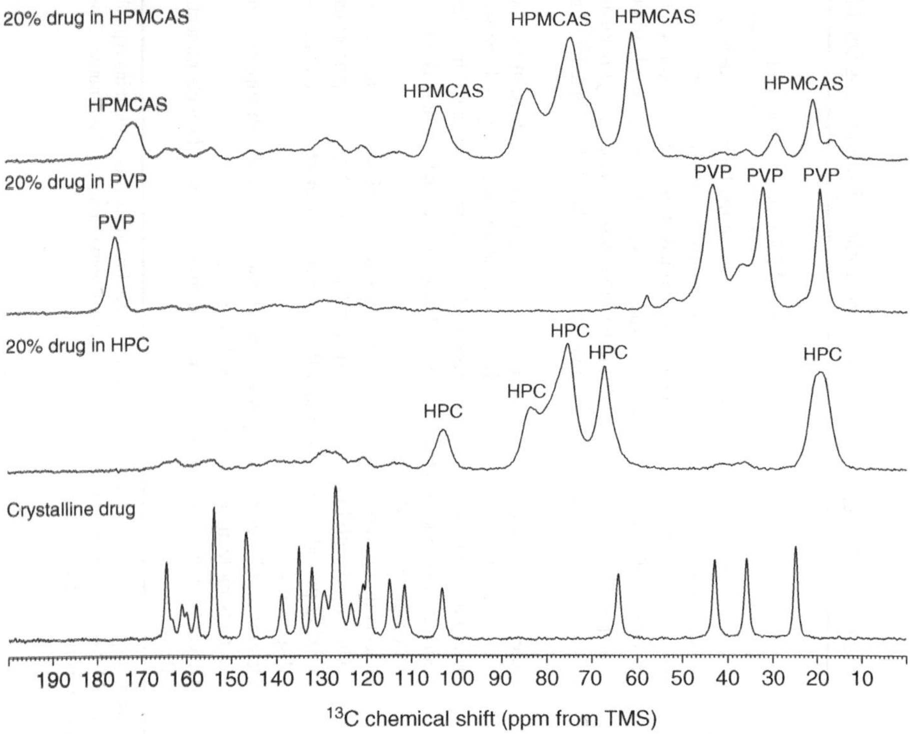

Figure 4.7. ^{13}C CP-TOSS spectra of 20% (w/w) dispersions of the drug 6-(2-(5-chloro-2-(2,4-difluorobenyzloxy)phenyl)cyclopent-1-enyl)picolinic acid in HPMCAS, HPC, and PVP prepared by acetone/methanol solvent evaporation [86]. The ^{13}C CP-TOSS spectrum of a crystalline form of this compound is shown for comparison. The ^{13}C resonances assigned to polymer sites are noted on each spectrum, and the spectra show the generally resolved nature of polymer resonances relative to resonances of a typical pharmaceutical compound that is obtained using this characterization technique.

CP-TOSS spectra of dispersions of the developmental compound 6-(2-(5-chloro-2-(2,4-difluorobenyzloxy)phenyl)cyclopent-1-enyl)picolinic acid in three common polymers are shown [86,109]. The polymers include PVP, HPC, and hydroxypropylmethylcellulose acetate succinate (HPMCAS). The results illustrate the typical sensitivity achieved using ^{13}C SSNMR with acquisition times in the vicinity of 5–10 h on dispersions containing 20% (w/w) of drug, which is sufficient to allow for detection of the carbon resonances of the drug. In Figure 4.7, the resonances assigned to the polymer sites are relatively well resolved from the resonances assigned to the drug molecule, which highlights a general feature of ^{13}C SSNMR relative to techniques such as IR and Raman spectroscopy (which can be seen in comparison with Figure 4.4). Broadened lineshapes are usually observed in the SSNMR spectra of amorphous solids relative to crystalline phases, as seen in Figure 4.7, because each nucleus experiences a range of chemical shift environments arising from differences in molecular conformation and local surroundings.

Conventional 1D CP-MAS and CP-TOSS methods are widely used to obtain basic characterization data for amorphous solid dispersions, like that shown in Figure 4.7. For example, 1D ^{13}C SSNMR methods have been applied to characterize dispersions such as troglitazone in PVP, tenoxicam–L-arginine complexes in PVP, ibuprofen and flurbiprofen in poly(methyl methacrylate)-based polymers, ketoprofen in poly(ethylene oxide), indomethacin–cyclodextrin complexes in PEG, and felodipine in PVP–VA [25,110–115]. Multivariate analysis methods can be applied to assist in the analysis of ^{13}C SSNMR spectra, such as to observe subtle trends that were predictive of recrystallization [115]. The ^{13}C chemical shift observed for carbonyl groups in 1D ^{13}C spectra often contains useful information about hydrogen bonding in amorphous solid dispersions that can be interpreted in a similar manner to carbonyl band shifts observed in IR spectra [57].

Experimental approaches that allow for measurement of 1D ^{1}H SSNMR spectra are also given in Table 4.1. Because of strong ^{1}H$-^{1}$H dipolar coupling, observation of high-resolution 1D ^{1}H spectra of amorphous solid dispersions often requires the use of relatively high ν_r settings (typically greater than 30 kHz), higher fields, and homonuclear decoupling [116,117]. Homonuclear decoupling can be accomplished using a variety of pulse sequences; for example, a commonly used family of sequences known by the acronym of DUMBO is highly effective at removing dipolar coupling, but yields a spectrum containing scaled ^{1}H chemical shifts and produces artifacts that must be avoided [116,117]. Unlike in solution, where rapid molecular tumbling averages dipolar coupling, ^{1}H SSNMR spectra remain broadened even when fast MAS, DUMBO, and high fields are used. However, useful information can still be obtained from the spectra for amorphous solid dispersions, primarily from the observation of deshielded ^{1}H environments (typically in the range of 9–15 ppm) that arise from protons engaged in hydrogen bonding [57]. The ^{1}H chemical shift of common hydrogen bonding protons in drugs, such as carboxylic acid protons, phenolic protons, and amide protons, is highly sensitive to hydrogen bond geometry. For example, the ^{1}H chemical shift of hydrogen bonding protons in carboxylic acids interacting with oxygen acceptors is deshielded by about 1 ppm with every 0.04 Å decrease in the H$\cdots$O interatomic distance [118]. Recently, several studies that utilize ^{1}H SSNMR have been reported that use this type of effect to probe hydrogen bonding, generally using DP-MAS, DUMBO, or related experiments with ν_r typically ranging from 30 to 70 kHz [79,119–121]. Hydrogen bonding resonances arising from protons engaged in longer hydrogen bonds can be overlapped with aromatic resonances in some cases of interest (e.g., with resonances in the 6–8 ppm region), but it is often possible to obtain ^{1}H chemical shift information indirectly using the 2D correlation experiments discussed below and summarized in Table 4.1.

1D SSNMR spectra that observe the ^{15}N nucleus are also highly useful in studies of dispersions wherein the drug and/or polymer contain nitrogen atoms. ^{15}N suffers from a low natural abundance (0.33%) and low sensitivity, but is still tractable for many applications to amorphous solid dispersions [105]. ^{15}N CP-MAS experiments on typical amorphous solid dispersions can require larger volume MAS probes (and hundreds of mg of material) as well as long acquisition times of a day or more. The results can be highly useful in studies of ionization state and hydrogen bonding involving nitrogen donors or acceptors, particularly in complex multicomponent amorphous dispersions. This was found to be the case in a study of dispersions of 1:2 tenoxicam:L-arginine in PVP, where detailed information about the ionization state of the drug and L-arginine was obtained using ^{15}N CP-MAS spectra [25]. Other recent SSNMR studies of dispersions have also employed observation of the ^{15}N nucleus [110]. A more recent approach using 2D SSNMR methods that indirectly observe the ^{14}N nucleus circumvents many of the challenges seen with ^{15}N SSNMR and is discussed below in more detail [121].

In contrast to ^{15}N SSNMR, studies of the ^{19}F nucleus benefit from high sensitivity and selectivity for fluorinated drugs [105]. As a large number of drugs currently in development contain fluorine, ^{19}F-based SSNMR techniques have seen widespread use [123]. ^{19}F SSNMR spectra of amorphous dispersions to date have typically been obtained using CP-MAS techniques [119]. The sensitivity of ^{19}F SSNMR allows for its use as an efficient bulk detection method for small amounts of crystalline content. For example, Figure 4.8 shows the ^{19}F CP-MAS spectrum of a dispersion containing <2% of drug and also containing a small amount (approximately 10% relative to the overall drug content) of crystalline drug physically mixed with the dispersion. Despite the low concentrations involved, the crystalline content can be seen as a feature on the broad signal obtained from the amorphous dispersion. The use of filtration methods based on the different relaxation properties of the amorphous dispersion and crystalline phase can be exploited to enhance the signal of the crystalline phase. In Figure 4.8, the results of adding a 40 ms spin-lock pulse prior to the CP transfer are shown. This approach enhances the signal of the crystalline phase because it has a much longer ^{1}H spin–lattice relaxation time in the rotating frame ($T_{1\rho}$) than the amorphous phase [124]. In addition to filtering using $T_{1\rho}$ differences for either ^{1}H nuclei or a heteronucleus such as ^{19}F or ^{13}C, a number of approaches based on spin–lattice relaxation times (T_1) or spin–spin relaxation times (T_2) are also available and can be used to enhance the signal of crystalline phases within amorphous solid dispersions. If enhancement of the amorphous signal is desirable in the presence of a crystalline phase, a simple approach is to shorten the relaxation delay or use a saturation recovery pulse sequence to suppress signal from the crystalline phase because of the typically shorter ^{1}H T_1 observed for the amorphous phase.

2D SSNMR experiments allow for observation of interactions between nuclei and provide information on nuclei that are near to one another in space (with distances in the range of a few Å) and nuclei that are covalently bonded or experience a significant amount

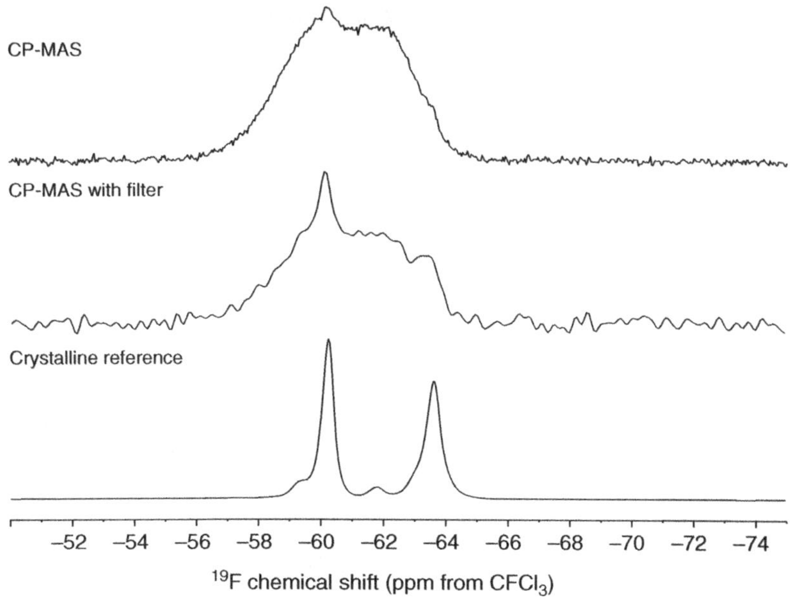

Figure 4.8. Detection of crystalline content by ^{19}F CP-MAS in a dispersion of a fluorinated drug containing a small amount of crystalline material introduced by physical blending. The top spectrum shows the centerband region of a conventional CP-MAS spectrum of the dispersion, while the middle spectrum shows a CP-MAS spectrum obtained using a 40 ms spin lock prior to the CP transfer on the ^1H channel, which enhances the signal from the crystalline fraction. The bottom spectrum shows a reference spectrum of the crystalline phase. Spectra were obtained at 11.7 T and 273 K.

of orbital overlap [95,125]. 2D SSNMR techniques are particularly valuable in studies of amorphous solid dispersions because they allow for direct, unambiguous detection of interactions between the drug and polymer as well as observation of structural effects such as hydrogen bonding, ionization state (for ionizable drugs, polymers, or additives), and molecular conformation [25,79,119,121,122]. Several 2D SSNMR pulse sequences that are particularly useful in the analysis of amorphous solid dispersions are summarized in Table 4.1. The experiments that allow for direct detection of drug–polymer interactions are generally based on through-space dipolar coupling interactions. Direct dipolar interactions between nuclei of interest in dispersions are typically limited to the 3–5 Å range because of the r^{-3} dependence of the dipolar coupling (where r is the internuclear distance) [95,96]. The heteronuclear correlation (HETCOR) class of pulse sequences are particularly useful because they correlate a heteronucleus, such as ^{13}C, ^{19}F, ^{23}Na, or ^{31}P, with ^1H nuclei directly through space or through a longer range process known as spin diffusion (see below) [119,126,127]. The use of ^1H nuclei in one of the spectral dimensions confers sufficient sensitivity on the experiment for it to be of practical use, while the use of a heteronucleus in the other spectral dimension enhances the specificity. In the study of dispersions, this increases the likelihood that, for example, a ^1H (or ^{13}C) resonance that can

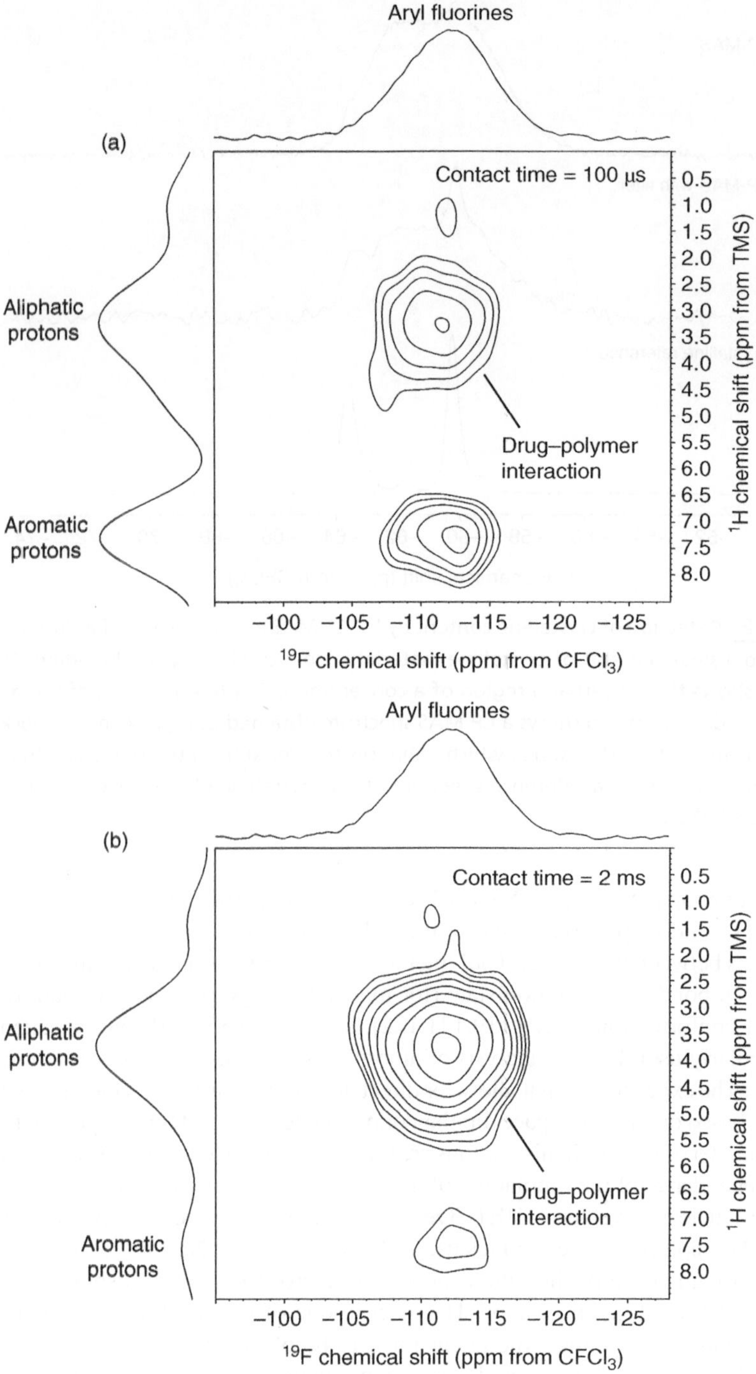
Aryl fluorines
(a)
Contact time = 100 μs
Aliphatic protons
Aromatic protons
Drug–polymer interaction
0.5
1.0
1.5
2.0
2.5
3.0
3.5
4.0
4.5
5.0
5.5
6.0
6.5
7.0
7.5
8.0
¹H chemical shift (ppm from TMS)
−100 −105 −110 −115 −120 −125
¹⁹F chemical shift (ppm from CFCl₃)
Aryl fluorines
(b)
Contact time = 2 ms
Aliphatic protons
Aromatic protons
Drug–polymer interaction
0.5
1.0
1.5
2.0
2.5
3.0
3.5
4.0
4.5
5.0
5.5
6.0
6.5
7.0
7.5
8.0
¹H chemical shift (ppm from TMS)
−100 −105 −110 −115 −120 −125
¹⁹F chemical shift (ppm from CFCl₃)

be definitely assigned to the drug (such as a set of aromatic protons) can be correlated to a ^{13}C (or ^{1}H) resonance assigned to the polymer [25,119]. When observed, such a correlation provides direct evidence of short-range interactions between the drug and the polymer and thus demonstrates miscibility.

The HETCOR-based approach to the study of drug–polymer miscibility can be demonstrated using $^{1}H-^{19}F$ CP-HETCOR experiments performed on a dispersion of diflunisal in PVP. Diflunisal lacks aliphatic protons, so its ^{1}H SSNMR spectrum has no resonances in the region between 1 and 5 ppm [119]. It also contains two fluorine atoms, which are not present in the polymer. A $^{1}H-^{19}F$ CP-HETCOR spectrum obtained with a short contact time of 100 μs of a 1% (w/w) dispersion of diflunisal in PVP is shown in Figure 4.9a. This spectrum shows a correlation between the drug and the polymer that can arise only if the two are in close proximity, that is, within several Å of each other, thus confirming that the two components are miscible at the molecular level [119]. The sensitivity of ^{19}F SSNMR allows for this analysis to be performed with only 1% (w/w) drug present in the sample in an experiment time of less than 2 h. Extending the contact time to 2 ms results in the spectrum shown in Figure 4.9b, which shows even greater intensity in the key correlation between the drug and the polymer. While this enhanced intensity could be caused by buildup of the direct dipolar interactions between ^{1}H and ^{19}F nuclei, the bulk of the enhancement is more likely caused by spin diffusion between ^{1}H nuclei prior to the CP transfer to ^{19}F nuclei. Spin diffusion results from multiple dipolar interactions between ^{1}H nuclei, and in the study of amorphous solid dispersions, spin diffusion extends the range over which correlations can be observed in 2D HETCOR and similar spectra [119,128,129].

The range of $^{1}H-^{1}H$ spin diffusion effects can be estimated from the maximum diffusion path length (L), which indicates the extent of $^{1}H-^{1}H$ spin diffusion that can occur within the ^{1}H spin lock during the contact time of a CP-HETCOR experiment (or during other periods where spin diffusion can occur). This quantity is related to the spin diffusion coefficient (D) and the time (t) over which the spin diffusion takes place (such as the CP contact time) by the following expression [130,131]:

$$\langle L^2 \rangle = 6Dt, \tag{4.8}$$

where the angle brackets denote an ensemble average. The spin diffusion coefficient (D) in Equation 4.8 may be estimated as

$$D = \langle l_0^2 \rangle / T_2, \tag{4.9}$$

Figure 4.9. $^{1}H-^{19}F$ CP-HETCOR spectra of a 1% (w/w) amorphous dispersion of diflunisal in PVP obtained using contact times of (a) 100 μs and (b) 2 ms [86]. The dispersion was prepared by solvent evaporation. The 1D ^{19}F CP-MAS spectrum (showing the centerband) is plotted along the F_2 (horizontal) axes, and a summation of columns is plotted along the F_1 (vertical) axes. The 2D spectra were obtained at 11.7 T and 273 K with $\nu_r = 12.5$ kHz using a Bruker Biospin Avance II+ spectrometer and a 4 mm HFX MAS probe. A total of 32 rows were acquired using 64 scans per row and a 3 s relaxation delay for a total experiment time of 1.7 h for each experiment.

where l_0 is the average distance between protons (typically about 0.1 nm in organic solids such as drug–polymer dispersions) and T_2 is typically in the range of several hundred 100 μs, which can be measured separately or estimated from spectral linewidths. Using Equation 4.8, L can be estimated for common contact times. For a 500 μs contact time, an L of about 5 Å is obtained, while for a 2 ms contact time an L of 11 Å is found [119]. This latter value is the source of the significantly enhanced drug–polymer correlation in Figure 4.9b, as it results in many neighboring protons within range to contribute to the intensity of this correlation.

The effects of spin diffusion in 2D ^{1}H–X CP-HETCOR spectra can be difficult to interpret in applications to amorphous solid dispersions because of the possibility that a particular 2D correlation can arise from (1) direct transfers and (2) direct transfers preceded by extensive spin diffusion [119,127]. The Lee–Goldburg CP (LGCP) method has been developed to suppress spin diffusion among the ^{1}H nuclei during the CP transfer, by applying a spin-lock pulse on the ^{1}H RF channel that also causes homonuclear ^{1}H dipolar decoupling [127]. The LGCP-HETCOR method (Table 4.1) is a 2D correlation method that utilizes LGCP transfers and can help assess the influence of spin diffusion on a particular 2D correlation. The effect of this pulse sequence is illustrated in Figure 4.10 for two amorphous solid dispersions of 6-(2-(5-chloro-2-(2,4-difluorobenyzloxy)phenyl)cyclopent-1-enyl)picolinic acid [86,109]. Figure 4.10 compares the extent of spin diffusion over a 2 ms contact period in 20% (w/w) dispersions of this compound in HPMCAS and HPC as observed by comparative ^{1}H$-^{19}$F CP-HETCOR and LGCP-HETCOR experiments [119]. The effects of spin diffusion are easily observed by comparing the spectra in Figure 4.10a because of the ability of LGCP to suppress spin diffusion. The correlation marked as arising from spin diffusion can then be easily assigned as shown. Furthermore, although the ^{1}H resolution in this system is less favorable than in the example in Figure 4.9, the level of spin diffusion observed using CP versus LGCP is consistent with the formation of a miscible solid solution [119]. In the corresponding spectra of the HPC dispersion in Figure 4.10b, noticeably less spin diffusion is observed despite a similar density of aliphatic protons in this polymer. This result can be interpreted as arising from a lesser degree of mixing between the drug and the polymer and is consistent with the formation of small domains of amorphous drug within the polymer.

The effect of miscibility on 2D CP-HETCOR spectra can be further illustrated by the example shown in Figure 4.11a. Here, a more concentrated 60% (w/w) dispersion of 6-(2-(5-chloro-2-(2,4-difluorobenyzloxy)phenyl)cyclopent-1-enyl)picolinic acid in PVP–VA is studied using a 2D ^{1}H$-^{19}$F CP-HETCOR experiment. In the dispersion prepared by hot melt mixing, clear correlations between the aryl fluoride groups from the drug and aliphatic groups either from the drug or more likely from PVP–VA (via spin diffusion) are observed. When a physical mixture of amorphous drug and PVP–VA is prepared by physical mixing and similarly analyzed as shown in Figure 4.11b, no such correlation is observed. The physical mixture began to recrystallize after 4 h (immediately after the ^{19}F SSNMR experiments shown in Figure 4.11b were complete). The physical mixture contains isolated domains of drug and polymer that are unable to undergo significant spin diffusion, leading to a lack of fluorine-to-aliphatic proton correlations. Even the intramolecular correlations to the small number of aliphatic

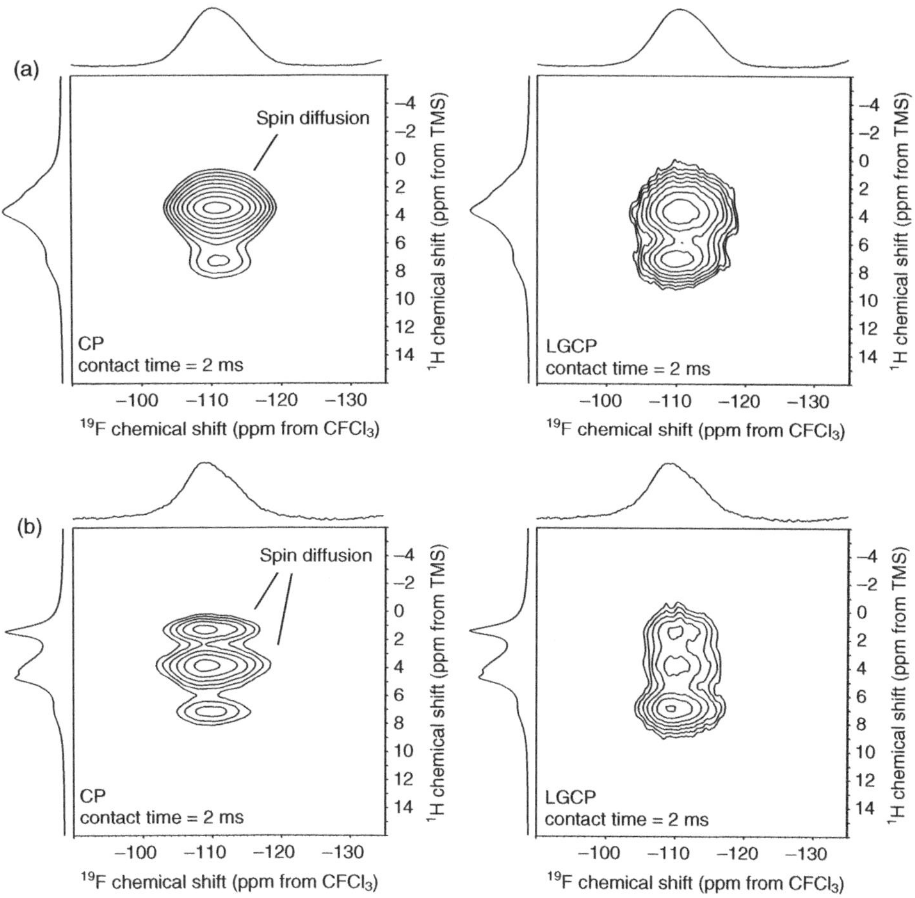

Figure 4.10. $^{1}H-^{19}F$ CP-HETCOR and corresponding LGCP-HETCOR spectra of 20% (w/w) dispersions of 6-(2-(5-chloro-2-(2,4-difluorobenyzloxy)phenyl)cyclopent-1-enyl)picolinic acid [86,109] in (a) HPMCAS and (b) HPC. Spectra were obtained with a 2 ms contact time using a Bruker Avance II+ spectrometer and a 4 mm HFX MAS probe. $^{1}H-^{1}H$ spin diffusion during the CP contact time prior to the $^{1}H-^{19}F$ transfer, while the use of LGCP suppresses $^{1}H-^{1}H$ spin diffusion and shows mainly the result of direct $^{1}H-^{19}F$ transfers. Spin diffusion from the drug to aliphatic polymer positions, denoted in each spectrum, significantly increases in the HPMCAS dispersion. This indicates closer contact between the polymer and the drug. 2D spectra were obtained at 11.7 T and 273 K with $\nu_r = 14$ kHz. The similarly obtained ^{19}F CP-MAS spectrum of each dispersion (showing the centerband) is plotted along the horizontal axes, and the ^{1}H DP-MAS spectrum obtained with $\nu_r = 35$ kHz using a 2.5 mm HFX probe is plotted along the vertical axes.

protons in the drug itself are minimal. Some intramolecular spin diffusion is observed in the crystalline form of this drug using a 2 ms contact time, as reported elsewhere [86]. The lack of a stronger correlation to aliphatic protons in Figure 4.11b is likely caused by mobility in the amorphous phase containing the drug, which has the effect of partially

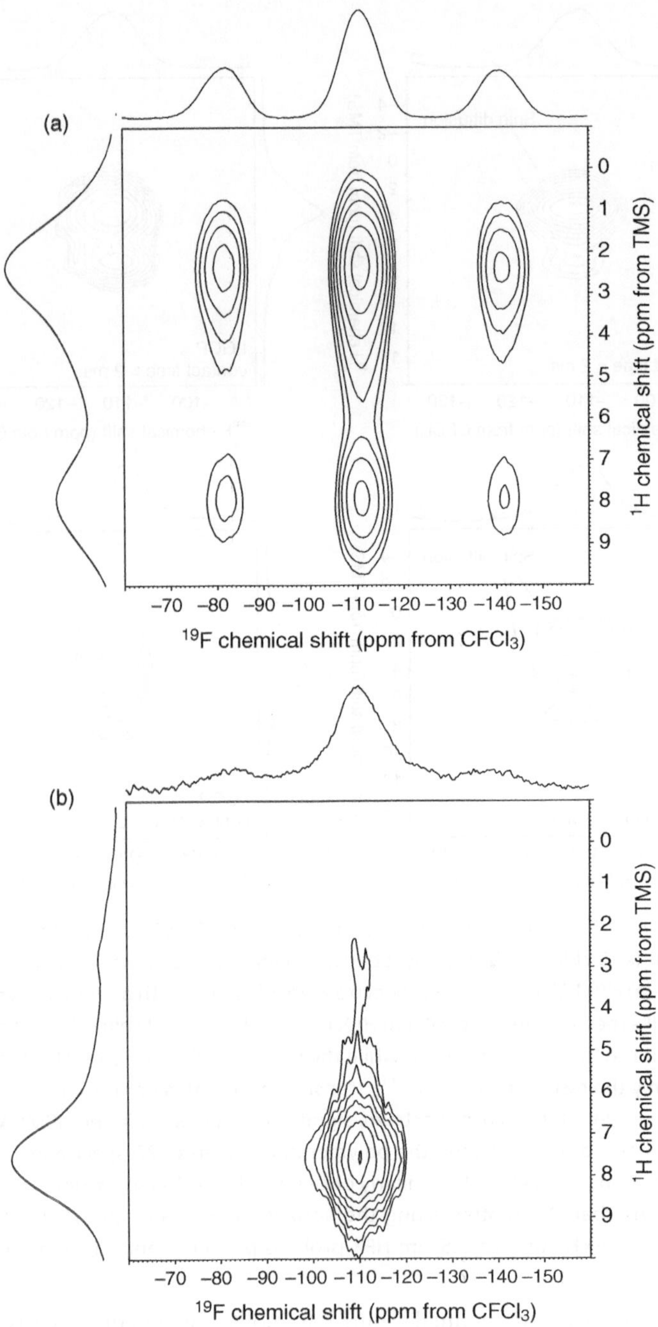
(a)
0 1 2 3 4 5 6 7 8 9
1H chemical shift (ppm from TMS)
−70 −80 −90 −100 −110 −120 −130 −140 −150
19F chemical shift (ppm from CFCl3)
(b)
0 1 2 3 4 5 6 7 8 9
1H chemical shift (ppm from TMS)
−70 −80 −90 −100 −110 −120 −130 −140 −150
19F chemical shift (ppm from CFCl3)

suppressing dipolar coupling and thus reduces CP. In Figure 4.11b, the spinning sidebands are also reduced in intensity compared with those in Figure 4.11a, and the spectrum is further broadened, which is also consistent with averaging of chemical shift anisotropy by molecular motion of the drug.

In addition to the SSNMR correlation methods that utilize heteronuclear dipolar coupling as their correlation mechanism, methods based on homonuclear dipolar coupling are also of use in some cases although limited ^{1}H resolution restricts their applicability [119,122]. The methods that have been applied to dispersions to date include ^{1}H double-quantum correlation spectroscopy using back-to-back excitation and reconversion pulses (DQ-BABA) and methods based on ^{1}H spin diffusion [119,122,132]. These methods allow for detection of interactions between the components of a dispersion in some cases and can be useful for assessment of miscibility. In addition, correlation methods based on through-bond *J*-coupling can be useful in spectral assignments for individual molecules in a dispersion, and in some cases may be able to probe strong intermolecular hydrogen bonding where significant orbital overlap and *J*-coupling occur [133].

2D SSNMR experiments have been applied to probe miscibility in a number of binary amorphous solid dispersion systems as well as more complex ternary and quaternary systems of pharmaceutical interest. These include dextran–trehalose, indomethacin–PVP, acetaminophen–PVP, acetaminophen with two methacrylate copolymers, telithromycin–PVP, voriconazole–PVP with sodium lauryl sulfate added as a surfactant, and tenoxicam:L-arginine–PVP wherein the L-arginine occurs in both mono-ionized and zwitterionized forms [25,119]. The SSNMR studies to date have largely focused on systems that contain modest drug concentrations (e.g., 10–50% (w/w)), but the SSNMR approach using a sensitive nucleus such as ^{19}F has been used to show miscibility on dispersions containing as low as 1% (w/w) of drug [86].

A recent addition to the catalog of 2D SSNMR methods useful for study of molecular association is the ^{1}H$-^{14}$N heteronuclear multiple-quantum coherence (HMQC) experiment that performs dipolar transfers using a rotational resonance approach [121]. This experiment utilizes the ^{14}N isotope, which has several advantages relative to the ^{15}N isotope [105,134]. The ^{1}H$-^{14}$N HMQC experiment generally performs best when ν_r is set to 45 kHz or above because of the need to suppress ^{1}H homonuclear coupling, and thus requires the use of MAS probes with rotor outer diameters in the range of 1.0–1.9 mm [121,135]. An initial application of the ^{1}H$-^{14}$N

Figure 4.11. Comparison of ^{1}H–^{19}F CP-HETCOR spectra of 6-(2-(5-chloro-2-(2,4-difluorobenyzloxy)phenyl)cyclopent-1-enyl)picolinic acid [86,109] in PVP–VA obtained from (a) a dispersion prepared by mixing a hot melt of the drug and PVP–VA and (b) a physical mixture of amorphous drug and polymer. The physical mixture of amorphous drug and polymer shown in (b) recrystallized rapidly in the rotor after this spectrum was obtained. Spectra were obtained using a 2 ms ramp CP contact time at 11.7 T and 273 K with ν_r = 14 kHz. The similarly obtained ^{19}F CP-MAS spectrum of each sample (showing the centerband and two sidebands) is plotted along the horizontal axes. A summation of columns is plotted along the F_1 (vertical) axes; because of the sensitivity of the physical mixture to recrystallization, ^{1}H DP-MAS spectra at a higher MAS rate could not be obtained.

HMQC experiment to a 50% (w/w) amorphous dispersion of acetaminophen in PVP demonstrated the effectiveness of this approach [121]. Although a direct, fully resolved correlation between a nitrogen resonance assigned to acetaminophen and protons assigned to PVP or vice versa could not be clearly observed because of spectral overlap in this particular dispersion, strong evidence of such a correlation was obtained by careful assignment of the $^{1}H-^{14}N$ HMQC spectrum and was further supported by changes in ^{1}H and ^{14}N chemical shifts [121]. As applications of this experiment are developed, the dipolar correlation information obtained should be augmented by valuable ^{14}N EFG and chemical shift parameters, given the sensitivity of nitrogen donors and acceptors to hydrogen bonding interactions in dispersions.

^{17}O SSNMR can provide insight into hydrogen bonding interactions in amorphous solid dispersions, and is particularly valuable in studies of oxygen hydrogen bond donor and acceptor groups such as carboxylic acids, esters, amides, alcohols, and phenols. In this role, it is complementary to ^{14}N and ^{15}N SSNMR experiments that are well suited to nitrogen hydrogen bond donors and acceptors. Like ^{14}N, ^{17}O is a quadrupolar nucleus that offers access to both EFG and chemical shift parameters. Unlike the ^{14}N nucleus, ^{17}O has a spin of 5/2, allowing direct observation of the quadrupolar central transition without the need for wide-line experiments or special indirect observation methods such as the aforementioned $^{1}H-^{14}N$ HMQC experiment. ^{17}O also requires synthetic labeling at oxygen sites of interest in order to facilitate observation because of its low (0.038%) natural abundance and the loss of sensitivity resulting from quadrupolar broadening [105]. ^{17}O SSNMR allows for direct study of hydrogen bonding effects and has been applied to detect hydrogen bonding differences in dispersions involving oxygen donor and acceptor sites that were much more difficult to assess using ^{1}H and ^{13}C SSNMR methods [79]. $^{1}H-^{17}O$ CP-HETCOR experiments can also be successfully performed on dispersions and provide specific information about proton environments in proximity to labeled oxygen sites [79].

The SSNMR methods described above have recently been extended to study an amorphous solid dispersion of ezetimibe in mesoporous silica [122]. Amorphous dispersions in silica systems can be characterized using a specially adapted set of SSNMR experiments, including ^{1}H spin diffusion and $^{1}H-^{29}Si$ CP-HETCOR experiments optimized for detection of interactions between silanol groups and the drug. In the case of a fluorinated drug such as ezetimibe, specialized experiments such as $^{19}F-^{29}Si$ CP can be used to directly detect interactions between the drug and the silica to which it is adsorbed [122]. Although not commonly thought of as a sensitive technique, the use of abundant isotopes such as ^{1}H, ^{19}F, and ^{31}P allows for SSNMR to achieve excellent sensitivity in a wide array of applications.

SSNMR methods that measure ^{1}H T_1 relaxation times are useful in studies of amorphous solid dispersions because of their ability to demonstrate phase dissociation [119,136]. The saturation recovery pulse sequence (Table 4.1) is commonly used to perform these measurements, and observation of ^{1}H T_1 relaxation via ^{13}C or ^{19}F heteronuclear detection is common [119,136]. This experiment is typically performed as a pseudo-2D experiment with 8–16 time increments, and exponential fitting, Laplace transforms or regularization, or other specialized fitting is used to extract or assess relaxation times [136]. In addition to measurements of ^{1}H T_1 relaxation, ^{1}H $T_{1\rho}$

relaxation measurements can also be useful in studies of phase heterogeneity and probe shorter range ^{1}H spin diffusion effects that occur during a ^{1}H spin-lock pulse [106]. A recent study observed both uniform ^{1}H T_1 and ^{1}H $T_{1\rho}$ relaxation times using ^{13}C detection for a series of amorphous solid dispersions between peaks assigned to the drug and peaks assigned to the polymer and other components, which was consistent with the molecular-scale mixing of the components observed by 2D SSNMR and DSC [25]. If a statistically significant difference in ^{1}H T_1 or ^{1}H $T_{1\rho}$ is obtained between ^{13}C resonances (or another heteronucleus) assigned to the polymer and those assigned to the drug, then the system is likely to be heterogeneous with respect to amorphous phases, which could result from an immiscible system, from the presence of a mixture of amorphous domains containing different drug/polymer ratios, or from a mesoporous dispersion that is fundamentally heterogeneous [122]. A uniform ^{1}H T_1 or ^{1}H $T_{1\rho}$ relaxation time is expected for a homogeneous phase because of strong ^{1}H spin diffusion effects at lower MAS rates, which equalize relaxation processes over longer timescales typically on the order of at least several milliseconds [136]. However, detection of a uniform ^{1}H T_1 or ^{1}H $T_{1\rho}$ relaxation time for a material does not provide the same level of proof of phase association as the 2D SSNMR experiments described above because of the possibility of degenerate (i.e., statistically indistinguishable) relaxation times between the phases. In the use of ^{1}H T_1 or ^{1}H $T_{1\rho}$ relaxation measurements on amorphous solid dispersions, it is important to note that the polymers in many dispersions have some permeability to paramagnetic O_2 gas and can be purged over time using the N_2 gas often used for spinning; as a result, it may be necessary to spin a rotor for a period of time to obtain consistent ^{1}H T_1 results (or otherwise purge the material of O_2 gas), or alternatively spin the sample using air as a bearing and drive gas instead of N_2. The use of ^{1}H T_1 or ^{1}H $T_{1\rho}$ relaxation time analysis for phase disassociation studies should be distinguished from their use to study mobility, which is discussed next.

Another major class of SSNMR experiments includes those useful for studying molecular mobility in solids. Information about molecular mobility over a wide range of timescales can be obtained using SSNMR methods. The nuclei accessed in SSNMR experiments undergo a variety of relaxation processes that can potentially report on mobility. SSNMR analysis of mobility in amorphous solid dispersions is therefore complementary to thermal analysis and dielectric spectroscopy techniques. Lineshape modeling can provide some information about specific dynamic motional models in solids, but in the typical case of an amorphous solid dispersion, the overall mobility of the solid is of interest. SSNMR relaxation methods are widely used to assess overall mobility in amorphous materials including dispersions, typically through measurements of ^{13}C or ^{19}F T_1 relaxation times [122,137,138]. Mobility in an amorphous phase measured by SSNMR relaxation time analysis can be related to the tendency to crystallize in some cases [137].

Measurements of ^{1}H $T_{1\rho}$ relaxation times provide a convenient measure of overall molecular mobility in the kHz amplitude regime in an amorphous solid dispersion, and thus are sensitive to the mobility increase that follows a glass transition. A study of amorphous solid dispersions containing 50% (w/w) nifedipine in PVP (exhibiting a single T_g with an onset at 67.1 °C by DSC) and 50% (w/w) indomethacin in PVP (exhibiting a single T_g with an onset at 71.6 °C by DSC) was made using variable-temperature ^{1}H $T_{1\rho}$ measurements to illustrate the ability of this approach to detect increases in motion below the T_g measured by

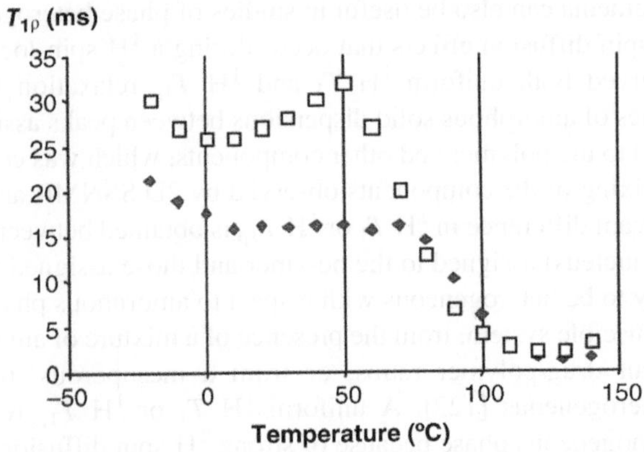

Figure 4.12. ^{1}H $T_{1\rho}$ values measured by SSNMR as a function of temperature for a 1:1 co-melt of nifedipine and PVP (open squares) and a 1:1 co-melt of indomethacin and PVP (closed diamonds). (Reproduced with permission from Ref. 140.)

DSC [139]. The results are shown in Figure 4.12, where a significant change in ^{1}H $T_{1\rho}$ times is observed near to the observed T_g for each system. The nifedipine–PVP dispersion shows significant increases in mobility starting at 50 °C, well below the measured T_g. Such a finding is consistent with the mobility expected above the Kauzmann temperature T_K. The ability to detect changes in mobility below the measured T_g (e.g., to assess relaxation behavior) may also be useful for studies of amorphous solid dispersions prepared by different processes. The pulse sequence used to measure ^{1}H $T_{1\rho}$ times involves a simple 90° RF pulse on the ^{1}H channel followed by a variable-duration spin-lock pulse and direct ^{1}H signal acquisition under moderate spinning rate MAS conditions. As with heteronuclear-detected experiments, a pseudo-2D data set is obtained, usually involving 8–16 spin-lock pulse time increments. The resulting exponential decay is fitted to obtain ^{1}H $T_{1\rho}$ times, and the experiment is repeated at different temperatures. Because of the sensitivity of ^{1}H detection, each ^{1}H $T_{1\rho}$ experiment can be completed in a matter of minutes, allowing a full temperature profile such as that shown in Figure 4.12 to be built in a reasonable time.

Finally, the use of ^{13}C as a detection nucleus for measurement of ^{1}H T_1, ^{1}H $T_{1\rho}$, ^{13}C T_1, or ^{13}C $T_{1\rho}$ relaxation times, which is usually seen in the literature to date, often results in long experimental durations because of the need to collect several hundred or thousand transients for each of 8–16 time points. Sensitivity is limited because the drug concentration in many dispersions of interest ranges from 10 to 50% (w/w). In some cases, more sensitive nuclei can be employed. For example, ^{19}F-detected ^{1}H T_1 measurements were employed for the efficient study of a dispersion containing 15% (w/w) voriconazole, while ^{13}C-detected ^{1}H T_1 measurements were used to separately obtain values for the high-concentration polymer [119]. Several different relaxation times of a solid dispersion of ezetimibe in mesoporous silica were observed via the ^{19}F nucleus, allowing for more sensitive experiments [122].

Dynamic nuclear polarization (DNP) using microwave irradiation of biradicals inserted into a sample is an emerging technique that can greatly enhance the sensitivity of SSNMR experiments for less sensitive nuclei [140]. In the past, DNP was applied to samples in a separate magnetic field outside of the NMR magnet, but more recently the approach has been streamlined by technical advances that allow for DNP to be performed within the NMR magnet. DNP has seen an initial successful application to an amorphous small organic molecule that is similar to an amorphous pharmaceutical system [141]. The results of a DNP-enhanced SSNMR experiment performed on an amorphous dispersion containing 16% (w/w) ezetimibe in mesoporous silica [122] are shown in Figure 4.13a. The DNP-enhanced spectrum is obtained using continuous microwave irradiation of an added radical from a 263 GHz continuous-wave gyrotron source that is coupled via a transmission line to a 3.2 mm MAS probe, with data acquired at a temperature of 100 K [142]. The rigid bis-TEMPO–bis-ketal (bTbK) biradical, where TEMPO is an acronym for (2,2,6,6-tetramethylpiperidin-1-yl)oxidanyl, was used for DNP enhancement in Figure 4.13a [143]. The biradical was soaked into the mesoporous solid dispersion using 1,1,2,2-tetrachloroethane (TCE), and the sample was analyzed with and without microwave irradiation to illustrate the signal enhancement as shown in Figure 4.13a. The strong enhancement allows for ^{13}C signals for ezetimibe to be observed in several minutes, when several hours are required for conventional methods [122]. The use of DNP enhancement in a typical polymeric amorphous solid dispersion is illustrated in Figure 4.13b, where spectra are shown for a dispersion of 30% (w/w) diflunisal prepared in HPMCAS by rapid solvent evaporation [79]. The biradical 1-(TEMPO-4-oxy)-3-(TEMPO-4-amino)propan-2-ol (TOTAPOL) was used for this experiment [143], with 16 mM of the biradical impregnated into the dispersion using deuterated dimethyl sulfoxide (d_6-DMSO) as a solvent. Again, signal enhancement is obtained for the spectrum obtained with continuous microwave irradiation. Although illustrated here for simple 1D ^{13}C CP-MAS spectra, the DNP approach is expected to be particularly useful in enhancing sensitivity for 2D SSNMR studies of amorphous solid dispersions as well as allowing increased access to 1D and possibly 2D spectra of less sensitive nuclei of interest such as ^{2}H, ^{15}N, or ^{17}O.

4.7 OTHER MOLECULAR SPECTROSCOPIC METHODS

In addition to SSNMR and vibrational spectroscopy, a number of other modes of molecular spectroscopy employed in pharmaceutical research also have applications to amorphous solid dispersions. Diffuse reflectance (DR) UV–visible spectroscopy measures electronic transitions in solids much as absorbance UV–visible spectroscopy does in solution [144]. DR UV–visible experiments on amorphous solid dispersions can be performed by fitting an accessory capable of collecting diffusely scattering light from a powder to a conventional UV–visible spectrometer [144]. DR UV–visible methods are particularly sensitive, since drugs often contain conjugated bonds that absorb UV radiation in the 200–300 nm range, while the polymers used to form dispersions often do not absorb as strongly in this region. Pharmaceutical molecules with a visible color offer even more specificity for the use of DR UV–visible spectroscopy. For example, amorphous solid dispersions of tenoxicam and L-arginine in PVP that exhibited a yellow color were readily characterized by DR UV–visible spectroscopy, and the resulting spectral differences were interpreted as ionization state changes arising from monoionization versus zwitterionization of

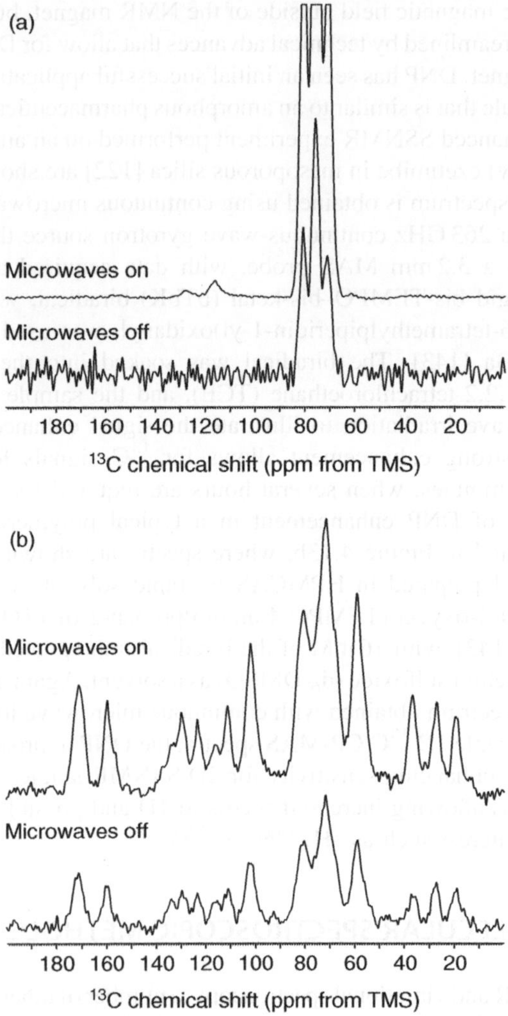

Figure 4.13. Examples of the enhancements possible using DNP in SSNMR analysis of amorphous dispersions. In (a), the DNP-enhanced ^{13}C CP-MAS spectrum of 16% (w/w) ezetimibe loaded onto mesoporous silica (Sylysia® SY350) is shown [122]. Spectra were obtained at $\nu_r = 12$ kHz. Each spectrum is the result of 64 scans requiring about 3 min. The use of a 3 s relaxation delay with continuous microwave irradiation (microwaves on) and 16 mM of bTbK radical leads to a significant DNP enhancement relative to the conventional CP-MAS spectrum (microwaves off). The enhancement allows detection of peaks in the 140 to 110 ppm region that are assigned to aromatic carbons in ezetimibe. Large signals from TCE solvent used to load the radical are observed between 85 and 65 ppm. Greater enhancements are possible using longer microwave irradiation periods. In (b), the DNP-enhanced ^{13}C CP-MAS spectrum of an amorphous dispersion containing 30% (w/w) of diflunisal in HPMCAS is shown [79]. Each spectrum required 16 scans and a 2 s relaxation delay was used. DNP enhancement was achieved using 16 mM of TOTAPOL radical impregnated using

tenoxicam [25]. Spectra measured by DR UV–visible spectroscopy can be reduced to color coordinates for assessment of the appearance of a batch using a set of values designed to correlate to human color perception [144]. Several different color coordinate systems are in use, including systems designed to be sensitive to the pale and off-white colors frequently encountered with pharmaceutical amorphous solid dispersions, allowing for straightforward quantitative appearance testing for quality control [144,145].

Fluorescence spectroscopy detects emitted fluorescence upon excitation using UV–visible radiation, and can be performed in several modes including emission scans with a constant excitation wavelength, excitation scans with a constant emission wavelength, synchronous scans of both monochromators, and total luminescence scans [146]. Solid-state fluorescence spectroscopy has potential in the study of amorphous solid dispersions, although only limited pharmaceutical applications have appeared to date primarily in studies of crystalline systems that undergo solvation phenomena [147]. Sampling of solid-state materials is typically accomplished using a front-facing sample holder offset at a slight angle (15–30°) relative to the excitation beam and reflected to the detector using a separate optical path than used for the normal 90° observation in solution-state fluorescence spectroscopy. The holder contains a small amount of packed solid material behind a quartz plate or similarly fluorescence-free, transparent material.

Solid-state fluorescence spectroscopy offers great sensitivity for applications in amorphous dispersions, but can suffer from limited specificity. In Figure 4.14a, the solid-state fluorescence emission spectra of amorphous dispersions of 30% (w/w) diflunisal in two polymers are shown [79]. Significant differences in intensity and emission maxima are observed between crystalline Form I and the dispersions in PVP and HPMCAS. The differences between the fluorescence spectra of the two dispersions are likely related to differences in the mobility of diflunisal in the glassy solid [147]. The overall lack of specificity limits the use of this technique for analysis of residual Form I content in either of the dispersions, however. This is not the case with the example shown in Figure 4.14b, where the fluorescence spectra of three dispersions of 1:2 tenoxicam:L-arginine in PVP are compared with the spectrum of crystalline Form III, because the latter exhibits much greater intensity and an emission maximum shift that would be useful for detection of residual Form III [25]. These dispersions were previously studied by DR UV–visible spectroscopy, as noted above. In Figure 4.14b, the solid-state fluorescence spectrum of Form III is seen to be distinctive relative to the spectra of the dispersions, with a Stokes-shifted emission maximum and a significant increase in intensity that is likely attributable to the zwitterionic state of tenoxicam in Form III [25]. The spectra of the dispersions show a more subtle shift in emission maximum compared with each other, as expected given the monoionic state of the drug, and also highlight the quantitative nature of fluorescence intensity versus drug concentration.

d_6-DMSO. Spectra were obtained at $\nu_r = 8$ kHz. A significant signal enhancement is again observed in the spectrum obtained with microwaves on. All spectra in (a) and (b) were obtained using a Bruker Avance III wide bore spectrometer operating at 9.4 T with a 263 GHz gyrotron and microwave transmission line, using a 3.2 mm MAS probe with sapphire rotors at a temperature of approximately 100 K. Other details of the system are described elsewhere [142].

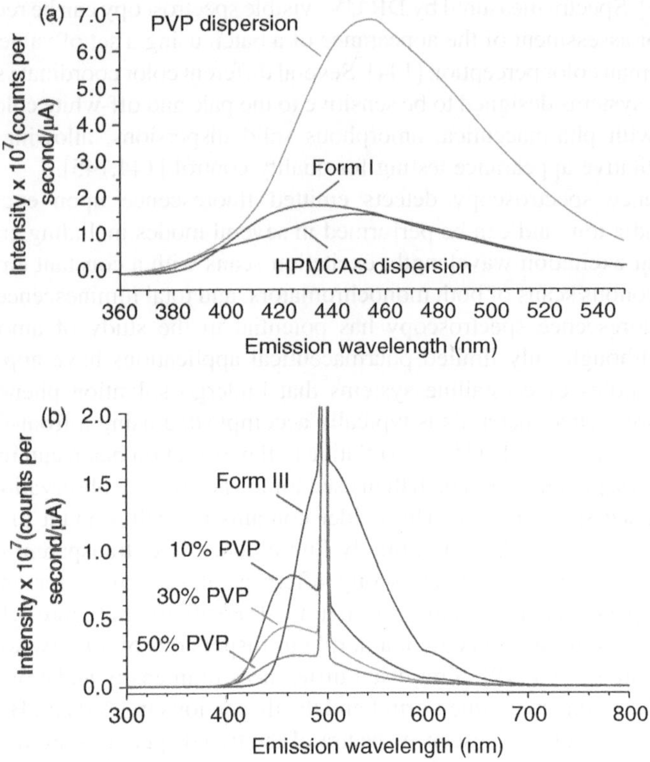

Figure 4.14. (a) Fluorescence emission spectra obtained with excitation at 340 nm of crystalline Form I of diflunisal, a 30% (w/w) amorphous dispersion of diflunisal in PVP, and a 30% (w/w) amorphous dispersion of diflunisal in HPMCAS [79]. (b) Fluorescence emission spectra obtained with excitation at 250 nm of crystalline Form III of tenoxicam and dispersions of tenoxicam:L-arginine containing 10, 30, and 50% (w/w) of PVP [25]. The vertical scale of the spectrum of Form III has been reduced by half. Spectra were obtained using a Horiba Jobin Yvon Fluorolog 3 spectrometer. The sharp feature at 500 nm arises from a second harmonic of the Rayleigh line. The excitation source was a 450 W Xe lamp, the signal detector was a photomultiplier tube, and the reference detector was a stabilized Si photodiode. Spectra were divided by the reference detector signal. A 1 nm slit width was used for both monochromators for the diflunisal materials, and a 2 nm slit width was used for the tenoxicam materials. Samples were packed in a solids holder covered with a quartz plate and analyzed using a front-facing geometry at an angle of 30°.

Fluorescence microscopy has also been applied to amorphous and semicrystalline solid dispersions designed for inhaled dosage forms [80]. In dosage forms where fluorescence differences between the drug and the polymer (or other carrier) are significant, fluorescence images can be used to assess the homogeneity of a dispersion in a very rapid manner (requiring only the short time needed to sample the image at video rates). Epifluorescence arrangements are often used for fluorescence imaging wherein fluorescence emission is selected using a filter that blocks wavelengths below 350 nm, while an excitation filter that selects wavelengths between 250 and 350 nm is also employed [146]. A conventional Hg lamp can be used as an excitation source. This technique can be used in conjunction with polarized light microscopic analysis for detection of small amounts of crystalline content using the same instrument, and can be further combined with confocal Raman microscopy.

4.8 X-RAY DIFFRACTOMETRY

PXRD techniques measure the diffraction of X-rays projected onto a solid sample [148]. The PXRD measurement is normally made in either reflection mode (known as the Bragg–Brentano geometry) or transmission mode (known as the Debye–Scherrer geometry) [148]. The lack of long-range order in amorphous solid dispersions limits the use of conventional applications of PXRD measurements toward crystalline phase identification and mixture analysis, which are widely used in other pharmaceutical applications to characterize crystalline drug substance polymorphic forms, salts, cocrystals, and dosage forms containing these species. PXRD analysis of a fully amorphous material generally results in broad diffuse scattering signals as opposed to sharp Bragg reflections. For example, Figure 4.15a shows the PXRD patterns obtained for three amorphous solid dispersions containing 20% (w/w) 6-(2-(5-chloro-2-(2,4-difluorobenyzloxy)phenyl)cyclopent-1-enyl) picolinic acid in HPMCAS, PVP, and HPC. Sharp Bragg peaks are absent from these patterns, which consist entirely of diffuse scattering. Nevertheless, at present PXRD provides at least two key results in studies of amorphous solid dispersions. First, PXRD methods are used to examine residual crystalline content in dispersions after manufacture or during stability studies. In this role, PXRD is a key tool for assessment of the potential for recrystallization and consequent loss of performance of a dispersion. Because it operates on bulk powders, PXRD is highly useful (e.g., in comparison with microscopic methods) for the overall quantitative analysis of crystalline content in a batch of a dispersion. The second key role that PXRD is beginning to fill more widely is in the assessment of drug–polymer miscibility using different approaches to modeling of the diffuse scattering pattern, such as those shown in Figure 4.15a.

A number of examples of the use of PXRD in the analysis of residual crystalline content in amorphous dispersions have been presented [3,149,150]. Typical detection limits for crystalline content are in the 1–5% (w/w) range using reflection methods. PXRD is widely used for quantitative and qualitative assessment of crystallinity when screening for amorphous solid dispersions as the drug content is varied and different polymers are explored. For example, in a study of the rate of recrystallization of felodipine from PVP and HPMCAS, PXRD was used to quantify the amount of drug that had recrystallized from dispersions containing 50–90% drug [150]. The drug was quantified at levels as low as 1.1%, and the quantitative analysis was used to assess the rate of recrystallization as dispersions were exposed to different RH levels. The quantitative analysis was performed using a PLS model that was built using calibration standards made by spiking crystalline felodipine into actual dispersions at different levels and cryomilling the physical mixtures to ensure homogeneity. This approach provides realistic calibration standards but can in some cases promote

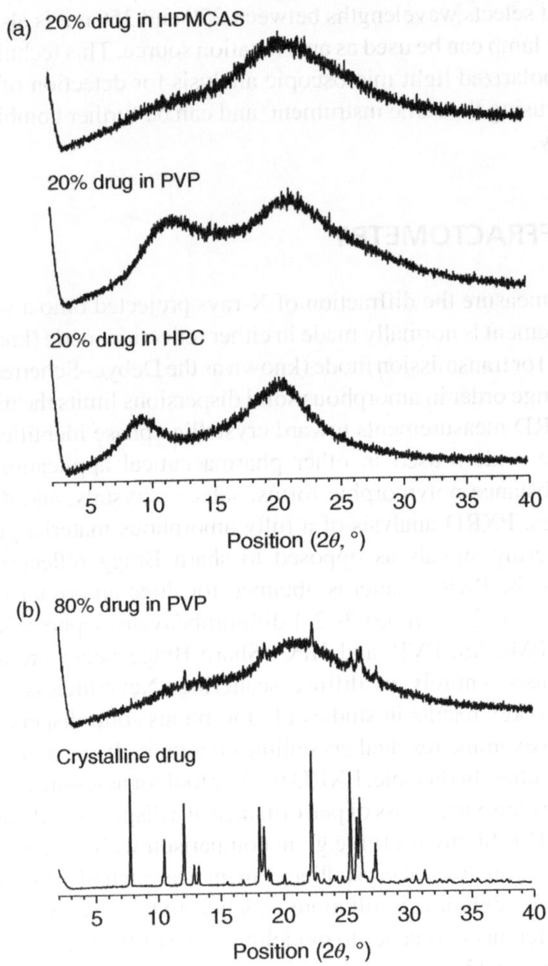

Figure 4.15. (a) Reflection PXRD patterns of amorphous dispersions consisting of 20% (w/w) 6-(2-(5-chloro-2-(2,4-difluorobenyzloxy)phenyl)cyclopent-1-enyl)picolinic acid in HPMCAS, PVP, and HPC, prepared by rapid evaporation from acetone/methanol solutions [86,109]. The patterns are illustrative of differences in diffuse scattering often observed between different amorphous dispersions. (b) Reflection PXRD pattern of an 80% (w/w) dispersion of the same drug in PVP, showing the presence of small Bragg peaks indicative of a small amount of crystalline content in comparison with the reference PXRD pattern of a stable crystalline form of this compound. PXRD patterns were collected using an X'Pert Pro diffractometer equipped with a real-time multistrip X'Celerator detector (Panalytical, Eindhoven, The Netherlands). The sample was scanned in continuous mode from 2° to 40° with a 2θ step size of 0.0167°, using Cu Kα radiation (1.54 Å) with a generator power and current of 40 kV and 40 mA, respectively.

recrystallization of the drug by providing seeds. Alternative methods such as SSNMR, which can study a bulk sample without the need for homogenization (e.g., by preparing a standard that fits entirely into a rotor but is not physically blended), can be used to check the standards before and after mixing to ensure recrystallization has not occurred. PXRD method optimization approaches, such as confining the scan range to a small 2θ range, can be used to further improve the sensitivity of a method for residual crystalline content if PXRD pattern of the form that can potentially crystallize during storage is known. PXRD methods can thus be used for quality control testing, including the release of batches after manufacture and the monitoring of batches for the appearance of crystalline content during stability studies. If multiple forms could crystallize during storage, or unknown forms might also appear, a full range PXRD scan from 5° to 40° 2θ is generally necessary at the cost of sensitivity. In Figure 4.15b, an example is shown for an 80% (w/w) dispersion of 6-(2-(5-chloro-2-(2,4-difluorobenyzloxy)phenyl)cyclopent-1-enyl) picolinic acid in PVP, where a small amount of recrystallization to a stable crystalline form has occurred [86]. This use of PXRD in the characterization of amorphous solid dispersions is extremely useful when screening dispersions produced in initial experiments and for assessing their stability, and is consequently widely used at present.

The second major use of PXRD in the characterization of amorphous solid dispersions is directed toward the characterization of miscibility and structure in the amorphous phase. Atomic pair distribution function (PDF) analysis is currently one of the most important tools used for analysis in these situations [151–153]. The PDF approach differs from conventional approaches to diffraction analysis. In SCXRD, detailed information about the structure of a crystalline phase is obtained from analysis of Bragg reflections in reciprocal space. In contrast, the PDF approach seeks to analyze the diffuse scattering component of the PXRD signal, which contains more information when amorphous materials are analyzed. Peaks in a PDF PXRD analysis correspond to the presence of pairs of atoms in the amorphous solid, and the position of the peak is related to the distance between the atoms [151–154]. The use of PDF PXRD can require extremely intense, monochromatic X-ray sources, which can in some cases limit its use.

The experimental PDF can be obtained by the sine FT of the experimentally determined total scattering structure function [155]. The calculated PDF $g(r)$ for a structure with atom pairs separated by a distance r is defined as [26,155]

$$g(r) = 4\pi r^2[\rho(r) - \rho_0], \tag{4.10}$$

where ρ_0 is the average atom number density of the structure and $\rho(r)$ is the atom pair density (or microscopic pair density) for X-ray scattering:

$$\rho(r) = \frac{1}{4\pi r^2}\sum_{pq}\frac{f_p f_q}{|\langle f\rangle|^2}\delta(r - r_{pq}). \tag{4.11}$$

In Equation 4.11, f_p and f_q are individual atomic form factors, $\langle f\rangle$ is the mean atomic form factor for the structure, r_{pq} is an atom pair distance, and δ represents a function that is infinity at zero and is zero elsewhere. Computation of a PDF from an atomistic structure can be performed for comparison with experimental data [156].

Although only recently introduced, PDF analysis has already been applied to probe miscibility in several amorphous solid dispersions of pharmaceutical interest, including dextran–trehalose, dextran–PVP, indomethacin–PVP, felodipine–PVP, nifedipine–PVP, ketoconazole–PVP, and felodipine–PAA [26,157]. The primary utility of PDF PXRD analysis in miscibility studies is the same as that for SSNMR, in that it provides another means besides DSC to demonstrate that an amorphous phase is homogeneously mixed. Like SSNMR, PDF PXRD can address situations where DSC does not provide sufficient clarity about the results, such as cases where two T_g values cannot be resolved or a T_g event cannot be detected at all [26].

In Figure 4.16a, the PDF PXRD analysis of a 50% (w/w) dispersion of dextran in PVP is shown [26]. MDSC analysis revealed two distinct T_g values for this system, which is

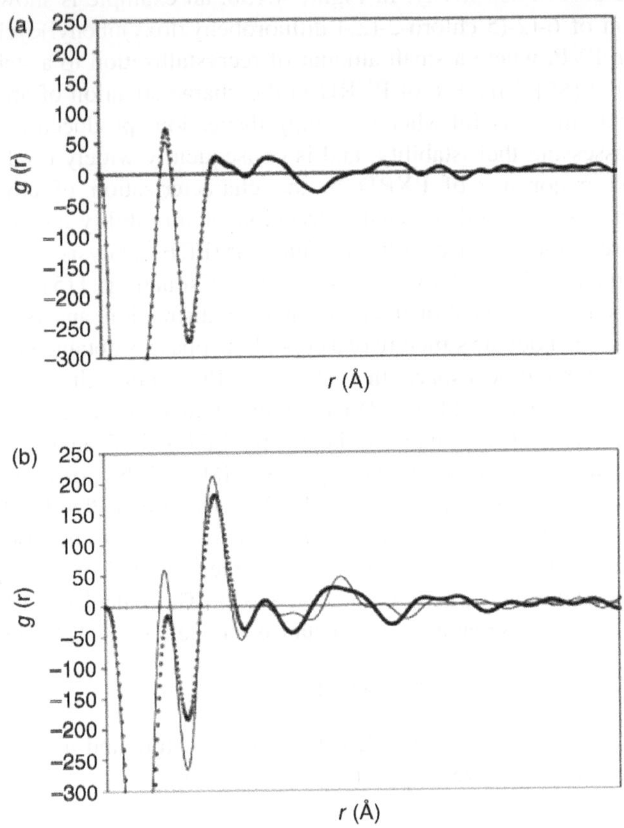

Figure 4.16. (a) PDF PXRD analysis of a 50% (w/w) dispersion of dextran in PVP. This system exhibits two independent T_g values by DSC, and the PDF analysis (black line) matches the experimental PXRD data (circles). The PDF is calculated using the separate amorphous reference materials, and hence the result for the 50% (w/w) dispersion of dextran in PVP indicates that this system is phase separated. (b) PDF PXRD analysis of a 50% (w/w) dispersion of indomethacin in PVP. This system exhibits a single T_g by DSC analysis. (Reproduced with permission from Ref. 26. Copyright 2008, Wiley-Liss, Inc. and the American Pharmacists Association.)

suggestive of phase separation with domain sizes on the order of 30 nm or greater [26]. The PDF-analyzed PXRD data (circles) of the dispersion match well with PDF PXRD results (black line) calculated using the separate amorphous reference materials. The agreement between the experimental and calculated PDF PXRD patterns is therefore indicative of a phase-separated system. In Figure 4.16b, PDF PXRD analysis of a 50% (w/w) dispersion of indomethacin in PVP is shown. In contrast to the dextran–PVP dispersion, this system exhibits a single T_g by DSC analysis, and the PDF PXRD analysis of the dispersion does not agree well with the patterns calculated from the separate amorphous reference materials [26]. PDF analysis can also be applied to neutron and electron diffraction data, although no applications of these alternative techniques to pharmaceutical amorphous solid dispersions have yet appeared in the literature [151]. As neutrons scatter from atomic nuclei (as opposed to X-rays, which scatter from electrons), they can offer an enhanced view of structure in many amorphous materials, but the limited accessibility of instrumentation prevents their wider usage.

Two approaches to multivariate analysis of PXRD data from amorphous dispersions have recently been presented [158]. The first approach is based on the use of a pure curve resolution method to qualitatively assess drug–polymer miscibility. The second method, based on alternate least squares analysis, is then used to quantify the degree of miscibility by determining the nearest-neighbor coordination number for the dispersion, which can provide an indication of the physical stability of the dispersion [158]. These multivariate methods provide another opportunity for the use of PXRD in studies of amorphous solid dispersions.

4.9 MICROSCOPIC AND SURFACE ANALYSIS METHODS

In addition to IR, Raman, and other molecular spectroscopy-based microscopy methods discussed above, several additional techniques based on microscopy and surface analysis are highly useful in the characterization of amorphous solid dispersions. Optical polarized light microscopy (PLM) is one of the best-known and most useful methods for detecting small amounts of crystalline materials through particle habit and birefringence [159,160]. Most crystalline pharmaceutical substances are birefringent, in that a single crystal will exhibit a different refractive index along each optical axis, which allows for rapid detection using PLM. Like MDSC, PLM is extensively applied in initial screening studies of dispersions, particularly on small evaporated films prepared on microscope slides. Hot-stage microscopy (HSM) can also be used to observe crystallization within dispersions at higher temperatures [159,160].

Scanning electron microscopy (SEM) is also widely used in the characterization of amorphous solid dispersions. SEM makes use of a monochromatic electron beam to probe the surface and near-surface areas of materials [160,161]. SEM analysis of amorphous solid dispersions can be used to probe particle morphology after dispersion formation by techniques such as spray drying, electrospinning, or milling of hot melt extrusion fragments. Images can be formed using both secondary electron (SE) and backscattered electron (BSE) detection, although the latter is often preferred for morphology studies because of its superior sensitivity to sample topology. In typical

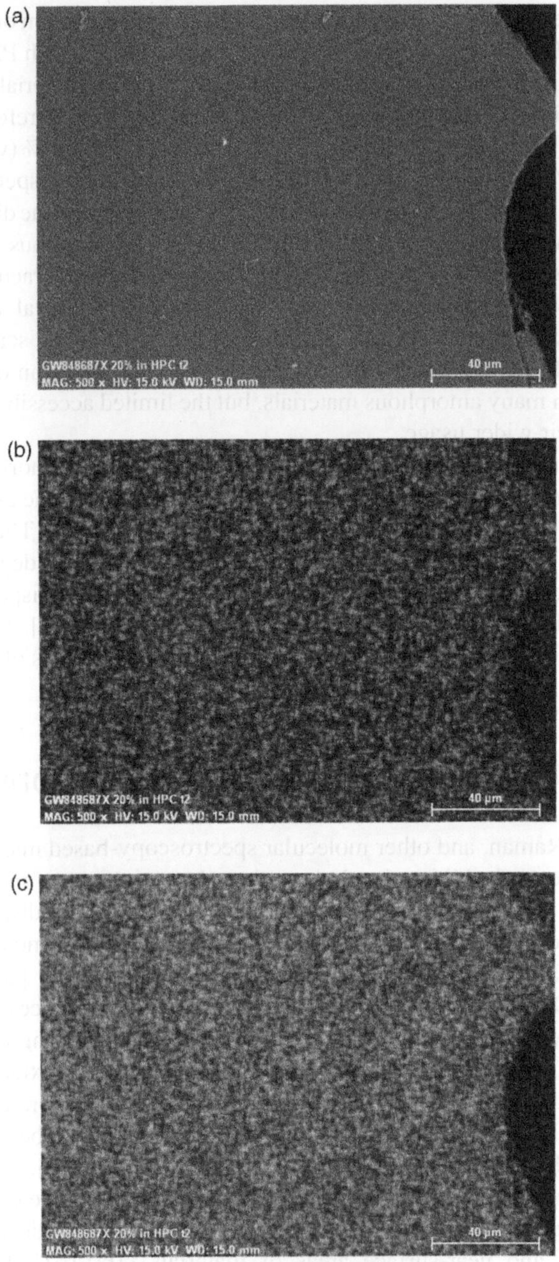

Figure 4.17. SEM–EDS mapping of a particle of an amorphous solid dispersion containing 20% (w/w) of 6-(2-(5-chloro-2-(2,4-difluorobenyzloxy)phenyl)cyclopent-1-enyl)picolinic acid in HPC [86,109]. Images were obtained using a Hitachi 3500 SEM equipped with a Bruker AXS Quantax Xflash AXS SDD energy-dispersive X-ray detector. In (a), a BSE topological image is shown. EDS maps of X-ray emission for (b) chlorine, (c) oxygen, and (d) fluorine are shown. More intense false-color map points in (b), (c), and (d) indicate a higher concentration of each element.

(d)

Figure 4.17. (*Continued*)

studies, SEM is used to image particle morphology of amorphous solid dispersions produced by methods such as spray drying [25]. Energy-dispersive X-ray spectroscopy (EDS) is often combined with SEM analysis and provides elemental information from the area probed by the electron beam. EDS measures X-ray emission from the inner shells of atoms that have been ionized by the SEM beam. Recent developments in silicon-drifted detector (SDD) technology have increased the speed of elemental mapping, allowing for efficient elemental area maps to be made over time periods of minutes to hours. X-ray photons are able to escape from depths of several μm within the sample upon excitation with a high-energy electron beam, so the EDS technique is considered to be sensitive to near-surface effects. Figure 4.17 shows elemental maps of a fractured piece of an evaporated film of a 20% (w/w) dispersion of 6-(2-(5-chloro-2-(2,4-difluorobenyzloxy)phenyl)cyclopent-1-enyl)picolinic acid in HPC. A generally even distribution of chlorine and fluorine (present only in the drug) and oxygen (present in both the drug and the polymer) is observed, and is consistent with a homogeneous dispersion at the spatial resolution limit of this technique of several μm. In addition to SEM, techniques such as transmission electron microscopy (TEM), which can provide alternative imaging modes as well as detection of electron diffraction from crystalline materials, can also be usefully applied to studies of amorphous solid dispersions [162].

X-ray photoelectronic spectroscopy (XPS) measures the energy of photoelectrons emitted from a material struck by a focused X-ray beam [163]. As photoelectrons can only escape from the first few nm of a solid material, XPS is a surface analysis technique and is capable of unique analyses of interfacial phenomena (in contrast to SEM–EDS, where X-rays can escape from much deeper regions of a solid). The XPS spectrum is specific for binding energies of most elements of interest in organic materials, and contains chemical shift information that can distinguish different environments for a particular element [163]. XPS has also been widely applied in chemical and elemental imaging applications [163]. To date, XPS has only seen limited use in the study of amorphous solid dispersions of pharmaceuticals. In an early study, XPS was used to probe dispersions of griseofulvin in crospovidone and in particular to study the loading mechanism when crospovidone particles were suspended in predissolved griseofulvin solutions [164]. XPS was found to be useful for imaging the distribution and homogeneity of griseofulvin loaded into particles of crospovidone with different morphology and also confirmed a diffusion mechanism for the loading process. More recently, XPS been applied in investigations of solvent interactions with amorphous solid dispersions of several drugs in HPMC and PVP to investigate a possible correlation between contact angle measurements, XPS results, and the results of other measurements including magnetic resonance imaging (MRI) techniques [165,166]. XPS instruments are available in an increasing number of laboratories, and with decreasing cost and technical advances in X-ray source designs and detectors, increased application of XPS in pharmaceutical applications is likely to occur.

A novel method based on second-harmonic generation (SHG) in intense light fields has been demonstrated as a more sensitive probe of crystal formation in amorphous materials than conventional optical microscopic techniques [167]. The method is known as second-order nonlinear optical imaging of chiral crystals (SONICC). In SHG, multiple photons are absorbed from an intense beam and are converted by a crystal into photons with twice the frequency [168]. In the initial study, SONICC was sensitive to the detection of small amounts of recrystallization from amorphous griseofulvin, and should have similarly sensitive applications to amorphous solid dispersions as the technique is further developed [167].

4.10 OTHER EMERGING ANALYTICAL METHODS

Rapid developments have occurred in the field of analytical chemistry, particularly in areas such as scanning probe microscopy, novel forms of spectroscopy making use of high-intensity radiation (such as to enable observation of nonlinear phenomena), and methods that utilize different radiation and particle sources. Some of these methods have already been demonstrated in initial applications to amorphous solid dispersions. For example, a study using nanoindentation was performed on dispersion films of clotrimazole in PVP–VA to allow for direct measurements of the mechanical properties of the solid, which are important in amorphous solid dispersion manufacturing because they can affect downstream processes such as particle size reduction (e.g., via milling), filling into capsules, and/or compression into tablets [169]. The dispersions were prepared as films with drug loading ranging from 0 to 100% (w/w). Interestingly, the addition of clotrimazole to the polymer reduced the glass transition temperature as observed by DSC, but the hardness, reduced elastic modulus, and storage modulus as measured by nanoindentation were found to increase to values greater than those of either pure component up to drug loadings of approximately 60% (w/w). Above this point, addition of clotrimazole to the system resulted in decreased hardness and moduli. This suggests a complex interplay between structural composition and properties that can be of significance in choosing the best formulation for manufacturing. Nanoindentation is also increasingly applied in studies of crystalline materials, again to measure elastic modulus and the onset of plastic deformation on cleaved surfaces, and may be useful in studies where amorphous solid dispersions are coformulated with crystalline excipients or when crystalline nanosuspensions form within an otherwise amorphous dispersion [170].

4.11 COMPUTATIONAL MODELS

Studies of crystalline pharmaceutical systems have benefitted greatly from computer simulations that primarily make use of density functional theory and force-field calculations. These models can assist with interpretation of data by allowing calculation of observables such as NMR spectral parameters, vibrational frequencies, optical properties, X-ray scattering, and bulk physical properties for comparison with experimental characterization data. Interpretation of the data obtained from many of the characterization techniques described above can similarly benefit from the availability of computational models. However, the modeling of amorphous systems presents additional challenges beyond those encountered in the crystalline state because of the lack of a crystal structure to use as input for calculations. Instead, more complex models of the amorphous state must be built. Periodic models using large cubic cells that contain initially random distributions of drug and polymer are often used to provide a starting point for structures of amorphous solid dispersions, which can then be progressed through time using a molecular dynamics (MD) calculation to create a set of models that can be used to obtain probabilistic information about structural motifs and properties in dispersions. Several recent studies have shown the potential of force-field calculations to model the properties of amorphous dispersions and closely related systems using periodic boundary condition models [171–174]. Atomistic force field-based MD calculations are generally preferred at the present time because of their balance of accuracy and performance in handling the large systems required for modeling the properties of amorphous dispersions relative to the more accurate but expensive quantum MD calculations (such as those based on density functional theory) or the faster but less accurate coarse-grained MD calculations. Although the present focus in the literature has largely been upon MD calculations, other techniques such as Monte Carlo calculations are also of importance [175].

An illustration of the structural model used in a periodic boundary condition MD calculation for a drug–polymer amorphous solid dispersion is given in Figure 4.18a, which represents a single initial frame in a larger ensemble of frames showing the evolution of the amorphous solid dispersion. To prepare the model for a MD simulation using the Dreiding force field and a constant volume and temperature (*NVT*) ensemble, two preparatory calculations were performed using the microcanonical constant volume and energy (*NVE*) ensemble (to stabilize the energy and relax the molecular structures) and constant pressure and temperature (*NPT*) ensemble (to allow the large cubic cell to relax and obtain a realistic density) [175]. A 200 ps *NVT* MD simulation was then executed after the preparation steps for collection of structural data for analysis. A common data analysis method is to utilize a PDF analysis of the MD trajectories as shown in Figure 4.18b for selected interatomic distances. In Figure 4.18b, two distances involving the drug and the polymer are compared. The PDF (also referred to as a radial distribution function (RDF)) involving the distance between the hydrogen donor in the

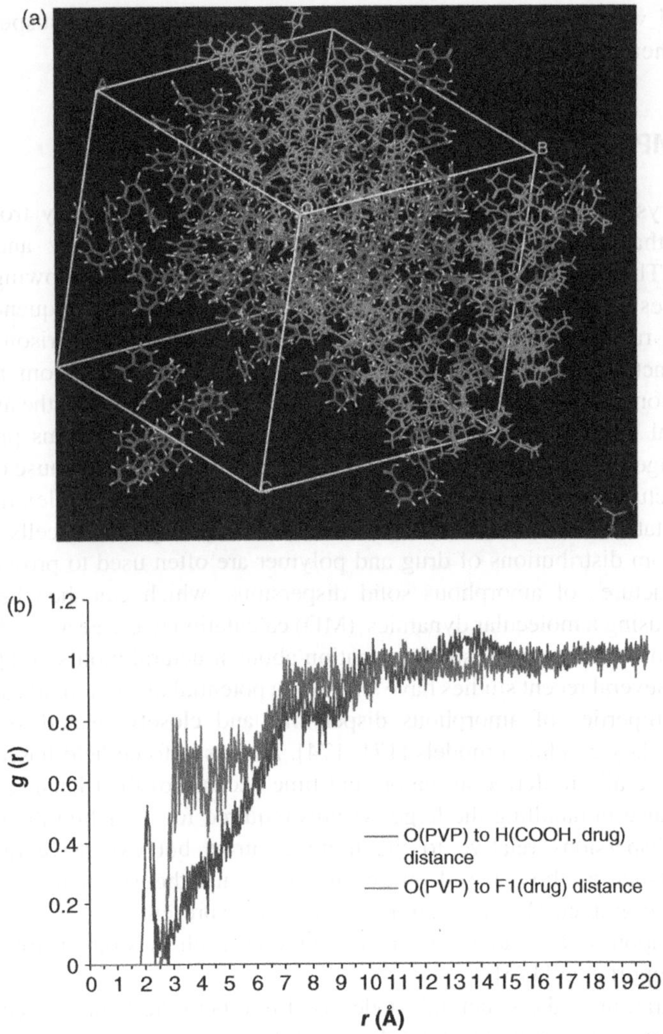

Figure 4.18. (a) View of a $40.13 Å^3$ amorphous cell calculated for an amorphous dispersion of 60% (w/w) 6-(2-(5-chloro-2-(2,4-difluorobenyzloxy)phenyl)cyclopent-1-enyl)picolinic acid in PVP. The drug is highlighted as a bright wireframe (and also appears in orange color in the online version) to distinguish it from the polymer. The structure was obtained after simulation using the *NVE* and *NPT* ensembles with the Dreiding force field to equilibrate energy and density. A cell of this type can be subjected to additional simulation to predict properties such as heat capacity and also provide distributions of structural parameters such as interatomic distances and PDFs. (b) An additional 200 ps *NVT* calculation with a 1.0 fs time step was performed to obtain 2000 frames, from which the PDFs shown were calculated. The PDFs are shown for two selected interatomic distances (see the text). Calculations were performed using the Amorphous Cell and Forcite software packages in Materials Studio version 6.0 (Accelrys, San Diego, CA, USA).

carboxylic acid group and the carbonyl oxygen in PVP (a hydrogen bond acceptor) shows evidence of a short ~2 Å contact consistent with some degree of hydrogen bonding occurring in the simulation. In contrast, the distance between a fluorine atom in the drug and the carbonyl oxygen in PVP shows no evidence of a short interaction. It is important to note that the choice of force-field methods used in these calculations varies in published studies and difference in force field, charging schemes, cutoffs, and the use of summation schemes such as the Ewald method can affect the results [175]. The results of MD calculations can be combined with the analytical techniques discussed previously to enhance the interpretation of results, particularly for techniques such as IR, Raman, and SSNMR spectroscopy (e.g., by modeling hydrogen bonding and other interactions of interest) and for diffraction techniques such as PXRD by assisting with models for PDF analysis [158].

The estimation of important properties such as T_g can also be accomplished computationally by MD methods [172,174]. For example, in a recent MD study of amorphous sucrose with varying amounts of water added as a plasticizer, T_g values of 367, 352, and 343 K were obtained for amorphous sucrose models containing 0, 3, and 5% (w/w) water, respectively [174]. The MD calculation was in reasonable agreement with experimental T_g values and also correctly predicted the influence of the plasticization effect of water on T_g [174]. In this study, MD simulations were performed using the COMPASS force field and the *NPT* ensemble, and the specific volumes of amorphous cells were computed in the temperature range of 265–440 K. The T_g was then defined from a characteristic "kink" obtained from volume–temperature plots using regression analysis. In addition to the analysis of T_g, PDF analysis of the MD simulation structures identified likely hydrogen bonding effects between sucrose hydroxyl oxygens and water oxygens [174].

Simulation methods may also allow investigation of molecular mobility at slow timescales that may be difficult to measure by the characterization techniques described here. For example, it is possible to relate MD simulations of polymers to nuclear spin relaxation measurements obtained by SSNMR, particularly when combined with field cycling methods, which has great potential for obtaining detailed insight into molecular mobility in amorphous solid dispersions [176]. The combination of MD simulations with dielectric relaxation methods and other techniques may also provide useful interpretative support for experimental findings.

4.12 CONCLUSIONS

Effective characterization plays a critical role in the development of pharmaceutical amorphous solid dispersions. The complexities of the amorphous state, the potential for heterogeneity over a wide range of spatial dimensions, and the multicomponent nature of these materials all present unique challenges for analytical characterization. A number of established techniques can be applied to obtain critical information about amorphous solid dispersions. Over the past five years in particular, significant progress has been made in adapting recently developed analytical techniques for the role of characterization of amorphous solid dispersions. This chapter has attempted to summarize both the basic methods widely used in the characterization of amorphous solid dispersions and many of the recently developed approaches that should continue to lend insight into the structure and performance of these systems.

ACKNOWLEDGMENTS

The author thanks Dr. Shane Pawsey (Bruker Biospin Corp.), Dr. Jochem Struppe (Bruker Biospin Corp.), Dr. Mark Strohmeier (Vertex Pharmaceuticals, Inc.), Dr. Robert A. Carlton (GlaxoSmithKline plc), Dr. Verne Hebard (Bruker Optics Corp.), Dr. Eunah Lee (Horiba Instruments, Inc.), and Dr. Glenn R. Williams (Rigaku Corp.) for providing data, assisting with experiments, and for helpful discussions.

REFERENCES

1. Threllfall, T.L. (1995) Analysis of organic polymorphs: a review. *Analyst*, 120:2435–2460.
2. Janssens, S. and Van den Mooter, G. (2009) Physical chemistry of solid dispersions. *J. Pharm. Pharmacol.*, 61:1571–1586.
3. Chiou, W.L. and Riegelman, S. (1971) Pharmaceutical applications of solid dispersion systems. *J. Pharm. Sci.*, 60:1281–1302.
4. Datta, S. and Grant, D.J.W. (2004) Crystal structures of drugs: advances in determination, prediction and engineering. *Nat. Rev. Drug Discov.*, 3:42–57.
5. Qian, K.K. and Bogner, R.H. (2012) Application of mesoporous silicon dioxide and silicate in oral amorphous drug delivery systems. *J. Pharm. Sci.*, 101:444–463.
6. Bissantz, C., Kuhn, B., and Stahl, M. (2010) A medicinal chemist's guide to molecular interactions. *J. Med. Chem.*, 53:5061–5084.
7. Hilden, L.R. and Morris, K.R. (2004) Physics of amorphous solids. *J. Pharm. Sci.*, 93:3–12.
8. Willart, J.F. and Descamps, M. (2008) Solid state amorphization of pharmaceuticals. *Mol. Pharm.*, 5:905–920.
9. Yu, L. (2001) Amorphous pharmaceutical solids: preparation, characterization, and stabilization. *Adv. Drug Deliv. Rev.*, 48:27–42.
10. Jäckle, J. (1986) Models of the glass transition. *Rep. Prog. Phys.*, 49:171–231.
11. Jones, R.A.L. (2002) *Soft Condensed Matter*, Oxford University Press, New York, pp. 18–23.
12. Shamblin, S.L., Tang, X., Chang, L., Hancock, B.C., and Pikal, M.J. (1999) Characterization of the time scales of molecular motion in pharmaceutically important glasses. *J. Phys. Chem. B*, 103:4113–4121.
13. Johari, G.P. and Goldstein, M. (1970) Viscous liquids and the glass transition II secondary relaxation in glasses of rigid molecules. *J. Chem. Phys.*, 53:2372–2388.
14. Hancock, B.C., Shalaev, E.Y., and Shamblin, S.L. (2002) Polyamorphism: a pharmaceutical science perspective. *J. Pharm. Pharmacol.*, 54:1151–1152.
15. Poole, P.H., Grandea, T., Sciortino, F., Stanley, H.E., and Angel, C.A. (1995) Amorphous polymorphism. *Comput. Mater. Sci.*, 4:373–382.
16. Shah, B., Kakumanu, V.K., and Bansal, A.K. (2006) Analytical techniques for quantification of amorphous/crystalline phases in pharmaceutical solids. *J. Pharm. Sci.*, 95:1641–1665.

17. Hair, J.F., Black, W.C., Babin, B.J., and Anderson, R.E. (2009) *Multivariate Data Analysis*, 7th edition, Prentice Hall, New York.
18. Baird, J.A. and Taylor, L.S. (2012) Evaluation of amorphous solid dispersion properties using thermal analysis techniques. *Adv. Drug Deliv. Rev.*, 64:396–421.
19. Clas, S.D., Dalton, C.R., and Hancock, B.C. (1999) Differential scanning calorimetry: applications in drug development. *Pharm. Sci. Technol. Today*, 2:311–320.
20. Richardson, M.J. and Savill, N.G. (1975) Derivation of accurate glass transition temperatures by differential scanning calorimetry. *Polymer*, 16:753–757.
21. Moynihan, C.T., Easteal, A.J., Wilder, J., and Tucker, J. (1974) Dependence of glass transition temperature on heating and cooling rate. *J. Phys. Chem.*, 78:2673–2677.
22. Gill, P.S., Sauerbrunn, S.R., and Reading, M. (1993) Modulated differential scanning calorimetry. *J. Therm. Anal.*, 40:931–939.
23. Hutchinson, J.M. (2003) Studying the glass transition by DSC and TMDSC. *J. Therm. Anal. Calorim.*, 72:619–629.
24. Schawe, J.E.K., Hütter, T., Heitz, C., Alig, I., and Lellinger, D. (2006) Stochastic temperature modulation: a new technique in temperature-modulated DSC. *Thermochim. Acta*, 446:147–155.
25. Patel, J.R., Carlton, R.A., Yuniatine, F., Needham, T.E., Wu, L., and Vogt, F.G. (2012) Preparation and structural characterization of amorphous spray-dried dispersions of tenoxicam with enhanced solubility. *J. Pharm. Sci.*, 101:641–663.
26. Newman, A., Engers, D., Bates, S., Ivanisevic, I., Kelly, R.C., and Zografi, G. (2008) Characterization of amorphous API:polymer mixtures using X-ray powder diffraction. *J. Pharm. Sci.*, 97:4840–4856.
27. Sakellariou, P., Rowe, R.C., and White, E.F.T. (1985) The thermomechanical properties and glass transition temperatures of some cellulose derivatives used in film coating. *Int. J. Pharm.*, 27:267–277.
28. Krause, S. and Iskander, M. (1977) Phase separation in styrene–α-methyl styrene block copolymers. In: Klempner, D. and Frisch, K.C. (Eds.), *Polymer Science and Technology*, Vol. 10, Plenum Press, New York, pp. 231–243.
29. Gordon, M. and Taylor, J.S. (1952) Ideal copolymers and the second-order transitions of synthetic rubbers 1. Non-crystalline copolymers. *J. Appl. Chem.*, 2:493–500.
30. Khougaz, K. and Clas, S.D. (2000) Crystallization inhibition in solid dispersions of MK-0591 and poly(vinylpyrrolidone) polymers. *J. Pharm. Sci.*, 89:1325–1334.
31. Couchman, P.R. and Karasz, F.E. (1978) Classical thermodynamic discussion of effect of composition on glass-transition temperatures. *Macromolecules*, 11:117–119.
32. Khougaz, K. and Clas, S.D. (2000) Crystallization inhibition in solid dispersions of MK-0591 and poly(vinylpyrrolidone) polymers. *J. Pharm. Sci.*, 89:1325–1334.
33. Ford, J.L. and Mann, T.E. (2012) Fast-scan DSC and its role in pharmaceutical physical form characterization and selection. *Adv. Drug Deliv. Rev.*, 64:422–430.
34. Mi, Y. and Zheng, S. (1998) A new study of glass transition of polymers by high pressure DSC. *Polymer*, 39:3709–3712.
35. O'Niell, M.A.A. and Gaisford, S. (2011) Application and use of isothermal calorimetry in pharmaceutical development. *Int. J. Pharm.*, 417:83–93.
36. Gaisford, S. (2012) Isothermal microcalorimetry for quantifying amorphous content in processed pharmaceuticals. *Adv. Drug Deliv. Rev.*, 64:431–439.

37. Alem, N., Beezer, A.E., and Gaisford, S. (2010) Quantifying the rates of relaxation of binary mixtures of amorphous pharmaceuticals with isothermal calorimetry. *Int. J. Pharm.*, 399:12–18.

38. Liu, J., Rigsbee, D.R., Stotz, C., and Pikal, M.J. (2002) Dynamics of pharmaceutical amorphous solids: the study of enthalpy relaxation by isothermal microcalorimetry. *J. Pharm. Sci.*, 91:1853–1862.

39. Smith, G., Duffy, A.P., Shen, J., and Olliff, C.J. (1995) Dielectric relaxation spectroscopy and some applications in the pharmaceutical sciences. *J. Pharm. Sci.*, 84:1029–1044.

40. Beckmann, P.A. (1988) Spectral densities and nuclear spin relaxation in solids. *Phys. Rep.*, 171:85–128.

41. Shmeis, R.A., Wang, Z., and Krill, S.L. (2004) A mechanistic investigation of an amorphous pharmaceutical and its solid dispersions, part I: a comparative analysis by thermally stimulated depolarization current and differential scanning calorimetry. *Pharm. Res.*, 21:2025–2030.

42. Shmeis, R.A., Wang, Z., and Krill, S.L. (2004) A mechanistic investigation of an amorphous pharmaceutical and its solid dispersions, part II: molecular mobility and activation thermodynamic parameters. *Pharm. Res.*, 21:2031–2039.

43. Correia, N.T., Alvarez, C., Moura-Ramos, J.J., and Descamps, M. (2001) The β-α branching in D-sorbitol as studied by thermally stimulated depolarization currents (TSDC). *J. Phys. Chem. B*, 105:5663–5669.

44. Correia, N.T., Campos, J.M., and Moura-Ramos, J.J. (1997) Molecular motions in solid chloropentamethylbenzene: a thermally stimulated depolarisation currents study. *J. Chem. Soc., Faraday Trans.*, 93:157–163.

45. Hancock, B.C. and Zografi, G. (1993) The use of solution theories for predicting water vapor absorption by amorphous pharmaceutical solids: a test of the Flory–Huggins and Vrentas models. *Pharm. Res.*, 10:1262–1267.

46. Zhou, L., Socha, J.M., Vogt, F.G., Chen, S., and Kord, A.S. (2010) A systematic method development strategy for water determinations in drug substance using Karl Fischer titrations. *Am. Pharm. Rev.*, 13:74–84.

47. Slade, L. and Levine, H. (1988) Non-equilibrium behavior of small carbohydrate–water systems. *Pure Appl. Chem.*, 60:1841–1864.

48. Kontny, M.J. and Zografi, G. (1995) Sorption of water by solids. In: Brittain, H.G. (Ed.), *Physical Characterization of Pharmaceutical Solids*, Marcel Dekker, New York, pp. 385–418.

49. Nyqvist, H. (1983) Saturated salt solutions for maintaining specified relative humidities. *Int. J. Pharm. Technol. Prod. Manuf.*, 4:47–48.

50. Mackin, L., Zanon, R., Park, J.M., Foster, K., Opalenik, H., and Demonte, M. (2002) Quantification of low levels (<10%) of amorphous content in micronised active batches using dynamic vapour sorption and isothermal microcalorimetry. *Int. J. Pharm.*, 231:227–236.

51. Anderson, R.B. (1946) Modifications of the Brunauer, Emmett and Teller equation. *J. Am. Chem. Soc.*, 68:686–691.

52. Zografi, G. and Kontny, M.J. (1986) The interactions of water with cellulose- and starch-derived pharmaceutical excipients. *Pharm. Res.*, 3:187–194.

53. Oksanen, C.A. and Zografi, G. (1990) The relationship between the glass transition temperature and water vapor absorption by poly(vinylpyrrolidone). *Pharm. Res.*, 7:654–657.

54. Young, P.M., Chiou, H., Tee, T., Traini, D., Chan, H.K., Thielmann, F., and Burnett, D. (2007) The use of organic vapor sorption to determine low levels of amorphous content in processed pharmaceutical powders. *Drug Dev. Ind. Pharm.*, 33:91–97.
55. Colthup, N., Daly, L., and Wiberly, S. (1990) *Introduction to Infrared and Raman Spectroscopy*, 3rd edition, Academic Press, New York.
56. Prasad, P.N., Swiatkiewicz, J., and Eisenhardt, G. (1982) Laser Raman investigation of solid state reactions. *Appl. Spectrosc. Rev.*, 18:59–103.
57. Jeffrey, G.A. (1997) *An Introduction to Hydrogen Bonding*, Oxford University Press, New York, pp. 220–225.
58. Bellamy, L.J. (1968) *Advances in Infrared Group Frequencies*, Meuthen & Co., New York, pp. 173–174.
59. Matsumoto, T. and Zografi, G. (1999) Physical properties of solid molecular dispersions of indomethacin with poly(vinylpyrrolidone) and poly(vinylpyrrolidone-*co*-vinyl-acetate) in relation to indomethacin recrystallization. *Pharm. Res.*, 16:1722–1728.
60. Taylor, L.S. and Zografi, G. (1997) Spectroscopic characterization of interactions between PVP and indomethacin in amorphous molecular dispersions. *Pharm. Res.*, 14:1691–1698.
61. Rumondor, A.C.F., Marsac, P.J., Stanford, L.A., and Taylor, L.S. (2009) Phase behavior of poly(vinylpyrrolidone) containing amorphous solid dispersions in the presence of moisture. *Mol. Pharm.*, 6:1492–1505.
62. Kazarian, S.G. and Martirosyan, G.G. (2002) Spectroscopy of polymer/drug formulations processed with supercritical fluids: *in situ* ATR-IR and Raman study of impregnation of ibuprofen into PVP. *Int. J. Pharm.*, 232:81–90.
63. Van den Mooter, G., Wuyts, M., Blaton, N., Busson, R., Grobet, P., Augustijns, P., and Kinget, R. (2001) Physical stabilisation of amorphous ketoconazole in solid dispersions with polyvinylpyrrolidone K25. *Eur. J. Pharm. Sci.*, 12:261–269.
64. Forster, A., Hempenstall, J., and Rades, T. (2001) Characterization of glass solutions of poorly water-soluble drugs produced by melt extrusion with hydrophilic amorphous polymers. *J. Pharm. Pharmacol.*, 53:303–315.
65. Fredericks, P., Rintoul, L., and Coates, J. (2004) Vibrational spectroscopy: instrumentation for infrared and Raman spectroscopy. In: Cazes, J. (Ed.), *Ewing's Analytical Instrumentation Handbook*, 3rd edition, Marcel Dekker, New York, pp. 163–238.
66. Zidan, A.S., Rahman, Z., Sayeed, V., Raw, A., Yu, L., and Khan, M.A. (2012) Crystallinity evaluation of tacrolimus solid dispersions by chemometric analysis. *Int. J. Pharm.*, 423, 341–350.
67. Onoue, S., Kojo, Y., Aoki, Y., Kawabata, Y., Yamauchi, Y., and Yamada, S. (2012) Physicochemical and pharmacokinetic characterization of amorphous solid dispersion of tranilast with enhanced solubility in gastric fluid and improved oral bioavailability. *Drug Metab. Pharmacokinet.*, 27:379–387.
68. Rahman, Z., Zidan, A.S., and Khan, M.A. (2010) Formulation and evaluation of protein-loaded solid dispersions by non-destructive methods. *AAPS J.*, 12:158–170.
69. Matthaus, C., Bird, B., Miljkovic, M., Chernenko, T., Romeo, M., and Diem, M. (2008) Infrared and Raman microscopy in cell biology. In: *Biophysical Tools for Biologists, Vol. 2*, Methods in Cell Biology, Vol. 89, Elsevier, New York, pp. 275–308.
70. Kazarian, S.G. and Chan, K.L.A. (2006) Applications of ATR-FTIR spectroscopic imaging to biomedical samples. *Biochim. Biophys. Acta*, 1758:858–867.

71. Chan, K.L.A., Elkhider, N., and Kazarian, S.G. (2005) Spectroscopic imaging of compacted pharmaceutical tablets. *Chem. Eng. Res. Des.*, 83:1303–1310.
72. Chan, K.L.A. and Kazarian, S.G. (2004) FTIR spectroscopic imaging of dissolution of a solid dispersion of nifedipine in poly(ethylene glycol). *Mol. Pharm.*, 1:331–335.
73. Kazarian, S.G. and Chan, K.L.A. (2003) "Chemical photography" of drug release. *Macromolecules*, 36:9866–9872.
74. Dazzi, A., Prater, C.B., Hu, Q., Chase, D.B., Rabolt, J.F., and Marcott, C. (2012) AFM–IR: combining atomic force microscopy and infrared spectroscopy for nanoscale chemical characterization. *Appl. Spectrosc.*, 66:1365–1384.
75. Van Eerdenbrugh, B., Lo, M., Kjoller, K., Marcott, C., and Taylor, L.S. (2012) Nanoscale mid-infrared imaging of phase separation in a drug–polymer blend. *J. Pharm. Sci.*, 101:2066–2073.
76. Van Eerdenbrugh, B., Lo, M., Kjoller, K., Marcott, C., and Taylor, L.S. (2012) Nanoscale mid-infrared evaluation of the miscibility behavior of blends of dextran or maltodextrin with poly(vinylpyrrolidone). *Mol. Pharm.*, 9:1459–1469.
77. Winnewisser, G. (1995) Spectroscopy in the terahertz region. *Vib. Spectrosc.*, 8:241–253.
78. Shen, Y.C. (2011) Terahertz pulsed spectroscopy and imaging for pharmaceutical applications: a review. *Int. J. Pharm.*, 417:48–60.
79. Vogt, F.G., Yin, H., Forcino, R.G., and Wu, L. (2013) ^{17}O solid-state NMR as a sensitive probe of hydrogen bonding in crystalline and amorphous solid forms of diflunisal. *Mol. Pharm.*, 10:3433–3446.
80. Breitenbach, J., Schrof, W., and Neumann, J. (1999) Confocal Raman spectroscopy: analytical approach to solid dispersions and mapping of drugs. *Pharm. Res.*, 16:1109–1113.
81. Park, C.W., Rhee, Y.-S., Vogt, F.G., Hayes, D., Zwischenberger, J., DeLuca, P.P., and Mansour, H.M. (2012) Advances in microscopy and complementary imaging techniques to assess the fate of drugs *ex vivo* in respiratory drug delivery. *Adv. Drug Deliv. Rev.*, 64:344–356.
82. Vogt, F.G. and Williams, G.R. (2012) Analysis of a nanocrystalline polymer dispersion of ebselen using solid-state NMR, Raman microscopy, and X-ray diffraction. *Pharm. Res.*, 29:1866–1881.
83. Padilla, A.M., Chou, S.G., Luthra, S., and Pikal, M.J. (2011) The study of amorphous phase separation in a model polymer phase-separating system using Raman microscopy and a low-temperature stage: effect of cooling rate and nucleation temperature. *J. Pharm. Sci.*, 100:1362–1376.
84. Padilla, A.M. and Pikal, M.J. (2011) The study of phase separation in amorphous freeze-dried systems, part 2: investigation of Raman mapping as a tool for studying amorphous phase separation in freeze-dried protein formulations. *J. Pharm. Sci.*, 100:1467–1474.
85. Widjaja, E., Kanaujiaa, P., Lau, G., Ng, W.K., Garland, M., Saal, C., Hanefeld, A., Fischbach, M., Maio, M., and Tan, R.B.H. (2011) Detection of trace crystallinity in an amorphous system using Raman microscopy and chemometric analysis. *Eur. J. Pharm. Sci.*, 42:45–54.
86. Vogt, F.G. and Strohmeier, M. (2013) Confocal UV and resonance Raman microscopic imaging of pharmaceutical products. *Mol. Pharm.*, 10:4216–4228.
87. Evans, C.L. and Xie, X.S. (2008) Coherent anti-Stokes Raman scattering microscopy: chemical imaging for biology and medicine. *Annu. Rev. Anal. Chem.*, 1:883–909.

88. Strachan, C.J., Windbergs, M., and Offerhaus, H.L. (2011) Pharmaceutical applications of non-linear imaging. *Int. J. Pharm.*, 417:163–172.

89. Slipchenko, M.N., Chen, H., Ely, D.R., Jung, Y., Carvajal, M.T., and Cheng, J.-X. (2010) Vibrational imaging of tablets by epi-detected stimulated Raman scattering microscopy. *Analyst*, 135:2613–2619.

90. Windbergs, M., Jurna, M., Offerhaus, H.L., Herek, J.L., Kleinebudde, P., and Strachan, C.J. (2009) Chemical imaging of oral solid dosage forms and changes upon dissolution using coherent anti-Stokes Raman scattering microscopy. *Anal. Chem.*, 81:2085–2091.

91. Kang, E., Robinson, J., Park, K., and Cheng, J.X. (2007) Paclitaxel distribution in poly (ethylene glycol)/poly(lactide-*co*-glycolic acid) blends and its release visualized by coherent anti-Stokes Raman scattering microscopy. *J. Control. Release*, 122:261–268.

92. Leproux, P., Couderc, V., de Angelis, A., Okuno, M., Kano, H., and Hamaguchi, H. (2011) New opportunities offered by compact sub-nanosecond supercontinuum sources in ultra-broadband multiplex CARS microspectroscopy. *J. Raman Spectrosc.*, 42:1871–1874.

93. Qian, K.K., Zhou, W., Xu, X., and Udovic, T. (2012) Characterization of medicinal compounds confined in porous media by neutron vibrational spectroscopy and first-principles calculations: a case study with ibuprofen. *Pharm. Res.*, 29:2432–2444.

94. Mitchell, P.C.H., Parker, S.F., Ramirez-Cuesta, A.J., and Tomkinson, J. (2005) *Vibrational Spectroscopy with Neutrons: With Applications in Chemistry, Biology, Materials Science, and Catalysis*, World Scientific, Singapore.

95. Slichter, C.P. (1990) *Principles of Magnetic Resonance*, 3rd edition, Springer, Heidelberg.

96. Gerstein, B.C. and Dybowski, C. (1985) *Transient Techniques in NMR of Solids*, Academic Press, Orlando, FL.

97. Eckert, H. (1992) Structural characterization of noncrystalline solids and glasses using solid state NMR. *Prog. Nucl. Magn. Reson. Spectrosc.*, 24, 159–293.

98. Schmidt-Rohr, K. and Spiess, H.W. (1994) *Multidimensional Solid-State NMR and Polymers*, Academic Press, London.

99. Chen, Q. and Schmidt-Rohr, K. (2004) ^{19}F and ^{13}C NMR signal assignment and analysis in a perfluorinated ionomer (Nafion) by two-dimensional solid-state NMR. *Macromolecules*, 37:5995–6003.

100. Harris, R.K. and Olivieri, A.C. (1992) Quadrupolar effects transferred to spin-½ magic-angle spinning spectra of solids. *Prog. Nucl. Magn. Reson. Spectrosc.*, 24:435–456.

101. VanderHart, D.L. (1996) Magnetic susceptibility and high resolution NMR of liquids and solids. In: Harris, R.K. and Grant, D.M. (Eds.), *The Encyclopedia of NMR*, Wiley, New York, pp. 2938–2946.

102. Sakellariou, D., Brown, S.P., Lesage, A., Hediger, S., Michel Bardet, M., Meriles, C.A., Pines, A., and Emsley, L. (2003) High-resolution NMR correlation spectra of disordered solids. *J. Am. Chem. Soc.*, 125:4376–4380.

103. Samoson, A., Tuherm, T., Past, J., Reinhold, A., Anupold, T., and Heinmaa, N. (2005) New horizons for magic-angle spinning NMR. *Top. Curr. Chem.*, 246:15–31.

104. Asano, A., Hori, S., Kitamura, M., Nakazawa, C.T., and Kurotsu, T. (2012) Influence of magic angle spinning on T_1^H of SBR studied by solid state ^{1}H NMR. *Polym. J.*, 44:706–712.

105. Harris, R.K., Becker, E.D., Cabral de Menezes, S.M., Goodfellow, R., and Granger, P. (2001) NMR nomenclature. Nuclear spin properties and conventions for chemical shifts (IUPAC recommendations 2001). *Pure Appl. Chem.*, 73:1795–1818.

106. Stejskal, E.O. and Memory, J.D. (1994) *High Resolution NMR in the Solid State*, Oxford University Press, New York.

107. Antzutkin, O.N. (1999) Sideband manipulation in magic-angle spinning NMR. *Prog. NMR Spectrosc.*, 35:203–266.

108. Hodgkinson, P. (2005) Heteronuclear decoupling in the NMR of solids. *Prog. NMR Spectrosc.*, 46:197–222.

109. Vogt, F.G., Wu, L., Olsen, M.A., and Clark, W.M. (2010) A spectroscopic and computational study of an unusual electrocyclized photodegradation product of 6-(2-(5-chloro-2-(2,4-difluorobenyzloxy)phenyl)cyclopent-1-enyl)picolinic acid. *J. Mol. Struct.*, 984:246–261.

110. Song, Y., Wang, L., Yang, P., Wenslow, R.M., Tan, B., Zhang, H., and Deng, Z. (2013) Physicochemical characterization of felodipine–Kollidon VA64 amorphous solid dispersions prepared by hot-melt extrusion. *J. Pharm. Sci.*, 102:1915–1923.

111. Geppi, M. and Guccione, S. (2005) Molecular properties of ibuprofen and its solid dispersions with Eudragit RL100 studied by solid-state nuclear magnetic resonance. *Pharm. Res.*, 22:1544–1555.

112. Mollica, G., Geppi, M., Pignatello, R., and Veracini, C.A. (2006) Molecular properties of flurbiprofen and its solid dispersions with Eudragit RL100 studied by high- and low-resolution solid-state nuclear magnetic resonance. *Pharm. Res.*, 23:2129–2140.

113. Schachter, D.M., Xiong, J., and Tirol, G.C. (2004) Solid state NMR perspective of drug–polymer solid solutions: a model system based on poly(ethylene oxide). *Int. J. Pharm.*, 281:89–101.

114. Wulff, M., Alden, M., and Tegenfeldt, J. (2002) Solid-state NMR investigation of indomethacin–cyclodextrin complexes in PEG 6000 carrier. *Bioconjug. Chem.*, 13:240–248.

115. Ito, A., Watanabe, T., Yada, S., Hamaura, T., Nakagami, H., Higashi, K., Moribe, K., and Yamamoto, K. (2010) Prediction of recrystallization behavior of troglitazone/polyvinylpyrrolidone solid dispersion by solid-state NMR. *Int. J. Pharm.*, 383, 18–23.

116. Salager, E., Stein, R.S., Steuernagel, S., Lesage, A., Elena, B., and Emsley, L. (2009) Enhanced sensitivity in high-resolution ^{1}H solid-state NMR spectroscopy with DUMBO dipolar decoupling under ultra-fast MAS. *Chem. Phys. Lett.*, 469:336–341.

117. Madhu, P.K. (2009) High-resolution solid-state NMR spectroscopy of protons with homonuclear dipolar decoupling schemes under magic-angle spinning. *Solid State Nucl. Magn. Reson.*, 35:2–11.

118. Zhou, D.H. and Rienstra, C.M. (2008) Rapid analysis of organic compounds by proton-detected heteronuclear correlation NMR spectroscopy with 40 kHz magic-angle spinning. *Angew. Chem., Int. Ed.*, 47:7328–7331.

119. Pham, T.N., Watson, S.A., Edwards, A.J., Chavda, M., Clawson, J.S., Strohmeier, M., and Vogt, F.G. (2010) Analysis of amorphous solid dispersions using 2D solid-state NMR and ^{1}H T_1 relaxation measurements. *Mol. Pharm.*, 7:1667–1691.

120. Tobyn, M., Brown, J., Dennis, A.B., Fakes, M., Gao, Q., Gamble, J., Khimyak, Y.Z., McGeorge, G., Patel, C., Sinclair, W., Timmins, P., and Yin, S. (2009) Amorphous drug–PVP dispersions: application of theoretical, thermal and spectroscopic analytical techniques to the study of a molecule with intermolecular bonds in both the crystalline and pure amorphous state. *J. Pharm. Sci.*, 98:3456–3468.

121. Tatton, A.S., Pham, T.N., Vogt, F.G., Iuga, D., Edwards, A.J., and Brown, S.P. (2013) Probing hydrogen bonding in cocrystals and amorphous dispersions using ^{14}N$-^{1}$H HMQC solid-state NMR. *Mol. Pharm.*, 10:999–1007.

122. Vogt, F.G., Roberts-Skilton, K., and Kennedy-Gabb, S. (2013) A solid-state NMR study of amorphous ezetimibe dispersions in mesoporous silica. *Pharm. Res.*, 30:2135–2331.

123. Vogt, F.G. (2010) The evolution of solid-state NMR in pharmaceutical analysis. *Future Med. Chem.*, 2:915–921.

124. Stejskal, E.O. and Memory, J.D. (1994) *High Resolution NMR in the Solid State*, Oxford University Press, New York.

125. Ernst, R.R., Bodenhausen, G., and Wokaun, A. (1987) *Principles of Nuclear Magnetic Resonance in One and Two Dimensions*, Oxford University Press, New York.

126. van Rossum, B.J., Förster, H., and de Groot, H.J.M. (1997) High-field and high-speed CP-MAS ^{13}C NMR heteronuclear dipolar-correlation spectroscopy of solids with frequency-switched Lee–Goldburg homonuclear decoupling. *J. Magn. Reson.*, 124:516–519.

127. van Rossum, B.J., de Groot, C.P., Ladizhansky, V., Vega, S., and de Groot, H.J.M. (2000) A method for measuring heteronuclear (^{1}H – ^{13}C) distances in high speed MAS NMR. *J. Am. Chem. Soc.*, 122:3465–3472.

128. Meier, B.H. (1994) Polarization transfer and spin diffusion in solid-state NMR. *Adv. Magn. Opt. Reson.*, 18:1–116.

129. Cheung, T.T.P. (1996) Spin diffusion in solids. In: Harris, R.K. and Grant, D.M. (Eds.), *The Encyclopedia of NMR*, Wiley, New York, pp. 4518–4524.

130. Schantz, S. and Ljungqvist, N. (1993) Structure and dynamics in polymer blends: a ^{13}C CPMAS NMR study of poly(3-octylthiophene)/poly(phenylene oxide). *Macromolecules*, 26:6517–6524.

131. Henrichs, P.M., Tribone, J., Massa, D.J., and Hewitt, J.M. (1988) Blend miscibility of bisphenol A polycarbonate and poly(ethylene terephthalate) as studied by solid-state high-resolution ^{13}C NMR spectroscopy. *Macromolecules*, 21:1282–1291.

132. Schnell, I. and Spiess, H.W. (2001) High-resolution ^{1}H NMR spectroscopy in the solid state: very fast sample rotation and multiple-quantum coherences. *J. Magn. Reson.*, 151:153–227.

133. Lesage, A., Charmont, P., Steuernagel, S., and Emsley, L. (2000) Complete resonance assignment of a natural abundance solid peptide by through-bond heteronuclear correlation solid-state NMR. *J. Am. Chem. Soc.*, 122:9739–9744.

134. O'Dell, L.A. (2011) Direct detection of nitrogen-14 in solid-state NMR spectroscopy. *Prog. Nucl. Magn. Reson. Spectrosc.*, 59:295–318.

135. Gan, Z., Amoureux, J.P., and Trebosc, J. (2007) Proton-detected ^{14}N MAS NMR using homonuclear decoupled rotary resonance. *Chem. Phys. Lett.*, 435:163–169.

136. Zumbulyadis, N., Antalek, B., Windig, W., Scaringe, R.P., Lanzafame, A.M., Blanton, T., and Helber, M. (1999) Elucidation of polymorph mixtures using solid-state ^{13}C CP/MAS NMR spectroscopy and direct exponential curve resolution algorithm. *J. Am. Chem. Soc.*, 121:11554–11557.

137. Aso, Y., Yoshioka, S., and Kojima, S. (2000) Relationship between the crystallization rates of amorphous nifedipine, phenobarbital, and flopropione, and their molecular mobility as measured by their enthalpy relaxation and ^{1}H NMR relaxation times. *J. Pharm. Sci.*, 89:408–416.

138. Aso, Y., Yoshioka, S., Miyazaki, T., and Kawanishi, T. (2009) Feasibility of ^{19}F\ for\ assessing the molecular mobility of flufenamic acid in solid dispersions. *Chem. Pharm. Bull.*, 57:61–64.

139. Forster, A., Apperley, D., Hempenstall, J., Lancaster, R., and Rades, T. (2003) Investigation of the physical stability of amorphous drug and drug/polymer melts using variable temperature solid state NMR. *Pharmazie*, 58:761–762.

140. Barnes, A.B., Paëpe, G.D., van der Wel, P.C., Hu, K.N., Joo, C.G., Bajaj, V.S., Mak-Jurkauskas, M.L., Sirigiri, J.R., Herzfeld, J., Temkin, R.J., and Griffin, R.G. (2008) High-field dynamic nuclear polarization for solid and solution biological NMR. *Appl. Magn. Reson.*, 34:237–263.

141. Ong, T.C., Mak-Jurkauskas, M.L., Walish, J.J., Michaelis, V.K., Corzilius, B., Albert, A., Smith, A.A., Clausen, A.M., Cheetham, J.C., Swager, T.M., and Griffin, R.G. (2013) Solvent-free dynamic nuclear polarization of amorphous and crystalline *ortho*-terphenyl. *J. Phys. Chem. B*, 117:3040–3046.

142. Rosay, M., Tometich, L., Pawsey, S., Bader, R., Schauwecker, R., Blank, M., Borchard, P.M., Cauffman, S.R., Felch, K.L., Weber, R.T., Temkin, R.J., Griffin, R.G., and Maas, W.E. (2010) Solid-state dynamic nuclear polarization at 263 GHz: spectrometer design and experimental results. *Phys. Chem. Chem. Phys.*, 12:5850–5860.

143. Matsuki, Y., Maly, T., Ouari, O., Karoui, H., Le Moigne, F., Rizzato, E., Lyubenova, S., Herzfeld, J., Prisner, T., Tordo, P., and Griffin, R.G. (2009) Dynamic nuclear polarization using a rigid biradical. *Angew. Chem., Int. Ed.*, 48:4996–5000.

144. Brittain, H.G. (2006) UV/Vis reflectance spectroscopy. In: *Drugs and the Pharmaceutical Sciences*, Vol. 160, Taylor & Francis, New York, pp. 121–149.

145. Zhou, L., Vogt, F.G., Overstreet, P.A., Dougherty, J.T., Clawson, J.S., and Kord, A.S. (2011) A systematic method development strategy for quantitative color measurement in drug substances, starting materials, and synthetic intermediates. *J. Pharm. Innov.*, 6:217–231.

146. Lakowicz, J.R. (2006) *Principles of Fluorescence Spectroscopy*, 3rd edition, Springer, New York.

147. Brittain, H.G. (2006) Luminescence spectroscopy. In: *Drugs and the Pharmaceutical Sciences*, Vol. 160, Taylor & Francis, New York, pp. 151–204.

148. Jenkins, R. and Snyder, R.L. (1996) *Introduction to X-Ray Powder Diffractometry*, Wiley, New York.

149. de Villiers, M.M., Wurster, D.E., Van der Watt, J.G., and Ketkar, A. (1998) X-ray powder diffraction determination of the relative amount of crystalline acetaminophen in solid dispersions with polyvinylpyrrolidone. *Int. J. Pharm.*, 163:219–224.

150. Rumondor, A.C.F., Stanford, L.A., and Taylor, L.S. (2009) Effects of polymer type and storage relative humidity on the kinetics of felodipine crystallization from amorphous solid dispersions. *Pharm. Res.*, 26:2599–2606.

151. Wagner, C.N.J. (1978) Direct methods for the determination of atomic-scale structure of amorphous solids (X-ray, electron, and neutron scattering). *J. Non-Cryst. Solids*, 31:1–40.

152. Billinge, S.J.L. and Kanatzidis, M.G. (2004) Beyond crystallography: the study of disorder, nanocrystallinity and crystallographically challenged materials with pair distribution functions. *Chem. Commun.*, 749–760.

153. Egami, T. and Billinge, S.J.L. (2003) *Underneath the Bragg Peaks: Structural Analysis of Complex Materials*, Plenum Press, Oxford.

154. Toby, B.H. and Egami, T. (1992) Accuracy of pair distribution function analysis applied to crystalline and noncrystalline materials. *Acta Crystallogr. A*, 48, 336–346.

155. Peterson, P.F., Bozin, E.S., Proffen, T., and Billinge, S.J.L. (2003) Improved measures of quality for the atomic pair distribution function. *J. Appl. Crystallogr.*, 36:53–64.

156. Cope, E.R. and Dove, M.T. (2007) Pair distribution functions calculated from interatomic potential models using the general utility lattice program. *J. Appl. Crystallogr.*, 40:589–594.
157. Rumondor, A.C.F., Ivanisevic, I., Bates, S., Alonzo, D.E., and Taylor, L.S. (2009) Evaluation of drug–polymer miscibility in amorphous solid dispersion systems. *Pharm. Res.*, 26:2523–2534.
158. Ivanisevic, I., Bates, S., and Chen, P. (2009) Novel methods for the assessment of miscibility of amorphous drug–polymer dispersions. *J. Pharm. Sci.*, 98:3373–3386.
159. McCrone, W.C., McCrone, L.B., and Delly, J.G. (1978) *Polarized Light Microscopy*, Ann Arbor Science Publishers, Ann Arbor, MI.
160. Carlton, R.A. (2011) *Pharmaceutical Microscopy*, Springer, New York.
161. Goldstein, J., Newbury, D., Joy, D., Lyman, C., Echlin, P., Lifshin, E., Sawyer, L., and Michael, J. (2003) *Scanning Electron Microscopy and X-Ray Microanalysis*, 3rd edition, Springer, New York.
162. Marsac, P.J., Rumondor, A.C., Nivens, D.E., Kestur, U.S., Stanciu, L., and Taylor, L.S. (2010) Effect of temperature and moisture on the miscibility of amorphous dispersions of felodipine and poly(vinyl pyrrolidone). *J. Pharm. Sci.*, 99:169–185.
163. Brundle, C.R., Watts, J.F., and Wolstenholme, J. (2005) X-ray photoelectron and Auger electron spectroscopy. In: Cazes, J. (Ed.), *Ewing's Analytical Instrumentation Handbook*, 3rd edition, Marcel Dekker, New York, pp. 399–427.
164. Carli, F. and Garbassi, F. (1985) Characterization of drug loading in crospovidone by X-ray photoelectron spectroscopy. *J. Pharm. Sci.*, 74:963–967.
165. Dahlberg, C., Millqvist-Fureby, A., Schuleit, M., and Furo, I. (2010) Polymer–drug interactions and wetting of solid dispersions. *Eur. J. Pharm. Sci.*, 39:125–133.
166. Dahlberg, C., Millqvist-Fureby, A., and Schuleit, M. (2008) Surface composition and contact angle relationships for differently prepared solid dispersions. *Eur. J. Pharm. Biopharm.*, 70:478–485.
167. Wanapun, D., Kestur, U.S., Kissick, D.J., Simpson, G.J., and Taylor, L.S. (2010) Selective detection and quantitation of organic molecule crystallization by second harmonic generation microscopy. *Anal. Chem.*, 82:5425–5432.
168. Fox, M. (2010) *Optical Properties of Solids*, 2nd edition, Oxford University Press, Oxford, UK, pp. 295–326.
169. Lamm, M.S., Simpson, A., McNevin, M., Frankenfeld, C., Nay, R., and Variankaval, N. (2012) Probing the effect of drug loading and humidity on the mechanical properties of solid dispersions with nanoindentation: antiplasticization of a polymer by a drug molecule. *Mol. Pharm.*, 9:3396–3402.
170. Ramos, K.J. and Bahr, D.F. (2007) Mechanical behavior assessment of sucrose using nanoindentation. *J. Mater. Res.*, 22:2037–2045.
171. Xiang, T.X. and Anderson, B.D. (2013) Molecular dynamics simulation of amorphous indomethacin–poly(vinylpyrrolidone) glasses: solubility and hydrogen bonding interactions. *J. Pharm. Sci.*, 102:876–891.
172. Xiang, T.X. and Anderson, B.D. (2013) Molecular dynamics simulation of amorphous indomethacin. *Mol. Pharm.*, 10:102–114.
173. Tantishaiyakul, V., Suknuntha, K., and Vao-Soongnern, V. (2010) Characterization of cimetidine–piroxicam coprecipitate interaction using experimental studies and molecular dynamic simulations. *AAPS PharmSciTech*, 11:952–958.

174. Gupta, J., Nunes, C., and Jonnalagadda, S. (2013) A molecular dynamics approach for predicting the glass transition temperature and plasticization effect in amorphous pharmaceuticals. *Mol. Pharm.*, 10:4136–4145.

175. Frenkel, D. and Smit, B. (2002) *Understanding Molecular Simulation*, 2nd edition, Oxford University Press/Academic Press, San Diego, CA.

176. Henritzi, P., Bormuth, A., and Vogel, M. (2013) Interpretation of ^{1}H and ^{2}H spin–lattice relaxation dispersions: insights from molecular dynamics simulations of polymer melts. *Solid State Nucl. Magn. Reson.*, 54:32–40.

5

PHYSICAL STABILITY AND CRYSTALLIZATION INHIBITION

Lynne S. Taylor

Purdue University, West Lafayette, IN, USA

5.1 INTRODUCTION

In order to successfully utilize amorphous solid dispersions (ASDs) to enhance drug delivery, it is essential to maintain the amorphous form of the drug during the manufacturing process and the shelf life of the formulated product. In addition, following oral delivery, the amorphous matrix will hydrate and the drug will dissolve. If the drug crystallizes within a short time frame upon exposure to biological fluids, it is clear that the solubility advantage of the amorphous formulation will not be realized. Therefore, it is essential to evaluate the crystallization tendency of the drug exposed to different environmental conditions and to develop strategies to modify crystallization kinetics through the addition of crystallization inhibitors; for pharmaceutical systems, polymers are the most widely used crystallization inhibitors. In this chapter, crystallization in the solid state will be discussed initially, including a consideration of the role of temperature, residual water, and additives. Subsequently, phase behavior during dissolution of an amorphous formulation will be explored. Figure 5.1 shows an overview of the factors that need to be understood and controlled for successful formulation of an amorphous solid dispersion.

Pharmaceutical Amorphous Solid Dispersions, First Edition. Edited by Ann Newman.

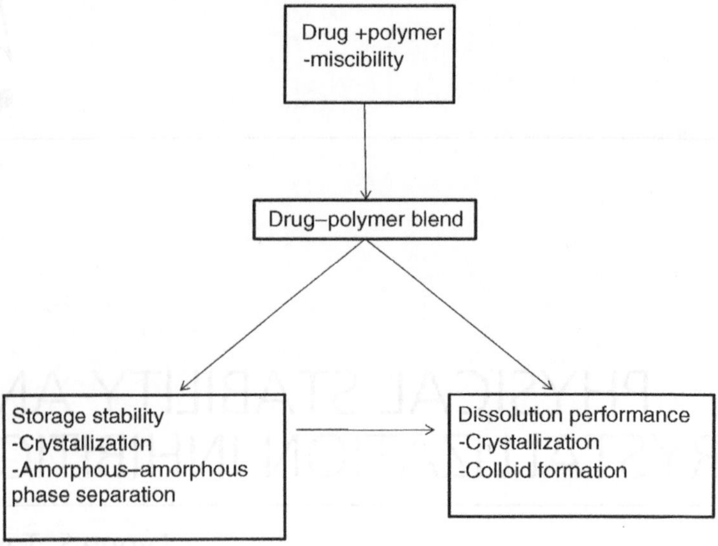

Figure 5.1. Schematic showing some of the factors important for the optimal performance of amorphous solid dispersions.

5.2 THEORY OF CRYSTALLIZATION IN THE SOLID STATE

The pure material in the amorphous state has a higher energy than the corresponding crystalline material, and thus will always have the potential to crystallize [1]. In order to understand the factors that impact the kinetics of crystallization from amorphous formulations, it is useful to briefly consider some theoretical concepts pertaining to melt crystallization. The crystallization process involves two events: nucleation to form an embryonic crystal followed by growth of that crystal; each of these processes will be briefly discussed.

5.2.1 Nucleation Theory

Crystallization begins when a small but critical number of molecules of the stable phase self-assemble to form a nucleus. Nucleation events can be classified as primary and secondary processes [2]. Primary nucleation occurs in a system that does not contain any crystalline matter, while secondary nucleation refers to the formation of nuclei in the vicinity of crystals already present in the system and is more relevant for solution crystallization where it is often seen. Primary nucleation may occur spontaneously (homogeneous) or can be induced by the presence of foreign surfaces (heterogeneous). For most pharmaceutical materials, it is likely that heterogeneous nucleation predominates.

The nucleation rate per unit volume, J, from a melt can be described by Equation 5.1:

$$J = A \exp\left[-\frac{\Delta G^{*} + \Delta G_{\mathrm{a}}}{k_{\mathrm{B}} T}\right]. \tag{5.1}$$

Equation 5.1 describes the nucleation rate in terms of the activation energy for molecular transport from the amorphous (undercooled) melt phase to the growing nucleus (ΔG_a) and the thermodynamic barrier for nucleation (ΔG^*). A is the pre-exponential factor, T is the temperature, and k_B is the Boltzmann constant. ΔG^* depends on the balance between the energy required to form a new interface (ΔG_s) and the bulk free energy difference between the crystalline and amorphous phases (ΔG_v):

$$\Delta G^* = \Delta G_s + \Delta G_v. \tag{5.2}$$

ΔG_s is a positive (unfavorable to nucleation) quantity and is proportional to r^2, the radius of the nucleus, and ΔG_v is a negative (favorable to nucleation) quantity proportional to r^3. For a spherical nucleus, Equation 5.2 can be expanded to

$$\Delta G^* = 4\pi r^2 \gamma + \frac{4}{3}\pi r^3 \Delta G_v, \tag{5.3}$$

where γ is the interfacial tension between the nucleus and the supercooled liquid. From Equation 5.3, it becomes apparent that there is a critical radius that must be exceeded for the nucleus to be thermodynamically favorable and thus to persist.

The difference in free energy between the liquid and the crystal can be estimated from enthalpy of fusion (ΔH_{fus}) and melting temperature (T_m) of the crystal [3]:

$$\Delta G_v = \frac{\Delta H_{fus}(T_m - T)T}{T_m^2}. \tag{5.4}$$

It can be seen from Equations 5.1–5.4 that several variables are important in determining the rate of nucleation from amorphous systems, including the heat of fusion of the crystalline phase and the crystal–amorphous interfacial energy. Temperature clearly also plays an extremely important role in determining the rate of nucleation, whereby an increase in undercooling ($T_m - T$) increases the driving force for nucleation. In practice, it is observed that the nucleation rate increases with the degree of undercooling, reaches a maximum, and then decreases. This behavior arises because of the decreased molecular mobility that occurs as the temperature decreases and the viscosity increases, as shown in Figure 5.2. This is accounted for in Equation 5.1 by the kinetic term ΔG_a that depends on the molecular mobility of the crystallizing species. Thus, the observed nucleation rate will be a complex interplay of thermodynamic, kinetic, and structural factors. In addition, since homogeneous nucleation is rarely encountered in practice, the presence of impurities, foreign particles, and surfaces can also impact the observed nucleation behavior [2].

5.2.2 Crystal Growth

Once nucleation has occurred, stable nuclei undergo crystal growth, which involves the successive transfer of molecules from the undercooled liquid phase to the surface of the growing nuclei. Crystal growth from the melt state is inherently very complex, and

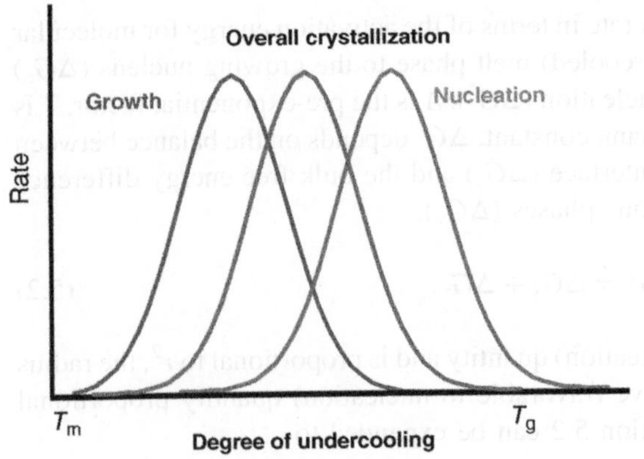

Figure 5.2. Schematic illustrating the temperature dependence of the nucleation, growth, and overall crystallization rates for undercooled melts.

several theories have been developed aimed at describing the mechanisms of crystal growth [2]. These have been extensively described and differ based on the structure and evolution of the interface. For smooth crystal interfaces, models such as the two-dimensional nucleation model and the screw dislocation model have been applied. At large undercooling, where growth rate is limited by mobility, as is the case for most pharmaceutical amorphous systems, Jackson [4] has proposed the following expressed for the growth rate ν, based on the original work of Frenkel [5]:

$$\nu = \frac{6aD}{\Lambda^2} f \exp\left(-\frac{\Delta S}{k}\right)\left[1 - \exp\left(-\frac{\Delta G}{kT}\right)\right], \tag{5.5}$$

where a is the molecular diameter, D is the diffusion coefficient, Λ is the diffusion jump distance, f is the fraction of interface sites that are active growth sites, ΔS is the entropy of difference between the two phases, k is the Boltzmann's constant, ΔG is the free energy difference between the two phases, and T is the temperature. Similar to the nucleation rate equation shown in Equation 5.1, it is apparent from Equation 5.5 that both the molecular mobility (i.e., the diffusion coefficient) and the thermodynamic driving force are key factors that determine the growth rate. Since these parameters show opposite temperature dependencies, the growth rate will first increase with decreasing temperature as the thermodynamic driving force increases, and then decrease as the molecular mobility decreases. Thus, the growth rate will show a maximum value, somewhere between the melting temperature and the glass transition temperature and crystal growth will be restricted at low temperatures by the reduced molecular mobility. The relationship with temperature is shown in Figure 5.2. The molecules also need to have the correct orientation to join the crystal lattice, and this factor is represented by the ΔS term.

In reality, crystal growth from supercooled liquids and glasses is more complicated and not always well described by Equation 5.5, in particular at temperatures close to and below the glass transition temperature (T_g). Thus, crystals at the surface grow much faster than crystals in the bulk [6] leading to biphasic crystallization patterns in particulate samples [7]. In addition, a new mode of growth, termed glass crystal (GC) growth has been observed to appear close to T_g for some crystal forms [8–11]. This mode of growth is thought to be diffusionless, and hence occurs much faster than the diffusion controlled growth described in Equation 5.5.

5.2.3 Overall Kinetics of Crystallization

The overall kinetics of crystallization from the melt at a particular temperature depends on the rate of nucleation, the induction or lag time for nucleation to occur, and the crystal growth rate. Based on the schematic shown in Figure 5.2, it would be unlikely that an Arrhenius-type relationship between the crystallization rate and temperature would hold for a very large temperature range. Thus, it can be very difficult to conduct accelerated stability testing at higher temperatures and extrapolate results to lower temperatures with good predictive success.

One of the most popular models to describe crystallization from amorphous systems is that of Kolmogorov–Johnson–Mehl–Avrami (KJMA) [12,13]. This theory is based on the assumption that changes in the degree of crystallization with time depend on the volume fraction of noncrystallized material and that rates of both nucleation and growth are time independent and that nucleation is homogeneous. If preseeding is absent, the lag time for nucleation is often significant and can be incorporated yielding the following equation:

$$x(t) = 1 - \exp[-K(t - \tau)^n], \tag{5.6}$$

where $x(t)$ is the fraction transformed into the crystalline phase at time t, K is a constant determined by the nucleation and growth constants, τ is the induction time (time before any crystalline material can be detected), and n is a constant characteristic of the experimental conditions called the Avrami exponent. The value of n can be used to understand possible mechanisms responsible for transformations. This model has been widely applied to pharmaceutical systems [14–16]. Zhou et al. [15] studied the crystallization of amorphous nifedipine. The experimental data obtained were fitted to different models. The KJMA model showed a good fit in the range of $x = 0.05$–0.8. At the early stages of crystallization (for $x = 0.05$–0.6), experimental values fit the KJMA model with $n = 3$ and changed to $n = 4$ for $x = 0.6$–0.8. The change in the exponent was attributed to the relative contributions of nucleation and crystal growth at different stages of the crystallization reaction.

The preceding discussion of nucleation and growth shows that crystallization is a complex phenomenon, influenced by both thermodynamic and kinetic factors. At lower temperatures, the undercooling is large (the system is further from the melting temperature), and there is a greater thermodynamic driving force for crystallization. At higher temperatures, mobility of the material will be higher and crystallization may proceed faster. Therefore, thermodynamic factors favor crystallization at lower temperatures, while kinetic factors favor crystallization at higher temperatures. Thus, the crystallization rate will be at a maximum somewhere between T_g and T_m, as shown in Figure 5.2, and will depend on the complex interplay of thermodynamic and kinetic factors. In order to understand the tendency of amorphous formulations to crystallize during storage, the inherent crystallization tendency of the compound will be considered first, and then factors that can impact crystallization behavior will be discussed.

5.3 FACTORS IMPACTING THE CRYSTALLIZATION TENDENCY OF ACTIVE PHARMACEUTICAL COMPOUNDS

It is intuitive that a compound that has a high tendency to crystallize will be more difficult to stabilize in an amorphous formulation than a compound with a correspondingly low crystallization tendency. Therefore, there has been some effort devoted to understand which compound properties are most indicative of crystallization tendency. Some of the factors that are thought to impact crystallization tendency of active pharmaceutical ingredients (APIs) include the free energy of crystallization, configurational entropy, interfacial free energy between the crystalline phase and the amorphous phase, molecular mobility, differences in hydrogen bonding patterns between amorphous and crystalline phases, and preparation method.

5.3.1 Thermodynamic and Structural Considerations

Configurational properties are often used to describe the thermodynamics of the amorphous state relative to the crystalline state. These properties are all a function of the configurational heat capacity ($C_{\mathrm{p,conf}}$), which is the difference in the heat capacity between the amorphous and crystalline states. $C_{\mathrm{p,conf}}$ can be measured experimentally by performing differential scanning calorimetry (DSC) experiments on both the amorphous and crystalline materials. The configurational entropy (S_{conf}), entropy of fusion (ΔS_{m}), configurational enthalpy (H_{conf}), and configurational free energy (G_{conf}) for a material can be expressed as

$$S_{\mathrm{conf}}(T) = \Delta S_{\mathrm{m}} - \int_{T}^{T_{\mathrm{m}}} \frac{C_{p,\mathrm{conf}}}{T}\, \mathrm{d}T, \tag{5.7}$$

$$H_{\mathrm{conf}}(T) = \Delta H_{\mathrm{m}} - \int_{T}^{T_{\mathrm{m}}} C_{p,\mathrm{conf}}\, \mathrm{d}T, \tag{5.8}$$

$$G_{\mathrm{conf}}(T) = H_{\mathrm{conf}}(T) - TS_{\mathrm{conf}}(T). \tag{5.9}$$

Such thermodynamic properties have been measured and attempts made to correlate these properties with the crystallization tendency. Zhou et al. [17] related the configurational properties (S_{conf}, H_{conf}, G_{conf}) and molecular mobility of five model pharmaceutical compounds to their crystallization behavior. It was observed that the compounds with the highest S_{conf} and lowest mobilities exhibited the lowest crystallization tendencies. The configurational entropy (S_{conf}) reflects the conformational flexibility of a molecule; thus, molecules with high S_{conf} can exist in multiple conformations in the undercooled liquid state and tend to have more complex structures. Thus, it would be expected that a system with a high S_{conf} would have more conformations to sample before finding the "right" conformation for nucleation, thus decreasing the nucleation rate and subsequent crystallization tendency [18]. On a somewhat related topic, similarities or differences in hydrogen bonding interactions in the amorphous and crystalline states may also influence the ability of the molecules to nucleate. Similar hydrogen bonding patterns and strengths in the glass and crystal may lead to more facile crystallization relative to instances where the strength and/or donor and acceptor group pairings show differences between the two phases [19]. Miyazaki et al. [20] found that for a series of structurally related dihydropyridines, the compounds with lower structural symmetry and bulkier side groups (nitrendipine and nilvadipine) exhibited lower crystallization rates compared with nifedipine and *m*-nifedipine, further supporting the important role of molecular structure.

In another study of the crystallization tendency of amorphous dihydropyridines [21], a higher crystallization tendency (higher nucleation rate) was observed for nifedipine relative to felodipine. This was correlated with the increase in thermodynamic driving force for crystallization, G_{conf}. In this instance, H_{conf} was the factor correlating with the crystallization tendency of these two compounds, not S_{conf} as observed in other studies. Another study evaluated the effect of heat of fusion (ΔH_m) and entropy on the crystallization tendencies of a series of hexitols (iditol, mannitol, sorbitol, and dulcitol) [22]. Here, the experimental observations suggested that the high crystallization tendencies of mannitol and dulcitol were due to their high ΔH_m and thus higher thermodynamic driving force for crystallization. Baird et al. [23] studied the crystallization tendency of a group of 50 compounds on cooling from the melt and found that, in general, rapidly crystallizing compounds had lower molecular weights (MWs) and a larger free energy difference between the crystal and the melt (on a per volume basis) than more slowly crystallizing compounds.

The interfacial energy is another important thermodynamic factor that will influence the crystallization tendency, although it is difficult to access experimentally. Carpentier et al. [24] measured the interfacial energies of a series of pentitols in an effort to delineate the relative contributions of both kinetic and thermodynamic driving forces to observed crystallization tendencies. The authors found that the crystallization tendencies observed did not correlate with the change in either the free energy or the molecular mobility of the compounds. For example, amorphous L-arabitol, which had the highest driving force for crystallization, did not recrystallize on heating, whereas adonitol, with the lowest driving force, crystallized upon heating. The differences in the crystallization tendencies were explained based on differences in the specific molecular conformations that led to different liquid–crystal interfacial energies.

5.3.2 Molecular Mobility

Crystallization from the amorphous state will be strongly affected by molecular motions [18]. Molecular mobility includes bending, rotation, and translation of molecules. Generally, crystallization rates will increase if the molecular mobility is increased, all other factors being equal. However, it is important to consider that supercooled liquids and glasses are spatially heterogeneous and that there are many different types of molecular motion occurring that will change with both temperature and time (in particular for glasses). While molecular mobility is much lower in materials stored below T_g, crystallization does indeed occur in glasses. Thus, there are numerous reports of crystallization of amorphous systems following storage at temperatures below T_g [25–27], and nucleation has been reported at temperatures as low as 55 °C below T_g [28]. Unfortunately, the types of molecular motion important for nucleation in glasses are not well understood. In addition, it has been observed that the relationship between bulk molecular diffusion and crystal growth rates also breaks down in glasses [8,9]. Another factor that should be considered is that the temperature dependence of nucleation and crystal growth rates differs. The maximum in the nucleation rate usually occurs at a lower temperature than that observed for the growth rate, as shown in Figure 5.2 [29]. Consequently, it is not currently possible to predict bulk crystallization kinetics using experimental measurements of molecular mobility, which tend to yield average values. Although it is not possible to predict long-term stability against crystallization from measurements of molecular mobility, correlations are frequently observed between these two factors.

Several studies have investigated the relationship between α-mobility and crystallization rates. α-Mobility, which arises from molecular rotation and translation, is generally associated with larger length scales and is the type of mobility associated with viscous flow and the glass transition. Aso et al. [30] probed the crystallization rate of three pharmaceutical compounds (nifedipine, phenobarbital, and flopropione) and measured the molecular mobility using ^{1}H nuclear magnetic resonance (NMR) relaxation times and enthalpy relaxation experiments. Crystallization rates for phenobarbital and flopropione were much lower below T_g relative to above T_g; in contrast, nifedipine showed much more modest differences. Relaxation experiments suggested that the molecular mobility of nifedipine was higher below T_g than that of either phenobarbital or flopropione, leading to faster crystallization kinetics. In another study, the molecular mobility of indomethacin and nifedipine was compared using ^{13}C and ^{1}H NMR [31]. The results showed that nifedipine had higher molecular mobility and lower activation energy for molecular motion in the glassy state compared with indomethacin, again correlating with the higher observed crystallization rate for nifedipine. Similar correlations between relaxation rates and crystallization rates were observed for amorphous indomethacin and salicin by Masuda et al. [32]. Using ^{13}C NMR, shorter relaxation times were observed for the more rapidly crystallizing salicin.

Many studies have also been conducted investigating relationships between β-mobility, a much faster motion that occurs on shorter length scales, and crystallization. A study conducted by Alie et al. [33] measured both the α- and β-relaxation processes as a function of temperature for a model compound using dynamic dielectric spectroscopy, and correlated the results with the observed characteristic crystallization time (τ). The temperature dependence of the crystallization time for amorphous SSR (an investigational compound) showed a correlation with the slope of the β_a-relaxation line, inferring that the crystallization kinetics was coupled with the β-relaxation process. Vyazovkin and Dranca [27] studied the α- and β-relaxation processes for amorphous indomethacin using DSC and proposed that at temperatures around T_g, the crystallization rate was controlled by α-relaxation in the material, while below T_g, the rate of crystallization was controlled by β-relaxation processes.

Viscosity measurements have also been used to determine the molecular mobility in glasses and supercooled liquids. The assumption here is that the Stokes–Einstein relationship holds, and that molecular diffusivity is inversely proportional to the viscosity. Thus, samples with a higher viscosity will crystallize more slowly. For example, samples with a higher viscosity at the melting point have been observed to be more likely to form glasses on cooling than compounds with a lower melt viscosity, which have a higher tendency to crystallize [34]. However, it has been observed that crystal growth rates do not always scale in the expected way with viscosity for some supercooled liquids. This is particularly true for fragile liquids such as APIs, where there is a breakdown in scaling between the viscosity and growth rate, particularly in the deeply supercooled region. In this region, the crystal growth rate is typically faster than would be expected based on considering the viscosity [35].

Despite the difficultly in making *a priori* predictions of crystallization from measurements of molecular mobility, it is clear that factors that increase molecular mobility will enhance the risk for crystallization. Of the various factors that enhance molecular mobility, the two of greatest practical importance are temperature and moisture content.

The impact of temperature on mobility has already been discussed to some extent. As temperature increases, the molecular mobility will increase; however, the typical Arrhenius relationship does not usually hold. The most common models to describe the temperature dependence of molecular mobility are the Adam–Gibbs–Vogel (AGV) [36], Vogel–Tammann–Fulcher (VTF) [37], Adam—Gibbs (AG) [38], and Williams–Landell–Ferry (WLF) [39] equations. The AGV equation relates the structural relaxation time τ of a glassy material below T_g:

$$\tau = \tau_0 \exp\left(\frac{DT_0}{T(1-(T_0/T_f))}\right), \tag{5.10}$$

whereas the VTF (Equation 5.11) is one of the most common models used to describe the temperature dependence of structural relaxation (or viscosity) above T_g:

$$\tau = \tau_0 \exp\left(\frac{DT_0}{T-T_0}\right), \tag{5.11}$$

where τ is the mean molecular relaxation time, τ_0 is the relaxation time constant for the unrestricted material, D is the strength parameter, T_f is the fictive temperature, and T_0 is the zero mobility or Kauzmann [40] temperature. These equations clearly show the strong dependence of the relaxation time on temperature.

Due to its very low T_g value (reported values range from 136 to 165 K) [41,42], the presence of water lowers the T_g of amorphous systems, and there is a corresponding increase in molecular mobility. The lowering of T_g is called a plasticizing effect and is well-known consequence of water absorption by amorphous systems [43,44].

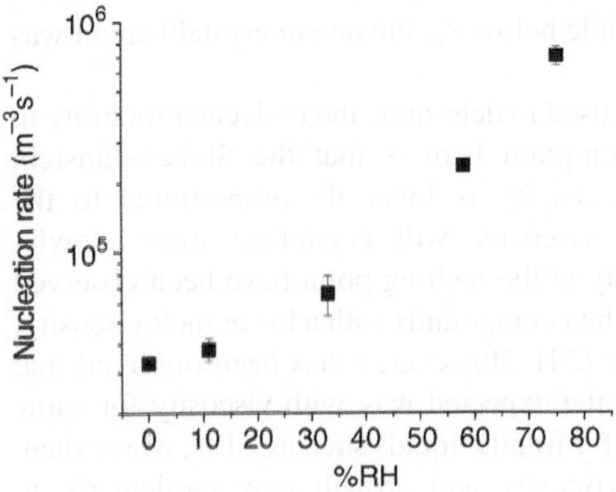

Figure 5.3. Nucleation rate of amorphous felodipine at 25 °C following storage at different relative humidities illustrating the rapid increase in nucleation rate as the storage relative humidity is increased.

The increased molecular mobility that results following water sorption will increase the crystallization tendency of an amorphous material. Andronis et al. [45] showed that increasing the storage relative humidity for amorphous indomethacin increased the amount of moisture sorbed by the system, resulting in a decrease in the measured T_g and increased crystallization rate. Another study reported the impact of moisture sorption on the crystallization of amorphous lamotrigine mesylate [46]. Increasing the amount of moisture sorbed in the system decreased the glass transition of the material and onset temperature for crystallization; hence, increasing the amount of water present increased the crystallization tendency from the amorphous state. Quantitative studies of crystal nucleation rates [47] and crystal growth rates [48] of felodipine exposed to different RH conditions illustrate the large impact of absorbed moisture on crystallization. Figure 5.3 shows the measured nucleation rate for the hydrophobic compound felodipine as a function of storage RH. Dry felodipine has a T_g of around 43 °C, and this decreases to about 35 °C following storage at 75% RH [47]. For this 10 °C reduction in T_g, the nucleation rate increases by more than an order of magnitude, illustrating the extremely potent impact of absorbed moisture on crystallization behavior.

5.3.3 Impact of Preparation Method and Processing Steps

It is well recognized that the properties of amorphous materials are highly dependent on both thermal and processing history. Common routes of preparing amorphous materials include rapid cooling from the melt, condensation from the vapor state, mechanical disruption of the crystalline lattice, and precipitation from solution and drying. Each of these different methods will yield a glass with different thermodynamic properties, whereby the various glasses have undergone different extents of structural relaxation. For example, the preparation method has been found to influence the enthalpic relaxation, crystallization tendency, and water vapor sorption of amorphous trehalose [49]. Based on isothermal crystallization studies, it was observed that the dehydrated material crystallized fastest, followed in order by the freeze-dried, spray-dried, and melt-quenched material. In addition to influencing the thermodynamic properties of the glass, the preparation method can impact the surface area with important implications for crystallization; it has been demonstrated that smaller particles crystallize more readily than larger particles [7,50], and this can be attributed to the faster surface crystal growth rates discussed earlier. Smaller particles with a larger surface area-to-volume ratio will thus be more susceptible to crystallization. The particle size and surface area of amorphous materials will be highly dependent on the preparation method, in particular if they are spray dried versus melt extruded. In addition, crystallization from glasses can be induced from surface heterogeneities induced, for example, by scratching [50], while milling of amorphous materials is known to promote crystallization [51]. The possibility of forming glasses with quenched-in nuclei, a common observation for metallic glasses [52–55] that has also been noted for organic glasses [56,57], also needs to be considered. Trapping of these nuclei in the glassy matrix can effect the crystallization tendency of a material in two ways. First, these nuclei can act as seeds for crystal growth below T_g over extended time periods as the glass undergoes either α- or β-relaxation and molecules diffuse to the nuclei [58,59]. Second, the presence of nuclei can greatly enhance the crystallization tendency during reheating of a glassy material. This occurs due to the fact that during cooling the nuclei were formed when the nucleation rate was maximum, but the growth rate was relatively low. Upon reheating above T_g, the quenched-in nuclei are exposed to the maximum growth rate resulting in recrystallization of the undercooled melt [52]. Hence, the presence of nuclei in the glassy state may play an important role in dictating crystallization tendency from the glassy state.

5.4 ROLE OF ADDITIVES IN MODIFYING SOLID-STATE CRYSTALLIZATION

Single-component amorphous drugs will typically crystallize much faster than the desired shelf life of the product, making it necessary to formulate them with crystallization inhibitors. It has been known for many years that combining an API with a polymer to form an amorphous solid dispersion can yield a solid with improved resistance to crystallization during processing and storage as well as enhanced dissolution properties. Other additives may also be present such as surfactants, which are often added to improve processing or drug release characteristics. In this section, the role of additives in modifying crystallization behavior during storage of amorphous solid dispersions will be discussed. However, before evaluating the mechanism of crystallization inhibition by polymers, it is first necessary to discuss miscibility in multicomponent systems.

5.4.1 Miscibility

For amorphous solid dispersions, it is very important to consider the miscibility of the various components since miscibility, or a lack of miscibility, can dramatically affect the crystallization tendency of the API [60,61]. Excipients can only modify crystallization behavior to a substantial degree when present in the same phase as the drug. Miscibility has been extensively investigated for polymer–solvent system blends of two or more polymers [62]. Following conventions developed in this arena, a miscible solid dispersion consisting of a drug and a polymer can be defined as consisting of a single chemically homogeneous phase where all of the components are intimately mixed at the molecular level, and the properties of the blend are different from the properties of the pure components. Hence, crystallization of the drug automatically leads to immiscibility as the drug is present in a separate crystalline phase. However, phase separation can also be manifested by the existence of two amorphous phases. A drug and a polymer may be only partially miscible, leading to a drug-rich amorphous phase and a polymer-rich amorphous phase. For example, felodipine is only partially miscible with polyacrylic acid [63], and the system readily crystallizes [61]. In contrast, felodipine is miscible with polyvinylpyrrolidone [64] and this polymer is a very effective crystallization inhibitor. Other examples of a partially miscible system are that of itraconazole and Eudragit E100 where the drug is miscible with the polymer only up to a drug loading of around 13%; at higher drug loadings, two phases can be observed [65]. It can be very difficult to determine if an amorphous solid dispersion is fully miscible or not. Typically, the presence of two glass transition events, for example, as measured using differential scanning calorimetry, would strongly suggest that the system contains two amorphous phases. However, the presence of a single T_g is not an unambiguous proof of miscibility. For example, a single T_g event was observed for a colyophilized dextran–trehalose mixture [66], but it was concluded that the system was actually poorly miscible based on the analysis of the pairwise distribution function of the X-ray diffraction data. It was suggested that a small domain size of phase separation may have resulted in only a single T_g event being observed. New approaches are constantly being sought to establish miscibility, with atomic force microscopy being one technique that appears to hold promise [67].

It has also been reported that demixing may be induced in a miscible binary drug–polymer combination by exposure to high relative humidity, which leads to absorption of water and the formation of a ternary system [60,68–71]. While the impact of absorbed water on molecular mobility is quite well known, for water to result in amorphous–amorphous phase separation, there also has to be a thermodynamic driving force. This is created by the change in composition on absorption of moisture [60,72], an effect well recognized in the polymer science field [73,74]. Amorphous solid dispersions that appear to be particularly susceptible to this phenomenon typically contain a very hygroscopic polymer, such as polyvinyl pyrrolidone (PVP), and a drug that is quite hydrophobic [75]. The extent of moisture-induced phase separation will depend on the exact combination of drug, polymer, and storage RH. Thus, exposure to high RH storage conditions can result in the conversion of a formerly miscible dispersion into a two amorphous phase system where it would be anticipated that the drug-rich amorphous phase will have a tendency to rapidly crystallize. Clearly, this type of phase behavior may confound interpretation of accelerated stability testing results and may not reflect the crystallization tendency of the product under less harsh environmental conditions.

5.4.2 Impact of Polymers on Crystallization

The crystallization rate of an amorphous solid dispersion is complex and will depend on several factors, including the crystallization tendency of the drug alone [21], storage temperature [76], water content [77], and the amount and type of polymer used [77,78]. Assuming that the polymer is mixed at a molecular level with the drug, it is of interest to consider what is known about the crystallization inhibitory effect of such additives. Polymers have been shown to decrease the crystallization tendency of amorphous APIs with explanations for their stabilizing ability including, reduction of molecular mobility [79,80], disruption of drug–drug interactions [81], and formation of specific polymer–drug interactions [81–83], while others have considered the effectiveness in the context of the crystallization tendency of the pure amorphous drug [84–86]. Each of these mechanisms will be discussed subsequently; however, it is apparent from the preceding discussions that crystallization is a complex phenomenon and it is unlikely that the stabilizing impact of a polymer can be attributed to a single mechanism.

Many polymers, including most of those used in amorphous solid dispersion formulations, have high glass transition temperatures, typically greater than 100 °C. The T_g of the drug is on the order of $2/3T_m$ (in Kelvin), and therefore drug T_g values typically range from about −45 °C (ibuprofen, T_m of 77 °C) [87] to greater than 100 °C (e.g., telaprevir with a T_g of 105 °C and a melting point of 246 °C [88]). Therefore, if a higher T_g polymer is combined with a drug with a lower T_g, the addition of a polymer can act as an antiplasticizer, decreasing the molecular mobility of the amorphous drug and restricting drug molecules from ordering to form stable nuclei, thus reducing the crystallization tendency. Thus, it would be anticipated that increasing the polymer MW, which will both increase T_g and decrease free volume, will result in more effective crystallization inhibition. In an investigation of the ability of various MW grades of PVP to inhibit the crystallization of a model compound MK-059, it was indeed observed that the crystallization inhibition (as judged from the size of the crystallization exotherm and onset of crystallization measured by DSC) increased with polymer MW [82]. From a theoretical perspective, the higher MW PVP contains fewer chain ends with decreased motion of the main chains, and thus should decrease the free volume and molecular mobility. These observations reinforce that decreasing the molecular mobility through the addition of a polymer can significantly decrease the crystallization tendency of the pure amorphous drug.

Specific interactions between the polymer and the drug also appear to be important in influencing the crystallization tendency from the amorphous state. Fourier transform infrared (FTIR) and Raman spectroscopy were used to show that indomethacin and PVP interacted through intermolecular hydrogen bonding in solid dispersions prepared by solvent evaporation [81]. It was postulated that drug–polymer hydrogen bonding interactions were important for stabilizing amorphous indomethacin by disrupting the self-association of indomethacin dimers, even at low polymer concentrations where the antiplasticizing effect of the polymer is minimal. For indomethacin, the dimeric interactions are critical to achieve the hydrogen bonding patterns present in the prevalent polymorphic forms. Thus, disruption of dimer–dimer interactions by the polymer prevents self-association of the drug molecules into the local structure found in the crystal. In a different study, the ability of two polymers (PVP, MW = 40,000, and polyacrylic acid (PAA), MW = 25,000) to inhibit crystallization of amorphous acetaminophen in solid dispersions prepared by melt quenching was investigated [76]. Following observations that PAA–drug solid dispersions exhibited much longer crystallization times compared with PVP–drug solid dispersions, it was speculated that this difference was caused by a stronger interaction between the carboxylic acid group on

PAA and the hydroxyl group of acetaminophen relative to the carbonyl group of PVP. These two examples suggest that the formation of specific interactions between the drug and the polymer appears to influence the polymer's ability to stabilize the drug in the amorphous state.

The stabilization ability of a particular polymer also appears to depend inherently on the crystallization tendency of the pure amorphous drug. Thus, it has been observed that amorphous ketoconazole does not recrystallize upon reheating until the melting point, and only by adding crystal seeds can recrystallization of the glass be initiated [84]. In the presence of PVP, recrystallization of the amorphous drug, even in the presence of seed crystals, could not be observed for up to 30 days of storage at room temperature and 52% RH. In this instance, no specific drug–polymer interactions could be detected, so the stability of PVP–ketoconazole solid dispersions was attributed to the stability of pure amorphous ketoconazole. Ritonavir, in a solid dispersion with polyethylene glycol (PEG) 8000, is another interesting example of a drug that forms a fairly stable glass. In this solid dispersion, the PEG is crystalline and therefore in a different phase from the drug. In spite of this phase separation, solid dispersions stored in the absence of moisture at 25 °C were stable for over 18 months [86], and this high stability can be attributed to the stability of pure amorphous ritonavir in the absence of the polymer. These studies show that polymers will be more effective stabilizers for compounds that are inherently stable in the glassy state. Hence, the higher the crystallization tendency of the pure amorphous drug, the greater the polymer must modify the amorphous phase of the drug to prevent devitrification.

5.5 ASSESSMENT OF PHYSICAL STABILITY

There are a number of challenges inherent in assessing the physical stability of amorphous solid dispersions. First, physical stability testing needs to be conducted in order to evaluate which is the best polymer from the perspective of inhibiting crystallization, as well as promoting dissolution. Not only must a specific polymer type and grade be selected, but the drug loading must also be determined. This can be a difficult task given the number of potential polymers available, as well as restrictions on the amount of drug substance available early in formulation development. A second challenge arises from the typically low drug loading found in amorphous solid dispersions, whereby significant analytical limitations exist in detecting crystallization when the drug comprises a low overall weight fraction of the dispersion. Third, due to the lack of first principle models that can accurately describe nucleation, growth, and overall crystallization kinetics in amorphous solid dispersion powders as a function of temperature and water content, it is very difficult to perform accelerated stability testing and extrapolate the results of these tests to predict stability under ambient conditions.

Crystallization can either be evaluated qualitatively or quantitatively. Qualitative determinations are useful during formulation design, when it is often sufficient to detect if any crystalline material is present. Quantitative determination of crystallization kinetics is important to set specifications for maximum crystallinity allowed, as well as to help extrapolate kinetic measurements to longer time periods to ensure adequate shelf life. In this section, different methods for physical stability assessment will be discussed, together with their strengths and limitations.

5.5.1 Optical and Polarized Light Microscopy

Most crystalline APIs are anisotropic and therefore show the phenomenon of birefringence. In contrast, amorphous materials are isotropic. Polarized light microscopy (PLM) is a contrast-enhancing technique that improves the image of birefringent materials. Typically, crystalline domains show up as colored and/or bright regions under cross polars, whereas amorphous materials show a similar color to the background, which is often black. Therefore, it is often easy to rapidly detect small quantities of crystalline material in an otherwise amorphous matrix. The advantage of PLM is that only small amounts of material are required; therefore, it is a useful screening technique. The disadvantage is that it can often be difficult to analyze fine powders, and particles can appear birefringent even though there is no crystallinity present due to the presence of strain. One way to circumvent these issues is to spin coat or melt thin films of the solid dispersion onto a glass substrate. Clear films indicate an amorphous sample, while colored regions under cross polarized light indicate crystallization. From initially transparent and colorless films, the evolution of crystal nucleation and growth can be readily monitored.

Microscopy has been used in a number of studies to evaluate the crystallization behavior of amorphous systems, as well as to evaluate optimum polymers for solid dispersions. In a small-scale screening study, the ability of seven different polymers to inhibit the crystallization of several readily crystallizable drugs was evaluated [56]. This involved spin coating a thin film of the drug and polymer onto an optically transparent substrate – essentially solvent casting a solution of the drug and polymer onto a cover slip, followed by rapid evaporation leading to an amorphous solid dispersion. The ability of the polymer to inhibit crystallization during solvent evaporation and short-term storage under different environmental conditions could be readily evaluated by examination of the films under a polarizing light microscope. It was found that compounds with a higher crystallization tendency were more difficult to stabilize requiring more polymer. In addition, the stabilizing ability of different polymers varied dramatically for the different compounds suggesting that the chemistry of the drug and the polymer are important for ensuring the formation of a stable dispersion. The small-scale screening approach and rapid sample evaluation enabled many samples to be studied and their relative crystallization tendency to be evaluated without providing quantitative information. However, it is also possible to study the crystallization of amorphous materials in a much more quantitative way using microscopic techniques. For example, optical microscopy was used to study the nucleation rate of pure amorphous indomethacin as a function of temperature [14]. The nucleation rate was estimated by counting the number of crystallites as a function of time in a particular view and depth of field. A complicated relationship between nucleation rate and temperature was observed, which in part could be accounted for by the formation of different polymorphs. This technique was subsequently applied to evaluate the impact of polymers on the nucleation rate of felodipine as a function of polymer amount and storage conditions [64,77]. It was found that the nucleation rate decreased dramatically with increasing polymer content, but exposure to high RH resulted in an increase in the nucleation rate. However, even though the polymer increased the hygroscopicity of the system, it still exerted an overall protective effect against crystallization by decreasing the nucleation rate relative to a system with no polymer, stored at a comparable RH. Such studies provide a quantitative comparison of the impact of polymers on one aspect of the crystallization process, namely, nucleation. Similar quantitative measurements have been made on crystal growth rates [6,10,57,89]. In one such study [10], the effect of 1 and 2% PVP on the linear growth rate of the β-polymorph of nifedipine as a function of temperature was measured. The polymer was able to decrease the crystal growth rate of nifedipine, with the inhibitory effect becoming more pronounced at lower temperatures, even though the polymer had no influence on T_g at these low concentrations. Using a similar method, the ability of different polymers to inhibit the crystal growth of felodipine as a function of temperature, at temperatures above T_g [90], was measured. It was found that PVP was a more efficient inhibitor of crystal growth than hydroxypropylmethyl cellulose acetate succinate (HPMCAS) for felodipine. Furthermore, the impact of different molecular weight grades of PVP on the growth rate was investigated whereby a small MW dependency of the inhibitory effect of PVP was observed, with higher MW grades being more effective inhibitors [91]. Crystal growth rate measurements provide a quantitative measurement of linear growth rates and enable polymer effectiveness to be directly compared, at least at low polymer concentrations where growth rates can be measured over experimentally accessible time-scales. However, in many instances it is of interest to determine the kinetics of the overall crystallization process and/or the amount of crystallized material.

5.5.2 X-Ray Diffraction

X-ray powder diffraction (XRPD) is one of the most common approaches used to detect and quantitate the presence of crystallinity in amorphous systems, and has been used for many years for this purpose [92]. Detection of crystallinity in amorphous solid dispersions is based on the appearance of diffraction peaks superimposed on the amorphous halo that are greater than the signal-to-noise ratio. With laboratory-based diffractometers, limits of detection are typically in the range of 1–5% crystallinity by weight [93–95]. With synchrotron-based systems, much lower detection limits, as low as 0.2%, have been reported [96]. When attempting to quantitate the extent of crystallinity in amorphous solid dispersions, it is important to prepare calibration samples that are as representative of the dispersions as possible. The best approach appears to be preparing the amorphous solid dispersion, and spiking this material with known amounts of crystalline material [94,97], rather than making a blend of amorphous and crystalline drug with the polymer. This is because interactions between the drug and the polymer in the dispersion impact the shape of the amorphous halo. Thus, a physical mixture of amorphous drug and polymer does not yield the same scattering pattern as the drug mixed at a molecular level with the polymer. Based on these differences, X-ray scattering has also been used to evaluate miscibility, sometimes in combination with pairwise distribution functions [63,66]. In addition, application of multivariate analysis has been found to be useful in improving the quantitative ability of X-ray methods [94,98].

There are a number of literature reports describing the use of XRPD to evaluate amorphous solid dispersions [98–101] and a few examples will be discussed. The crystallization kinetics of felodipine solid dispersions prepared at different weight ratios with either PVP or HPMCAS were evaluated as a function of RH storage conditions. It was found that the PVP dispersions were very stable when stored at low RH conditions, but rapidly crystallized when stored above 75% RH, 25 °C. In contrast, the HPMCAS dispersions were much more robust to the effects of absorbed moisture, and showed fairly similar crystallization kinetics regardless of the storage conditions [102]. Trace crystallinity in tacrolimus amorphous dispersions was evaluated with a variety of techniques including XRPD [97]. For this system, near-infrared spectroscopy was found to be a superior technique to XRPD. Amorphous solid dispersions of resveratrol formulated with different polymers were subjected to a variety of harsh storage conditions including elevated temperatures and high RH, in order to compare the ability of different polymers to prevent crystallization [103]. The evolution of crystallinity in the samples was followed using XRPD. It was found that PVP and Eudragit E100 formed the most stable dispersions with resveratrol and these systems were remarkably stable to crystallization, remaining amorphous for several months. In another study, crystallinity in several drug polymer blends immediately following melt extrusion was evaluated using a combination of XRPD and DSC [104]. For some samples, diffraction peaks were observed with XRPD and a melt endotherm was present in the DSC thermogram. The presence of crystalline material for the samples was thought to arise from incomplete melting of the drug during the extrusion process.

5.5.3 Solid-State Nuclear Magnetic Resonance Spectroscopy

Solid-state nuclear magnetic resonance spectroscopy (SSNMR) has been used to investigate a number of aspects of amorphous solid dispersions, including miscibility [105,106], molecular mobility [107], intermolecular interactions [108], and ionization state [109], and for the detection [106] and quantitation of crystallinity [110] in dispersions.

SSNMR has been applied in a number of studies to evaluate mobility in single-component and multicomponent amorphous systems [16,107,111–114]. ^{19}F SSNMR, which can be used if the drug or polymer contains one or more fluorine atoms, was used to compare and contrast the mobility of flufenamic solid dispersions prepared either with PVP or hydroxypropylmethyl cellulose (HPMC) at a 20% drug loading. The crystallization kinetics were also evaluated using NMR spectroscopy and correlated with differences in the observed molecular mobility. By measuring parameters relating to the rotational molecular mobility, it was observed that the molecular mobility of flufenamic acid was higher in a HPMC dispersion relative to the PVP dispersion. The stability of flufenamic acid to crystallization was found to be better in the PVP dispersion compared with the HPMC dispersion, and this was attributed to the lower molecular mobility in the former dispersion.

One advantage of SSNMR spectroscopy is the ability to extract information about the local environment of the drug, which can enable miscibility to be determined at different length scales [105,106]. Thus, by measuring different relaxation times, some estimate of the length scales of drug and polymer domains can be achieved, based on the assumption that the relaxation time of the solid dispersion will be monoexponential but different from the monoexponential relaxation profiles of the individual amorphous materials, whereas a phase-separated system will be biexponential. Using nifedipine as the model drug, Aso et al. found that the T_1 and $T_{1\rho}$ relaxation times were monoexponential for HMPC and PVP dispersions [105]. In contrast, dispersions with α,β-poly (*N*-5-hydroxypentyl)-L-aspartamide were monoexponential for the T_1 relaxation times, but biexponential for $T_{1\rho}$ relaxation measurements. From these results, it was concluded that the latter system was phase separated into two amorphous phases whereby the size of the phase-separated domains was between 5 and 50 nm.

SSNMR has been used to evaluate drug–polymer hydrogen bonding interactions [106]. For a dispersion containing 30% acetaminophen blended with PVP, it was concluded that the phenol group in acetaminophen formed a hydrogen bond with the carbonyl group of the polymer, with a hydrogen bond distance of approximately 1.75 ± 0.04 Å [106]. Based on additional measurements, it was further demonstrated that acetaminophen and PVP are mixed on a molecular level.

SSNMR can be a very sensitive technique for the detection of crystallinity in amorphous solid dispersions. Using ^{19}F SSNMR, a detection limit of around 3% crystalline material in a solid dispersion of an experimental drug with HPMCAS was observed [110]. Using this method, it was found that incomplete crystallization of the drug occurred from the amorphous dispersion. It was postulated that there is a composition of drug and polymer for which crystallization does not occur because the drug is below its solubility limit in the polymer.

5.5.4 Calorimetric Methods

Calorimetric methods, in particular DSC, are widely used for the characterization of amorphous solid dispersions. Due to the nonisothermal character of DSC and hence the potential for sample changes during heating, some care must be exercised when interpreting results. One of the most common applications of DSC is to determine the T_g or T_g values of the solid dispersion. It is of interest to know the T_g since crystallization is known to be faster above T_g than below T_g [14]. In addition, samples above T_g become sticky and may be difficult to handle. Therefore, for amorphous dispersions, it is desirable to have a dispersion T_g that is above the temperature range of storage and processing conditions. While it would be nice to be able to predict the likelihood of crystallization of the dispersion from a knowledge of the T_g, it appears that crystallization does not correlate well with T_g for systems that contain more than one component [77,103].

Another major application of DSC is to determine if crystallization occurs during heating, or if a melting endotherm (indicative of the presence of crystalline material arising from processing or storage) is present. Experiments can be performed in different ways. For example, the solid dispersion can be held at a certain temperature using the DSC in isothermal mode to evaluate the crystallization kinetics at this temperature. Yoshihashi et al. [115] used this approach to characterize the crystallization kinetics in an amorphous solid dispersion. They first heated the solid dispersions to above the melting point of the drug to eliminate any residual crystalline material. This heating stage was followed by rapid cooling and then the samples were heated to the temperature of interest. By holding the samples isothermally until crystallization could be detected from the appearance of an exothermic peak, the induction times for the crystallization process could be evaluated. For flurbiprofen and tolbutamide dispersions with PVP, it was observed that the induction time for crystallization increased with an increasing amount of PVP; in contrast, increasing the temperature decreased the induction time. By assuming that the induction time is inversely proportional to the nucleation rate, it was concluded that PVP decreased the nucleation rate. PVP was more effective at increasing the induction times for tolbutamide compared with flurbiprofen, and this is thought to be due to better specific interactions between tolbutamide and PVP relative to those between flurbiprofen and PVP. Using the same method, induction times for solid dispersions of naproxen and HPMCAS were also measured [116]. In this instance, it was observed that the induction time for naproxen crystallization decreased with temperature until a minimum value was obtained at around 335 K and then started to increase with increasing temperature. This result highlights the complex temperature dependence of nucleation whereby it seems that the nucleation rate reached a maximum at 335 K. For these examples, induction times were relatively short, making it possible to study them with DSC. Unfortunately, for many systems of interest, the long induction times preclude direct measurement of the crystallization event using this approach.

An alternative approach, which does not require long periods of equilibration in the instrument, is to store samples under certain conditions, and then analyze them by DSC using a more typical scanning mode. The crystallization of solid dispersions of PVP with either nifedipine or phenobarbital was studied using this approach [30]. Solid dispersions were stored between 25 and 80 °C in a dry environment, and at certain time intervals samples were analyzed using DSC. Nifedipine dispersions crystallized when heated above T_g and the magnitude of this exotherm was used to estimate the amount of amorphous material remaining. With increasing storage times, more material crystallized and thus the size of the recrystallization exotherm decreased. The fraction of amorphous nifedipine remaining was calculated from the ratio of the exotherm obtained following storage at a particular time to that of a freshly prepared amorphous sample. For this experimental approach, the inherent assumption is that all of the remaining amorphous nifedipine in the solid dispersion will be able to crystallize during heating. For the phenobarbital–PVP dispersions, an alternative approach was used. For these systems, the change in the heat capacity as the system passed through T_g was evaluated using modulated DSC. For a pure compound, it has been demonstrated that the change in heat capacity at T_g correlates well with the extent of crystallinity [117]. This is because, as the drug crystallizes, the ΔC_p value for the T_g event of the remaining amorphous fraction decreases in a manner proportional to the loss of the amorphous material due to crystallization. By assuming that a similar relationship holds for a solid dispersion, the fraction of amorphous drug remaining at time $R(t)$ can be calculated [30]:

$$R(t) = \frac{\Delta C_{p_{T_g}}(t)}{\Delta C_{p_{T_g}}(0)}, \tag{5.12}$$

where $\Delta C_{p_{T_g}}(t)$ is the change in heat capacity of the glass transition event after time t and $\Delta C_{p_{T_g}}(0)$ is the corresponding heat capacity change at T_g for the freshly prepared sample with no crystalline material. However, it should be noted that this expression does not account for the contribution of the polymer to the $\Delta C_{p_{T_g}}$ term and would therefore only be an appropriate approximation for dispersions containing a minor quantities of polymer. For these types of experiments, the time taken for 10% of the material to crystallize (t_{90}) is often determined by plotting the fraction of amorphous material remaining as a function of time. This value can then be compared for systems subjected to different storage temperatures or formulated with different polymers. Using this approach, it was observed that for a given storage temperature, t_{90} increased by a factor of 100–1000 times in the presence of 10% PVP relative to the value for pure drug. This result again highlights how effective some polymers are as crystallization inhibitors.

Crystallization rates have also been estimated using the magnitude of the melt endotherm observed for a solid dispersion subjected to different storage conditions and times. This approach involves comparing the melt endotherm of the partially crystallized solid dispersion (normalized to percentage of drug present) with the melt endotherm of the pure drug with the ratio yielding the fraction of drug crystallized. An investigation of efavirenz–PVP solid dispersions [118] that employed this method found that the maximum extent of drug crystallization was only about 40% of the total drug present. It should be recognized that there are potential caveats associated with using the melting transition of a drug substance when there is polymer present, particularly when intimately mixed as in an amorphous solid dispersion. The first issue is that melting point depression can occur in the presence of a polymer [119,120]. A large change in melting temperature will affect the magnitude of the enthalpy value measured. The next issue is that mixing of the drug and the polymer during melting (which can alternatively be viewed as the dissolution of the drug into the polymer) can reduce the magnitude of the observed melting endotherm, sometimes by an appreciable amount. Clearly, this will result in an underestimation of the percent crystallinity and, in some cases, even though crystalline drug is present at room temperature, no melting endotherm will be observed [121]. A final concern is that a different polymorphic form may crystallize from the solid dispersion, with a different enthalpy of fusion from the initial crystalline material.

5.5.5 Vibrational Spectroscopy

Both Raman and IR spectroscopy have been used to provide information about solid dispersions, including measurements of crystallization rates, as well as for determination of the presence of drug–polymer intermolecular interactions. Raman spectroscopy is useful for discriminating between crystalline and amorphous materials since the Raman spectra are generally sharper and less susceptible to sample preparation compared with IR spectra, which typically show broader peaks. In addition, the different selection rules for IR and Raman spectroscopy often result in the drug having a stronger Raman spectrum than the excipient, while excipients and drugs both typically have strong IR absorption peaks. Thus, Raman spectroscopy is a useful technique to evaluate crystallization in amorphous solid dispersions. A very dramatic example of the enhanced sensitivity of Raman spectroscopy to the drug versus the excipients is provided by ibipinabant, a cannabinoid receptor antagonist [122]. This substance was formed into an amorphous solid dispersion with PVP and then the dispersion was formulated into a tablet so that the weight fraction of the drug present in the tablet was only 1%. Despite the very low weight fraction of total drug present in the tablet, using Raman spectroscopy it was possible to detect the presence of 5% crystalline material, which equates to less than 0.05% of the total mass of the sample. This was possible due to the extremely strong Raman scattering of ibipinabant with distinct peaks in a spectral region where the excipients did not have any signal, as well as discernable differences in the spectra of the amorphous and crystalline forms. By constructing calibration curves for the percent crystallinity, it was also possible to quantitatively monitor the crystallization kinetics as a function of time. Another advantage of Raman spectroscopy is that it can be coupled with a microscope, enabling high spatial resolution chemical images to be obtained. Raman mapping has been used to discriminate between amorphous and crystalline domains in solid dispersions of troglitazone with PVP [123]. This system is quite complex since troglitazone has two asymmetric carbons and normally crystallizes as a mixture of two diastereomeric pairs, namely, RR/SS and RS/SR. The RR/SS crystals could be distinguished from the RS/SR crystals using Raman spectroscopy, as could the amorphous solid. Using Raman mapping, the distribution of the crystalline and amorphous materials in the solid dispersions subjected to different heating regimes could be evaluated. It was found that the amorphous form of troglitazone coexisted with the crystalline RR/SS and RS/SR forms in some of the solid dispersions. Thus, by using Raman mapping, this complex system could be characterized in terms of the presence and distribution of different diastereomeric pairs/amorphous materials in the presence of PVP.

Raman spectroscopy has also been used to monitor crystallization of amorphous solids on contact with aqueous media [124], and thus can be used to help assess the risk for crystallization during dissolution. By coupling a fiber optic probe to a Raman spectrometer, it is possible to remotely sample the system of interest. This enables Raman spectra to be obtained from systems subjected to the experimental conditions of interest. The crystallization kinetics of amorphous felodipine powder, slurried in a pH 6.8 buffer, were monitored using a Raman immersion probe that was inserted into the slurry. By obtaining spectra at frequent time intervals, the crystallization kinetics of felodipine could be readily monitored.

It is more difficult to use IR spectroscopy for quantitation of crystallinity in amorphous solid dispersion for several reasons, including sampling, the typically strong IR absorption by the excipient, and the broader peaks present in the IR spectrum. However, FTIR spectroscopy is a very useful technique to investigate the state of mixing between the drug and the polymer in an amorphous solid dispersion and to evaluate if specific drug–polymer interactions are present. As already discussed, an amorphous drug can be either molecularly dispersed in the polymer matrix or, if the system is only partially miscible, present as separate drug- and polymer-rich domains. By evaluating drug–polymer interactions, FTIR spectroscopy can be used to help understand if the drug is interacting with the polymer at the molecular level, or if the system shows little interaction. It is very important to obtain the appropriate control spectra when undertaking such measurements. FTIR spectroscopy was used to characterize interactions in amorphous solid dispersions of PVP and indomethacin [81]. In this study, it was observed that hydrogen bonds were formed between the carboxylic acid function of the drug and the carbonyl acceptor function of the polymer. These drug–polymer hydrogen bonds led to the disruption of the carboxylic acid dimers formed in the pure amorphous drug; in other words, the drug–polymer interactions disrupted the drug–drug interactions. The interpretation of the solid dispersion spectra was aided by obtaining reference spectra of both the pure amorphous drug (rather than the crystalline forms) and the (dry) polymer in order to elucidate the presence or absence of drug–polymer hydrogen bonds. It was noted that comparing spectra of solid dispersions with those of physical mixtures (of amorphous drug and polymer) of the same composition proved to be the easiest way to identify the presence of specific interactions. Drug–polymer interactions in amorphous dispersions prepared with felodipine have been extensively probed using IR spectroscopy [64]. The strength of drug–polymer hydrogen bonding for dispersions prepared with three polymers (PVP, HPMC, and HPMCAS)) was compared. It was found that PVP formed the strongest hydrogen bonds with felodipine; this was inferred from the shift in the position of the drug NH group in the solid dispersions using the position of the peak in the pure amorphous drug as the reference point. Furthermore, the presence of concentration-dependent specific interactions between the drug and the polymer indicates that the two components are mixed, at least to some extent, at the molecular level, in agreement with the other experimental measurements, including T_g measurements. IR spectroscopic data can also be obtained following manipulation of the environmental conditions, for example, as a function of temperature. This may be relevant for amorphous solid dispersions subjected to elevated temperatures, for example, during the melt extrusion process. In the case of a felodipine and PVP dispersion, it was found that the system remained miscible even when heated to the melting temperature of the drug, with somewhat weakened drug–polymer hydrogen bonding interactions persisting at higher temperatures [68]. More sophisticated data analysis methods have also been applied in order to evaluate solid dispersion miscibility. In one example, principal components analysis on IR spectra of solid dispersion and physical mixture was used to evaluate miscibility of the solid dispersions [63]. In this study, it was found that the score values of the physical mixtures were similar to those of the dispersions of equivalent concentrations for immiscible systems. In contrast, for the miscible systems, much larger differences were seen between the score values obtained from the solid dispersions and those of the physical mixtures.

One important parameter that needs to be controlled when trying to infer the presence of drug–polymer hydrogen bonds using IR spectroscopy is the potential contamination by absorbed water. It is essential to eliminate the effects of sorbed water through rigorous drying of the sample prior to IR analysis since water absorbed into an amorphous solid dispersion may obscure absorption bands arising from drug–polymer interactions, making them much more difficult to characterize [68]. In addition, it has been found that for some drug–polymer combinations, absorbed water can actually disrupt drug–polymer hydrogen bonding, resulting in amorphous–amorphous phase separation [60,102]. This would clearly be undesirable since the polymer concentration in the drug phase would be reduced, and thus it would be anticipated that the system would be much more susceptible to crystallization.

There are some new IR-based methodologies under development that are of potential interest for the characterization of amorphous solid dispersions. In photothermal induced resonance spectroscopy, an atomic force probe is used to measure the local thermal expansion following irradiation by IR light produced from a tunable laser [125]. Because an AFM probe is used to detect the signal, the spatial resolution is much improved compared with conventional IR imaging, and it is possible to obtain spectra from submicrometer domains as well as chemical images with submicrometer resolution. In a study of felodipine and polyacrylic acid dispersions, the AFM topographical image showed the presence of the domains, but could not provide any chemical information about their chemical composition. However, nano-IR imaging showed that the discrete domains were drug rich, and these domains were embedded in a polymer-rich continuous phase [126]. This methodology thus appears to be an emerging technique that offers the potential for improved understanding of the microstructure of amorphous solid dispersions.

5.5.6 Second-Order Nonlinear Optical Imaging of Chiral Crystals

As discussed previously, detection of small amounts of crystalline material in solid dispersions can be challenging, in particular when the drug loading is low and the dispersion is predominantly comprised of a polymer. A promising technique for extending the limits of detection is second-order nonlinear optical imaging of chiral crystals (SONICC) that has recently been reported to show very high sensitivity for detecting tiny amounts of crystalline material present in an otherwise amorphous sample [127]. The detection limits for crystalline griseofulvin and chlorpropamide in predominantly amorphous sample of pure drug were estimated to be a factor of 10^5 better than with conventional methods such as lab XRPD. SONICC is dependent on second harmonic generation (SHG), which is a nonlinear optical effect that is possible only with noncentrosymmetrically ordered materials, and hence can usually only detect crystals with chiral space groups. Thus, this technique is not universally applicable to all APIs, for example, compounds that are racemic or achiral. However, it holds promise for detecting small amounts of crystalline material of SHG-active compounds in a solid dispersion since the carrier polymers employed typically will have low-to-zero SHG activity, potentially allowing the detection of nanoscopic domains (on the order of 100 nm) of crystalline material. In a study with naproxen–HPMCAS dispersions, the ability to detect crystalline material with Raman spectroscopy, XRPD, and SONICC was compared [128]. Crystalline material could be detected in the amorphous solid dispersion immediately after production using the SONICC technique, but could not be detected based on the XRPD patterns or the Raman spectra. A comparison of the images obtained from the SONICC studies and the XRPD are shown in Figure 5.4. The bright areas in the SONICC images depict crystalline material, while the dark areas arise from amorphous sample. It is clear that crystallinity can be sensitively detected with the SONICC imaging method. Based on these results, it is certainly of interest to further explore the application of SONICC for the detection of the early stages of crystallization in amorphous solid dispersions.

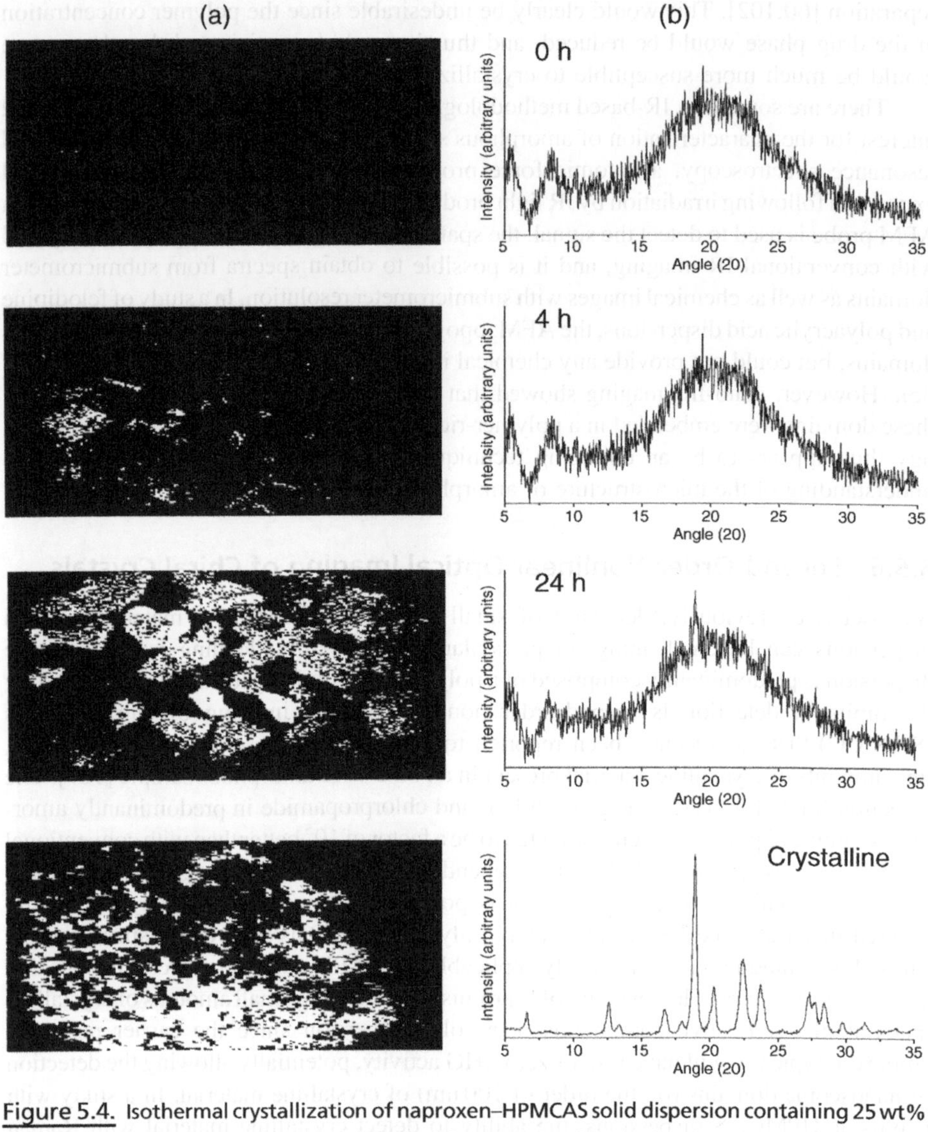

Figure 5.4. Isothermal crystallization of naproxen–HPMCAS solid dispersion containing 25 wt% naproxen monitored at 55 °C using SONICC (a) and PXRD (b). The bright areas in the SONICC image are areas of naproxen crystallinity and the width of the SONICC images are 400 μm. For reference, data obtained from pure crystalline naproxen powder are also shown.

5.6 CRYSTALLIZATION IN AQUEOUS ENVIRONMENTS

It is, of course, essential that the amorphous solid dispersion does not crystallize as the matrix hydrates during the dissolution process, and that the supersaturated solution that is produced by dissolution of the amorphous formulation is maintained for a biologically

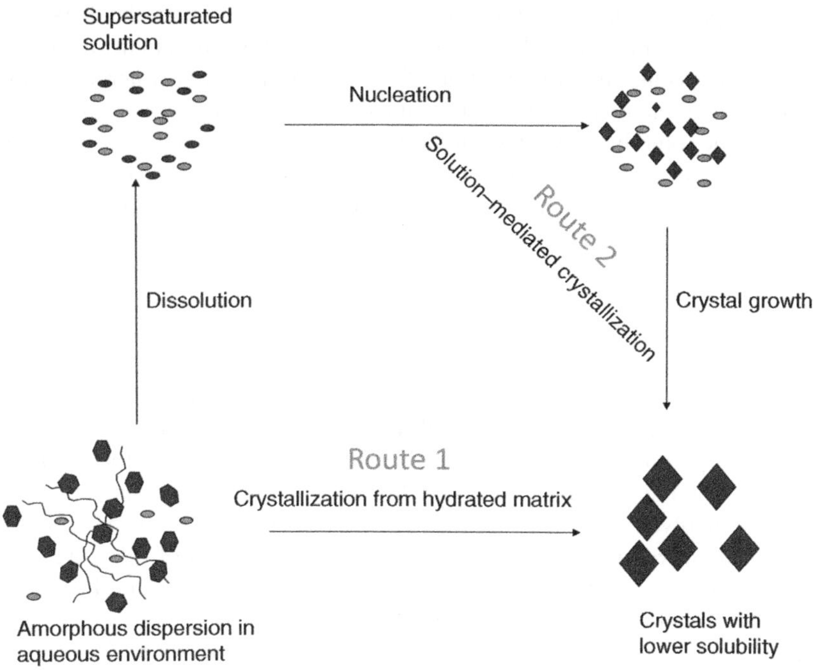

Figure 5.5. Schematic showing the different possible routes of crystallization following the addition of an amorphous solid dispersion to an aqueous medium.

relevant time frame. Thus, there are two routes to crystallization in an aqueous environment that must be evaluated: matrix crystallization and solution crystallization, as summarized in Figure 5.5. Crystallization of the matrix during the dissolution process will be considered first.

5.6.1 Matrix Crystallization

The dissolution rate of an amorphous solid dispersion is likely to be extremely complex and will depend on, among other factors, the drug:polymer ratio, the solubility and viscosity of the polymer, and the solubility of the drug (which will depend on whether it remains amorphous or crystallizes), as well as on other factors normally affecting dissolution processes, for example, surface area. At high drug loadings, it would be anticipated that the drug will control the rate of release, while at low drug loadings, the polymer will have a controlling role [129]. For many solid dispersions, the drug is the minor component, present at a level of <25% by weight, and hence the polymer plays a determining role in impacting the release rate. For such cases, the rate of drug release relative to the rate of matrix crystallization needs to be considered. For example, in the case of HPMCAS dispersions, the polymer shows low solubility at gastric pH conditions, and thus release of drug from the dispersion will be slow until the higher pH environment of the small intestine is encountered. Hence, during transit through the stomach, which may take a couple of hours depending on if the state is fed or fasted, the dispersion will undergo hydration, with the possibility of crystallization of the drug in the matrix. Due to the increase in molecular mobility resulting from the absorption of water, crystallization will be more favorable than during storage under low humidity conditions. If crystallization occurs, the solubility advantage of the amorphous dispersion will be lost. Thus, the dissolution behavior will be more similar to that of the crystalline material rather than that of the amorphous form. It is therefore essential to understand the propensity for an amorphous dispersion to crystallize when added to an aqueous environment, in order to determine if the matrix crystallization kinetics are faster than the dissolution kinetics. There are relatively few studies that have investigated crystallization kinetics in amorphous solid dispersions exposed to dissolution media. In one study, undissolved amorphous solid dispersion was removed at different time points, dried, and analyzed by DSC in order to determine if any crystalline material was present [130]. Two different solid dispersions, prepared with either HPMCAS or PVP–VA, and containing 40% by weight of an experimental drug, were evaluated. It was noted that for the PVP–VA dispersion, the polymer dissolved faster than the drug, increasing the concentration of the drug in the undissolved material. This system was observed to crystallize within 15 min. The crystallization was attributed to a decreased protective effect of the polymer due to its dissolution, as well as to the plasticization effect of the absorbed water. In contrast, for the HPMCAS dispersion, the polymer and the drug released at similar rates, and the undissolved dispersion did not contain detectable crystalline drug. Clearly, matrix crystallization is an important consideration and warrants further study.

5.6.2 Solution Crystallization

If the drug is released into solution faster than the rate of matrix crystallization, then it is possible to generate supersaturated solutions, in particular if release is delayed until the small intestine where the fluid volume is low [131]; in other words, dissolution is occurring under nonsink conditions. Therefore, it is important to evaluate the impact of the polymer on inhibiting crystallization from supersaturated solutions. At this juncture, it is important to differentiate between higher solution concentrations achieved by solubilization, where the equilibrium solubility of the crystalline drug is increased, versus supersaturated solutions, in which there is a thermodynamic driving force for crystallization because the saturation solubility is exceeded. At higher concentrations, polymers, similar to other additives such as cyclodextrins and surfactants, can potentially increase the equilibrium solubility of a compound [132]. For example, a concentration of 0.25% PVP was found to double the aqueous solubility of acetazolamide [132]. Solubilization by a polymer can be readily determined by measuring the solubility of a crystalline solid in equilibrium with a solution containing different concentrations of the polymer of interest. Once equilibrium is reached and if more of the crystalline material has been dissolved, then it implies that the polymer has led to an increase in the equilibrium solubility. For solubilized solutions, there is no thermodynamic driving force for crystallization, and hence the solubility increases will be maintained as long as the additive is not diluted. Alternatively, dissolution of an amorphous formulation can lead to a supersaturated solution, where the solution concentration of the compound exceeds the equilibrium solubility of the crystalline material, and thus crystallization is favorable from a thermodynamic perspective. While it might be thought that it is better to achieve solubilization rather than supersaturation due to the risk for crystallization and thus the loss of the solubility advantage in supersaturated solutions, flux across a membrane is enhanced for supersaturated solutions, but not for solubilized solutions [133,134]. Therefore, there is a great deal of interest in creating and maintaining supersaturated solutions. Polymers can act as crystallization inhibitors, even when present at low concentrations, sustaining supersaturation without altering the driving force for crystallization. While polymers are often used to fulfill this role, there is little mechanistic understanding of the properties of a polymer that make it a good inhibitor of crystallization for a given compound. It should also be noted that many studies evaluate the ability of a given additive to inhibit precipitation rather than crystallization. Precipitation is the appearance of a new phase, and most of the methods used to evaluate precipitation, for example, turbidity, do not provide information about the structure of the new phase, for example, if it is crystalline or amorphous. This consideration is important because if the new phase that evolves in the supersaturated solution is an amorphous phase, then supersaturation can be maintained [135,136]. In contrast, if the new phase is crystalline, supersaturation will be depleted.

5.6.2.1 Influence of Additives on Precipitation. There have been a number of studies that have screened the impact of various additives on the precipitation behavior of compounds of interest. The methodology used for such studies has an impact on whether such studies should be considered as quantitative evaluations of the impact of additives on supersaturation duration or qualitative investigations. For quantitative investigations, it is necessary to first determine the equilibrium crystalline solubility in the exact test medium in which supersaturation duration is to be evaluated; this will provide information about the solubility-enhancing potential of the excipients and will allow the actual degree of supersaturation to be determined. Second, the initial degree of supersaturation (rather than concentration of drug added) should be constant for the different additives if the additive effectiveness is to be compared. Following precipitation, the precipitated material needs to be immediately separated from the solution and assayed using a quantitative method. Ideally, the separated phase would also be analyzed to better understand the desupersaturation process. Unfortunately, such quantitative studies are often challenging to perform experimentally. For example, separating the precipitated material from the solution phase by filtration can be difficult since material can adsorb to the filter, the sample handling needed for the filtration process can induce crystallization, and very small precipitated species can pass through the filter. Challenges also exist with methods such as centrifugation. This has led to the development of more qualitative high-throughput screening methods, where the precipitate is evaluated *in situ* using a nonspecific method such as turbidity measurements. Turbidity measurements enable the time when a new phase evolves to be evaluated without sample handling. Turbidity measurements were used in one study to evaluate the precipitation behavior of danazol in the presence and absence of many different types of polymers, added at concentrations of 0.1–0.001% [137]. It was observed that certain cellulose derivatives appeared to be effective precipitation inhibitors; however, the turbidity profiles as a function of time were quite complex for the different systems examined, suggesting that a more detailed examination of the mechanism of the turbidity evolution could yield additional insight into the role of the polymers. There are also a number of studies that have specifically probed the impact of polymers on both crystal nucleation and growth and examples will be discussed in the following sections.

5.6.2.2 Impact of Polymers on Crystal Nucleation. Nucleation is a first-order phase transition that involves the formation of a new phase from a supersaturated solution. According to classical nucleation theory, the nucleation rate J is given by the following expression [2,138]:

$$J = A\exp\left(-\frac{B}{\ln^2 S}\right), \tag{5.13}$$

where A and B are constants for a given system at a particular temperature and S is the supersaturation ratio equal to c/c^*, where c is the actual solution concentration and c^* is the saturation solubility in the medium of interest. The exponent $B/\ln^2 S$ represents the energy barrier for nucleation. Equation 5.13 clearly shows that the nucleation rate is highly sensitive to the supersaturation and, therefore, when evaluating the crystallization tendency of supersaturated solutions generated from amorphous solid dispersions, it is very important to have knowledge of the extent of the supersaturation of the solution.

Because it is very difficult to measure the nucleation rate directly, induction time measurements are typically used to evaluate the impact of additives on nucleation. The experimentally observed induction time is the time taken for the system to nucleate and also includes the additional time taken for the material to grow to a detectable size. The induction time is typically regarded as being inversely proportional to the nucleation rate. Nucleation rates are notoriously sensitive to experimental factors, including volume of the test solution, the presence of impurities, and stirring rate. Therefore, when comparing the impact of different polymers on the nucleation behavior, it is important not only to compare systems at the same supersaturation but also to maintain similar experimental conditions.

Polymers have been shown to have very dramatic effects on nucleation rates; however, not all polymers are equally effective and the most effective polymer can vary from compound to compound. The impact of polymers on the induction times of celecoxib, efavirenz, and ritonavir was studied at the supersaturation expected to be generated by dissolution of the amorphous form [139]. The induction times in the presence of low concentrations of polymers (5 ppm) increased by a factor of at least 2 and up to a factor of 5 in the presence of the most effective polymers. For ritonavir and efavirenz, the best polymers were similar for the two compounds; however, for celecoxib, different polymers were found to be effective at modifying the nucleation behavior. The effectiveness of the various polymers appeared to depend on the hydrophobicity of the polymer relative to that of the drug. A strong dependence of the induction time on supersaturation was also observed in the presence of HPMC for celecoxib. At the highest supersaturation measured, the induction time for crystallization was about 20 min in the presence of the polymer, compared with less than 5 min in the absence of the polymer. However, when the supersaturation was halved, no crystallization was observed in the presence of the polymer over a 6 h period. In another study, the impact of HPMC on the nucleation rate of felodipine was investigated, again at the supersaturation expected to be generated upon dissolution of the amorphous solid, assuming no crystallization has occurred [140]. HPMC was found to be effective at inhibiting both crystal nucleation and growth. However, the impact on crystal growth was marginal (factor of 2 reduction), whereas the nucleation rate was reduced by a factor of 1000. Ensuring that nucleation does not occur in this system over a biologically relevant time period is clearly important for this system. Although difficult from an experimental perspective, evaluating the impact of polymers on crystal nucleation is clearly an area of investigation that warrants more attention. Some of the challenges that will need to be met to tackle this topic include evaluating the impact of endogenous substances such as surfactants on nucleation rates, further investigation of the formation of colloidal species [136,141] in highly supersaturated solutions and their impact on nucleation behavior, and selecting correct hydrodynamic conditions for nucleation experiments that mimic *in vivo* conditions.

5.6.2.3 Inhibition of Crystal Growth. It is also important to evaluate the impact of polymers on crystal growth. Crystals will grow in supersaturated solutions following the formation of nuclei of a critical size, or may be present in the formulation having been generated during storage or the manufacturing process. Therefore, it is useful to consider how effective polymers are at inhibiting crystal growth, and thus preventing the depletion of the supersaturation caused by dissolution of the amorphous formulation. Equation 5.14 describes an empirical relationship between the overall rate of growth (R_g) of a crystal and the supersaturation ratio (S), where the growth rate constant is k_G and g is the overall order [142]:

$$R_G = k_G S^g. \tag{5.14}$$

For the overall order (g), a value of 1 indicates diffusion-controlled growth, whereas a value of 2 indicates integration-controlled growth [143].

In some cases, polymers can be extremely efficient inhibitors of crystal growth. Take, for example, sulfathiazole and PVP [144]. In water, the linear growth rate of sulfathiazole crystals was observed to be a linear function of the supersaturation ratio less than 2. However, in the presence of PVP in solution, the crystal growth rate was altered, as was the growth morphology. When a certain level of PVP was added, the growth was completely blocked. The amount of PVP needed to block complete growth increased as the supersaturation increased and the lower molecular weight grades of PVP were found to be more effective on a per weight basis than the higher molecular weight grades. This was attributed to their faster diffusion rates that enabled polymer molecules to reach the surface of the sulfathiazole crystal and block the diffusion-controlled growth process. Once the polymer molecules were adsorbed onto the surface of the crystal, the process was found to be largely irreversible. This was demonstrated by placing a crystal with adsorbed PVP into a supersaturated sulfathiazole solution that did not contain polymer. No crystal growth was observed for this system, suggesting that PVP does not readily desorb from the crystal surface, and thus continues to inhibit growth even when no longer present in the solution phase.

The studies conducted with sulfathiazole only extended to supersaturation ratios of around 2. This is not necessarily representative of the levels of supersaturation that would be anticipated following dissolution of an amorphous solid dispersion. The crystal growth rates of felodipine were investigated as a function of *S* in the presence and absence of different amounts of HPMC [140]. HPMC was found to adsorb to felodipine crystals, with a plateau in the adsorption isotherm at a polymer concentration of around 1 μg/ml. Quantitative crystal growth rate studies revealed that the impact of HPMC on felodipine crystal growth rate plateaued at around this polymer concentration; thus, increasing the polymer concentration much beyond this value did not lead to further inhibition of crystal growth. The ability of HPMC to inhibit crystal growth was found to be highly dependent on *S*. At low values of *S*, HPMC was quite effective at slowing crystal growth. However, at a *S* of 10, the supersaturation expected to be reached by dissolving an amorphous solid dispersion, the highest concentration of HPMC reduced the crystal growth rate by only a factor of about 2. Thus, in the presence of crystal seeds, HPMC was not able to prevent the rapid desupersaturation of supersaturated felodipine solutions.

Studies with ritonavir have also shown that polymers are much more effective at low *S*, but lose effectiveness as the driving force for crystallization increases [145]. Thus, at a supersaturation corresponding to the amorphous solubility of ritonavir, polymers could not reduce desupersaturation rates to a large extent in solutions to which seed crystals were added. Furthermore, it was observed that the chemistry of the polymers was very important for determining how effectively they can alter the kinetics of crystal growth when present at comparable concentrations. For ritonavir, the more hydrophilic and very hydrophobic polymers were found to have little impact on crystal growth rates, while a group of moderately hydrophobic polymers was found to be most effective at reducing the crystal growth rate. The extent of inhibition was also dependent on the ionic strength of the medium, as well as on the pH when the polymer was ionizable [146]. As the ionic strength increased, the polymers became relatively more effective at inhibiting growth. For anionic polymers, it was noted that they are better modifiers of crystal growth when they are ionized, relative to when no charge is present. It should also be noted that a polymer that is effective for one compound may not be a useful inhibitor for another compound. For example, PVP/VA was observed to be one of the better inhibitors evaluated for celecoxib and efavirenz crystal growth, but it was completely ineffective for ritonavir [147]. Finally, it has been observed that the optimum polymer for inhibiting nucleation may differ from the optimum polymer for delaying crystal growth [147]. This observation suggests that polymers may interfere with different crystallization processes via different mechanisms and underscores further the general lack of understanding of how polymers disrupt crystallization.

5.7 SUMMARY AND OUTLOOK

With the increasing number of developmental compounds exhibiting poor aqueous solubility, there is little doubt that amorphous solid dispersions will be more widely used for enhancing oral drug delivery. Some of the main challenges that remain to be addressed include a mechanistic understanding of crystallization in the solid material so that predictive physical stability models can be developed, the detection of the early stages of crystallization in low drug loading formulations, and understanding the tendency for phase transformations in the gastrointestinal tract. There are considerable opportunities for research in the area of amorphous solid dispersions, in particular in understanding failure mechanisms and in the optimization of excipients for the purpose of inhibiting crystallization.

REFERENCES

1. Hancock, B.C. and Zograf, G. (1997) Characteristics and significance of the amorphous state in pharmaceutical systems. *J. Pharm. Sci.*, 86(1):1–12.
2. Mullin, J.W. (2001) *Crystallization*, 4th edition, Butterworths, London.
3. Hoffman, J.D. (1958) Thermodynamic driving force in nucleation and growth processes. *J. Chem. Phys.*, 29:1192–1193.
4. Jackson, K.A. (1969) On the theory of crystal growth: the fundamental rate equation. *J. Cryst. Growth*, 5:13–18.
5. Frenkel, J. (1932) Note on a relation between the speed of crystallization and viscosity. *Phys. Z. Sowjetunion*, 1:498–500.
6. Wu, T. and Yu, L. (2006) Surface crystallization of indomethacin below T_g. *Pharm. Res.*, 23(10):2350–2355.
7. Crowley, K.J. and Zografi, G. (2003) The effect of low concentrations of molecularly dispersed poly(vinylpyrrolidone) on indomethacin crystallization from the amorphous state. *Pharm. Res.*, 20(9):1417–1422.
8. Wu, T. and Yu, L. (2006) Origin of enhanced crystal growth kinetics near T_g probed with indomethacin polymorphs. *J. Phys. Chem. B*, 110(32):15694–15699.
9. Sun, Y., Xi, H., Ediger, M.D., and Yu, L. (2008) Diffusionless crystal growth from glass has precursor in equilibrium liquid. *J. Phys. Chem. B*, 112(3):661–664.
10. Ishida, H., Wu, T.A., and Yu, L.A. (2007) Sudden rise of crystal growth rate of nifedipine near T_g without and with polyvinylpyrrolidone. *J. Pharm. Sci.*, 96(5):1131–1138.
11. Hikima, T., Adachi, Y., Hanaya, M., and Oguni, M. (1995) Determination of potentially homogeneous-nucleation-based crystallization in *o*-terphenyl and an interpretation of the nucleation-enhancement mechanism. *Phys. Rev. B*, 52(6):3900–3908.
12. Avrami, M. (1939) Kinetics of phase change: I. General theory. *J. Chem. Phys.*, 7(12):1103–1112.
13. Avrami, M. (1940) Kinetics of phase change: II. Transformation–time relations for random distribution of nuclei. *J. Chem. Phys.*, 8(2):212–224.
14. Andronis, V. and Zografi, G. (2000) Crystal nucleation and growth of indomethacin polymorphs from the amorphous state. *J. Non-Cryst. Solids*, 271(3):236–248.
15. Zhou, D., Schmitt, E.A., Zhang, G.G., Law, D., Vyazovkin, S., Wight, C.A., and Grant, D.J.W. (2003) Crystallization kinetics of amorphous nifedipine studied by model-fitting and model-free approaches. *J. Pharm. Sci.*, 92(9):1779–1792.
16. Aso, Y., Yoshioka, S., and Kojima, S. (2000) Relationship between the crystallization rates of amorphous nifedipine, phenobarbital, and flopropione, and their molecular mobility as

measured by their enthalpy relaxation and ^{1}H NMR relaxation times. *J. Pharm. Sci.*, 89(3):408–416.

17. Zhou, D.L., Zhang, G.G.Z., Law, D., Grant, D.J.W., and Schmitt, E.A. (2002) Physical stability of amorphous pharmaceuticals: importance of configurational thermodynamic quantities and molecular mobility. *J. Pharm. Sci.*, 91(8):1863–1872.
18. Bhugra, C. and Pikal, M.J. (2008) Role of thermodynamic, molecular, and kinetic factors in crystallization from the amorphous state. *J. Pharm. Sci.*, 97(4):1329–1349.
19. Tang, X.C., Pikal, M.J., and Taylor, L.S. (2001) A spectroscopic investigation of hydrogen bond patterns in crystalline and amorphous phases in dihydropyridine calcium channel blockers. *Pharm. Res.*, 19(4):477–483.
20. Miyazaki, T., Yoshioka, S., Aso, Y., and Kawanishi, T. (2007) Crystallization rate of amorphous nifedipine analogues unrelated to the glass transition temperature. *Int. J. Pharm.*, 336(1):191–195.
21. Marsac, P.J., Konno, H., and Taylor, L.S. (2006) A comparison of the physical stability of amorphous felodipine and nifedipine systems. *Pharm. Res.*, 23(10):2306–2316.
22. Siniti, M., Carre, J., Bastide, J.P., Letoffe, J.M., and Claudy, P. (1993) The thermal behavior of hexitols: 2. Polymorphism of dulcitol. *Thermochim. Acta*, 224:105–110.
23. Baird, J.A., Van Eerdenbrugh, B., and Taylor, L.S. (2010) A classification system to assess the crystallization tendency of organic molecules from undercooled melts. *J. Pharm. Sci.*, 99(9):3787–3806.
24. Carpentier, L., Desprez, S., and Descamps, M. (2003) Crystallization and glass properties of pentitols: xylitol, adonitol, arabitols. *J. Therm. Anal. Calorim.*, 73(2):577–586.
25. Alig, I., Braun, D., Lagendorf, R., Voigt, M., and Wendorff, J.H. (1997) Simultaneous ageing and crystallization processes within the glassy state of a low molecular weight substance. *J. Non-Cryst. Solids*, 221(2–3):261–264.
26. Bhugra, C., Shmeis, R., Krill, S.L., and Pikal, M.J. (2008) Prediction of onset of crystallization from experimental relaxation times: II. Comparison between predicted and experimental onset times. *J. Pharm. Sci.*, 97(1):455–472.
27. Vyazovkin, S. and Dranca, I. (2005) Physical stability and relaxation of amorphous indomethacin. *J. Phys. Chem. B*, 109(39):18637–18644.
28. Vyazovkin, S. and Dranca, I. (2007) Effect of physical aging on nucleation of amorphous indomethacin. *J. Phys. Chem. B*, 111(25):7283–7287.
29. Okamoto, N. and Oguni, M. (1996) Discovery of crystal nucleation proceeding much below the glass transition temperature in a supercooled liquid. *Solid State Commun.*, 99(1):53–56.
30. Aso, Y., Yoshioka, S., and Kojima, S. (2004) Molecular mobility-based estimation of the crystallization rates of amorphous nifedipine and phenobarbital in poly(vinylpyrrolidone) solid dispersions. *J. Pharm. Sci.*, 93(2):384–391.
31. Apperley, D.C., Forster, A.H., Fournier, R., Harris, R.K., Hodgkinson, P., Lancaster, R.W., and Rades, T. (2005) Characterisation of indomethacin and nifedipine using variable-temperature solid-state NMR. *Magn. Reson. Chem.*, 43(11):881–892.
32. Masuda, K., Tabata, S., Sakata, Y., Hayase, T., Yonemochi, E., and Terada, K. (2005) Comparison of molecular mobility in the glassy state between amorphous indomethacin and salicin based on spin-lattice relaxation times. *Pharm. Res.*, 22(5):797–805.

33. Alie, J., Menegotto, J., Cardon, P., Duplaa, H., Caron, A., Lacabanne, C., and Bauer, M. (2004) Dielectric study of the molecular mobility and the isothermal crystallization kinetics of an amorphous pharmaceutical drug substance. *J. Pharm. Sci.*, 93(1):218–233.
34. Baird, J.A., Santiago-Quinonez, D., Rinaldi, C., and Taylor, L.S. (2012) Role of viscosity in influencing the glass-forming ability of organic molecules from the undercooled melt state. *Pharm. Res.*, 29(1):271–284.
35. Ediger, M.D., Harrowell, P., and Yu, L. (2008) Crystal growth kinetics exhibit a fragility-dependent decoupling from viscosity. *J. Chem. Phys.*, 128(3):034709.
36. Hodge, I.M. (1987) Effects of annealing and prior history on enthalpy relaxation in glassy polymers: 6. Adam–Gibbs formulation of nonlinearity. *Macromolecules*, 20:2897–2908.
37. Bohmer, R., Ngai, K.L., Angell, C.A., and Plazek, D.J. (1993) Nonexponential relaxations in strong and fragile glass formers. *J. Phys. Chem.*, 99:4201–4209.
38. Adam, G. and Gibbs, J.H. (1965) On temperature dependence of cooperative relaxation properties in glass-forming liquids. *J. Chem. Phys.*, 43(1):139–146.
39. Williams, M.L., Landel, R.F., and Ferry, J.D. (1955) The temperature dependence of relaxation mechanisms in amorphous polymers and other glass-forming liquids. *J. Am. Chem. Soc.*, 77:3701–3707.
40. Kauzmann, W. (1948) The nature of the glassy state and the behavior of liquids at low temperatures. *Chem. Rev.*, 43(2):219–256.
41. Johari, G.P., Astl, G., and Mayer, E. (1990) Enthalpy relaxation of glassy water. *J. Chem. Phys.*, 92(1):809–810.
42. Velikov, V., Borick, S., and Angell, C.A. (2001) The glass transition of water, based on hyperquenching experiments. *Science*, 294(5550):2335–2338.
43. Hancock, B.C. and Zografi, G. (1994) The relationship between the glass transition temperature and the water content of amorphous pharmaceutical solids. *Pharm. Res.*, 11(4):471–477.
44. Duddu, S.P. and Sokoloski, T.D. (1995) Dielectric analysis in the characterization of amorphous pharmaceutical solids: 1. Molecular mobility in poly(vinylpyrrolidone)–water systems in the glassy state. *J. Pharm. Sci.*, 84(6):773–776.
45. Andronis, V., Yoshioka, M., and Zografi, G. (1997) Effects of sorbed water on the crystallization of indomethacin from the amorphous state. *J. Pharm. Sci.*, 86(3):346–357.
46. Schmitt, E., Davis, C.W., and Long, S.T. (1996) Moisture-dependent crystallization of amorphous lamotrigine mesylate. *J. Pharm. Sci.*, 85(11):1215–1219.
47. Marsac, P.J., Konno, H., Rumondor, A.C.F., and Taylor, L.S. (2008) Recrystallization of nifedipine and felodipine from amorphous molecular level solid dispersions containing poly (vinylpyrrolidone) and sorbed water. *Pharm. Res.*, 25(3):647–656.
48. Rumondor, A.C.F., Jackson, M.J., and Taylor, L.S. (2010) Effects of moisture on the growth rate of felodipine crystals in the presence and absence of polymers. *Cryst. Growth Des.*, 10(2):747–753.
49. Surana, R., Pyne, A., and Suryanarayanan, R. (2004) Effect of preparation method on physical properties of amorphous trehalose. *Pharm. Res.*, 21(7):1167–1176.
50. Kestur, U.S., Ivanesivic, I., Alonzo, D.E., and Taylor, L.S. (2013) Influence of particle size on the crystallization kinetics of amorphous felodipine powders. *Powder Technol.*, 236 197–204.

51. Trasi, N. and Byrn, S. (2012) Mechanically induced amorphization of drugs: a study of the thermal behavior of cryomilled compounds. *AAPS PharmSciTech*, 13(3):772–784.
52. Schroers, J., Masuhr, A., Johnson, W.L., and Busch, R. (1999) Pronounced asymmetry in the crystallization behavior during constant heating and cooling of a bulk metallic glass-forming liquid. *Phys. Rev. B*, 60(17):11855–11858.
53. Muniz, P., De Toro, J.A., and Riveiro, J.M. (2004) Influence of the quenched-in nuclei on the crystallisation of amorphous Ni80B20. *J. Magn. Magn. Mater.*, 272:E1129–E1130.
54. Wang, J.Q., Zhang, H.W., Gu, X.J., Lu, K., Sommer, F., and Mittemeijer, E. (2004) Kinetics of primary nanocrystallization in Al-rich metallic glass with quenched-in nuclei. *Mater. Sci. Eng. A*, 375:980–984.
55. Scudino, S., Eckert, J., and Schultz, L. (2004) Possible influence of quenched-in nuclei on quasicrystal formation in mechanically alloyed Zr57Ti8Nb2.5Cu13.9Ni11.1Al7.5 glassy powder. *J. Mater. Res.*, 19(8):2211–2215.
56. Van Eerdenbrugh, B. and Taylor, L.S. (2010) Small scale screening to determine the ability of different polymers to inhibit drug crystallization upon rapid solvent evaporation. *Mol. Pharm.*, 7(4):1328–1337.
57. Strydom, S., Liebenberg, W., Yu, L., and de Villiers, M. (2009) The effect of temperature and moisture on the amorphous-to-crystalline transformation of stavudine. *Int. J. Pharm.*, 379(1):72–81.
58. Swiatkowski, W. (2000) Effect of different sizes of subcritical nuclei on kinetics of transformation in the solid state. *J. Non-Cryst. Solids*, 262(1–3):162–168.
59. Tomitaka, S., Mizukami, M., Paladi, F., and Oguni, M. (2005) Thermal and dielectric studies of 2,2′-dihydroxybenzophenone: progress of crystal nucleation and growth below the glass transition temperature. *J. Therm. Anal. Calorim.*, 81(3):637–643.
60. Rumondor, A.C.F., Marsac, P.J., Stanford, L.A., and Taylor, L.S. (2009) Phase behavior of poly(vinylpyrrolidone) containing amorphous solid dispersions in the presence of moisture. *Mol. Pharm.*, 6(5):1492–1505.
61. Ivanisevic, I. (2010) Physical stability studies of miscible amorphous solid dispersions. *J. Pharm. Sci.*, 99(9):4005–4012.
62. Patterson, D. (1982) Polymer compatibility with and without a solvent. *Polym. Eng. Sci.*, 22:64–73.
63. Rumondor, A.C.F., Ivanisevic, I., Bates, S., Alonzo, D.E., and Taylor, L.S. (2009) Evaluation of drug–polymer miscibility in amorphous solid dispersion systems. *Pharm. Res.*, 26(11):2523–2534.
64. Konno, H. and Taylor, L.S. (2006) Influence of different polymers on the crystallization tendency of molecularly dispersed amorphous felodipine. *J. Pharm. Sci.*, 95(12):2692–2705.
65. Six, K., Murphy, J., Weuts, I., Craig, D.Q.M., Verreck, G., Peeters, J., Brewster, M., and Van den Mooter, G. (2003) Identification of phase separation in solid dispersions of itraconazole and EUDRAGIT E100 using microthermal analysis. *Pharm. Res.*, 20(1):135–138.
66. Newman, A., Engers, D., Bates, S., Ivanisevic, I., Kelly, R.C., and Zografi, G. (2008) Characterization of amorphous API: polymer mixtures using X-ray powder diffraction. *J. Pharm. Sci.*, 97(11):4840–4856.
67. Lauer, M., Siam, M., Tardio, J., Page, S., Kindt, J., and Grassmann, O. (2013) Rapid assessment of homogeneity and stability of amorphous solid dispersions by atomic force microscopy: from bench to batch. *Pharm. Res.*, 30(8):2010–2022.

68. Marsac, P.J., Rumondor, A.C.F., Nivens, D.E., Kestur, U.S., Stanciu, L., and Taylor, L.S. (2010) Effect of temperature and moisture on the miscibility of amorphous dispersions of felodipine and poly(vinyl pyrrolidone). *J. Pharm. Sci.*, 99(1):169–185.

69. Rumondor, A.C.F. and Taylor, L.S. (2010) Effect of polymer hygroscopicity on the phase behavior of amorphous solid dispersions in the presence of moisture. *Mol. Pharm.*, 7(2):477–490.

70. Rumondor, A.C.F., Wikstrom, H., Van Eerdenbrugh, B., and Taylor, L.S. (2011) Understanding the tendency of amorphous solid dispersions to undergo amorphous–amorphous phase separation in the presence of absorbed moisture. *AAPS PharmSciTech*, 12(4):1209–1219.

71. Vasanthavada, M., Tong, W.Q., Joshi, Y., and Kislalioglu, M.S. (2004) Phase behavior of amorphous molecular dispersions: I. Determination of the degree and mechanism of solid solubility. *Pharm. Res.*, 21(9):1598–1606.

72. Marsac, P.J., Rumondor, A.C.F., Nivens, D.E., Kestur, U.S., Stanciu, L., and Taylor, L.S. (2010) Effect of temperature and moisture on the miscibility of amorphous dispersions of felodipine and poly(vinyl pyrrolidone). *J. Pharm. Sci.*, 99(1):169–185.

73. Pouchly, J. and Patterson, D. (1976) Polymers in mixed solvents. *Macromolecules*, 9(4):574–579.

74. Robard, A., Patterson, D., and Delmas, G. (1977) Delta-chi effect and polystyrene poly(vinyl methyl ether) compatibility in solution. *Macromolecules*, 10(3):706–708.

75. Rumondor, A.C.F. and Taylor, L.S. (2010) Effect of polymer hygroscopicity on the phase behavior of amorphous solid dispersions in the presence of moisture. *Mol. Pharm.*, 7(2):477–490.

76. Miyazaki, T., Yoshioka, S., Aso, Y., and Kojima, S. (2004) Ability of polyvinylpyrrolidone and polyacrylic acid to inhibit the crystallization of amorphous acetaminophen. *J. Pharm. Sci.*, 93(11):2710–2717.

77. Konno, H. and Taylor, L.S. (2008) Ability of different polymers to inhibit the crystallization of amorphous felodipine in the presence of moisture. *Pharm. Res.*, 25(4):969–978.

78. Yoshioka, M., Hancock, B.C., and Zografi, G. (1995) Inhibition of indomethacin crystallization in poly(vinylpyrrolidone) coprecipitates. *J. Pharm. Sci.*, 84(8):983–986.

79. Crowley, K.J. and Zografi, G. (2001) The use of thermal methods for predicting glass-former fragility. *Thermochim. Acta*, 380(2):79–93.

80. Hancock, B.C. and Shamblin, S.L. (2001) Molecular mobility of amorphous pharmaceuticals determined using differential scanning calorimetry. *Thermochim. Acta*, 380(2):95–107.

81. Taylor, L.S. and Zografi, G. (1997) Spectroscopic characterization of interactions between PVP and indomethacin in amorphous molecular dispersions. *Pharm. Res.*, 14(12):1691–1698.

82. Khougaz, K. and Clas, S.D. (2000) Crystallization inhibition in solid dispersions of MK-0591 and poly(vinylpyrrolidone) polymers. *J. Pharm. Sci.*, 89(10):1325–1334.

83. Tantishaiyakul, V., Kaewnopparat, N., and Ingkatawornwong, S. (1996) Properties of solid dispersions of piroxicam in polyvinylpyrrolidone k-30. *Int. J. Pharm.*, 143(1):59–66.

84. Van den Mooter, G., Wuyts, M., Blaton, N., Busson, R., Grobet, P., Augustijns, P., and Kinget, R. (2001) Physical stabilisation of amorphous ketoconazole in solid dispersions with polyvinylpyrrolidone K25. *Eur. J. Pharm. Sci.*, 12(3):261–269.

85. Weuts, I., Kempen, D., Decorte, A., Verreck, G., Peeters, J., Brewster, M., and Van den Mooter, G. (2004) Phase behaviour analysis of solid dispersions of loperamide and two

structurally related compounds with the polymers PVP-K30 and PVP-VA64. *Eur. J. Pharm. Sci.*, 22(5):375–385.

86. Law, D., Krill, S.L., Schmitt, E.A., Fort, J.J., Qiu, Y.H., Wang, W.L., and Porter, W.R. (2001) Physicochemical considerations in the preparation of amorphous ritonavir–poly (ethylene glycol) 8000 solid dispersions. *J. Pharm. Sci.*, 90(8):1015–1025.
87. Dudognon, E., Danède, F., Descamps, M., and Correia, N. (2008) Evidence for a new crystalline phase of racemic ibuprofen. *Pharm. Res.*, 25(12):2853–2858.
88. Kwong, A.D., Kauffman, R.S., Hurter, P., and Mueller, P. (2011) Discovery and development of telaprevir: an NS3-4A protease inhibitor for treating genotype 1 chronic hepatitis C virus. *Nat. Biotechnol.*, 29(11):993–1003.
89. Zhu, L., Wong, L., and Yu, L. (2008) Surface-enhanced crystallization of amorphous nifedipine. *Mol. Pharm.*, 5(6):921–926.
90. Kestur, U.S. and Taylor, L.S. (2010) Role of polymer chemistry in influencing crystal growth rates from amorphous felodipine. *CrystEngComm*, 12(8):2390–2397.
91. Kestur, U.S., Lee, H., Santiago, D., Rinaldi, C., Won, Y.Y., and Taylor, L.S. (2010) Effects of the molecular weight and concentration of polymer additives, and temperature on the melt crystallization kinetics of a small drug molecule. *Cryst. Growth Des.*, 10(8):3585–3595.
92. Black, D.B. and Lovering, E.G. (1977) Estimation of the degree of crystallinity in digoxin by X-ray and infrared methods. *J. Pharm. Pharmacol.*, 29:684–687.
93. Saleki-Gerhardt, A. (1991) *Estimation of Percent Crystallinity in Milled Samples of Sucrose*, University of Wisconsin-Madison.
94. Rumondor, A.C.F. and Taylor, L.S. (2010) Application of partial least-squares (PLS) modeling in quantifying drug crystallinity in amorphous solid dispersions. *Int. J. Pharm.*, 398(1–2):155–160.
95. Surana, R. and Suryanarayanan, R. (2000) Quantitation of crystallinity in substantially amorphous pharmaceuticals and study of crystallization kinetics by X-ray powder diffractometry. *Powder Diffr.*, 15(1):2–6.
96. Nunes, C., Mahendrasingam, A., and Suryanarayanan, R. (2005) Quantification of crystallinity in substantially amorphous materials by synchrotron X-ray powder diffractometry. *Pharm. Res.*, 22(11):1942–1953.
97. Zidan, A.S., Rahman, Z., Sayeed, V., Raw, A., Yu, L., and Khan, M.A. (2012) Crystallinity evaluation of tacrolimus solid dispersions by chemometric analysis. *Int. J. Pharm.*, 423(2):341–350.
98. Palermo, R., Short, S., Anderson, C., Tian, H., and Drennen, J., III. (2012) Determination of figures of merit for near-infrared, Raman and powder X-ray diffraction by net analyte signal analysis for a compacted amorphous dispersion with spiked crystallinity. *J. Pharm. Innov.*, 7(2):56–68.
99. Palermo, R., Anderson, C., and Drennen, J., III. (2012) Review: use of thermal, diffraction, and vibrational analytical methods to determine mechanisms of solid dispersion stability. *J. Pharm. Innov.*, 7(1):2–12.
100. Shah, S., Maddineni, S., Lu, J., and Repka, M.A. (2013) Melt extrusion with poorly soluble drugs. *Int. J. Pharm.*, 453(1):233–252.
101. Paudel, A. and Mooter, G. (2012) Influence of solvent composition on the miscibility and physical stability of naproxen/PVP K 25 solid dispersions prepared by cosolvent spray-drying. *Pharm. Res.*, 29(1):251–270.

102. Rumondor, A.C.F., Stanford, L.A., and Taylor, L.S. (2009) Effect of polymer type and storage relative humidity on the kinetics of felodipine crystallization from amorphous solid dispersions. *Pharm. Res.*, 26(12):2599–2606.

103. Wegiel, L.A., Mauer, L.J., Edgar, K.J., and Taylor, L.S. (2013) Crystallization of amorphous solid dispersions of resveratrol during preparation and storage: impact of different polymers. *J. Pharm. Sci.*, 102(1):171–184.

104. Forster, A., Hempenstall, J., Tucker, I., and Rades, T. (2001) The potential of small-scale fusion experiments and the Gordon–Taylor equation to predict the suitability of drug/polymer blends for melt extrusion. *Drug Dev. Ind. Pharm.*, 27(6):549–560.

105. Aso, Y., Yoshioka, S., Miyazaki, T., Kawanishi, T., Tanaka, K., Kitamura, S., Takakura, A., Hayashi, T., and Muranushi, N. (2007) Miscibility of nifedipine and hydrophilic polymers as measured by ^{1}H NMR spin-lattice relaxation. *Chem. Pharm. Bull.*, 55(8):1227–1231.

106. Pham, T.N., Watson, S.A., Edwards, A.J., Chavda, M., Clawson, J.S., Strohmeier, M., and Vogt, F.G. (2010) Analysis of amorphous solid dispersions using 2D solid-state NMR and ^{1}H T_1 relaxation measurements. *Mol. Pharm.*, 7(5):1667–1691.

107. Aso, Y. and Yoshioka, S. (2006) Molecular mobility of nifedipine–PVP and phenobarbital–PVP solid dispersions as measured by ^{13}C-NMR spin-lattice relaxation time. *J. Pharm. Sci.*, 95(2):318–325.

108. Tatton, A.S., Pham, T.N., Vogt, F.G., Iuga, D., Edwards, A.J., and Brown, S.P. (2013) Probing hydrogen bonding in cocrystals and amorphous dispersions using ^{14}N-^{1}H HMQC solid-state NMR. *Mol. Pharm.*, 10(3):999–1007.

109. Patel, J.R., Carlton, R.A., Yuniatine, F., Needham, T.E., Wu, L.M., and Vogt, F.G. (2012) Preparation and structural characterization of amorphous spray-dried dispersions of tenoxicam with enhanced dissolution. *J. Pharm. Sci.*, 101(2):641–663.

110. Calahan, J.L., Zanon, R.L., Alvarez-Nunez, F., and Munson, E.J. (2013) Isothermal microcalorimetry to investigate the phase separation for amorphous solid dispersions of AMG 517 with HPMC-AS. *Mol. Pharm.*, 10(5):1949–1957.

111. Aso, Y., Yoshioka, S., and Kojima, S. (1998) Determination of the diffusion coefficient of insulin and lysozyme in crosslinked dextran hydrogels by pulsed-field-gradient NMR. *Chem. Pharm. Bull.*, 46(11):1836–1839.

112. Aso, Y., Yoshioka, S., Zhang, J., and Zografi, G. (2002) Effect of water on the molecular mobility of sucrose and poly(vinylpyrrolidone) in a colyophilized formulation as measured by ^{13}C-NMR relaxation time. *Chem. Pharm. Bull.*, 50(6):822–826.

113. Yoshioka, S., Aso, Y., and Kojima, S. (2003) Molecular mobility of lyophilized poly (vinylpyrrolidone) and methylcellulose as determined by the laboratory and rotating frame spin-lattice relaxation times of ^{1}H and ^{13}C. *Chem. Pharm. Bull.*, 51(11):1289–1292.

114. Aso, Y., Yoshioka, S., Miyazaki, T., and Kawanishi, T. (2009) Feasibility of ^{19}F-NMR for assessing the molecular mobility of flufenamic acid in solid dispersions. *Chem. Pharm. Bull.*, 57(1):61–64.

115. Yoshihashi, Y., Iijima, H., Yonemochi, E., and Terada, K. (2006) Estimation of physical stability of amorphous solid dispersion using differential scanning calorimetry. *J. Therm. Anal. Calorim.*, 85(3):689–692.

116. Yoshihashi, Y., Yonemochi, E., Maeda, Y., and Terada, K. (2010) Prediction of the induction period of crystallization of naproxen in solid dispersion using differential scanning calorimetry. *J. Therm. Anal. Calorim.*, 99(1):15–19.

117. Guinot, S. and Leveiller, F. (1999) The use of MTDSC to assess the amorphous phase content of a micronised drug substance. *Int. J. Pharm.*, 192(1):63–75.

118. Yang, J., Grey, K., and Doney, J. (2010) An improved kinetics approach to describe the physical stability of amorphous solid dispersions. *Int. J. Pharm.*, 384(1–2):24–31.

119. Forster, A., Hempenstall, J., Tucker, I., and Rades, T. (2001) Selection of excipients for melt extrusion with two poorly water-soluble drugs by solubility parameter calculation and thermal analysis. *Int. J. Pharm.*, 226(1–2):147–161.

120. Marsac, P.J., Shamblin, S.L., and Taylor, L.S. (2006) Theoretical and practical approaches for prediction of drug–polymer miscibility and solubility. *Pharm. Res.*, 23(10):2417–2426.

121. Sun, Y., Tao, J., Zhang, G.G.Z., and Yu, L. (2010) Solubilities of crystalline drugs in polymers: an improved analytical method and comparison of solubilities of indomethacin and nifedipine in PVP, PVP/VA, and PVAc. *J. Pharm. Sci.*, 99(9):4023–4031.

122. Sinclair, W., Leane, M., Clarke, G., Dennis, A., Tobyn, M., and Timmins, P. (2011) Physical stability and recrystallization kinetics of amorphous ibipinabant drug product by Fourier transform Raman spectroscopy. *J. Pharm. Sci.*, 100(11):4687–4699.

123. Furuyama, N., Hasegawa, S., Hamaura, T., Yada, S., Nakagami, H., Yonemochi, E., and Terada, K. (2008) Evaluation of solid dispersions on a molecular level by the Raman mapping technique. *Int. J. Pharm.*, 361(1–2):12–18.

124. Alonzo, D.E., Gao, Y., Zhou, D.L., Mo, H.P., Zhang, G.G.Z., and Taylor, L.S. (2011) Dissolution and precipitation behavior of amorphous solid dispersions. *J. Pharm. Sci.*, 100(8):3316–3331.

125. Kjoller, K., Felts, J.R., Cook, D., Prater, C.B., and King, W.P. (2010) High-sensitivity nanometer-scale infrared spectroscopy using a contact mode microcantilever with an internal resonator paddle. *Nanotechnology*, 21(18):1–6.

126. Van Eerdenbrugh, B., Lo, M., Kjoller, K., Marcott, C., and Taylor, L.S. (2012) Nanoscale mid-infrared imaging of phase separation in a drug–polymer blend. *J. Pharm. Sci.*, 101(6):2066–2073.

127. Wanapun, D., Kestur, U.S., Kissick, D.J., Simpson, G.J., and Taylor, L.S. (2010) Selective detection and quantitation of organic molecule crystallization by second harmonic generation microscopy. *Anal. Chem.*, 82(13):5425–5432.

128. Kestur, U.S., Wanapun, D., Toth, S.J., Wegiel, L.A., Simpson, G.J., and Taylor, L.S. (2012) Nonlinear optical imaging for sensitive detection of crystals in bulk amorphous powders. *J. Pharm. Sci.*, 101(11):4201–4213.

129. Simonelli, A.P., Mehta, S.C., and Higuchi, W.I. (1976) Dissolution rates of high energy sulfathiazole–povidone coprecipitates II: characterization of form of drug controlling its dissolution rate via solubility studies. *J. Pharm. Sci.*, 65(3):355–361.

130. Qian, F., Wang, J., Hartley, R., Tao, J., Haddadin, R., Mathias, N., and Hussain, M. (2012) Solution behavior of PVP–VA and HPMC–AS-based amorphous solid dispersions and their bioavailability implications. *Pharm. Res.*, 29(10):2766–2776.

131. Schiller, C., Fröhlich, C.P., Giessmann, T., Siegmund, W., Mönnikes, H., Hosten, N., and Weitschies, W. (2005) Intestinal fluid volumes and transit of dosage forms as assessed by magnetic resonance imaging. *Aliment. Pharmacol. Ther.*, 22(10):971–979.

132. Loftsson, T., Friðriksdóttir, H., and Guðmundsdóttir, T.K. (1996) The effect of water-soluble polymers on aqueous solubility of drugs. *Int. J. Pharm.*, 127(2):293–296.

133. Davis, A.F. and Hadgraft, J. (1991) Effect of supersaturation on membrane transport: 1. Hydrocortisone acetate. *Int. J. Pharm.*, 76(1–2):1–8.

134. Theeuwes, F., Gale, R.M., and Baker, R.W. (1976) Transference: a comprehensive parameter governing permeation of solutes through membranes. *J. Memb. Sci.*, 1(1):3–16.

135. Hsieh, Y.L., Ilevbare, G.A., Van Eerdenbrugh, B., Box, K.J., Sanchez-Felix, M.V., and Taylor, L.S. (2012) pH-induced precipitation behavior of weakly basic compounds: determination of extent and duration of supersaturation using potentiometric titration and correlation to solid state properties. *Pharm. Res.*, 29(10):2738–2753.

136. Ilevbare, G.A. and Taylor, L.S. (2013) Liquid–liquid phase separation in highly supersaturated aqueous solutions of poorly water-soluble drugs: implications for solubility enhancing formulations. *Cryst. Growth Des.*, 13(4):1497–1509.

137. Warren, D.B., Benameur, H., Porter, C.J.H., and Pouton, C.W. (2010) Using polymeric precipitation inhibitors to improve the absorption of poorly water-soluble drugs: a mechanistic basis for utility. *J. Drug Target.*, 18(10):704–731.

138. Davey, R.J., Schroeder, S.L.M., and ter Horst, J.H. (2013) Nucleation of organic crystals: a molecular perspective. *Angew. Chem. Int. Ed.*, 52(8):2166–2179.

139. Ilevbare, G.A., Liu, H., Edgar, K.J., and Taylor, L.S. (2012) Maintaining supersaturation in aqueous drug solutions: impact of different polymers on induction times. *Cryst. Growth Des.*, 13(2):740–751.

140. Alonzo, D.E., Raina, S., Zhou, D., Gao, Y., Zhang, G.G.Z., and Taylor, L.S. (2012) Characterizing the impact of hydroxypropylmethyl cellulose on the growth and nucleation kinetics of felodipine from supersaturated solutions. *Cryst. Growth Des.*, 12(3):1538–1547.

141. Ilevbare, G.A., Liu, H., Pereira, J., Edgar, K.J., and Taylor, L.S. (2013) Influence of additives on the properties of nanodroplets formed in highly supersaturated aqueous solutions of ritonavir. *Mol. Pharm.*, 10(9):3392–3403.

142. Mohan, R. and Myerson, A.S. (2002) Growth kinetics: a thermodynamic approach. *Chem. Eng. Sci.*, 57(20):4277–4285.

143. Judge, R.A., Johns, M.R., and White, E.T. (1995) Protein purification by bulk crystallization: the recovery of ovalbumin. *Biotechnol. Bioeng.*, 48(4):316–323.

144. Simonelli, A.P., Mehta, S.C., and Higuchi, W.I. (1970) Inhibition of sulfathiazole crystal growth by polyvinylpyrrolidone. *J. Pharm. Sci.*, 59(5):633–638.

145. Ilevbare, G.A., Liu, H., Edgar, K.J., and Taylor, L.S. (2012) Understanding polymer properties important for crystal growth inhibition: impact of chemically diverse polymers on solution crystal growth of ritonavir. *Cryst. Growth Des.*, 12(6):3133–3143.

146. Ilevbare, G.A., Liu, H., Edgar, K.J., and Taylor, L.S. (2012) Inhibition of solution crystal growth of ritonavir by cellulose polymers: factors influencing polymer effectiveness. *CrystEngComm.* 14:6503–6514.

147. Ilevbare, G.A., Liu, H., Edgar, K.J., and Taylor, L.S. (2013) Impact of polymers on crystal growth rate of structurally diverse compounds from aqueous solution. *Mol. Pharm.*, 10(6):2381–2393.

6

SOLUBILITY AND DISSOLUTION CONSIDERATIONS FOR AMORPHOUS SOLID DISPERSIONS

Grace A. Ilevbare,[1] Wei Xu,[2] Christopher T. John,[3] James D. Ormes,[1] Jesse L. Kuiper,[4] Allen C. Templeton,[5] and Annette Bak[6]

[1]*Discovery Pharmaceutical Sciences, Merck & Co., Rahway, NJ, USA*
[2]*Formulation Sciences, Merck & Co., West Point, PA, USA*
[3]*Discovery Pharmaceutical Sciences, Merck & Co., West Point, PA, USA*
[4]*Analytical Sciences, Merck & Co., West Point, PA, USA*
[5]*Analytical Sciences, Merck & Co., Kenilworth, NJ, USA*
[6]*Discovery Pharmaceutical Sciences, Merck & Co., Kenilworth, NJ, USA*

6.1 SOLUBILITY AND DISSOLUTION: AN OVERVIEW

Solubility- and dissolution-limited absorption have been a problem for pharmaceutical scientists for decades. Key advances have been made on many fronts; however, at the same time, new chemical entities have become less soluble and the problem is still prominent in discovery and preclinical and early pharmaceutical development. With these challenging molecules, pharmaceutical scientists are faced with meeting high exposure demands for early toxicology and clinical studies [1].

Solubility and dissolution are interconnected but not identical properties of a drug molecule. Solubility is the thermodynamic parameter measured by the concentration in solution of a compound in equilibrium with excess solid compound. Solubility is

Pharmaceutical Amorphous Solid Dispersions, First Edition. Edited by Ann Newman.

dependent on drug (e.g., pK_a, lipophilicity, and intermolecular forces in the solid state) and solvent properties (e.g., pH, polarity, and solvent interactions) [1–3]. Due to these multifactorial dependencies, it is difficult to describe solubility accurately in a mathematical fashion across a broad class of compounds and solvents. The reader is encouraged to consult comprehensive texts on the topic such as Refs [2,4,5]. One equation that has been used to describe solubility in nonideal solutions is based on the Hildebrand solubility approach and is shown in Equation 6.1:

$$-\log X_2 = \frac{\Delta H_f}{2.303RT}\left(\frac{T_0 - T}{T_0}\right) + \frac{V_2\Phi_1}{2.303RT}(\delta_1 - \delta_2)^2, \tag{6.1}$$

$$\delta^2_{(\mathrm{Total})} = \delta^2_\mathrm{D} + \delta^2_\mathrm{P}\delta^2_\mathrm{H} \text{ (D: nonpolar effects; P: polar effects; H: hydrogen bonding effects)},$$

where

- X_2 is the solubility expressed in mole fraction,
- ΔH_f is the heat of fusion,
- R is the gas constant,
- T_0 is the melting point,
- T is the temperature of the solution,
- Φ is the volume fraction occupied by the solid,
- V_2 is the molar volume of the liquid solute, and
- δ_1 is the solubility parameter of the solvent and δ_2 of the solute.

Although the equation cannot be applied to predict solubility for all solutes in all solvents, it gives us valuable insight into what factors influence solubility. Accordingly, factors such as higher melting point, higher heat of fusion, and lower solution temperature will translate into lower solubility. The solubility parameters (δ) are energy contributions reflecting intermolecular and solute–solvent interactions and how these relate to solubility is also reflected in Equation 6.1.

An inference from Equation 6.1 is that solubility is difficult to accurately model without experimentation. According to Ref. [6], one of the main challenges in early development is collecting high-quality, accurate drug-relevant solubility data. Solubility measurement methodology must cover the wide diversity of chemical space and account for the relevant parameters. Obtaining high-quality solubility data experimentally requires careful attention to experimental details and may require several iterations.

Dissolution is solubility's kinetic cousin and the first theories describing dissolution in a quantitative sense were published in 1897 by Noyes and Whitney [7]. These now classical equations have since been modified to include elements of Fick's law of diffusion [8], as well as other parameters, leading to the most extant modified Noyes–Whitney equation [9,10]:

$$\mathrm{DR} = \frac{\mathrm{d}X_\mathrm{d}}{\mathrm{d}t} = \frac{A \times D}{\delta}\left(\frac{C_\mathrm{S} - X_\mathrm{d}}{V}\right), \tag{6.2}$$

where

- DR is the dissolution rate,
- A is the effective surface area,
- D is the drug diffusion coefficient,
- δ is the effective diffusion boundary next to the dissolving particle,
- C_s is the equilibrium solubility under luminal conditions,
- X_d is the amount of drug in solution, and
- V is the volume of the dissolution medium [11].

Equation 6.2 illustrates that the dissolution rate will increase with increasing surface area (usually approximated with particle size), diffusivity, and drug solubility. It is also evident that dissolution rate will decrease with increasing thickness of the aqueous boundary and the concentration of drug already in solution.

Why are we concerned about solubility and dissolution in the context of oral absorption? As illustrated in Figure 6.1, a drug generally needs to be in solution to be absorbed, which involves diffusing from the luminal liquid, across the unstirred water layer, and permeating the intestinal mucosa into the blood circulatory system [12]. Generally speaking, dissolution will drive the initial solubilization of solid drug, while intestinal absorption removes solubilized drug from the luminal liquid. The drug can potentially precipitate if the drug concentration in the luminal liquid exceeds the drug solubility limit [13]. In other words, oral absorption is controlled by dynamic processes such as dissolution, diffusion, permeation, and precipitation.

The importance of solubility, dissolution, and permeability is further illustrated by Equation 6.3:

$$F_{\text{absorption}} = C_b \times P \times \text{SA}, \tag{6.3}$$

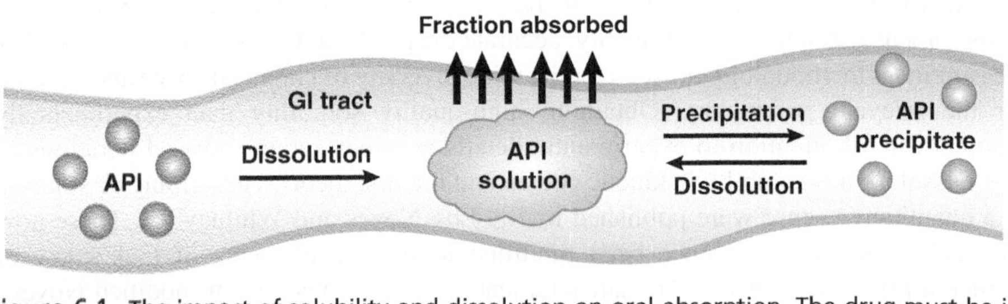

Figure 6.1. The impact of solubility and dissolution on oral absorption. The drug must be in solution to be absorbed. Dissolution will drive the initial solubilization of the API and solubility in the gastrointestinal media will drive potential precipitation events. At the same time, the intrinsic permeability of the compound will drive the flux across the mucosa or, in other words, absorption [13,17].

where $F_{absorption}$ is the mass transport of free dissolved active pharmaceutical ingredient (API) molecules across the intestinal membrane proportional to

- P, the permeability,
- C_b, the free concentration of drug in the unstirred water layer, and
- SA, the surface area available for absorption [14].

In addition to drug concentration in solution and permeability, it can be seen that gastrointestinal (GI) surface area is also important. Surface area in the human GI is highest in the small intestine (jejunum > ileum > duodenum) and permeability generally decreases as a function of distance down the GI tract [14,15]. As a result, there is a need to keep as much drug in solution in the upper GI tract as possible. Therefore, the transit time of the upper GI tract (jejunum ~2 h and ileum ~1.5 h) in terms of desired time in solution before precipitation and the pH-dependent solubility profile of the drug (jejunum pH~6.2–6.4 and ileum pH~6.6–7.4) become important considerations when designing formulations to achieve optimal bioperformance. This is especially important for formulations designed to drive absorption via supersaturation such as amorphous solid dispersions.

How does suboptimal solubility or dissolution manifest itself *in vivo*? Dissolution, in a very general sense, becomes rate limiting when the amount of time required for dissolution is longer than the window of time that the compound is available for absorption across the relevant GI membranes. In this case, the drug either remains undissolved as it is transported further down the GI tract or ultimately dissolves but in a region of the GI in which absorption is poor. For many drugs (but not all) the absorption window will mainly be the small intestine, and the small intestinal transit time is ~3–4 h in humans as mentioned previously. Solubility-limited absorption, on the other hand, mainly occurs when the dose is so high that the gut fluids are saturated with the drug(s) and therefore demonstrate a less than proportional increase in AUC with dose [16]. Stated differently, the excess compound in the GI that cannot be dissolved for lack of solubility remains undissolved and is excreted with the feces, not entering systemic circulation.

Moreover, a number of additional processes such as chemical degradation in the GI tract, active transport, substrate-specific efflux, and first pass metabolism can either facilitate and/or limit systemic exposure. These important processes are beyond the scope of this brief introduction, but the reader is encouraged to consult recent reviews on these topics [16–18].

With these principles in mind, how do we make our solubility and dissolution studies relevant for oral absorption? There are generally two types of experiments described in the literature: dissolution testing as a quality control (QC) measure and/or as a bioperformance test (QC dissolution experiments exist, but is not a focus of this chapter) [19]. Comprehensive books [20] and official guidelines [21–24] describe important dissolution studies that are to be performed as quality control on dosage form batches to ensure consistency, limit batch-to-batch variability, and ensure drug release on stability. These methods are typically developed using USP (United States Pharmacopeia) apparatus 1 (basket) or 2 (paddle) and are robust, well accepted, and therefore suited for routine quality control testing [11]. However, surfactants are often added to the buffer systems in nonphysiological amounts to ensure dissolution of insoluble and/or poorly wetting drugs. The addition of surfactant in these quality methods is often not representative of the contents of actual GI media, thus their relevance to actual drug absorption is usually either not quantified or suspected to be poor. Media that more closely represent GI physiological composition (e.g., pH, viscosity, and surface tension) can be used however, but present a number of challenges for quality control methodology and, as a result, are usually avoided in quality control tests. There are a number of studies that have deployed media mimicking actual physiological GI conditions and these have been found to be useful in developing *in vitro–in vivo* relations (IVIVR) or *in vitro–in vivo* correlation (IVIVC) [11,16,25,26] as well as in directing formulation development decision-making to obtain desired bioperformance. These biorelevant studies are the focus of this chapter.

Absorption in the lower part of the GI (i.e., the colon) is relevant to modified release dosage forms, which are beyond the scope of this brief introduction and chapter. The reader is encouraged to consult Ref. [26] for additional details on simulated colonic fluids. For the proximal portion of the GI (i.e., the stomach and the small intestine), the USP suggests chemical compositions for biorelevant media based on pH (SGF at pH 1.2 and SIF at pH 6.8) with/without Pepsin added (SGF) or with/without pancreatin added (SIF) [27]. The *in vivo* situation is more complex than depicted by these media, so they cannot be expected to adequately represent the full picture on *in vivo* solubility and dissolution for a drug in question. In other words, USP biorelevant media may not give rise to IVIVR or IVIVC. Biorelevant media that more closely resemble fluids found in the proximal GI were proposed by Dressman et al. in 1998 [9] and recently revised [28–30], as summarized in Table 6.1. Adjustments to classical biorelevant media reflected in Table 6.1 include the FaSSGF pH value, adjustment of the pepsin concentration, and inclusion of bile salts, plus a reduction of the media's surface tension. The FeSSGF medium proposed in Table 6.1 is meant to represent a time frame of 75–165 min after a meal. The addition of a 1:1 mixture of milk is to provide an appropriate amount of fat absent from previous media. In the revised version of FaSSIF, the osmolality has been adjusted, the phosphatidylcholine concentration decreased, and the buffer has been changed from phosphate to maleate to increase physiological relevance. Finally, the newest version of FeSSIF was developed to better reflect postprandial changes in pH, buffer capacity, osmolality, and bile salt concentrations, while utilizing concentrations of lipolysis products that reflect the values in the micelle phase of fed intestinal contents. These changes are based on data from human aspirations [16,26].

Compared with typical quality control dissolution method media or previous versions of simulated gastric/intestinal fluids, these more advanced media can be complex, expensive, and have batch-to-batch variability. Therefore, the authors of this chapter advocate considering the purpose of the solubility and dissolution studies to be conducted carefully prior to selection of a media for testing. The relatively simple simulated fluids recommended by USP can be sufficient for rank-ordering compound series in discovery, but researchers may consider investigating more complex media when trying to predict bioperformance of specific formulations intended for important clinical investigations. An additional set of considerations are those involving differences in GI physiology, such as the GI fluid composition of various preclinical and clinical species used in pharmaceutical drug development. This brief introduction covers

TABLE 6.1. Recommended Biorelevant Media Compositions [11,25,26,28–30]

Media Component	Units	FaSSGF-V2	FeSSGF[a]	FaSSIF-V2	FeSSIF-V2
Sodium taurocholate	mmol/l	0.08		3	10
Lecithin	mmol/l	0.02		0.2	2
Pepsin	mg/ml	0.1			
Glyceryl monooleate	mmol/l				5
Sodium oleate	mmol/l				0.8
Pancreatin	U/l				100
Acetic acid	mmol/l		17.2		
Maleic acid	mmol/l			19.2	55.02
Sodium hydroxide	mmol/l			34.8	81.65
Sodium acetate	mmol/l		29.75		
Sodium chloride	mmol/l	68	237.02	68.62	125.5
pH		1.6	5	6.5	5.8
Surface tension	mN/m	42.6			
Buffer capacity	mmol/ΔpH		25	10	25
Osmolality	mOsmol/kg		400	180	390

FaSSGF: fasted-state simulated gastric fluid; FeSSGF: fed-state simulated gastric fluid; FaSSIF: fasted-state simulated intestinal fluid; FeSSIF: fed-state simulated intestinal fluid.
[a]Mixed 1:1 with temperature-treated milk.

human bioperformance as the end target for drug development, but as highlighted in a recent review by Reppas et al. [26], animal GI physiology can be significantly different and species-appropriate models should be considered for toxicology formulations or when developing formulations for animal health purposes.

Another important consideration when conducting biorelevant solubility and dissolution studies is to mimic the fluid dynamics of the GI tract. Biopharmaceutical Classification System (BCS) class I and III drugs (see Table 6.2 for an overview of the BCS [31,32]) are soluble drugs at their projected doses, and hence dissolution experiments are not as sensitive to physiologically relevant sink or hydrodynamics. For BCS class II drugs, where solubility and dissolution are rate limiting for absorption and also where high permeability may create an *in vivo* sink, special attention to the volumes and hydrodynamics of the dissolution test is required. The fasted and fed volumes in the human small intestine have been determined by magnetic resonance imaging to be 45–319 and 20–156 ml, respectively [25]. To create sink for a BCS class II

TABLE 6.2. The Biopharmaceutical Classification System [31,32]

Class I	Class II	Class III	Class IV
High solubility High permeability Generally well absorbed	Low solubility High permeability Solubility/dissolution-limited absorption	High solubility Low permeability Permeability limited absorption	Low solubility Low permeability Generally poorly absorbed

High-solubility drugs have dose/solubility ratio >250 across pH 1–8; high-permeability drugs have permeability $>20 \times 10^{-6}$ cm/s in Caco-2 cell model.

drug, the volume used in dissolution testing will need to be increased or sink simulated by other means such as the equipment (e.g., other apparatus). For BCS class IV drugs, where the drug is removed more slowly by permeation into the bloodstream, a sink condition may not be required for establishing IVIVR.

An interest in making solubility and dissolution studies biorelevant in the hope that the data will correlate with *in vivo* PK data in humans has been established. Why is this important? The drivers are primarily time and cost. On a very simplistic level, it is quite apparent that a dissolution study takes shorter time and costs less than a human PK study to investigate product bioperformance. Without good biorelevant prediction tools, the researcher is left to use expensive and time-consuming trial-and-error to solve the problem of requisite human bioperformance. In addition, if IVIVC can be attained, a biowaiver can be sought with regulatory bodies [33,34], which allows costly and time-consuming *in vivo* bioavailability and bioequivalence studies to be avoided on filing and may also help avoid unnecessary quality testing. IVIVC for immediate release dosage forms can be categorized in four levels. These levels and criteria are summarized in Table 6.3 [25,33–36].

While it is not possible to manipulate GI physiology, chemists and pharmaceutical scientists are in a position to influence absorption through drug properties (e.g., salts, cocrystals, and amorphous forms) and formulation engineering (e.g., particle size control, lipid and self-emulsifying formulations, and amorphous dispersions). Most of these approaches follow the strategy of boosting absorption by solubilizing the drug and also maintaining the drug in solution state by slowing precipitation sufficiently (illustrated in Figure 6.1.) so that the drug can be absorbed before it precipitates, thereby prolonging the supersaturated state [37–40]. Over the last 10–15 years, amorphous solid dispersions have been increasingly popular for boosting solubility and keeping the drug in a supersaturated state to enhance absorption. This chapter seeks to provide important context on the following topics:

- Differences between dispersions, amorphous materials, and crystalline forms (Section 6.2)
- Impact of excipient selection, especially polymers (Section 6.3)
- Provide guidance for solubility and dissolution testing of amorphous dispersions (Section 6.4)

TABLE 6.3. The Four Levels of IVIVC [25,33–36]

Level	Conditions
A	Most desirable and describes a generally linear correlation between *in vitro* dissolution and the fraction of drug absorbed as deconvoluted from *in vivo* plasma levels
B	Describes a correlation between *in vitro* dissolution and mean residence or *in vivo* dissolution time
C	Describes correlation between a single-point dissolution parameter and a pharmacokinetic parameter
Multiple C	Describes multiple point C correlations

6.2 DIFFERENCES BETWEEN CRYSTALLINE API, AMORPHOUS MATERIALS, AND AMORPHOUS DISPERSIONS AS IT PERTAINS TO SOLUBILITY AND DISSOLUTION

As already discussed, scientists are able to influence absorption of poorly soluble molecules by crystal form selection and formulation engineering [39]. Figure 6.2 shows an outline of some of the various API crystal forms available for selection in discovery/early development. The majority of drugs in the 1990s and early 2000s were delivered in preclinical and clinical studies, as well as in marketed products, as a crystalline API form. Crystalline solids are mostly used in pharmaceutical drug formulations because of their physical and chemical stabilities [35]. However, the superior stability of crystalline solids can have a significant adverse impact on dissolution and solubility, especially for BCS class II and IV compounds. Low solubility is a significant barrier to the effective delivery of therapeutic agents; consequently, high-energy forms such as amorphous solids can be utilized to improve solubility and delivery. An amorphous solid may be described as a liquid that has solidified by the removal of thermal energy or solvent, in a way that avoids crystallization and results in material with lack of ordered morphology. Amorphous solids are commonly described as condensed phases that lack the long-range translational order typical of a crystalline solid, although the molecules may have short-range order. Thus, they are sufficiently immobilized to behave as a solid on the macroscopic scale.

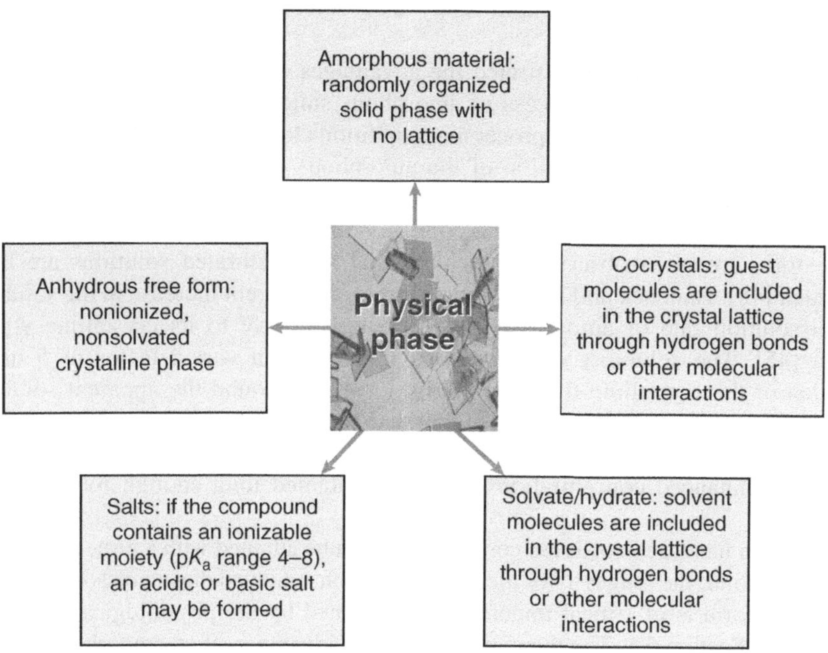

Figure 6.2. Overview of API physical phases available to late-discovery/early-development scientists for API form selection.

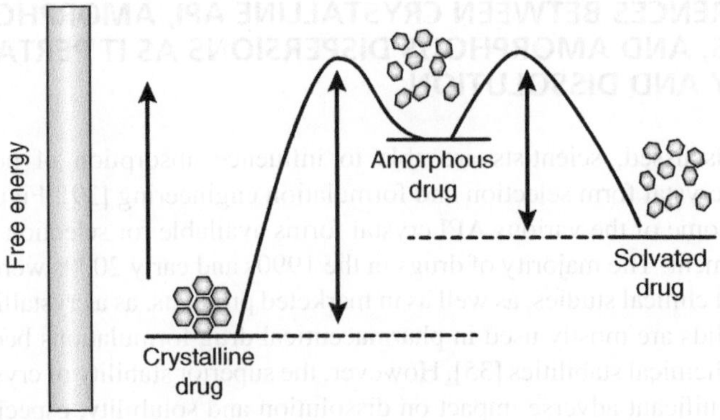

Figure 6.3. Theoretical energy differences between crystalline, solvated, and amorphous materials. These energy differences give rise to very different physicochemical properties and *in vivo* performance.

Due to the lack of long-range order, amorphous materials are more energetic than their crystalline counterparts (see Figure 6.3 for an illustration). This gives rise to certain advantages of amorphous materials relative to crystalline materials, including higher specific volume, higher solubility, and bioavailability, but also certain disadvantages, including risk for lower physical and chemical stability. Conversion of the amorphous phase to a crystalline form with significantly lower solubility can lead to bioperformance loss [41,42].

Many studies have demonstrated the advantages of formulating drugs as amorphous forms [43–53]. The success of amorphous solids as a supersaturating dosage form depends on the choice of processing conditions to yield a pure amorphous solid with no or minimal crystallization of the amorphous drug during storage and upon dosing. In dissolution studies of amorphous solids, API solution concentrations much higher than those achieved using the corresponding thermodynamically stable crystalline form are often observed, indicating that supersaturated solutions are being generated [48]. Hancock and Parks demonstrated an apparent increase in the solubility and dissolution rate of amorphous indomethacin relative to its crystalline γ-polymorph [45]. The solubility of amorphous indomethacin was a factor of 5 higher than that of the crystalline form. Similarly, Law et al. found the apparent solubility of amorphous ritonavir to be 10-fold higher than that of its crystalline counterpart [50]. Typically, these increases in solubility will translate into an increase in oral absorption if the enhanced concentrations can be maintained long enough for absorption to occur.

While an increase in solution concentration can be attained with a neat, unstabilized amorphous solid, the longevity of the supersaturation achieved during dissolution of the amorphous form is of critical importance, as dictated by the physiological conditions discussed in Section 6.1. The theoretical solubility advantage of an amorphous material (illustrated by Equation 6.4) [54] can be estimated using the experimentally determined crystalline solubility (C_{eq}), the free energy difference between the crystalline and amorphous forms ($\Delta G_{a\rightarrow c}$), and the activity of the amorphous solute saturated with water ($\exp[-I(a_2)]$), as shown in the following equation:

$$C_{\text{amorphous}} = C_{\text{eq}} \times \exp[-I(a_2)] \times \exp\left[\frac{\Delta G_{a\rightarrow c}}{RT}\right], \tag{6.4}$$

where R is the gas constant, T is the temperature in Kelvin, and $\exp[-I(a_2)]$ and $\Delta G_{a\rightarrow c}$ can be estimated using the methods presented in Refs [55–57], respectively. The theoretical enhancement in solubility can be several orders of magnitude [45]. However, when comparing predicted solubility values for several amorphous drugs with experimentally reported values in the literature, there is typically a poor correlation. This inconsistency has been attributed, in part, to the rapid crystallization to a lower solubility drug form from the amorphous material upon contact with dissolution fluids or crystallization from the supersaturated solution generated upon dissolution of the amorphous solid [45,48].

Crystallization from solution is a two-part process by which crystals are formed. It involves nucleation and growth, and both processes require a thermodynamic driving force to occur. This thermodynamic driving force is created by the presence of a supersaturated solution. The driving force for crystallization is the difference in chemical potential ($\Delta\mu$) of a molecule in the supersaturated and saturated solution [58]:

$$\Delta\mu = kT \ln\left(\frac{c}{c_0}\right), \tag{6.5}$$

where

- k is the Boltzmann constant,
- T is the absolute temperature,
- c is the solute concentration, and
- c_0 is the equilibrium solute concentration at equilibrium temperature T_0.

For dilute solutions, the supersaturation ratio S is given by the ratio of concentrations [59]:

$$S = \frac{c}{c_0}. \tag{6.6}$$

The magnitude of S is an important factor in determining how long supersaturation can be maintained. A system is always stable (i.e., no crystallization will occur) when $c - c_0 < 0$. A system is considered to be metastable when $c - c_0 > 0$; the concentration is above the equilibrium solubility and an increase in bioavailability can be observed since spontaneous crystallization is improbable. Under these conditions, nucleation, precipitation, deposition, and growth of a new phase are possible. Spontaneous nucleation occurs when $c - c_0 \gg 0$. It is possible to prolong the metastable state using certain additives [58]. This observation is of great interest since it means that the solution concentration can be increased and maintained with a low risk of crystallization. Guzmán

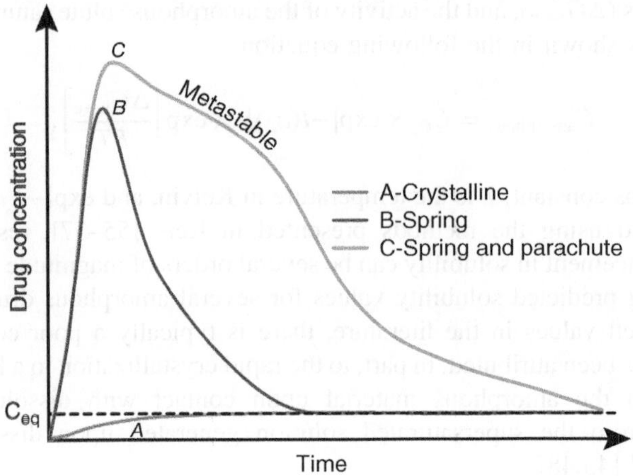

Figure 6.4. Graphical illustration of spring and parachute effect [60].

et al. [60] described the generation and stabilization of a metastable supersaturated state using the "spring and parachute" model (Figure 6.4). Higher energy forms such as amorphous solids that transiently increase solubility relative to that of the stable crystalline form were likened to the "spring." The higher energy state of amorphous solids can translate into an increase in solubility and drug absorption, but rapid crystallization back to the more stable and less soluble crystalline form can limit this benefit [60]. Additives that inhibit crystallization were described as "parachutes"; they maintain supersaturation for an extended time period. The enhanced and prolonged solubilization could lead to further increases in absorption and bioavailability.

The use of amorphous solid dispersions has become a well-known strategy for inhibiting crystallization, as the amorphous solid dispersions can have increased physical stability over neat amorphous material. For example, the crystallization tendency of AMG-517, an insoluble compound, was tested on a neat amorphous drug and various drug loads in a HPMC-AS (hydroxypropyl methyl cellulose acetate succinate) solid dispersion. It was found that the HPMC-AS polymer slows the crystallization rate of amorphous AMG-517 [61]. Polymers can prevent or delay crystallization of amorphous drug during dissolution (*in vitro* and in the GI) as well as stabilizing amorphous drug against recrystallization in solid dosage forms to provide adequate shelf life. Higher concentration–time profiles may be attained with amorphous solid dispersions compared with those achieved with the pure amorphous API, indicating that certain polymer excipients function to further enhance solution concentrations by inhibiting API crystallization in solution. Using beagle dogs as an animal model, Law et al. studied the absorption of poly(ethylene glycol) (PEG)–ritonavir amorphous solid dispersions with different drug loadings [51]. The PEG–ritonavir solid dispersions were prepared using the solvent evaporation–fusion method. Amorphous solid dispersions containing 10–30% drug exhibited significant increases in exposure over crystalline drug. For example, a 10% amorphous solid dispersion exhibited increases of 22- and 13.7-fold in

AUC and C_{max}, respectively. The increase in solution concentration was higher than that reported for pure amorphous ritonavir (based on *in vitro* study results) [50]. Similarly, an animal bioavailability study in dogs was performed to compare a capsule containing 1:2 (w/w) itraconazole:HPMC-P (hydroxypropylmethyl cellulose phthalate) amorphous solid dispersion, prepared by spray-drying, with those filled with crystalline itraconazole [62]. The itraconazole:HPMC-P amorphous solid dispersion exhibited significantly higher C_{max} and bioavailability than the crystalline material alone. The C_{max} was ~17–35-fold higher compared with crystalline material, while the systemic exposure was 511–1911% of that observed from crystalline itraconazole. Finally, an oral bioavailability study was performed in fasted cynomolgus monkeys using a spray-dried amorphous solid dispersion of a poorly water-soluble drug molecule (AMG-517) and HPMC-AS [47]. Pharmacokinetic studies of AMG-517 in capsule [15 wt% AMG-517 in HPMC-AS blended with 5 wt% SDS (sodium dodecyl sulfate)] in monkeys increased AUC and C_{max} by 163 and 145%, respectively, in comparison to an OraPlus suspension control, indicating a significant improvement in exposure. In addition, this study provides a view of the power of amorphous dispersions over crystal engineering approaches, as AMG-517 is a sorbic acid cocrystal in the OraPlus vehicle that was used for early clinical studies of the compound [63].

The above-mentioned case studies demonstrate that amorphous solid dispersions provide improved physical stability over neat amorphous material and enhanced oral bioavailability for poorly soluble compounds over crystalline APIs. We also illustrated the physiological importance of solubility, dissolution, and supersaturation in formulation design and characterization. The next section will discuss the impact of polymers on dissolution and solubility of amorphous solids.

6.3 THE RELATIONSHIP OF POLYMER PROPERTIES WITH SOLUBILITY, DISSOLUTION, AND SUPERSATURATION

Numerous pharmaceutically acceptable polymers have been used as carriers to prepare solid dispersions in the literature [64]. These polymers help maintain the amorphous state of the drug upon storage and are critical to the dissolution and solubilization of the drug upon *in vitro/in vivo* release. Although not an exhaustive list, the following five common classes of polymers are found in the literature:

- Cellulose-derived polymers [65] (i.e., methyl cellulose, hydroxyl propyl methyl cellulose, and hydroxyl propyl methyl cellulose acetate succinate)
- PEGs (polyethylene glycols of various chain lengths) and pegylated materials such as surfactants [66] (i.e., poloxamer [67,68], vitamin E TPGS, Solutol™/Soluplus™)
- Acrylate-based polymers (i.e., EUDRAGITS™)
- Polyvinyl pyrrolidine-derived polymers/copolymers [69,70] (i.e., PVP or PVP VA)
- Biodegradable polylactide/glycolide polymers

The general structure of each polymer class can be found in Figure 6.5.

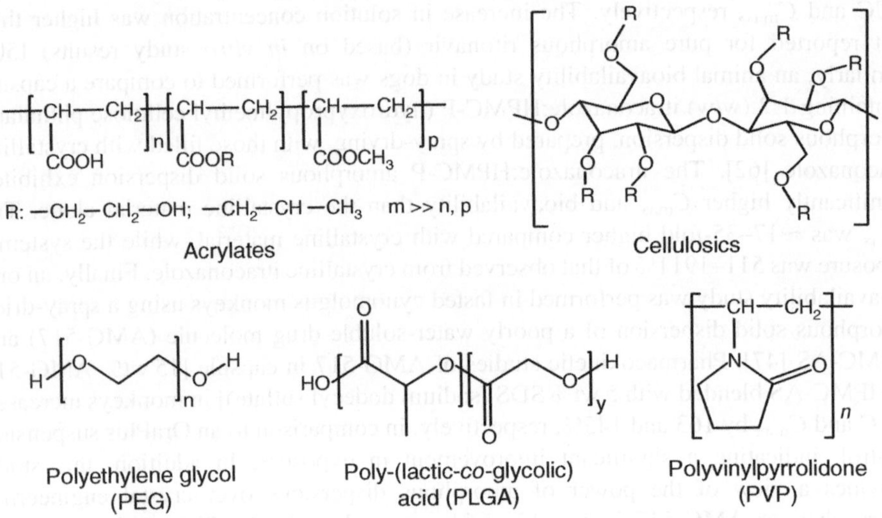

Figure 6.5. Structures of common polymers used for solid dispersions.

Typically, physical properties and structure are used to classify polymers into different classes. Polymer properties can affect the storage conditions, shelf life, packaging/processing considerations, dissolution, and solubility of amorphous dispersions. Some of the key polymer properties include size, molecular weight, types of functional groups, ionization state, hygroscopicity, and hydrophilicity/hydrophobicity balance. These properties, presented in Table 6.4, should be considered when selecting a polymer to improve solubilization, dissolution rate and supersaturation. The following are some of the key considerations:

1. Selection of a polymer with an optimum hydrophobicity for the drug substance. For instance, hydrophobic molecules may have enhanced interactions with polymers containing a similar hydrophobicity. The drug and polymer may form nonspecific hydrophobic interactions that can hinder nucleation and crystal growth [59].
2. Selection of a polymer with functional groups favorable for acid/base or hydrogen bonding interactions with the drug substance. Examples include selecting PVP [66] as a hydrogen bond acceptor for an acidic drug or HPMC-AS [65] as a hydrogen bond donor to a basic drug. Often, crystalline lattices form by hydrogen bond donor/acceptor interactions. By selecting a polymer with functional groups capable of hydrogen bonding with the drug, a polymer may disrupt the crystal nucleation and/or growth processes, thus leading to an increase in supersaturation.
3. Screening for polymers, which provide solubility enhancement or supersaturation in relevant *in vitro* experiments [65]. Screening efforts are required in part due to the lack of detailed physical understanding of the process of crystallization (nucleation or growth) for a given molecule. Careful consideration of polymer properties will help guide selection and rank order screening experiment priority.

TABLE 6.4. Important Polymer Properties

Polymer Property	Key Considerations Around Each Parameter
Size/molecular weight	A descriptor of the average size of the polymer. Other parameters, such as glass transition temperature, are affected by molecular weight
Polydispersity	A parameter for describing the polymer size distribution
Functional groups present	A consideration for molecular interactions such as hydrogen bonding, which can affect physical stability, dissolution rate, and supersaturation properties
Impurity profile	A parameter affecting drug substance chemical stability and dispersion safety profile
Hydrophilic/lipophilic balance	A parameter for matching polarity with the drug substance. Matching polarity may be necessary for favorable physicochemical interactions affecting physical stability, dissolution rate, and supersaturation
Ionization state/apparent pK_a	A parameter for pH-dependent applications and is predictive of potential acid/base interactions with the drug substance. Given the pH gradient that occurs in the digestive tract, ionization state can be critical to consider for *in vivo* performance
Glass transition temperature	This thermal parameter is closely linked to the molecular mobility of an amorphous material
Aqueous solubility and dissolution rate	Parameters affecting aqueous dissolution, solubilization, and supersaturation performance *in vitro* and *in vivo*
Organic solvent solubility	A parameter to consider for processing via the spray-drying process
Thermal stability	A parameter to consider for processing via the hot melt extrusion process
Hygroscopicity	A parameter affecting storage conditions of dispersions and potential for moisture uptake
Biocompatibility	Parameters affecting the safety/tolerability profile of the polymer and ability for polymer to be metabolized *in vivo*

6.3.1 The Impact of Polymer Properties on Dissolution, Solubility, and Supersaturation

Dissolution of a solid dispersion depends on the proportional quantities and properties of both the drug and the carrier [71]. Craig presented a simplified model of the process by which drug particles are released during dissolution [72]. This model assumes that there is a highly concentrated polymer layer at the dissolving surface (at least at high polymer content) through which the drug must pass before it is released into the bulk solution phase. The two main mechanisms for drug release from solid dispersions that have been described in the literature are carrier drug release and carrier-controlled release [72–75]. For the carrier drug release, the drug is molecularly dispersed within the concentrated polymer layer and the rate-limiting step for dissolution of drug is the release of polymer itself. For the carrier-controlled release, the dissolution of the drug particles into the polymer-rich layer is a slow process. These two release mechanisms may occur simultaneously if the drug particles are partially dissolved in the polymer-rich diffusion layer before being released. It is important to note that at low drug concentrations, the release mechanism could be either carrier drug release or carrier-controlled release [72]. Using a similar model, Simonelli et al. described the rate of release of sulfathiazole from compressed tablets containing PVP. The apparent solubility and the rate of dissolution of sulfathiazole were significantly increased when sulfathiazole was coprecipitated with PVP, because coprecipitation resulted in the formation of a higher energy (amorphous) form [76,77]. The proposed model describes a situation in which there is an external layer of sulfathiazole at lower PVP weight fractions and a controlling PVP external layer at higher PVP weight fractions. At low PVP fractions, the relative rates of dissolution appear to fall on a plateau, indicating that in these cases the sulfathiazole dissolution rates were not affected by the presence of PVP in the tablets. This implies that the surface is covered by a sulfathiazole layer that is controlling the rate of dissolution. Although this model is oversimplified, the data from the study agreed well with the model and allowed a detailed characterization of the dissolution process.

Numerous attempts have been made to enhance apparent solubility and prolong supersaturation of poorly water-soluble drug compounds. Although a number of mechanisms by which polymers inhibit drug crystallization have been proposed in the literature, there is no consensus on the polymer property that is primarily responsible for inhibiting crystallization and thus for maintaining supersaturation. It has been suggested that molecular forces, which are critical to holding together a crystal lattice, can be disrupted by the use of a polymer and the onset of crystallization (nucleation) or the growth of a crystal can be delayed [78]. Important intermolecular interactions between a polymer and a drug molecule include specific interactions such as hydrogen bonding and nonspecific interactions such as hydrophobic interactions. Examples of each of the above-mentioned interactions and their importance are as follows.

- *Specific Intermolecular Interactions:* An important interaction in the crystal lattices of many molecules is hydrogen bonding. Typically, a hydroxyl or carboxyl functional group (hydrogen bond donor) will interact with a functional group such as an amide or a sulfonamide (hydrogen bond acceptor). Since these interactions are critical to forming and maintaining a crystalline lattice, if these interactions can be disrupted, the result is a delay to either crystal nucleation or growth. Naringenin/Hesperetin (common flavonoids) amorphous solid dispersions have been shown to have favorable hydrogen bonding characteristics, in which phenolic hydroxyls (hydrogen bond donors) from each of the flavonoids are able to hydrogen bond with PVP (a hydrogen bond acceptor), which in turn leads to prolonged supersaturation in *in vitro* dissolution studies. A number of studies have demonstrated the importance of specific interactions in preventing crystallization of amorphous solid dispersions [79–81]. Although most of these studies were performed in the solid state (in the absence of aqueous media), these interactions may also be important for solution crystallization inhibition. For instance, Kestur et al. studied the effect of different polymers (PVP, PAA, and PVAc) on the crystalline growth rates of amorphous bifonazole and nimesulide [82]. It was found that crystalline growth rates in the solid state are highly dependent on hydrogen bonding interactions between the drug and the polymer. In aqueous solution, Ilevbare et al. studied the effectiveness of a group of chemically diverse polymers on crystal growth of three structurally diverse compounds (celecoxib, efavirenz, and ritonavir) [83]. Using FTIR spectroscopy, the authors demonstrated that the formation of specific interactions is in part responsible for the crystal growth inhibition of ritonavir by poly(*N*-isopropylacrylamide) [83,84] and efavirenz by PVP [83]. Specific intermolecular interactions (such as hydrogen bonding) may be important in cases in which polymer and the crystallizing drug have accessible donor and acceptor groups.

- *Hydrophobic–Hydrophilic Interactions:* In addition to acid/base and hydrogen bonding interactions, hydrophobic/hydrophilic interactions can play a critical role in crystal nucleation and growth. Selecting a polymer with optimum hydrophobic/hydrophilic balance to the dispersed molecule may be critical to maintaining prolonged supersaturation. In a structure–property study, Ilevbare et al. found that the hydrophobicity of the polymer, as well as the presence of ionized functional groups, was important in determining the effectiveness of a polymer as a crystal growth inhibitor for three model compounds—celecoxib, efavirenz, and ritonavir [83,84]. Similarly, the effectiveness of the various polymers in inhibiting nucleation was found to depend on the hydrophobicity of the polymer relative to that of the drug [59].

The selection of appropriate *in vitro* dissolution conditions is an important step for evaluating the performance of an amorphous solid dispersion. The following section will focus on various factors that need to be considered for solubility and dissolution studies.

6.4 SOLUBILITY AND DISSOLUTION FACTORS TO CONSIDER FOR DISPERSIONS

6.4.1 Solubility

Aqueous solubility is a thermodynamic concept, and fairly straightforward to understand for crystalline compounds, as the concentration in solution of a solid phase in equilibrium with a solubilized form of the drug (illustrated in Equation 6.7) [85]:

$$\text{Solute}_{(\text{solid, crystalline, excess})} + \text{solvent} \overset{K}{\Leftrightarrow} \text{solute}_{\text{solution}}. \tag{6.7}$$

Theoretically, determining solubility would involve measuring the solubility as a function of time, while analyzing the solids, until the solubility and the composition of solids remain constant [85]. This well-known definition of solubility for crystalline materials will not apply for amorphous materials as they are an inherently metastable phase [86]. The theoretical solubility advantage that can be attained for amorphous material has already been discussed in Section 6.2 and has also been the subject of a recent publication [54]. However, one major issue entailed in determining amorphous solubility is the inherent tendency of these materials to crystallize, and hence the actual measured

TABLE 6.5. A Method for Determining Solubility of Amorphous Material Adapted from Ref. [85]

(1) Ensure material has a uniform particle size to avoid multiple dissolution rates. 75–150 μm recommended
(2) Add a known excess of amorphous material to 250 ml medium in a USP type II apparatus. At least 10× the crystalline solubility is recommended
(3) In order to improve powder dispersion, a paddle speed of at least 300 rpm is recommended
(4) Draw samples at select time points and filter using an appropriate filter
(5) Dilute samples immediately using organic solvent suitable for analysis method to prevent precipitation and/or crystallization of drug
(6) Determine concentration
(7) Pull an unfiltered sample at each time point and analyze for appearance of crystallinity
(8) Determine the concentration versus time profile and note the first appearance of crystalline material

solubility may not reach the maximum attainable amorphous solubility. It is therefore recommended to use the term "apparent solubility" for the maximum solubility attained in a solubility versus time experiment, and use this number as a practical designation of amorphous solubility. This recommendation is in line with recent publications and reviews [38,85,87].

One of these recent reviews has also outlined a procedure for measuring amorphous solubility, which has been summarized in Table 6.5 [85]. This method is a good general starting point for developing solubility assays. We recommend keeping the following aspects in mind: First, in discovery there may be insufficient material available to conduct the solubility experiment at 250 ml scale. The assay can be scaled down to a beaker or small vial, while still maintaining the same principles outlined in Table 6.5. Next, 300 rpm in a USP apparatus is a high stir speed. The authors selected this speed to avoid wetting issues. As recently published, the most common stir speed selected for dispersions is 100 rpm [87]. Therefore, it is recommended to select a lower stir speed if the vehicle contains surfactants or the material is otherwise not expected to have wetting issues. Finally, the isolation of the material is important. As described later in this section, submicrometer particles can form when amorphous materials are dissolving. Filters may not be able to separate these, and hence these particles may give rise to incorrect results.

As an illustration of the solubility enhancement that can be attained by amorphous material, in one study the apparent solubility of amorphous thiazide diuretics (prepared by spray-drying) was determined at 37 °C in 0.01 N HCl with 1% PVP. The increase in apparent solubility of the amorphous phase over the crystalline solid ranged from 10-fold for polythiazide to 1.1-fold for hydrochlorothiazide [88]. On the other hand, Gupta et al. predicted the theoretical solubility advantage of amorphous celecoxib from thermal data, but found that the measured value, although higher than the crystalline solubility, was significantly less than the predicted amorphous value [89].

There are additional caveats for dispersions. One factor to consider is the amount of excess material added. The amount of solid material in the experiment can drive whether dissolution of amorphous material or crystallization is the main driving factor for the maximum point on the solubility versus time profile [85]. Another consideration for dispersions is the impact of the carrier or polymer. As reviewed in Section 6.3, polymers are frequently incorporated with amorphous material to create amorphous solid dispersions. The polymer, if selected judiciously, tends to stabilize the amorphous material from converting to a crystalline form in the solid state and prolongs the supersaturated state [38,87]. However, the polymer may impact the solubility experiment and can create an even higher supersaturated state than what should theoretically be attainable with pure amorphous material because the polymer dissolution rate will impact the results [85].

6.4.2 Dissolution

For oral solid dosage forms to be effective, the active pharmaceutical ingredient typically reaches its target through systemic circulation. In order to assess the performance of a drug product with respect to its dissolution in the human intestinal tract, dissolution methods designed to mimic the conditions found *in vivo* have been developed. As discussed in the introduction to this chapter, these are called "biorelevant" dissolution methods. They are *in vitro* methods that are designed to characterize how an API is extracted out of a solid dosage form under physiologically relevant conditions. The biorelevant dissolution studies are typically carried out in a simulated biological medium using an apparatus that provides agitation at a controlled temperature. Biorelevant dissolution is often used by formulation scientists during the early development stage to understand the dissolution mechanism and predict the relative bioperformance of different formulations [9,10,90]. When used in conjunction with animal PK studies, biorelevant dissolution can be an effective tool to select formulations with good *in vivo* performance. The biorelevant dissolution method explored during the development of an amorphous solid dispersion formulation is usually, if not always, different from a quality control dissolution test that is mainly designed to verify the consistency of a drug product as well as the process robustness and long-term stability. In addition to measuring the drug concentration in solution, it is sometimes necessary to perform other characterization studies to investigate the dissolution mechanism and evaluate the performance of an amorphous solid dispersion formulation.

A variety of biorelevant dissolution tests have been published in the literature for amorphous solid dispersion formulations, ranging from a single medium dissolution study in USP dissolution apparatus to multimedia, multicompartment dissolution systems to dissolution–membrane permeation tests. Newman et al. [87] recently surveyed the dissolution testing strategy for amorphous solid dispersions based on 40 research papers. The survey shows that USP apparatus II (74%) was used in the majority of the studies, followed by USP type I apparatus (9%), USP type IV apparatus (6%), and other (12%) that includes multistage dissolution systems. USP type I and II apparatuses are robust, easy to operate, and widely available. The USP chapters on dissolution testing discuss the different types of USP apparatuses in detail [91–93]. The multicompartment dissolution system and the dissolution–membrane permeation tests will be discussed in more detail in Section 6.4.2.2.

6.4.2.1 Traditional Dissolution Protocols The selection of the dissolution medium is dependent on the nature of the drug and the polymer. The drug molecule has a pH-dependent solubility profile dependent on its pK_a values. In addition, the solubility of some drug molecules can be very sensitive to bile micelle concentration. Commonly used polymers for amorphous solid dispersions have different solubility as a function of pH. PEG, PVP, PVP–PVA, HPC, and HPMC are soluble at all pH conditions, while HPMC-AS, HPMC-P, and EUDRAGIT L are enteric polymers that have very low solubility under acidic conditions (<pH 5.5). EUDRAGIT E is soluble in buffer solutions at pH values up to 5 and swells at higher pH values. Based on Newman et al.'s survey, the most commonly used acidic mediums are 0.1 HCl and simulated gastric fluid (SGF) without enzymes [87]. Water and pH 6.8 buffer are most commonly used to characterize the dissolution of amorphous solid dispersions at neutral pH, trailed by simulated intestinal fluids (SIF) and surfactant solutions [87]. For an amorphous solid dispersion containing a free base drug and a pH-independent polymer, an acidic solution such as SGF or HCl may be an appropriate dissolution medium, while a pH 6.8 buffer may provide better prediction for an amorphous dispersion containing nonionized drug in an enteric polymer. If the solubility of the drug is highly dependent on micelle concentration, SIF or a surfactant solution is recommended.

The volume of the dissolution medium and the agitation speed are also important factors for a dissolution test. The survey by Newman et al. shows that a wide range of volumes from 25 to 1000 ml were reported, although the majority of the studies use typical dissolution volumes of 500 or 900 ml for USP I and II [87]. The survey also showed that a majority of the studies used an agitation speed between 50 and 100 rpm. For early screening of amorphous solid dispersion intermediates, miniaturization of the dissolution experiment is desired due to limited drug availability. Instead of standard USP apparatus, small-scale dissolution studies in beakers or glass vials with magnetic stirring could be considered to characterize and compare the dissolution of different amorphous solid dispersions.

The sink condition is a critical component for a biorelevant dissolution method; however, it appears to be underappreciated. Sink can be defined as utilizing a volume of dissolution media that is 3–10× greater than the volume required for a drug to be saturated. It was reported in a recent review that "only 8% specified sink conditions, 6% specified non-sink conditions, and 86% did not specify the conditions used" [87]. The dissolution behavior of amorphous drug is governed by the degree of supersaturation. Drug precipitation can take place under supersaturated conditions and the rate of precipitation is affected by the degree of supersaturation, drug physiochemical properties, and the interaction between the drug and the polymer. In the early stage of amorphous solid dispersion development, it is important to characterize the solubility of the crystalline and amorphous forms of the drug of interest in dissolution media. The total drug concentration in a biorelevant dissolution study is often dose dependent. For example, a total drug concentration of 0.2 mg/ml may be chosen to evaluate the dissolution of a formulation containing 50 mg of drug in 250 ml SGF. For many insoluble drugs, 0.2 mg/ml could be above the amorphous solubility limit. Understanding the sink condition is therefore critical to the proper interpretation of dissolution data. It also helps in deciding if additional studies should be conducted to further understand the dissolution mechanism. Dissolution studies under nonsink condition at a drug load relevant to dose can be more discriminating than studies conducted under sink condition, since the ability for a formulation to maintain supersaturation can only be tested under nonsink conditions. In late-stage formulation development, the QC test is usually conducted under sink conditions for the solid dosage form and is provided to regulatory agencies [94,95].

Some published studies rely on biorelevant dissolutions alone to evaluate the performance of amorphous solid dispersions. Tho et al. used pH 7.3 phosphate buffer modified with 1 mM sodium taurocholate to evaluate the dissolution and release performance of ritonavir melt extrudates, comparing the performances with water and aqueous buffer media [74]. In another study by Boetker et al., achlorhydric media (pH 3.5 acetate buffer) were used to demonstrate the advantage of amorphous amlodipine over crystalline forms [96]. Mehanna et al. studied the application of pluronic block copolymers in the creation of amorphous solid dispersions of tadalafil. Ozaki et al. [97] used dissolution in FaSSIF media to gauge the performance of these amorphous solid dispersions versus the crystal form, showing a significant increase in solubility and dissolution rate for the solid solutions versus the free tadalafil and tadalafil comixed with pluronic F-127 [98]. Lust et al. studied the efficacy of multiple polymers as carriers for amorphous solid dispersions of piroxicam with SGF as the dissolving media [99]. Amorphous dispersions of piroxicam showed enhancement of both the rate and extent of dissolution versus the crystalline form. In Ref. [97], the authors used FaSSIF to determine the supersaturation and rate of crystallization of poorly soluble APIs formulated as amorphous solid dispersions versus the crystalline forms.

6.4.2.2 Newly Developed Dissolution Protocols For amorphous formulations that form drug precipitates (nano- or micrometer-sized particles) during dissolution, it is of great interest to determine solution concentration without interference from these particulates. The dynamic dialysis method has been widely used to study the drug release rate of nanoparticulate systems [100–105]. For example, Leo et al. studied the *in vitro* release of doxorubicin from doxorubicin–gelatin nanoparticles (submicrometer particles, $<1\ \mu m$) in the absence as well as presence of a proteolytic enzyme that degraded the gelatin carrier [102]. Nagarwal et al. performed dynamic dialysis studies to measure the release of 5-fluorouracil (5-FU) from chitosan-based nanoparticles targeted for ophthalmic delivery of 5-FU [103]. In this study, a membrane with a specific molecular weight cutoff (MWCO), typically 12 KDa or less, was used to separate the donor and acceptor compartments. The membrane served as a physical barrier to prevent diffusion of nanoparticles from the donor to the acceptor side and drug concentration in the acceptor compartment was monitored with time. To maintain sink conditions and prevent reverse drug diffusion, the volume of the dissolution medium in the donor compartment was about 10 times less than that in the acceptor compartment. The diffusion rate of the drug from the donor to the acceptor compartment was determined using the concentration of solubilized drug in the acceptor compartment, the permeability of the drug across the membrane, and the area of the membrane. If the solution concentration in the donor compartment is not replenished quickly by undissolved drug particles, diffusion becomes dissolution limited.

The dynamic dialysis method can be used to characterize and compare the *in vitro* dissolution rate of different amorphous solid dispersion formulations of the same drug. Upon dissolution of an amorphous solid dispersion in a biorelevant media, drug can exist in a number of forms: free drug, drug in bile salt micelles, drug–polymer nanoaggregates, micrometer-sized or larger drug precipitates [106]. Without having to centrifuge the dissolution samples at high velocity (high *G* force), the dissolution rates of different types of formulations can be compared based on the diffusion rate, if drug permeation across the dialysis membrane is not rate limiting. Using the dynamic dialysis method, Alonzo et al. investigated the dissolution mechanism and impact of drug–polymer composition on the degree of supersaturation generated upon dissolution of amorphous solid dispersions of HPMC–felodipine [107]. The flux (diffusion rate normalized by membrane surface area) of pure amorphous felodipine and HPMC–felodipine amorphous solid dispersions (10 and 50% drug loading) were compared with crystalline felodipine (1 μg/ml solubility in the dissolution medium) and supersaturated solutions of felodipine at 5 and 10 μg/ml, generated using the cosolvent method. Using UV fiber optic probes, the felodipine solution concentration was monitored in both the donor and acceptor chambers. Although the "apparent" solution concentration of the amorphous solid dispersion with 10% drug in the donor compartment was 25 μg/ml (based on UV measurements), the diffusion rate of the compound from the donor to the acceptor compartment was similar to that of the 10 μg/ml supersaturated solution, indicating that only 10 μg/ml of drug dissolved. This apparent increase in solution concentration was attributed to the presence of 200–300 nm particulates in solution. Characterization of the particulates generated during dissolution of amorphous solid dispersions will be discussed later in this chapter. In contrast to the USP I or II dissolution apparatuses, the dynamic dialysis technique can be used to determine the degree of solubility enhancement achieved by an amorphous solid dispersion without the interference from submicrometer particulates generated *in situ* during dissolution.

Although the dynamic dialysis method has been extensively used to study *in vitro* drug release, care should be taken when interpreting the diffusion data. The assumption made in most studies is that the drug diffusion is controlled by the concentration gradient between the solution in the donor and acceptor compartments and the drug permeation across the membrane is not rate limiting. Zambito et al. conducted dialysis studies to understand the rate-limiting step in the diffusion of diclofenac- or ofloxacin-loaded chitosan nanoparticles across the dialysis membrane [108]. The authors found that the diffusion process was controlled by drug permeation across the membrane for both drug-loaded chitosan nanoparticles. For drug molecules that exhibit low permeability through the dialysis membrane, the diffusion rate may not reflect the dissolution rate accurately and the dialysis method may not be an appropriate choice.

Babcock et al. published a patent on an *in vitro* membrane permeation test to characterize pharmaceutical compositions [109]. This test was designed to be well suited for evaluating the absorption rate of insoluble drugs ($\log P > 2$) from amorphous solid dispersion formulations. It has several major differences compared with the dynamic dialysis method. The membrane permeation test utilizes a hydrophobic microporous membrane with high porosity (preferably >40%). The preferred membrane was the Accurel® polypropylene membrane with a pore size of 0.1 μm and thickness of 90 μm. Similar to the dialysis method, the flux of the drug across the membrane was determined by measuring the concentration of drug in the receiving medium as a function of time. A term called "maximum flux" was used to compare different amorphous solid dispersion formulations of the same drug; the higher the amount of dissolved drug, the higher the magnitude of the maximum flux. Babcock et al. demonstrated that the membrane permeation test is a useful method for identifying amorphous formulations that can improve bioavailability or predict food effect [109].

Precipitation from a supersaturated solution is driven by degree of supersaturation. In the presence of an absorption surface such as the intestinal membrane, the degree of supersaturation as a function of time could be significantly different than that in a nonabsorption environment such as a dissolution vessel. Although the dialysis membrane or the porous polypropylene membrane offers an absorption interface, these membranes are not necessarily biorelevant. Bevernage et al. recently studied the precipitation kinetics of loviride from supersaturated solutions at the apical side of CACO-2 cell monolayers and found that the data do not correlate with precipitation kinetics in nonabsorption environment [110]. In addition, the effectiveness of HPMC-AS as an antiprecipitation agent for loviride was also different in the two environments. This study demonstrated the power of "implementing transepithelial permeation into *in vitro* precipitation assays" to predict the bioperformance of supersaturating drug delivery systems. The incorporation of cell membranes in the dissolution test poses some limitations on the types of samples that can be examined. It typically requires a presuspended powder sample or a presupersaturated solution. In addition, the study is not applicable if the formulation contains excipient(s) toxic to the CACO-2 cells.

In addition to the dissolution membrane tests, a number of multicompartment dissolution techniques have emerged to more closely mimic the process of gastric emptying and fluid transport along the human GI tract. These methods are used to understand complex drug release mechanisms and begin building *in vitro/in vivo* correlations. One of these methods is the artificial stomach duodenal model, which is a two-compartment system to represent stomach and duodenum [111,112]. API or formulation is placed in the stomach chamber at the beginning of the experiment. The gastric contents are transferred to the duodenum chamber to be mixed with simulated intestinal fluid and then transferred out of the duodenum. The stomach and duodenum secretions are controlled programmatically by pumps. Drug concentration is measured in each of the chambers using a UV fiber optic probe to evaluate formulation performance. Polster et al. used this method to study the effect of stomach pH on the inter- and intrapatient plasma exposure variability of a development compound LCX and to select a dosing solution that could potentially reduce the exposure variability of the capsule formulation. The model is well suited to study amorphous solid dispersions since it can capture dynamic phenomena such as supersaturation, precipitation, and redissolution.

The simulated gastrointestinal tract model 1 (TIM-1) is another multicompartmental, dynamic, computer-controlled system developed to simulate the conditions in the human upper GI tract [113,114]. It consists of compartments mimicking stomach, duodenum, jejunum, and ileum, and can simulate many aspects of the physiological conditions and dynamic processes such as secretion of bile salts and enzymes, transit from one compartment to another, drug absorption, and food intake. Dissolution studies with TIM-1 are more time consuming compared with other methods and often not readily available.

Animal and human PK data are available in some of the published research to compare with the dissolution data. There are two types of predictions for *in vitro* and *in vivo* data: IVIVR and IVIVC as briefly described in Section 6.1. The former is a broader, semiquantitative association between the two data sets and the latter is a more rigorous, predictive mathematical model that describes the predictability of *in vitro* data for *in vivo* performance. The vast majority of early developmental work with biorelevant media focuses on IVIVR [22,33,115]. For example, modified fasted duodenal (MFD) fluid was used to determine the dissolution of several APIs from amorphous solid dispersions of HPMC and HPMC-AS. Friesen et al. demonstrated a clear solubility advantage for solid dispersions of drug made with varying grades of HPMC-AS versus the crystalline API. Also described in the same publication is a comparison of the *in vitro* data with the *in vivo* data and demonstration of IVIVR [106]. Finally, modified SGF was used by Boghra et al. to determine the release rate of irbesartan from HPMC solid dispersions. The data were compared with the dissolution rates of crystalline API from suspension in the same media, showing an advantage for the amorphous solid dispersion. Bioavailability enhancement in rabbit model (New Zealand gray) was also demonstrated, establishing an IVIVR for the dissolution and animal PK [116].

6.4.3 Characterization of Species Generated during Dissolution of Amorphous Solids and Their Impact on Performance

Many times when amorphous solid dispersion is dissolved in aqueous media, the resultant solution is observed to be cloudy with a slight bluish color, characteristic of the presence of a second light scattering phase. This observation is consistent with studies in which supersaturating dosage forms, consisting of amorphous solid dispersions, are found to have a rapid dissolution rate [54,64,117,118] and submicrometer particles are formed *in situ* upon dispersion in the aqueous media [119]. In particular, the formation of drug nanosuspensions (submicrometer colloidal dispersions of pure particles of drug) is of interest in the context of better understanding of how amorphous solid dispersions lead to enhanced bioavailability [119]. By investigating the properties of solutions formed from the dissolution of amorphous solid dispersions of PVP and beta carotene, Tachibana and Nakamura first reported that amorphous solid dispersions can lead to the formation of colloidal particles [120]; since then, numerous researchers have demonstrated that amorphous solid dispersions are capable of generating colloidal-sized species when dissolved in aqueous solution [64,106,120–122].

Although the formation of colloidal dispersions from amorphous solid dispersions has been widely reported, the formation mechanism and the factors influencing the size and stability of these species are not well understood. In recent studies, it has been demonstrated that when sufficiently high supersaturated solutions are generated in aqueous media, poorly water-soluble compounds can undergo liquid–liquid phase separation (LLPS) [54,117,119]. When a certain concentration has been exceeded, a dispersed phase that consists of submicrometer particles is produced, provided crystallization does not occur first. The LLPS concentration is consistent for a given medium and temperature, is predictable, and can be related to the theoretical amorphous "solubility" [119]. Since amorphous solid dispersions dissolve to generate supersaturated drug solutions, it is very likely that the colloidal species observed upon dissolution of the rapidly dissolving amorphous solid dispersions originate from LLPS. The LLPS concentration, which is comparable to the amorphous "solubility", represents the upper limit of solution concentration of "free" drug that can be achieved following the dissolution of an amorphous solid dispersion; exceeding the LLPS concentration leads to the formation of a colloidal phase of submicrometer size [119]. This observation is consistent with the study by Friesen et al., where it was found that colloidal particles began to form for an amorphous solid dispersion containing a poorly water-soluble drug only at concentrations above the amorphous solubility [106]. For a more detailed discussion on LLPS, colloidal dispersions, and the theoretical amorphous "solubility," the reader is referred to Refs [54,106,117,123].

Colloidal particles generated during dissolution of amorphous solid dispersions may be crystalline or noncrystalline (amorphous or supercooled) in nature. The supersaturation behavior of drug compounds is highly correlated with their inherent crystallization tendency [117,124]. Since colloidal dispersions are generated from the supersaturated solution upon dissolution of amorphous solid dispersions, it is expected that the nature of the species formed will also depend on the crystallization tendency of the API present in the amorphous solid dispersion or any supersaturating dosage form. Compounds have different crystallization tendencies and, depending on the behavior shown during the solidification process (e.g., cooling from the melt and solvent evaporation), they can be classified as slow, intermediate, and fast crystallizers [125]. Fast crystallizing compounds tend to be low molecular weight, simple structures with fewer number of rotatable bonds compared with slow crystallizing species, which tend to be more structurally complex. In general, poorly water-soluble drug compounds that yield noncrystalline colloidal particles have been shown to belong to the group of slow crystallizing compounds [54,117,119]. However, by using a suitable polymer, the crystallization of intermediate or fast crystallizing compounds can be significantly delayed and, therefore, amorphous blends of these types of compounds with water-soluble crystallization inhibitors may generate noncrystalline particulates upon dissolution. It is also expected that the crystallization tendency of colloidal species from aqueous solutions would depend on additional factors, including the presence of water, which will promote molecular mobility of amorphous species generated. Thus, the duration of time the generated particulates remain amorphous in aqueous solution may also be related to the ability to form stable amorphous formulations that possess long shelf life stability [106,117,123] and, when dissolved, yield supersaturated solutions that can be maintained for a physiologically relevant time period.

Several methods can be used to characterize the size and the physical nature (crystalline or amorphous) of the colloidal species generated upon dissolution. Some of the methods that have been employed in the literature include turbidity measurements [54,119,122,123], dynamic light scattering [54,64,106,118,119,126] and fluorescence spectroscopy [54,119], scanning electron microscopy (SEM) [106,120,123], cryogenic transmission electron microscopy (cryo-TEM) [106,127], optical microscopy [54,64,117], asymmetrical flow field–flow fractionation (AFIFFF), [122] powder X-Ray diffraction (PXRD) [117], and differential scanning calorimetry (DSC) [54,117,119]; these methods and their applications are summarized in Table 6.6. Customarily, colloidal particles are separated from

TABLE 6.6. Characterization Methods for Particulates Formed During Dissolution of Amorphous Solid Dispersions

Method	Data Measured	Property Measured/ Information Obtained	Remarks	References
Turbidity measurement	Transmission, absorbance, extinction, or visual detection of cloudiness	Light scattering. Determination of onset of colloidal particle formation	Sensitivity depends on the method of detection. No information on structure and nature of species	[53,66,70]
Dynamic light scattering (DLS)	Particle size (*Z*-average diameter) with time	Size distribution of the colloidal species in solution	High-throughput method. Should be used simultaneously with another method (e.g., fluorescence) for thoroughness. DLS provides no information on structure and nature of species	[53,59,104, 118,126]
Fluorescence spectroscopy	Peak intensity versus wavelength	Emission peak intensity. Determination of crystalline nature of particles	Fluorescence probe investigates the local environment in the solution	[53]
Optical microscopy	Microscopy under the influence of cross-polarized light	Morphology, surface structure, and size of the particles	Limited size information	[53,59,66,117]
Scanning electron microscopy (SEM)	Microscopy under the influence of electron radiation	Morphology, surface structure, and size of particles	Sample preparation required	[104,120,123]
Cryogenic-transmission electron microscopy (Cryo-TEM)	Microscopy under the influence of electron radiation with cooling	Morphology, surface structure, and size of particles	Sample preparation required	[104,127]

Asymmetrical flow field-flow fractionation (AFIFFF) w/UV/vis	Fractogram	Size, UV absorbance, refractive index, and separation of species	Interfaces with modern analytical methods such as MALS and MALDI-TOF. Fractions can be collected and used for offline analysis	[122]
Differential scanning calorimetry (DSC)	Heat flow versus temperature	Thermal event. Determination of physical nature (crystalline or amorphous) of material	Requires separation of supernatant from precipitate prior to analysis	[53,117]
Powder X-ray diffraction (PXRD)	Diffractogram	Crystallographic properties. Determination of physical nature (crystalline or amorphous) material	Requires separation of supernatant from precipitate prior to analysis	[117]

supernatant using centrifugation or filtration and then the species are characterized using characterization methods such as PXRD and DSC to verify their nature. The colloidal particles may also be characterized by lyophilizing the colloidal suspension and analyzing the lyophilized material [106]. To completely characterize the species in solution, a combination of two or more of the above-mentioned techniques should be utilized. A combination of DLS and fluorescence spectroscopy methods allows high-throughput screening of the species in solution; DLS data do not provide information about structure of the disperse phase; thus, another technique should be used to investigate the local environments in the solution. Time-resolved DLS provides information on size and colloidal stability, while fluorescence spectroscopy provides information on the crystalline or noncrystalline nature of the species [54]. For fluorescence spectroscopy, a fluorescence probe, which exhibits different fluorescence characteristics depending upon the properties of the solubilizing medium [128,129], may be used. If the colloidal species formed have a noncrystalline nature, then it would be expected that a hydrophobic probe molecule such as pyrene would partition into this phase and consequently undergo a change in emission characteristics relative to the emission spectrum in water. In contrast, if the disperse phase is crystalline, no change in probe emission spectrum would be expected since pyrene will not penetrate the crystal lattice [128,129]. Imaging techniques such as optical microscopy, cryo-TEM, and SEM may also be used to confirm the nature of the colloidal species. For optical microscopy, the presence of birefringent particles is indicative of crystallinity, while for noncrystalline particles, SEM and TEM micrographs reveal smooth, spherical liquid-like species [120,123]; the structure of these noncrystalline colloidal particles is typically a sharp contrast to the particles characteristic of crystalline counterpart. The use of optical microscopy as a size quantification method is limited, while TEM and SEM will provide size, as well as surface morphology information, of the colloidal species [84].

The drug precipitates can come in a wide range of sizes, which poses different challenges for solution concentration measurements. An online UV-vis probe is widely used to monitor drug concentration in solution during a dissolution study. In general, it is assumed that the drug particles generated during dissolution of amorphous solid dispersions simply scatter light, which in most cases leads to a vertical baseline shift in the absorption spectrum that can have a slope when scattering is strongly dependent on wavelength. These vertical shifts will lead to inaccurate concentration estimates if based on absorption peaks alone; so baseline corrections or first derivatives of the calibration and sample spectra are necessary for proper analysis. In the case of a system where there is wavelength-dependent scattering and the baseline shift is linearly sloped, a second derivative is used. However, when particles in solution are present on a nanometer scale both absorption and scattering contributes to the overall signal [130]. This has an important implication for performing dissolution experiments. When the particles present in the solution are of a relatively large size, baseline corrections and second derivatives are generally sufficient for predicting accurate solution concentrations. However, in the presence of organic nanospecies (<1 μm) generated *in situ*, the effectiveness of derivatives and baseline corrections is limited [130]. In many cases, nanoparticles are not present in the dissolution medium and conventional detection methods, such as UV-vis, will provide accurate results. However, care should be taken to ensure that there are no submicrometer species present in a solution during dissolution experiments, particularly when using a UV-vis spectrometer coupled to a fiber optic probe to monitor solution concentrations. Ultracentrifugation or filtration of the solution with submicrometer membrane pore size (prior to solution concentration measurements) to separate colloidal particles from the supernatant may be necessary to accurately determine the drug concentration in the solution phase [54,106,117,118,126,131]. As discussed in Section 6.4.2.2, the dynamic dialysis method may be used to determine drug concentration without notable interference from these particulates.

The formation of drug-rich submicrometer particles upon dissolution of amorphous solid dispersions also provides an alternative method of generating nanospecies and might be important not only for oral drug delivery but also for other routes of delivery where it is desirable to form drug-rich nanoparticles, such as pulmonary drug delivery [119,132]. It has been suggested that colloidal species are important to the bioavailability enhancement often seen with amorphous solid dispersions [106,133,134]. Frenkel et al. proposed that the formation of colloidal particles (especially particles with hydrodynamic radii less than 100 nm) may be responsible for the enhanced bioavailability of nonnucleoside reverse-transcriptase inhibitors (NNRTIs) by promoting their absorption by particle-recognizing microvilli cells in Peyer's patches of mucosa-associated lymphoid tissue [133]. In addition, Serajuddin et al. hypothesized that colloidal particles could dissolve during GI transit, thus increasing bioavailability [134]. However, in order to take advantage of the properties of colloidal dispersions, the submicrometer particles must be stabilized. Colloidal particles may undergo coalescence since colloidal particles are generated through the formation of highly supersaturated solutions. Polymers have been used to stabilize colloidal dispersions [106,120,121,127,135] and can also serve as crystallization inhibitors [59,84]. It has been proposed that the size and stability of colloidal particles depend on solution conditions (such as pH, buffer, and ionic strength) and the structure and properties of the molecules, including the number of polar atoms, number of hydrogen bond donors and acceptors, conformational flexibility, and presence of ionized groups [119,133]. Analysis of aggregate size distributions for NNRTI APIs under various physiological conditions demonstrated dependence of aggregation size on solution pH [133]; at constant ionic strength, aggregate size was found to increase with increasing pH (decreasing extent of ionization). This dependence of particle size on drug and/or additive ionization, and thus on pH, may have important implications for drug absorption since pH varies along the GI tract.

In summary, developing a biorelevant dissolution method is a critical activity for amorphous formulation development. Different dissolution methods can be applied based on the development stage to guide formulation selection and process development. Since the number of dissolution conditions is broad, scientists should consider what set of conditions best matches the needs for the questions they intend to answer. Understanding the properties of the drug molecule and the polymers are critical to the choice of dissolution apparatus, testing conditions, and analytical methods. At early stages of development, a simple one-stage dissolution experiment in UPS type I/II apparatus or beaker/glass vials can be used to explore different formulation compositions for solubility enhancement and provide a qualitative trend. At later stages, a more sophisticated and resource-intensive dissolution system may be explored if a simple test does not correlate with bioperformance in human or there is a desire for the *in vitro* test to provide more quantitative information for *in vivo* prediction. Multiple techniques should be explored to investigate the dissolution mechanism at dose-relevant concentrations. This often requires the characterization of insoluble drug particles. Submicrometer drug particles can form during the dissolution study under certain conditions and present additional challenges for determining truly dissolved drug concentration. Having the appropriate sample clarification method should be a critical component for any dissolution method for amorphous solid dispersion formulations. Animal PK studies can be leveraged to test the biorelevance of the *in vitro* dissolution method before human studies to derisk the clinical formulation. For amorphous solid dispersion formulations, physical changes such as crystallization or amorphous–amorphous phase separation can take place under stability conditions. The availability of a biorelevant dissolution method could be extremely useful to define the failure mode without resorting to extensive animal or human PK studies. Finally, dissolution of amorphous solid dispersions in the human GI tract is a highly dynamic process that is affected by many factors that can be difficult to mimic simultaneously. Although developing a predictive *in vitro* method or establishing IVIVC may remain a challenging task for pharmaceutical scientists, progress has been made with new tools and techniques that provide more quantitative information.

6.5 SOLUBILITY AND DISSOLUTION MEASUREMENTS FOR AMORPHOUS DISPERSIONS: SUMMARY, CONCLUSIONS, AND RECOMMENDATIONS

It has been estimated that ~40% of the world's top oral drugs are classified as BCS II and IV compounds, and therefore have low solubility [136], and that the problem is even worse in discovery pipelines [137]. Consequently, there has been an increasing interest in crystal (e.g., salts, cocrystals, and amorphous forms) and formulation engineering (e.g., particle size control, nanoformulations, and amorphous dispersions) strategies to mitigate poor solubility. The increased interest in these techniques needs to be seen in light of the tight link between solubility, dissolution, and good oral absorption. The latter is one (out of several) important prerequisite for a commercial oral drug. The majority of crystal and formulation engineering approaches follow the strategy of enhancing the time span between dissolution and precipitation (illustrated in Figure 6.1) sufficiently to enable drug absorption before precipitation [37–40]. Among these approaches, amorphous solid dispersions have proven especially successful, and as recently reviewed, improved bioavailability in 82% of 40 amorphous dispersion case studies [87].

As presented in this chapter, solubility and dissolution are intimately interconnected but not identical properties. Solubility is a thermodynamic parameter and is defined by the concentration in solution in equilibrium with excess solids. As recently reviewed, a major consideration when conducting solubility measurements is determining when equilibrium is reached. Therefore, theoretically, measuring solubility would involve sampling the solubility experiment as a function of time until the solubility and the solid phase is constant. This is not a feasible method for amorphous materials as they are inherently metastable phases [85].

The theoretical solubility advantage that can be attained for amorphous materials has already been discussed in Section 6.2 and has recently been the subject of a literature report [54]. Importantly, one major issue entailed in determining amorphous solubility is the inherent tendency of these materials to crystallize and hence the actual measured solubility may not reach the maximum attainable amorphous solubility [85]. One additional caveat for amorphous dispersions is the impact of the carrier or polymer. As reviewed in Sections 6.2 and 6.3, polymers are frequently added to amorphous materials to create amorphous solid dispersions. The polymer, if selected appropriately, can stabilize the amorphous material, delay the conversion to crystalline drug form in the solid state, and prolong the supersaturated state in an aqueous environment. The polymer can and will however impact the solubility experiment, and as discussed in detail in Section 6.4, it will influence the particle engineering and therefore the interaction of the medium with the amorphous material [85]. Given this complexity, what is then the best and recommended way to measure solubility of amorphous material and dispersions? In the context of this chapter, which takes a late-discovery/early-development view, the main objective of solubility experiments is to attain IVIVR (as discussed in Sections 6.1 and 6.4) and therefore the biorelevance of the measurement. The recommendation is therefore to consider the maximum value of a biorelevant dissolution curve (solubility value and time) in the context of the transit time through the small intestine of the species targeted for bioperformance. This recommendation is in line with recent publications and reviews [38,85,87].

This leaves the question: How to conduct a dissolution experiment for an amorphous dispersion? The steps involved have been discussed in Sections 6.1 and 6.4 and recently reviewed [38,85,87]. A flowchart for the suggested steps is presented in Figure 6.6. The authors recommend starting with the end in mind; in other words, the desired outcome. Figure 6.6 outlines three main outcomes for dissolution studies: (i) early characterization of discovery batches for dispersion optimization, (ii) early development studies feeding into IVIVR or maybe even IVIVC considerations, and (iii) studies aimed at quality control aspects of formulations, such as batch-to-batch variability. The latter dissolution studies have not been the focus of this chapter and the reader is encouraged to read FDA and ICH guidelines for more details on this important area of dissolution study design [21–24].

In terms of determining the study objective (the next step in Figure 6.6), which is the desired outcome best served by a (i) solubility study, (ii) biorelevant dissolution study, or (iii) QC dissolution study? Focusing on the former two and since solubility of an amorphous dispersion is best determined by the maximum concentration on the dissolution curve, these studies may be conducted in an identical fashion and information collected to conclude on both aspects. However, the experimental design could be different according to the stage of discovery or development. The discovery or development stage usually dictates the availability of resources and material. As recently reviewed, biorelevant dissolution and solubility experiments may be conducted at small scale in plates, tubes, and beakers rather than large-scale traditional dissolution equipment to save time and material [85,87].

The next step is selection of the appropriate testing equipment. As already mentioned, a variety of plates, tubs, and vessels may be used for early-development

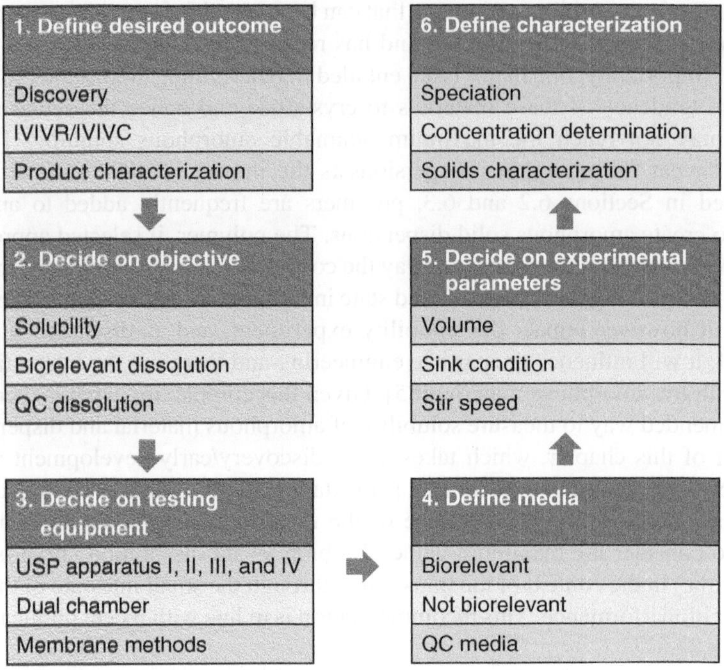

Figure 6.6. Flowchart and decision process for designing dissolution experiments for an amorphous dispersion.

evaluations. However, focusing on early-development biorelevant dissolution studies, there are several equipment types to consider: (i) USP apparatuses I–IV, (ii) dual chamber methods, or (iii) membrane methods. As recently reviewed, the USP apparatuses are most frequently used with USP II (paddle method) being used in 74% of 40 reviewed cases [87]. The USP approach offers the advantages of common use, simplicity, and translatability. Dual chamber or phase methods offer the advantage of better simulating pH changes from the stomach to the small intestine and membrane methods add the biological sink of intestinal permeability as an additional component [87]. The latter two are more complex but may give rise to better IVIVR.

As Figure 6.6 indicates, the next step is to select media and three main options are usually considered: (i) biorelevant, (ii) nonbiorelevant, and (iii) a variety of QC media. Focusing on the former two, biorelevant media were described in Section 6.1 and in recent literature [9,28–30]. The authors are advocating a stage-dependent approach to media selection. In discovery, where the objective is to rank order compounds, the more simple USP media [27] may be used. Later in development, when the objective is to attain IVIVR, more complex media [9,28–30] may give rise to better correlation. Recently, 40 dispersion case studies were reviewed showing the low pH media and buffers did not give rise to IVIVR [87].

In selecting experimental conditions, three main considerations are listed in Figure 6.6: media volume, sink conditions, and stir speed. The standard dissolution vessel on a USP dissolution apparatus is 900 ml. Therefore, not surprisingly, 60% of 40 surveyed dispersion case studies used 900 ml media [87]. However, this may not reflect the physiological conditions within human as intestinal volumes are much smaller. According to a recent study, volumes average 45 ml (fasted)/686 ml (fed) in the stomach, 105 ml (fasted)/54 ml (fed) in the small intestine, and 13 ml (fasted)/11 ml (fed) in the large intestine [138]. Therefore, considering smaller and more physiologically relevant volumes is recommended for biorelevant studies. A related consideration to volume is whether or not sink is established under the experimental conditions. The USP defines sink as three times the media required for saturation [95], which is a rule that should be adhered to when conducting QC dissolution experiments. *In vivo* sink may, for certain drugs, be established by permeation across the gastrointestinal mucosa; however, for insoluble drugs such as BCS class II (poor solubility and high permeability) or especially BCS class IV (low solubility and low permeability), sink may not be attained and lower volumes should be considered in biorelevant experiments. Stir speed is important as it will affect the unstirred water layer around the dissolving particle and thereby dissolution rate. It was described in Section 6.4 and recently published that the most common stir rate used in a USP apparatus is 100 rpm [87]. Stir speed is a difficult parameter to make biorelevant as mixing in the *in vivo* situation will depend on a variety of factors, such as species, feeding state, hydration state, GI motility, and therefore also age and medication-related factors. The authors recommend that stir rate is adjusted empirically as IVIVR correlations are developed. Stir speed may not be important for early studies, where rank ordering is the objective, and in these cases it is recommended to keep it at a constant rate.

Finally, the dissolution design flowchart outlined in Figure 6.6 recommends considering characterization that usually will involve (i) speciation or, in other words, detecting if any nanosized particles or aggregates are formed in the experiments, (ii) concentration determination, and (iii) characterization of any solids remaining in the experiment. Concentration determination is usually conducted via HPLC, UV, or other similar methods [90] and any remaining solids isolated are usually analyzed for crystalline/amorphous content by standard solid-state methods such as DSC and PXPD. Recently, it has been published that submicrometer particles can form during dissolution of amorphous dispersion, as discussed in detail in Section 6.4. These particles can have significant impact on bioperformance and can make analysis of dissolution experiments difficult. They can give erroneous results when using UV detection due to particle scattering and falsely high results when using HPLC, if the particles and the medium are not separated appropriately. As outlined in Section 6.4, ultracentrifugation is recommended for isolating the solids if submicrometer particles are being formed in the dissolution experiment and Table 6.6 outlines analysis methods to detect and characterize these submicrometer species.

In conclusion, we have illustrated that *in vitro* solubility and dissolution studies represent an important link between formulation design and *in vivo* bioperformance. This is especially true for supersaturating formulations such as amorphous dispersions. These formulations are based on high-energy and metastable API forms and may therefore crystallize *in vivo*, resulting in less than desired bioperformance. Well-designed biorelevant solubility and dissolution studies could serve as a resource-sparing alternative to increase the probability of success of desired *in vivo* bioperformance. Therefore, it is important that a tight collaboration and flawless communication exist between researchers who are designing formulations, performing the biorelevant solubility and dissolution testing, and the researchers who are responsible for conducting and evaluating bioperformance.

ACKNOWLEDGMENTS

The authors would like to thank Maria Cueto (Scientific Knowledge Discovery, Merck & Co.) and Michelle Morrisseau (Formulation and Basic Pharmaceutical Sciences, Merck & Co.) for help with the literature search that was a crucial part of writing this chapter. In addition, we would like to acknowledge Paul Harmon (Analytical Sciences, Merck & Co) and Caroline McGregor (Discovery Pharmaceutical Sciences) and Michael H. Kress (Process Research, Merck & Co.) for an insightful review and comments. Finally, Ann Marie Norris (Creative Studios-Scientific Presentations, Merck & Co.) is acknowledged for assistance with figures and tables.

REFERENCES

1. Avdeef, A. (2001) Physicochemical profiling (solubility, permeability, charge state). *Curr. Topics Med. Chem.*, 1(4):277–354.
2. Martin, A. (1993) *Physical Pharmacy*, 4th ed., Lea & Febiger, Philadelphia.
3. Martin, A., Newburger, J., and Adjei, A. (1980) Extended Hildebrand solubility approach: solubility of theophylline in polar binary solvents. *J. Pharm. Sci.*, 69(5):487–491.
4. Yalkowsky, S.H. (1999) *Solubility and Solubilization in Aqueous Media*, Oxford University Press, New York.
5. James, K.C. (1986) *Solubility and Related Properties*, Marcel Dekker, New York.
6. Wang, J. and Hou, T. (2011) Recent advances on aqueous solubility prediction. *Comb. Chem. High Throughput Screen.*, 14(5):328–338.
7. Noyes, A.A. and Whitney, W.R. (1897) The rate of solution of solid substances in their own solutions. *J. Am. Chem. Soc.*, 19(12):930–934.
8. Nernst, W. (1904) Theorie der Reaktionsgeschwindigkeit in heterogenen Systemen. *Z. Phys. Chem.*, 47:52–55.
9. Dressman, J.B., Amidon, G.L., Reppas, C., and Shah, V.P. (1998) Dissolution testing as a prognostic tool for oral drug absorption: immediate release dosage forms. *Pharm. Res.*, 15(1):11–22.
10. Horter, D. and Dressman, J.B. (1997) Influence of physicochemical properties on dissolution of drugs in the gastrointestinal tract. *Adv. Drug Deliv. Rev.*, 25(1):3–14.
11. Klein, S. (2010) The use of biorelevant dissolution media to forecast the *in vivo* performance of a drug. *AAPS J.*, 12(3):397–406.
12. Dressman, J.B. and Reppas, C. (2010) *Oral Drug Absorption: Prediction and Assessment*, Vol. 193. 2nd ed. Informa Healthcare, New York.
13. Dressman, J.B. and Reppas, C. (2000) *In vitro–in vivo* correlations for lipophilic, poorly water-soluble drugs. *Eur. J. Pharm. Sci.*, 11(2):73–80.

14. Sjögren, E., Westergren, J., Grant, I., et al. (2013) *In silico* predictions of gastrointestinal drug absorption in pharmaceutical product development: application of the mechanistic absorption model GI-Sim. *Eur. J. Pharm. Sci.*, 49(3):679–698.

15. He, Y-L., Murby, S., Warhurst, G., et al. (1998) Species differences in size discrimination in the paracellular pathway reflected by oral bioavailability of poly(ethylene glycol) and D-peptides. *J. Pharm. Sci.*, 87(5):626–633.

16. Dressman, J.B., Thelen, K., and Jantratid, E. (2008) Towards quantitative prediction of oral drug absorption. *Clin. Pharm.*, 47(10):655–667.

17. Fagerholm, U. (2007) Prediction of human pharmacokinetics: gastrointestinal absorption. *J. Pharm. Pharmacol.*, 59(7):905–916.

18. Yu, L.X. (1999) An integrated model for determining causes of poor oral drug absorption. *Pharm. Res.*, 16(12):1883–1887.

19. Shah, V.P. (2001) Dissolution: a quality control test vs. a bioequivalence test. *Dissolut. Technol.*, 8(4):6–7.

20. Miller, J. (2000) *Analytical Chemistry in a GMP Environment: A Practical Guide*, 1st ed., Wiley-Interscience, New York.

21. U.S. Department of Health and Human Services (1997) *Guidance for Industry: Dissolution Testing of Immediate Release Solid Oral Dosage Forms*, Center for Drug Evaluation and Research (CDER), Food and Drug Administration, Rockville, MD.

22. U.S. Department of Health and Human Services (1997) *Guidance for Industry: Extended Release Oral Dosage Forms—Development, Evaluation, and Application of* In Vitro/In Vivo *Correlations*, Center for Drug Evaluation and Research (CDER), Food and Drug Administration, Rockville, MD.

23. ICH (1999) ICH Harmonised Tripartite Guideline: Specifications: Test Procedures and Acceptance Criteria for New Drug Substances and New Drug Products: Chemical Substances: Q6A. http://www.ich.org/fileadmin/Public_Web_Site/ICH_Products/Guidelines/Quality/Q6A/Step4/Q6Astep4.pdf.

24. ICH (1999) ICH Harmonised Tripartite Guideline: Specifications: Test Procedures and Acceptance Criteria for New Drug Substances and New Drug Products: Chemical Substances: Q6A: Decision Trees. http://www.ich.org/fileadmin/Public_Web_Site/ICH_Products/Guidelines/Quality/Q6A/Step4/Decision_Trees.pdf.

25. Kleberg, K., Jacobsen, J., and Müllertz, A. (2010) Characterising the behaviour of poorly water soluble drugs in the intestine: application of biorelevant media for solubility, dissolution and transport studies. *J. Pharm. Pharmacol.*, 62(11):1656–1668.

26. Reppas, C. and Vertzoni, M. (2012) Biorelevant *in-vitro* performance testing of orally administered dosage forms. *J. Pharm. Pharmacol.*, 64(7):919–930.

27. USP (2006) *United States Pharmacopeia and National Formulary (USP 30-NF 25)*, United States Pharmacopeial Convention, Rockville, MD.

28. Jantratid, E., Janssen, N., Reppas, C., and Dressman, J.B. (2008) Dissolution media simulating conditions in the proximal human gastrointestinal tract: an update. *Pharm. Res.*, 25(7):1663–1676.

29. Vertzoni, M., Dressman, J., Butler, J., Hempenstall, J., and Reppas, C. (2005) Simulation of fasting gastric conditions and its importance for the *in vivo* dissolution of lipophilic compounds. *Eur. J. Pharm. Biopharm.*, 60(3):413–417.

30. Vertzoni, M., Pastelli, E., Psachoulias, D., Kalantzi, L., and Reppas, C. (2007) Estimation of intragastric solubility of drugs: in what medium? *Pharm. Res.*, 24(5):909–917.

31. Amidon, G.L., Lennernas, H., Shah, V.P., and Crison, J.R. (1995) Theoretical basis for a biopharmaceutical drug classification: correlation of *in vitro* drug product dissolution and *in vivo* bioavailability. *Pharm. Res.*, 12:413–420.
32. Martinez, M.N. and Amidon, G.L. (2002) A mechanistic approach to understanding the factors affecting drug absorption: a review of fundamentals. *Clin. Pharmacol.*, 42(6): 620–643.
33. U.S. Department of Health and Human Services (2000) *Guidance for Industry: Waiver of* In Vivo *Bioavailability and Bioequivalence Studies for Immediate-Release Solid Oral Dosage Forms Based on a Biopharmaceutics Classification System*, Center for Drug Evaluation and Research (CDER), Food and Drug Administration, Rockville, MD.
34. U.S. Department of Health and Human Services (2000) *Guidance for Industry: Waiver on* In Vivo *Bioavailability and Bioequivalence Studies for Immediate-Release Solid Oral Dosage Forms Based on a Biopharmaceutics Classification Systems*, Center for Drug Evaluation and Research (CDER), Food and Drug Administration, Rockville, MD.
35. Li, S., He, H., Lakshman, P.J., Hequn, Y., and Serajuddin, A.Y.M. (2005) IV–IVC considerations in the development of immediate-release oral dosage form. *J. Pharm. Sci.*, 94(7):1396–1416.
36. Lu, Y., Kim, S., and Park, K. (2011) *In vitro–in vivo* correlation: perspectives on model development. *Int. J. Pharm.*, 418(1):142–148.
37. Brouwers, J., Brewster, M.E., and Augustijns, P. (2009) Supersaturating drug delivery systems: the answer to solubility-limited oral bioavailability? *J. Pharm. Sci.*, 98(8): 2549–2572.
38. Gao, P. and Shi, Y. (2012) Characterization of supersaturatable formulations for improved absorption of poorly soluble drugs. *AAPS J.*, 14(4):703–713.
39. Kawakami, K. (2012) Modification of physicochemical characteristics of active pharmaceutical ingredients and application of supersaturatable dosage forms for improving bioavailability of poorly absorbed drugs. *Adv. Drug Deliv. Rev.*, 64(6):480–495.
40. Singh, A., Worku, Z.A., and Van Den Mooter, G. (2011) Oral formulation strategies to improve solubility of poorly water-soluble drugs. *Expert Opin. Drug Deliv.*, 8(10):1361–1368.
41. Hancock, B.C. (2002) Disordered drug delivery: destiny, dynamics and the Deborah number. *J. Pharm. Pharmacol.*, 54(6):737–746.
42. Mosharraf, M. and Nyström, C. (1999) The effect of dry mixing on the apparent solubility of hydrophobic, sparingly soluble drugs. *Eur. J. Pharm. Sci.*, 9(2):145–156.
43. Bollag, G., Hirth, P., Tsai, J., et al. (2010) Clinical efficacy of a RAF inhibitor needs broad target blockade in BRAF-mutant melanoma. *Nature*, 467(7315):596–599.
44. Goldberg, A.H., Gibaldi, M., and Kanig, J.L. (1966) Increasing dissolution rates and gastrointestinal absorption of drugs via solid solutions and eutectic mixtures III: experimental evaluation of griseofulvin–succinic acid solid solution. *J. Pharm. Sci.*, 55(5):487–492.
45. Hancock, B.C. and Parks, M. (2000) What is the true solubility advantage for amorphous pharmaceuticals? *Pharm. Res.*, 17(4):397–404.
46. Joshi, H.N., Tejwani, R.W., Davidovich, M., et al. (2004) Bioavailability enhancement of a poorly water-soluble drug by solid dispersion in polyethylene glycol–polysorbate 80 mixture. *Int. J. Pharm.*, 269(1):251–258.
47. Kennedy, M., Hu, J., Gao, P., et al. (2008) Enhanced bioavailability of a poorly soluble VR1 antagonist using an amorphous solid dispersion approach: a case study. *Mol. Pharm.*, 5(6):981–993.

48. Konno, H., Handa, T., Alonzo, D.E., and Taylor, L.S. (2008) Effect of polymer type on the dissolution profile of amorphous solid dispersions containing felodipine. *Eur. J. Pharm. Biopharm.*, 70(2):493–499.
49. Kwong, A.D., Kauffman, R.S., Hurter, P., and Mueller, P. (2011) Discovery and development of telaprevir: an NS3-4A protease inhibitor for treating genotype 1 chronic hepatitis C virus. *Nat. Biotechnol.*, 29(11):993–1003.
50. Law, D., Krill, S.L., Schmitt, E.A., et al. (2001) Physicochemical considerations in the preparation of amorphous ritonavir–poly(ethylene glycol) 8000 solid dispersions. *J. Pharm. Sci.*, 90(8):1015–1025.
51. Law, D., Schmitt, E.A., Marsh, K.C., et al. (2004) Ritonavir–PEG 8000 amorphous solid dispersions: *in vitro* and *in vivo* evaluations. *J. Pharm. Sci.*, 93(3):563–570.
52. Leuner, C. and Dressman, J. (2000) Improving drug solubility for oral delivery using solid dispersions. *Eur. J. Pharm. Biopharm.*, 50(1):47–60.
53. Six, K., Verreck, G., Peeters, J., Brewster, M., and Van Den Mooter, G. (2004) Increased physical stability and improved dissolution properties of itraconazole, a class II drug, by solid dispersions that combine fast- and slow-dissolving polymers. *J. Pharm. Sci.*, 93(1):124–131.
54. Ilevbare, G.A. and Taylor, L.S. (2013) Liquid–liquid phase separation in highly supersaturated aqueous solutions of poorly water-soluble drugs: implications for solubility enhancing formulations. *Cryst. Growth Des.*, 13(4):1497–1509.
55. Murdande, S., Pikal, M., Shanker, R., and Bogner, R. (2010) Solubility advantage of amorphous pharmaceuticals: II. Application of quantitative thermodynamic relationships for prediction of solubility enhancement in structurally diverse insoluble pharmaceuticals. *Pharm. Res.*, 27(12):2704–2714.
56. Murdande, S.B., Pikal, M.J., Shanker, R.M., and Bogner, R.H. (2010) Solubility advantage of amorphous pharmaceuticals: I. A thermodynamic analysis. *J. Pharm. Sci.*, 99(3): 1254–1264.
57. Hoffman, J.D. (1958) Thermodynamic driving force in nucleation and growth processes. *J. Chem. Phys.*, 29(5):1192–1193.
58. Mullin, J.W. (2001) *Solutions and Solubility*, 4th ed., Butterworth-Heinemann, Oxford, pp. 86–134.
59. Ilevbare, G.A., Liu, H., Edgar, K.J., and Taylor, L.S. (2012) Maintaining supersaturation in aqueous drug solutions: impact of different polymers on induction times. *Cryst. Growth Des.*, 13(2):740–751.
60. Guzmán, H.R., Tawa, M., Zhang, Z., et al. (2007) Combined use of crystalline salt forms and precipitation inhibitors to improve oral absorption of celecoxib from solid oral formulations. *J. Pharm. Sci.*, 96(10):2686–2702.
61. Calahan, J.L., Zanon, R.L., Alvarez-Nunez, F., and Munson, E.J. (2013) Isothermal microcalorimetry to investigate the phase separation for amorphous solid dispersions of AMG 517 with HPMC-AS. *Mol. Pharm.*, 10(5):1949–1957.
62. Engers, D., Teng, J., Jimenez-Novoa, J., et al. (2008) Screening and animal bioavailability studies on itraconazole amorphous dispersions. *AAPS Annual Meeting, Atlanta, Georgia.*
63. Bak, A., Gore, A., Yanez, E., et al. (2008) The co-crystal approach to improve the exposure of a water-insoluble compound: AMG 517 sorbic acid co-crystal characterization and pharmacokinetics. *J. Pharm. Sci.*, 97(9):3942–3956.
64. Alonzo, D.E., Gao, Y., Zhou, D., Mo, H., Zhang, G.G., and Taylor, L.S. (2011) Dissolution and precipitation behavior of amorphous solid dispersions. *J. Pharm. Sci.*, 100(8):3316–3331.

65. Ilevbare, G.A., Liu, H., Edgar, K.J., and Taylor, L.S. (2012) Inhibition of solution crystal growth of ritonavir by cellulose polymers: factors influencing polymer effectiveness. *CrystEngComm*, 14(20):6503–6514.
66. Kanaze, F.I., Kokkalou, E., Niopas, I., Georgarakis, M., Stergiou, A., and Bikiaris, D. (2006) Dissolution enhancement of flavonoids by solid dispersion in PVP and PEG matrixes: a comparative study. *J. Appl. Polym. Sci.*, 102(1):460–471.
67. Karekar, P., Vyas, V., Shah, M., Sancheti, P., and Pore, Y. (2009) Physicochemical investigation of the solid dispersion systems of etoricoxib with poloxamer 188. *Pharm. Dev. Technol.*, 14(4):373–379.
68. Kolašinac, N., Kachrimanis, K., Homšek, I., Grujić, B., Đurić, Z., and Ibrić, S. (2012) Solubility enhancement of desloratadine by solid dispersion in poloxamers. *Int. J. Pharm.*, 436(1–2):161–170.
69. Andrews, G.P., Abudiak, O.A., and Jones, D.S. (2010) Physicochemical characterization of hot melt extruded bicalutamide- polyvinylpyrrolidone solid dispersions. *J. Pharm. Sci.*, 99(3):1322–1335.
70. Karavas, E., Ktistis, G., Xenakis, A., and Georgarakis, E. (2006) Effect of hydrogen bonding interactions on the release mechanism of felodipine from nanodispersions with polyvinyl-pyrrolidone. *Eur. J. Pharm. Biopharm.*, 63(2):103–114.
71. Corrigan, O.I. (1985) Mechanisms of dissolution of fast release solid dispersions. *Drug Dev. Ind. Pharm.*, 11(2-3):697–724.
72. Craig, D.Q.M. (2002) The mechanisms of drug release from solid dispersions in water-soluble polymers. *Int. J. Pharm.*, 231(2):131–144.
73. Albers, J., Alles, R., Matthée, K., Knop, K., Nahrup, J.S., and Kleinebudde, P. (2009) Mechanism of drug release from polymethacrylate-based extrudates and milled strands prepared by hot-melt extrusion. *Eur. J. Pharm. Biopharm.*, 71(2):387–394.
74. Tho, I., Liepold, B., Rosenberg, J., Maegerlein, M., Brandl, M., and Fricker, G. (2010) Formation of nano/micro-dispersions with improved dissolution properties upon dispersion of ritonavir melt extrudate in aqueous media. *Eur. J. Pharm. Sci.*, 40(1):25–32.
75. Vasconcelos, T., Sarmento, B., and Costa, P. (2007) Solid dispersions as strategy to improve oral bioavailability of poor water soluble drugs. *Drug Discov. Today*, 12 (23–24):1068–1075.
76. Simonelli, A.P., Mehta, S.C., and Higuchi, W.I. (1969) Dissolution rates of high energy polyvinylpyrrolidone (PVP)-sulfathiazole coprecipitates. *J. Pharm. Sci.*, 58(5):538–549.
77. Simonelli, A.P., Mehta, S.C., and Higuchi, W.I. (1976) Dissolution rates of high energy sulfathiazole–povidone coprecipitates II: characterization of form of drug controlling its dissolution rate via solubility studies. *J. Pharm. Sci.*, 65(3):355–361.
78. Erdemir, D., Lee, A.Y., and Myerson, A.S. (2009) Nucleation of crystals from solution: classical and two-step models. *Acc. Chem. Res.*, 42(5):621–629.
79. Marsac, P., Konno, H., and Taylor, L. (2006) A comparison of the physical stability of amorphous felodipine and nifedipine systems. *Pharm. Res.*, 23(10):2306–2316.
80. Tang, X.C., Pikal, M.J., and Taylor, L.S. (2002) A spectroscopic investigation of hydrogen bond patterns in crystalline and amorphous phases in dihydropyridine calcium channel blockers. *Pharm. Res.*, 19(4):477–483.
81. Wegiel, L.A., Mauer, L.J., Edgar, K.J., and Taylor, L.S. (2013) Mid-infrared spectroscopy as a polymer selection tool for formulating amorphous solid dispersions. *J. Pharm. Pharmacol.*, 66(2):244–255.

82. Kestur, U.S., Van Eerdenbrugh, B., and Taylor, L.S. (2011) Influence of polymer chemistry on crystal growth inhibition of two chemically diverse organic molecules. *CrystEngComm*, 13(22):6712–6718.
83. Ilevbare, G.A., Liu, H., Edgar, K.J., and Taylor, L.S. (2013) Impact of polymers on crystal growth rate of structurally diverse compounds from aqueous solution. *Mol. Pharm.*, 10(6):2381–2393.
84. Ilevbare, G.A., Liu, H., Edgar, K.J., and Taylor, L.S. (2012) Understanding polymer properties important for crystal growth inhibition: impact of chemically diverse polymers on solution crystal growth of ritonavir. *Cryst. Growth Des.*, 12(6):3133–3143.
85. Murdande, S.B., Pikal, M.J., Shanker, R.M., and Bogner, R.H. (2011) Aqueous solubility of crystalline and amorphous drugs: challenges in measurement. *Pharm. Dev. Technol.*, 16(3):187–200.
86. Byrn, S.R., Pfeiffer, R.R., and Stowell, J.G. (1999) *Solid State Chemistry of Drugs*, Ssci Inc., West Lafayette, IN.
87. Newman, A., Knipp, G., and Zografi, G. (2011) Assessing the performance of amorphous solid dispersions. *J. Pharm. Sci.*, 101(4):1355–1377.
88. Corrigan, O.I., Holohan, E.M., and Sabra, K. (1984) Amorphous forms of thiazide diuretics prepared by spray drying. *Int. J. Pharm.*, 18(1–2):195–200.
89. Gupta, P., Kakumanu, V.K., and Bansal, A.K. (2004) Stability and solubility of celecoxib–PVP amorphous dispersions: a molecular perspective. *Pharm. Res.*, 21(10): 1762–1769.
90. Wang, Q., Ma, D., and Higgins, J.P. (2006) Analytical method selection for drug product dissolution testing. *Dissolut. Technol.*, 13(3):6–13.
91. USP (2004) *United States Pharmacopeia and National Formulary (USP 27-NF 22): Dissolution*, United States Pharmacopeial Convention, Rockville, MD.
92. USP (2004) *United States Pharmacopeia and National Formulary (USP 27-NF 22): Drug Release*, United States Pharmacopeial Convention, Rockville, MD.
93. Cohen, J.L., Hubert, B.B., Leeson, L.J., et al. (1990) The development of USP dissolution and drug release standards. *Pharm. Res.*, 7(10):983–987.
94. USP (1998) *British Pharmacopeia: Dissolution of Solid Oral Dosage Forms*, Vol. 2, United States Pharmacopeial Convention, Rockville, MD.
95. USP (2006) *United States Pharmacopeia and National Formulary (USP 41-NF 34): The Dissolution Procedure—Development and Validation*, United States Pharmacopeial Convention, Rockville, MD.
96. Boetker, J.P., Savolainen, M., Koradia, V., et al. (2011) Insights into the early dissolution events of amlodipine using UV imaging and Raman spectroscopy. *Mol. Pharm.*, 8(4):1372–1380.
97. Ozaki, S., Minamisono, T., Yamashita, T., Kato, T., and Kushida, I. (2012) Supersaturation–nucleation behavior of poorly soluble drugs and its impact on the oral absorption of drugs in thermodynamically high-energy forms. *J. Pharm. Sci.*, 101(1):214–222.
98. Mehanna, M.M., Motawaa, A.M., and Samaha, M.W. (2010) Insight into tadalafil–block copolymer binary solid dispersion: mechanistic investigation of dissolution enhancement. *Int. J. Pharm.*, 402(1–2):78–88.
99. Lust, A., Laidmäe, I., Palo, M., et al. (2013) Solid-state dependent dissolution and oral bioavailability of piroxicam in rats. *Eur. J. Pharm. Sci.*, 48(1–2):47–54.

100. Essa, S., Rabanel, J.M., and Hildgen, P. (2011) Characterization of rhodamine loaded PEG-g-PLA nanoparticles (NPs): effect of poly(ethylene glycol) grafting density. *Int. J. Pharm.*, 411 (1–2):178–187.
101. Kulhari, H., Pooja, D., Prajapati, S.K., and Chauhan, A.S. (2011) Performance evaluation of PAMAM dendrimer based simvastatin formulations. *Int. J. Pharm.*, 405(1–2):203–209.
102. Leo, E., Cameroni, R., and Forni, F. (1999) Dynamic dialysis for the drug release evaluation from doxorubicin–gelatin nanoparticle conjugates. *Int. J. Pharm.*, 180(1):23–30.
103. Nagarwal, R.C., Singh, P.N., Kant, S., Maiti, P., and Pandit, J.K. (2011) Chitosan nanoparticles of 5-fluorouracil for ophthalmic delivery: characterization, *in-vitro* and *in-vivo* study. *Chem. Pharm. Bull.*, 59(2):272–278.
104. Saremi, S., Atyabi, F., Aklaghi, S.P., Ostad, S.N., and Dinarvand, R. (2011) Thiolated chitosan nanoparticles for enhancing oral absorption of docetaxel: preparation, *in vitro* and *ex vivo* evaluation. *Int. J. Nanomed.*, 6:119–128.
105. Tan, Y.-L. and Liu, C.-G. Preparation and characterization of self-assembled nanoparticles based on folic acid modified carboxymethyl chitosan. *J. Mater. Sci. Mater. Med.*, 22(5): 1213–1220.
106. Friesen, D.T., Shanker, R., Crew, M., Smithey, D.T., Curatolo, W.J., and Nightingale, J.A.S. (2008) Hydroxypropyl methylcellulose acetate succinate-based spray-dried dispersions: an overview. *Mol. Pharm.*, 5(6):1003–1019.
107. Alonzo, D.E., Raina, S., Zhou, D., Gao, Y., Zhang, G.G.Z., and Taylor, L.S. (2012) Characterizing the impact of hydroxypropyl methyl cellulose on the growth and nucleation kinetics of felodipine from supersaturated solutions. *Cryst. Growth Des.*, 12(3):1538–1547.
108. Zambito, Y., Pedreschi, E., and Di Colo, G. (2012) Is dialysis a reliable method for studying drug release from nanoparticulate systems?: a case study. *Int. J. Pharm.*, 434(1–2):28–34.
109. Babcock, W.C., Friesen, D.T., Mccray, S.B., Babcock, W.C., Friesen, D.T., Mccray, S.B., Babcock, W.C., Friesen, D.T., and Mccray, S.B.S. (2005) Method and device for evaluation of pharmaceutical compositions. WO2005095950 A1, October 13, 2005.
110. Bevernage, J., Brouwers, J., Annaert, P., and Augustijns, P. (2012) Drug precipitation–permeation interplay: supersaturation in an absorptive environment. *Eur. J. Pharm. Biopharm.*, 82(2):424–428.
111. Carino, S.R., Sperry, D.C., and Hawley, M. (2006) Relative bioavailability estimation of carbamazepine crystal forms using an artificial stomach–duodenum model. *J. Pharm. Sci.*, 95(1):116–125.
112. Polster, C.S., Atassi, F., Wu, S.-J., and Sperry, D.C. (2010) Use of artificial stomach–duodenum model for investigation of dosing fluid effect on clinical trial variability. *Mol. Pharm.*, 7(5):1533–1538.
113. Dickinson, P., Abu Rmaileh, R., Ashworth, L., et al. (2012) An investigation into the utility of a multi-compartmental, dynamic, system of the upper gastrointestinal tract to support formulation development and establish bioequivalence of poorly soluble drugs. *AAPS J.*, 14(2):196–205.
114. Mainville, I., Arcand, Y., and Farnworth, E.R. (2005) A dynamic model that simulates the human upper gastrointestinal tract for the study of probiotics. *Int. J. Food Microbiol.*, 99(3):287–296.
115. Brown, C.K., Chokshi, H.P., Nickerson, B., Reed, R.A., Rohrs, B.R., and Shah, P.A. (2004) Acceptable analytical practices for dissolution testing of poorly soluble compounds. *Pharm. Technol.*, 28(12):56–65.

116. Boghra, R.J., Kothawade, P.C., Belgamwar, V.S., Nerkar, P.P., Tekade, A.R., and Surana, S.J. (2011) Solubility, dissolution rate and bioavailability enhancement of irbesartan by solid dispersion technique. *Chem. Pharm. Bull.*, 59(4):438–441.
117. Hsieh, Y.L., Ilevbare, G.A., Van Eerdenbrugh, B., Box, K.J., Sanchez-Felix, M.V., and Taylor, L.S. (2012) PH-induced precipitation behavior of weakly basic compounds: determination of extent and duration of supersaturation using potentiometric titration and correlation to solid state properties. *Pharm. Res.*, 29(10):2738–2753.
118. Karavas, E., Georgarakis, E., and Bikiaris, D. (2006) Felodipine nanodispersions as active core for predictable pulsatile chronotherapeutics using PVP/HPMC blends as coating layer. *Int. J. Pharm.*, 313(1–2):189–197.
119. Ilevbare, G.A., Liu, H., Pereira, J., Edgar, K.J., and Taylor, L.S. (2013) Influence of additives on the properties of nanodroplets formed in highly supersaturated aqueous solutions of ritonavir. *Mol. Pharm.*, 10(9):3392–3403.
120. Tachibana, T. and Nakamura, A. (1965) A method for preparing an aqueous colloidal dispersion of organic materials by using water-soluble polymers: dispersion B-carotene by polyvinylpyrrolidone. *Kolloid-Zeit. Zeit. Polym.*, 203(2):130–133.
121. Aisha, A.F.A., Ismail, Z., Abu-salah, K.M., and Majid, A.M.S.A. (2012) Solid dispersions of α-mangostin improve its aqueous solubility through self-assembly of nanomicelles. *J. Pharm. Sci.*, 101(2):815–825.
122. Frank, K.J., Westedt, U., Rosenblatt, K.M., et al. (2012) The amorphous solid dispersion of the poorly soluble ABT-102 forms nano/microparticulate structures in aqueous medium: impact on solubility. *Int. J. Nanomed.*, 7:5757–5768.
123. Brick, M.C., Palmer, H.J., and Whitesides, T.H. (2003) Formation of colloidal dispersions of organic materials in aqueous media by solvent shifting. *Langmuir*, 19(16):6367–6380.
124. Van Eerdenbrugh, B., Baird, J.A., and Taylor, L.S. (2010) Crystallization tendency of active pharmaceutical ingredients following rapid solvent evaporation: classification and comparison with crystallization tendency from undercooled melts. *J. Pharm. Sci.*, 99(9): 3826–3838.
125. Baird, J.A., Van Eerdenbrugh, B., and Taylor, L.S. (2010) A classification system to assess the crystallization tendency of organic molecules from undercooled melts. *J. Pharm. Sci.*, 99(9):3787–3806.
126. Murdande, S.B., Pikal, M.J., Shanker, R.M., and Bogner, R.H. (2011) Solubility advantage of amorphous pharmaceuticals: Part 3. Is maximum solubility advantage experimentally attainable and sustainable? *J. Pharm. Sci.*, 100(10):4349–4356.
127. Douroumis, D. and Fahr, A. (2007) Stable carbamazepine colloidal systems using the cosolvent technique. *Eur. J. Pharm. Sci.*, 30(5):367–374.
128. Goddard, E.D., Turro, N.J., Kuo, P.L., and Ananthapadmanabhan, K.P. (1985) Fluorescence probes for critical micelle concentration determination. *Langmuir*, 1(3):352–355.
129. Zhao, C.L., Winnik, M.A., Riess, G., and Croucher, M.D. (1990) Fluorescence probe techniques used to study micelle formation in water-soluble block copolymers. *Langmuir*, 6(2):514–516.
130. Van Eerdenbrugh, B., Alonzo, D.E., and Taylor, L.S. (2011) Influence of particle size on the ultraviolet spectrum of particulate-containing solutions: implications for *in-situ* concentration monitoring using UV/vis fiber-optic probes. *Pharm. Res.*, 28(7):1643–1652.
131. Yamashita, K., Nakate, T., Okimoto, K., et al. (2003) Establishment of new preparation method for solid dispersion formulation of tacrolimus. *Int. J. Pharm.*, 267(1–2):79–91.

132. Azarmi, S., Roa, W.H., and Lobenberg, R. (2008) Targeted delivery of nanoparticles for the treatment of lung diseases. *Adv. Drug Deliv. Rev.*, 60(8):863–875.

133. Frenkel, Y.V., Clark, A.D., Das, K., et al. (2005) Concentration and pH dependent aggregation of hydrophobic drug molecules and relevance to oral bioavailability. *J. Med. Chem.*, 48(6):1974–1983.

134. Serajuddin, A.T.M. (1999) Solid dispersion of poorly water-soluble drugs: early promises, subsequent problems, and recent breakthroughs. *J. Pharm. Sci.*, 88(10):1058–1066.

135. Moribe, K., Pongpeerapat, A., Tozuka, Y., and Yamamoto, K. (2006) Drug nanoparticle formation from drug/HPMC/SDS ternary ground mixtures. *Die Pharm. Int. J. Pharm. Sci.*, 61(2):97–101.

136. Dahan, A., Miller, J.M., and Amidon, G.L. (2009) Prediction of solubility and permeability class membership: provisional BCS classification of the world's top oral drugs. *AAPS J.*, 11(4):740–746.

137. Lipinski, C. (2002) Poor aqueous solubility: an industry wide problem in drug discovery. *Am. Pharm. Rev.*, 5(3):82–85.

138. Fadda, H.M., McConnell, E.L., Short, M.D., and Basit, A.W. (2009) Meal induced acceleration of tablet transit through the human small intestine. *Pharm. Res.*, 26(2):356–360.

7

TRANSLATIONAL DEVELOPMENT OF AMORPHOUS DISPERSIONS

Patrick R. Connelly,[1] Brian Patrick Quinn,[2] Steve Johnston,[1] Philip Bransford,[1] Praveen Mudunuri,[1] Andrey Peresypkin,[1] Majed Fawaz,[1] Setu Roday,[1] Anuj Kuldipkumar,[1] Hong-Ren Wang,[1] Phillip Snyder,[1] Jeff Katstra,[3] Tapan Sanghvi,[4] Bill Rowe,[4] and Patricia Hurter[5]

[1]*Materials Discovery and Characterization, Vertex Pharmaceuticals Incorporated, Boston, MA, USA*

[2]*Parthenon - Ernst & Young, New York, NY, USA*

[3]*Formulation Development, Vertex Pharmaceuticals Incorporated, Cambridge, MA, USA*

[4]*Formulation Development, Vertex Pharmaceuticals Incorporated, Boston, MA, USA*

[5]*Global Pharmaceutical Development and Regulatory Affairs, Vertex Pharmaceuticals Incorporated, Boston, MA, USA*

7.1 INTRODUCTION: TRANSLATIONAL DRUG DEVELOPMENT

7.1.1 The Evolution of Early Development

The basic objective of early pharmaceutical development—to develop a composition and a process that maximize a certain molecule's performance in nonclinical, preclinical,

Pharmaceutical Amorphous Solid Dispersions, First Edition. Edited by Ann Newman.

and, later, human clinical studies—has been the same for decades. However, research and development (R&D) in the drug industry has grown steadily more sophisticated over the past 20 years. The field has evolved in its scientific depth, of course—but also, notably, in the organizational models it has followed. These organizational developments are only irregularly recognized; nonetheless, they have enabled the industry to take fuller advantage of its scientific growth than would otherwise be possible. Crucially, the most recent model—one that we term *translational drug development*, or TDD—has proven well suited to the task of developing poorly soluble compounds into amorphous pharmaceutical materials.

To gain a sense of the evolution that has led to translational development, it is useful to compare three different organizational models that early development has followed over time.

Before the early 1990s, the pharmaceutical R&D process was, broadly, unidirectional. Discovery groups identified molecular scaffolds that proved to be strong and specific ligands for the protein target of interest. The typical output of discovery, at the time, was a molecule that would be dosed in animals to gather data on *in vivo* pharmacokinetics (PK), efficacy, and safety. These molecules were passed along to a development group, charged with studying their physicochemical properties and composing a drug product that could be manufactured, would be stable, and would demonstrate optimal disposition (i.e., absorption, distribution, metabolism, and excretion) in humans. This product would, in turn, be tested in clinical trials, while the development group turned its attention to the manufacturing process that eventually produced the drug at commercial scale [1]. For the sake of taxonomy, we can call this model Development 1.0.

The shortcomings of this process had become apparent by the mid-1990s, and the Development 1.0 model came in for extensive criticism [2]. Its principal drawback was inefficiency: an alarmingly large fraction of drug candidates were not making it all the way through the process (see below). As various observers sought out the root of this problem, the transition from discovery to development earned special attention. This step was, as these practitioners realized, less of a transition and more of a discontinuity. Molecules were being chosen for development before their physicochemical parameters were well understood [2]; by the time properties such as stability, aqueous solubility, and polymorphism were quantified, the choice of molecule for a given target was usually *fait accompli*. As a result, development groups were sometimes left to formulate a molecule with properties that made for an unavoidably poor drug product. A number of clinical programs were delayed or, worse, derailed altogether.

In an effort to reduce this attrition, some organizations implemented what we could call Development 2.0. Briefly, this model consisted of bringing the core competencies of scientists in preclinical development—including physical chemistry, process chemistry, engineering, pharmaceutics, and materials science—into the discovery stage. Venkatesh and Lipper [2] were among the first to report on this expanded role for development scientists in the discovery phase. Before a single molecular scaffold was decided on, preclinical specialists would study physicochemical and biopharmaceutical properties, and even assess manufacturability and potential for scale-up. Only then would a compound be advanced to full preclinical development. In other words, the discovery-to-development transition became bidirectional: molecules moved forward into development, as before—but first, the expertise gained further downstream was fed back into discovery groups and brought to bear on go/no-go decisions at (or shortly before) the candidate selection stage.

Thanks to the Development 2.0 approach, the transition from discovery to development is much less fraught than it used to be. The idea that these two stages require some collaborative, bidirectional information flow to succeed has gained widespread acceptance [2]. More recently, a number of firms have begun to scale up these studies with high-throughput, miniaturized, or automated physicochemical screening during the hit-to-lead and lead optimization phases of discovery [3–8]. Toxicity, solubility, pharmacokinetics, ease of manufacture, and drug disposition are among the areas studied in these "multivariate" programs. The data collected have been assembled into databases that, in turn, guided the development of several *a posteriori* empirical rules of thumb, like Lipinski's famous "Rule of 5" [9] and the Biopharmaceutical Classification System [10]. Strategically speaking, in each case, access to development stage expertise gives discovery teams and managers advanced notice of problems and opportunities downstream. With their improved awareness, they can make better informed decisions about which compounds to move into development.

Still, the later stages of preclinical and clinical development can pose further problems that early development expertise may not recognize. For example, biopharmaceutical properties in humans are notoriously difficult for even specialists to predict before they are observed in clinical trials. Likewise, scale-up at the manufacturing plant (even at clinical scale, never mind commercial) often proves much more complex than just running the same reactions in a bigger kettle. Crystallizations, in particular, often become problematic at scale. Finally, while pharmaceutical firms under Development 1.0 once spent tremendous effort trying, in vain, to develop undevelopable molecules, the opposite problem may apply in Development 2.0. In other words, molecules that were otherwise promising—that could, for instance, have high binding affinity and specificity for their target—have been written off because of their poor stability, or biopharmaceutical properties, or scale-up potential. Some of these failures may have been premature. However, even in a Development 2.0 model, expertise in materials science, biopharmaceutics, and manufacturing is rarely made available to decision makers early in R&D.

This shortcoming led some of those in preclinical development to a conclusion: pharmaceutical R&D organizations stand to benefit from taking the salutary developments of Development 2.0 to their next logical level. In other words, the binary relationship of preclinical and early development groups should be expanded to include expertise in *every* phase of drug development. It should also be applied earlier in discovery research, including the lead optimization phase when molecules are crafted and selected for further testing. Decision makers should be able to recognize not only whether a molecule can be readily rendered as a drug product, but also whether an alternate material form could be a path forward for an otherwise undevelopable scaffold. These same decision makers should be able to estimate biopharmaceutical properties and be prepared in advance for any difficulties in manufacturing scale-up. R&D managers must, in short, think translationally. They must be mindful of the many evolutions that a discovery stage molecule will have to undergo—as it is formulated, manufactured, stored, dosed, absorbed, metabolized, and transported—before it reaches its target. They must ask: Can this molecule make it to *every* subsequent stage of its life cycle, and not just into an early clinical formulation? Can it become a pharmaceutical material, such as a capsule of crystalline active pharmaceutical ingredient (API) or a tablet of amorphous dispersion? What will it do on the shelf and in the body? And finally, can it be manufactured at scale? Translating a drug candidate effectively into a drug product, and then into the plant, and ultimately into human patients demands that know-how gathered at *all* stages of development informs the earlier stages [11]. For this reason, we have called the process of Development 3.0 *translational drug development*.

Translational drug development should not be confused with well-established translational, or bench-to-bedside, medicine, which seeks to adapt scientific developments into clinical therapies. TDD is, instead, a process of reducing the likelihood that therapeutically promising compounds will go to waste by designing in downstream developability. Such a thorough rethinking of the relationship between discovery and development is urgently necessary. Some problems in developability may be corrected during preclinical development, but by and large this approach is far from optimal. Costs increase dramatically as drugs progress toward and through the clinic. Moreover, the risk of expensive failures only increases as an API or formulation, especially a challenging material with a complex synthesis, is ramped up to clinical and eventually commercial-scale manufacture. Yet, for all of its expense, attrition during preclinical drug development is extraordinarily common. A recent report prepared for the industry group PhRMA estimates that only 1 out of every 250 early development molecules will ever become an FDA-approved drug [12], while other approximations run from 1 approval in 100 molecules to 1 in 5000 [13]. According to another recent analysis, pharmaceutical firms advance eight drugs to Phase I clinical trials for every one approved [14]. There is, therefore, little reason to believe that current models of developing new drugs will be sustainable forever [2]. Reducing developmental inefficiency, on the other hand, would produce substantial benefits, including a shorter time to market, a lower capitalized cost of new drugs, and ultimately more cost-effective healthcare.

Ultimately, reducing inefficiency is a principal motivation for adopting TDD. The process reduces the likelihood of unwelcome surprises downstream in development, allowing management teams to make the most prudent, well-informed decisions about which molecules to advance. In TDD, molecules fail faster, if they fail at all.

However, in our experience, TDD can also keep promising molecules from failing in the first place. In some cases, early awareness of late development concerns may send medicinal chemists back to the drawing board in search of a different scaffold. In other cases, TDD rescues promising compounds that would otherwise be dismissed. For instance, a potent compound with poor aqueous solubility may still become a clinically viable drug if it can be delivered in an alternate material form. Such alternatives include cocrystals, nanocrystals, and amorphous dispersions, among others.

To date, our organization has developed and commercialized amorphous dispersions twice with drugs for treating cystic fibrosis and hepatitis C infection requiring the navigation of global marketing approval paths for amorphous materials through agencies such as the FDA and EMA as well as the regulatory authorities of Canada and Australia. In both cases, the TDD process led us to use an alternate material form to maximize the clinical value of a promising compound. Below, we describe the TDD process in general. In later sections, we discuss in greater detail the steps we take to develop amorphous dispersions—with references to our first commercialized drug, telaprevir (Incivek®), as a case study.

7.1.2 How TDD Works

Our organization has implemented Development 3.0 by tasking a single department—the Materials Discovery and Characterization (MDC) department—with collaborating with the other pharmaceutical development line functions—analytical, formulation, and chemical development—as well as discovery research and our commercial pharmaceutical operations unit, to support translation throughout the entire development process. The benefit of an integrated department like this one primarily accrues over time. As translational experience grows, a substantial body of knowledge is produced that perhaps could, eventually, allow development groups to predict the physical properties of a molecule able to bind a specific target.

Because MDC's personnel are involved in projects from lead optimization, through preclinical and clinical development, into commercial manufacture and ultimately line extension, the group is necessarily interdisciplinary. It includes physical chemists, such as specialists in thermodynamics, crystallography, and spectroscopy; analytical chemists; materials scientists and engineers; protein biochemists; and biopharmaceutics experts. Although the department's disciplinary base is broad, its name, Materials Discovery and Characterization, reflects a common focus. The group considers the whole drug product to be a single unit of analysis—the API, in a given crystalline or amorphous form, with all its excipients. As a material, it has certain properties, for instance, solubility, chemical stability, physical stability (i.e., polymorphic and amorphous-to-crystalline interconversion), ease of manufacture, and disposition in the body. The department's mission is to understand these properties as rigorously and as early in development as possible and, thereby, support other line functions in their objectives as needed.

This orientation evokes the "drugs as materials" approach described by Gardner et al. [15]. However, it differs in two respects. First, as this chapter will demonstrate, personnel involved in TDD are integrated into each stage of discovery and development, serving as regular members of working groups, projects teams, and major steering committees. Second, as described below, the basic tool of these TDD personnel is rational, structure-based drug design, rather than a high-throughput approach.

In practice, the department begins its involvement at the lead optimization stage of discovery, when a disease area management team defines what is known as a target product profile (TPP). A TPP includes the pathophysiology of the disease and the clinical needs of the patient population. Working backward from that goal, the profile defines a desired route of administration, relying on an understanding of the patient population and its needs provided by commercialization experts. Available scaffolds can then be considered. On the basis of the ultimate goal and on the resources available to meet it, the management team defines acceptable ranges for physical and chemical properties.

Defining a TPP is common practice in the pharmaceutical industry. What changes in a TDD paradigm is that early and late development experts are available to contribute to planning even at this early stage. This flow of information—later stage expertise guiding earlier stage decision making—is central to TDD, as it was (in more limited fashion) to the innovations of Development 2.0. More specifically, from this early stage on, TDD closely follows the quality by design (QbD) recommended by global regulatory authorities including the FDA and EMA. Under QbD, "quality should be built into a product with a thorough understanding of the product and [the] process by which it is developed and manufactured[,] along with a knowledge of the risks involved . . . and how best to mitigate those risks." [14] In fact, the entire purpose of TDD is to achieve the most thorough possible understanding at the earliest possible date—and, thereby, to mitigate risks downstream.

In offering their perspective on lead selection, development scientists are able to discuss the full range of options for downstream development—spotting both possible problems and opportunities to solve them. For instance, they can suggest material forms that may make a compound with poor aqueous solubility into an orally bioavailable one. This information, in turn, may contribute to the selection of a molecule.

The case of telaprevir is exemplary here because it demonstrates several of the principles just described. The next section describes a particular conundrum seen with telaprevir, some of the solutions that were available for it, and the decisions that led us to move forward with an amorphous spray-dried dispersion (SDD).

7.1.3 The Potency–Insolubility Conundrum and Telaprevir

Telaprevir is a small-molecule peptidomimetic inhibitor of the NS3:4a protease found in the hepatitis C virus [16]. The protease is essential to HCV replication [17]; hence, when combined with the preexisting standard of care (pegylated interferon and ribavirin), inhibiting NS3:4a might, it was hypothesized, improve a patient's ability to clear the virus *vis-à-vis* the standard of care alone.

Telaprevir acts on NS3:4a competitively, by binding to the protease's active site with high affinity ($K_i = \sim 7$ nM at 37 °C [18]). On the basis of binding affinity alone, it might perform well as an inhibitor. However, the nature of the target complicated matters. The active site of NS3:4a is largely apolar; hence, only other apolar compounds are likely to bind to it tightly. The stability of apolar–apolar interactions is due to their low free energy of interaction, created by dehydration and packing effects; in addition, a hydrophobic microenvironment shelters hydrogen bonds from interference by water, lowering the surrounding dielectric and eventuating tighter bonds [19–21]. Regardless of the mechanism, though, the apolar target poses a conundrum. An apolar ligand may be able to bind tightly to the target site, but may not be highly soluble in water or in physiologically relevant aqueous media. Telaprevir was highly hydrophobic and, as a result, displayed a very low aqueous solubility. In water, it is less soluble than marble [14].

Aqueous solubility alone is neither a necessary nor a sufficient condition for an effective, orally delivered drug, but it is one critical parameter that development scientists have to account for. As most oral compounds are absorbed in the small intestine, the API has to reach a concentration in the intestinal lumen sufficient to drive it into the systemic circulation. For compounds with higher gut permeability, a lower solubility may be acceptable. Any compound, however, has to have a certain minimum solubility given its permeability—or no level of binding affinity even matters, because it will never reach adequate levels at its target. Figure 7.1a illustrates this phenomenon for a fictional Compound X. F_a, or fraction absorbed, is limited by both solubility (logarithmically scaled on the x-axis) and permeability (measured in a Caco-2 cell culture model and scaled on the y-axis). Within a certain narrow range, in the case of the compound depicted, low solubility can still result in acceptable F_a if the compound proves to permeate well. However, in most of the range of possible solubilities, solubility may affect whether Compound X will be absorbed sufficiently or not. Hence, the low solubility of crystalline telaprevir meant that, effective as it was *in vitro*, it would be difficult to dose the drug in its crystalline state.

Later research, following the eventual commercialization of telaprevir, would investigate this Catch-22 as a generalizable trend. The problem, which we came to call the *potency–insolubility conundrum*, has long been widely recognized in the pharmaceutical industry, which has responded with empirical attempts to understand and manage it. For instance, the Biopharmaceutical Classification System offers a 2×2 matrix of compounds, describing them as having high or low gut permeability and high or low solubility, with different guidelines for each quadrant. Lipinski's Rule of 5 offers useful rules of thumb to guide development decisions [9]. Only recently, however, have explanations for the fundamental structural and thermodynamic connection between potency and insolubility been elucidated [22]. The investigation that led to these findings echoed the type of studies that could be conducted at the lead optimization phase of discovery, in which the two possible explanations for a compound's insolubility—its hydrophobicity and its propensity to form tight crystal lattices—were analyzed separately. The free energy of solvation—the transfer of the molecule from gas to solution—was distinguished from the free energy of sublimation—the transfer of the molecule from crystal to gas. In Figure 7.1b, hydrophobicity is depicted on the vertical axis and tightness of crystal lattice on the horizontal axis. Therefore, a compound in the lower-left quadrant of the graph would readily dissolve in water. A compound in the upper quadrants would likely be too hydrophobic for much aqueous solubility. But for compounds in the lower-right quadrant, the cause of insolubility is not aversion to water but the strength of the crystal lattice. Hence, disrupting this lattice could increase overall solubility. Hence, the solubilities of itraconazole as well as telaprevir can be increased by two orders of magnitude once the strength of the crystal lattice is eliminated as an obstacle. (Compounds X, Y, and Z are fictional data for illustrative purposes.)

These findings, in turn, suggest a possible framework for addressing this problem with alternate material forms that interrupt the normal, tight formation of a crystal lattice. For telaprevir, cocrystals and amorphous dispersions were initially investigated, but R&D management decided to move only amorphous dispersion into development.

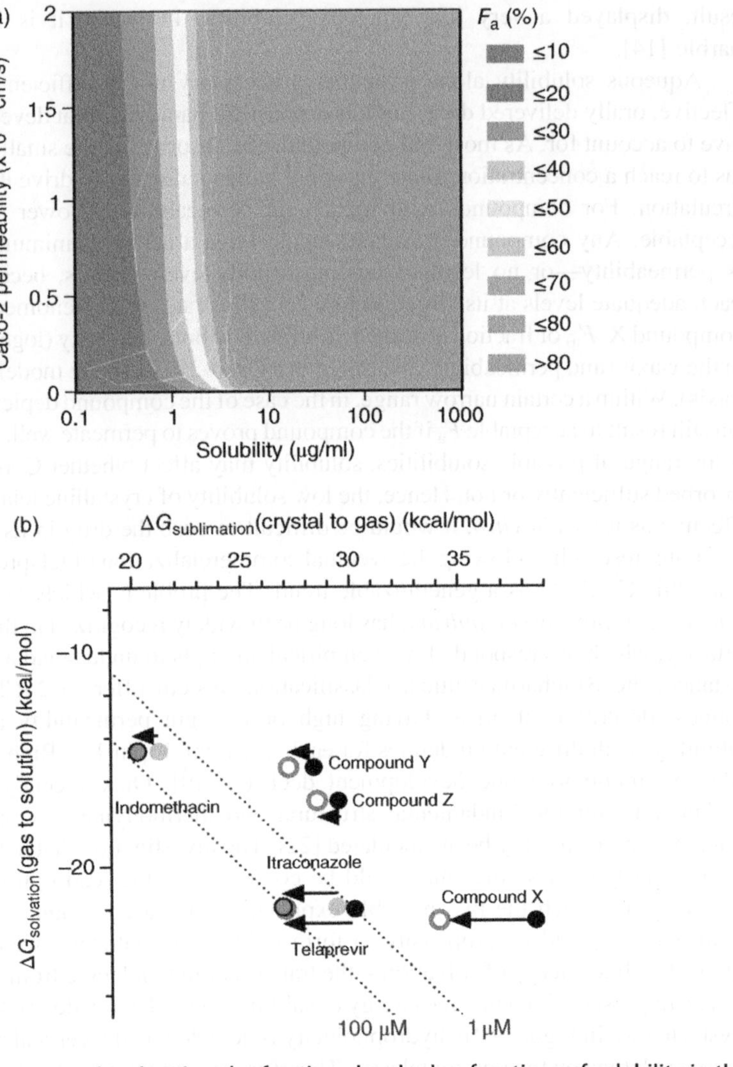

Figure 7.1. (a) A contour plot showing the fraction absorbed as functions of solubility in the fluid of the intestinal lumen and the effective permeability as quantified from the flux in a Caco-2 cell permeability assay. (b) A thermodynamic contribution diagram showing the sublimation and solvation contributions to the free energy of solution.

The following two sections describe the process of TDD, both in general and as applied to telaprevir. First, we describe the efforts undertaken at the discovery stage, which focused on measuring and modeling absorption, optimally preparing and characterizing the amorphous dispersion, and developing a suspension formulation. Next, we discuss the translational activities that are required as the dispersion moves forward into development. The objective here is to prepare the material for GLP preclinical investigations, early GMP manufacture, and first-in-human (FIH) clinical studies. These efforts focus, in part, on assessing the potential for future solid dosage forms. The balance of the work assesses stability both *in vivo* and on storage (in the course of manufacture and on the shelf). We conclude the chapter with a brief summary.

7.2 TRANSLATIONAL DEVELOPMENT AT THE DISCOVERY STAGE

For our organization, laboratory studies of a new candidate drug begin during late pharmacology, before dose range-finding (DRF) studies are conducted. By the time a compound moves forward into development—a shift marked by GLP toxicology studies and, subsequently, a FIH trial—development scientists are responsible for delivering one physical product: a suspension formulation that will be serviceable for initial animal and human dosing. This is fairly typical for preformulation groups, and their equivalents, across a number of pharmaceutical companies. Under a TDD or Development 3.0 paradigm, however, the same group is responsible for a number of scientific or intellectual deliverables as well—results aimed at elucidating mission-critical information about a new compound as early as possible and guiding discovery stage decision making. In addition, when the drug substance in question takes the form of an amorphous dispersion, this group has the further task of developing the process to produce the dispersion at laboratory scale, characterizing the resulting dispersion, and ensuring that it remains amorphous during preparation and suspension dosing. Our procedures for all of these tasks are described in this section in roughly chronological order.

7.2.1 Absorption Modeling

The first studies undertaken by development scientists concern the absorption of an API across the intestinal wall, through the hepatic portal vein to the liver, and ultimately into the systemic circulation. As the Biopharmaceutical Classification System reflects, absorption is driven by both permeability (of the intestinal wall to a given API) and solubility (of the API in the aqueous media of the stomach and intestine). Relatively higher permeability may make up for relatively lower solubility; the reverse is also true. However, these properties differ in one crucial respect. Solubility can be increased in the laboratory by rendering a compound in a thermodynamically higher energy state—either a higher energy crystalline polymorph or, as here, an amorphous dispersion. Permeability, however, is difficult to change for a given compound: it is dependent on structure and is only an issue once the molecule is already in solution. This is not to say that permeability is static: a compound's "effective permeability," a parameter that takes into account the effect of cellular transporters as well as passive uptake mechanisms, may vary along the length of the gastrointestinal tract. Nonetheless, while a compound's solubility is within the capacities of development scientists to change, permeability is not. For this reason, it is important to understand, as early in discovery as possible, both what a compound's absorption is likely to be and how much of that absorption depends on solubility, as opposed to permeability.

One model currently used at our organization is as follows. The states in the model include the mass of undissolved drug in the GI tract (M), the concentration of dissolved drug (C), and the fraction of the dose absorbed (A). The dissolution rate scales with the amount of undissolved drug in the GI tract and the concentration gradient between the diffusion boundary layer around the undissolved particles—a zone that is taken to be saturated with respect to drug—and bulk C. The solubility of the API, which is also its concentration in the boundary layer, is s, and its mass transfer rate is the constant h. V is the volume of the fluid in the GI tract. D is the dose. The dissolution rate admits two fixed points: $C = s$ and $D = 0$. (In other words, dissolution ceases when the concentration in the bulk reaches the thermodynamic solubility of the material or when all of the solid has dissolved.) Finally, the model assumes first-order absorption with a constant permeation rate k. The last equation below defines the cumulative amount of drug absorbed.

$$\frac{\mathrm{d}M}{\mathrm{d}t} = -hM(s - C),$$

$$V\frac{\mathrm{d}C}{\mathrm{d}t} = hM(s - C) - kC,$$

$$D\frac{\mathrm{d}A}{\mathrm{d}t} = kC.$$

Initial conditions are

$$M(0) = D,$$
$$C(0) = A(0) = 0.$$

Next, we can introduce the following nondimensionalization scheme:

$$\hat{M} = \frac{M}{D}, \quad \hat{C} = \frac{CV}{D}, \quad \hat{A} = \frac{A}{D}, \quad \hat{t} = \frac{t}{T},$$

$$\sigma = \frac{s}{D/V}, \quad \eta = hsT, \quad \xi = \frac{kT}{V},$$

where T is the transit time of the drug through the GI tract, σ is solubility normalized by the maximum theoretical concentration in the GI tract, η is the dimensionless mass transfer rate, and ζ is a dimensionless permeation rate. The absorption model becomes

$$\frac{\mathrm{d}\hat{M}}{\mathrm{d}\hat{t}} = -\eta\hat{M}\left(1 - \frac{\hat{C}}{\sigma}\right),$$

$$\frac{\mathrm{d}\hat{C}}{\mathrm{d}\hat{t}} = \eta\hat{M}\left(1 - \frac{\hat{C}}{\sigma}\right) - \xi\hat{C},$$

$$\frac{\mathrm{d}\hat{A}}{\mathrm{d}\hat{t}} = \xi\hat{C}.$$

$$\hat{M}(0) = 1,$$
$$\hat{C}(0) = \hat{A}(0) = 0.$$

The parameter σ is the solubility relative to maximum theoretical concentration, η is the rate of mass transfer relative to the rate of transport through the GI, and ζ is the rate of permeation relative to the rate of transport through the GI.

This model is illustrated by Figure 7.2. Figure 7.2a depicts a simulated absorption model, using first-order absorption, that applies over a wide dose range (100–800 mg). As Figure 7.2b reflects, the bioavailability of a theoretical soluble material ($\sigma \gg 1$) is more sensitive to dissolution rate than an insoluble material ($\sigma \ll 1$). Imagine starting from the upper left of a graph and traversing from left to right; one crosses contours of constant fraction absorbed ($\%F_a$) at a nearly perpendicular angle. If one were to instead begin from the lower-left corner, the contours of constant $\%F_a$ are nearly parallel, and therefore difficult to bisect. As a comparison of the four panels indicates, permeability affects the spacing of the contours such that a highly permeable compound is more sensitive to dissolution rate and solubility than a low-permeability compound.

7.2.2 Preparing and Characterizing an Amorphous Dispersion: Spray Drying and Glass Transition

Once absorption and the relative contributions of dissolution and permeability are adequately understood, subsequent discovery stage work—which includes preclinical formulation development—requires a group to have the desired physical form of the API in hand. If the target product profile (see Section 7.1) calls for a crystalline compound, and the solubility of the molecule's crystalline form (or forms) allows it, then various, extensive solid form studies should be done. These studies would determine whether any other substance could cocrystallize with the API to form a salt, solvate, clathrate, or cocrystal, and would attempt to induce form, or polymorph, changes in the neat API. Although it is not currently possible to know whether all potential polymorphs have been discovered, a rigorous polymorph screen substantially derisks later development—and even commercialization. Ritonavir—an HIV protease inhibitor developed by Abbott that suddenly began converting to an alternate, less soluble form in the commercial manufacturing plant and had to be redesigned [23]—is an example of how tragic such an event can be for the patients affected.

However, if the strategic decision has been made to move forward with an amorphous dispersion, the next step is to prepare it—in other words, to render a crystalline drug amorphous and keep it that way. Several options exist for producing an amorphous dispersion at laboratory scale, including melt quenching, rotovapping, and simply grinding it. (These options are discussed in detail by our colleagues in Chapter 10.) For the examples described in this chapter, we used spray drying.

Spray drying produces an amorphous dispersion, but two important considerations remain. To a greater or lesser degree, they represent conflicting priorities. First, a spray-dried dispersion generally retains some amount of residual solvent. Since spray drying solvents tend to be organic substances that are unacceptable above trace levels in a human or animal formulation, a secondary drying step is necessary to gently remove it. Second, the SDD, however, could be a metastable state with higher free energy than a crystalline form; hence, the dispersion will, on some timescale, revert to the crystalline state. The kinetics of this recrystallization in the solid state can be temperature dependent.

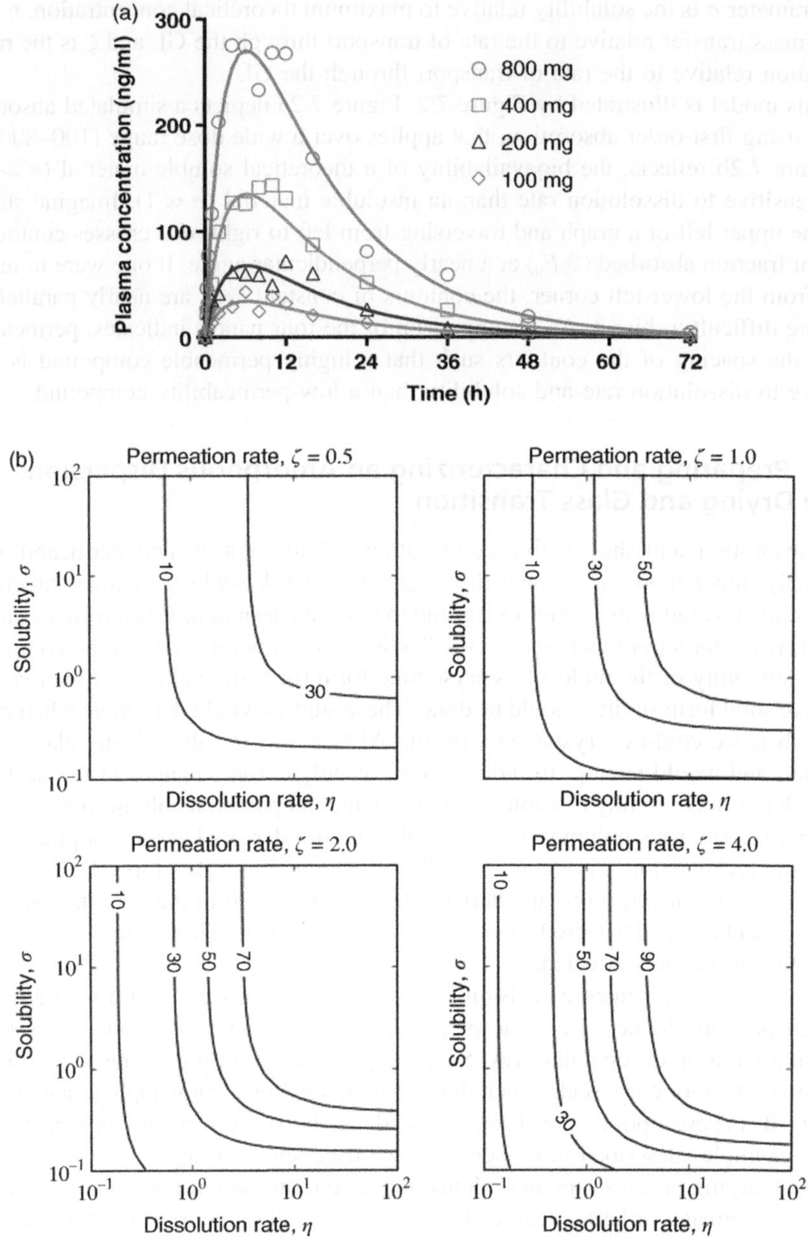

Figure 7.2. (a) A simulated single ascending dose panel reflecting first-order absorption. (b) Contour plots showing the influences that dissolution rate, solubility, and permeability have on bioavailability.

Hence, until the material is fully characterized, it is unclear whether the elevated temperatures required for secondary drying may increase the risk of recrystallization.

Development groups can obtain an early gauge of this risk by determining the material's glass transition temperature, or T_g. The SDD, as an amorphous solid, is a supercooled liquid (or glass); higher temperatures increase molecular mobility, up to the point (called the T_g) at which the material is able to nucleate and revert to the lower energy crystalline state. Below this temperature, the material is still expected to eventually crystallize with temperature-dependent kinetics; the question, which we discuss below, is whether this will occur in a time frame relevant to manufacture, storage, and dosing of the drug. One source posits that a stable dispersion will have a T_g 40–50 °C above the temperature at which it will likely be stored [24]. At 50 °C below T_g, crystallization kinetics are expected to be slow enough to ensure a 2-year shelf life. The first issue, however, is to determine T_g, because above or near that temperature the material is at risk of *sudden* crystallization, sometimes termed "catastrophic crystallization" in regulatory documents.

This temperature is measured by differential scanning calorimetry (DSC), which detects and quantifies thermal events in the material as heat is added. Once the T_g is measured, it can be used as a guide to experiment with drying temperature protocols, with the general knowledge that the risk of sudden recrystallization is lower as long as the material is below that temperature.

It is important to note, though, that T_g is not a fixed property of a given dispersion. Residual solvent lowers the T_g, in some cases quite substantially. Solvents generally have extremely low T_g (e.g., that of water is 136 K); hence, they are highly efficient plasticizers. Plasticizing an amorphous dispersion increases its molecular mobility. If the dispersion encounters temperatures at or above the plasticized T_g, the risk of crystallization may rise.

The simplest way to model the effect of solvent content on T_g is to treat the solid component (the API and any polymers (see below) in the SDD) as one phase and the solvent or solvent system as another phase. Then, the Gordon–Taylor equation can be used to describe the mixing of these components:

$$T_{g,mix} = \frac{w_1 T_{g,1} + K w_2 T_{g,2}}{w_1 + K w_2},$$

where

$$K = \frac{\rho_1 \Delta\alpha_2}{\rho_2 \Delta\alpha_1},$$

given that ρ is the density and $\Delta\alpha$ is the change in the thermal expansion coefficient. The Gordon–Taylor equation has the advantage of being able to account for the behavior of polymer blends using known properties of the polymer components, such as true density and change in thermal expansivity. To simplify the calculation further by assuming comparable densities and $\Delta\alpha$, one can also use the Fox equation:

$$\frac{1}{T_{g,mix}} = \frac{w_1}{T_{g,1}} + \frac{w_2}{T_{g,2}}.$$

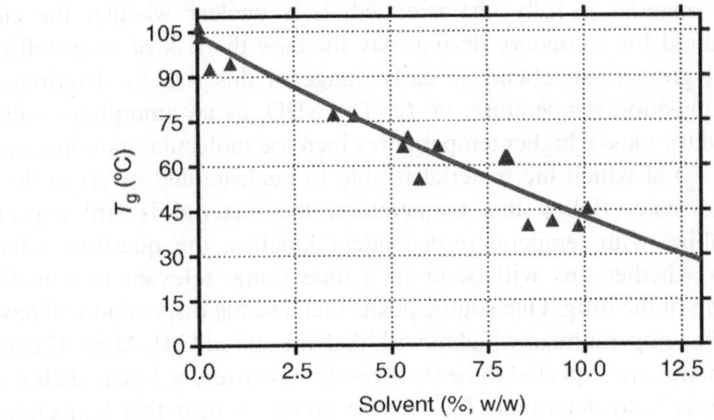

Figure 7.3. The influence of solvent content on glass transition temperature.

With this model, the T_g of the SDD can be estimated over a range of solvent levels. However, this approach is insufficient to rigorously describe the T_g. Solvent–solute interactions, which are difficult to predict, are likely to cause some degree of deviation from the predictions of the Fox equation. Hence, a combination of experimental (DSC-based) and data-fitting approaches may be useful. With the T_g of both dry and plasticized SDDs known, the Fox equation allows one to work from what we call an "effective T_g," essentially the product of the Fox equation's prediction and an interaction parameter that accounts for solvent–solute interactions. (For more on interaction parameters, which become especially salient in blended dispersions, please see Section 7.3.1.) This T_g, graphed as a function of solvent content (%, w/w), can help determine secondary drying temperatures used to remove residual solvent from the SDD (Figure 7.3). Figure 7.4 depicts the change in T_g (squares) as solvent 1 (triangles) and solvent 2 (circles) are removed.

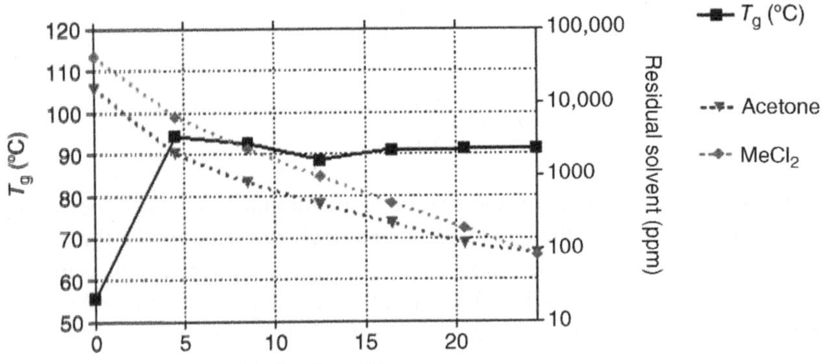

Figure 7.4. Time course of drying solvents from telaprevir SDD indicating the glass transition temperature as a function of residual solvent content.

Naturally, solvent plasticization poses a Catch-22 for early formulation development: the dispersion has to be dried to remove the residual solvent, but the solvent lowers the T_g and restricts the range of temperatures that can be used to dry the material. On the other hand, as the material dries, its T_g increases, meaning that the drying temperature can safely be raised and the drying process thereby accelerated. Hence, modeling the effects of solvent on T_g for a given SDD allows a development team to select the highest possible drying temperature without compromising the product.

This approach was used in developing telaprevir and not only gave us an awareness of the T_g, but also illuminated the molecular mechanism of the drying process. Spray drying crystalline telaprevir produced an amorphous dispersion that contained ~7–13% (w/w) residual solvent. This material was then dried in a biconical dryer at a 50 °C jacket temperature for up to 24 h. To determine the effects of drying on the residual solvent levels and on T_g, samples were taken every 2 h and examined by gas chromatography (GC) (to determine residual solvent levels) and DSC (to measure T_g). GC showed that about 24 h were required to reduce residual solvent to below 100 ppm. DSC thermograms of samples collected at different drying times are shown in Figure 7.5. As the figure reflects, a second T_g emerges and increases in magnitude as drying progresses; by the end, only a single T_g is visible. To further investigate these results, the 12 h sample was analyzed by thermal cycling, in which it was heated above T_g, cooled below that temperature, and heated again. Two T_g values were observed in the first heating cycle and a single, intermediate T_g in the second and third cycles.

Thus, it appeared that two distinct regions were emerging in the material. At the surface, solvent evaporation occurs quickly; this cools the surface, slowing down further evaporation and causing a solvent gradient to develop. When the material is dried further, the gradient disappears due to diffusion of the solvent, with the remaining residual solvent distributing evenly throughout the material. This is the point at which we observe only a single T_g.

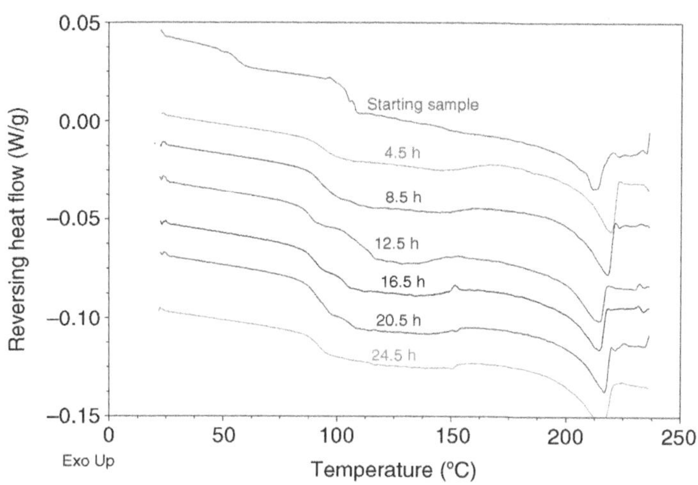

Figure 7.5. DSC thermograms of telaprevir SDD at various stages of secondary drying.

While the mechanism of drying, and the observed second T_g, may be unusual, measuring the plasticized T_g gave us actionable information. SDDs that had been thoroughly dried (per GC) had a T_g of 105 °C. When the same material first came off the spray dryer, its T_g was about 60 °C. While the common rule of thumb would not recommend long-term room-temperature storage of fully plasticized telaprevir SDD—since its T_g is not quite 40 °C above the storage temperature—this result also gave confidence that the drying temperature of 50 °C would not cause the material to suddenly crystallize on drying.

As noted above, an SDD is a thermodynamically higher energy state that would, given sufficient time, be expected to revert to its crystalline form. Our experience has been that some degree of more sophisticated formulation development is required, even at this early stage. Measuring, modeling, and predicting the kinetics of recrystallization during distribution and shelf storage are described in Section 7.3. The next section describes the development and characterization of the first formulation required from an SDD—a suspension for use in preclinical studies, including pharmacology and toxicology, and possibly even for FIH.

7.2.3 Preparation and Characterization of SDD Suspension Formulations: Vehicle and Aqueous Stability

The first suspension formulation is meant to be a rough-and-ready way to get the API into the systemic circulation and, ideally, to its target. Among the key parameters to measure for a given formulation is its maximal exposure *in vivo*; stability of the suspension and solubility and dissolution rate in simulated biological media are also key parameters to evaluate *in vivo* in attempts to maximize exposure and formulation robustness.

To determine the optimal vehicle, several are typically tested, drawing on experience to select excipients to test. These suspension formulations can include surfactants, as they tend to increase the solubility of the SDD. The drawback to any suspension formulation, however, is the possibility of disappointment later in development: it may produce such excellent exposure that no solid form is ever able to equal it.

In addition to the surfactant TPGS, a large majority of the vehicles we screen also include polymers. These are included to improve the physical stability of the SDD in suspension and prevent an overly rapid solvent-mediated phase transition (SMPT)—a process in which the SDD solubilizes, and then recrystallizes in its lower energy, lower solubility crystalline state and precipitates out. While stability is, again, secondary to solubility and dissolution rate, a certain stability in the vehicle is required to ensure consistent exposure from the suspension formulations. If the API precipitates inconsistently from suspensions before or during dosing, the resultant exposures will be too variable to be of any scientific use. It is worth noting that a suspension formulation is, by definition, supersaturated with respect to the crystalline form; thus, the SDD needs to resist recrystallizing not only in the vehicle, but specifically when supersaturated in the vehicle. For a description of how we determine more detailed and rigorous parameters for acceptable stability, please see Section 7.3.4.

Since the suspension will be dosed orally, the SDD also needs to remain in suspension at physiological temperature (i.e., 37 °C) in simulated fluids—specifically,

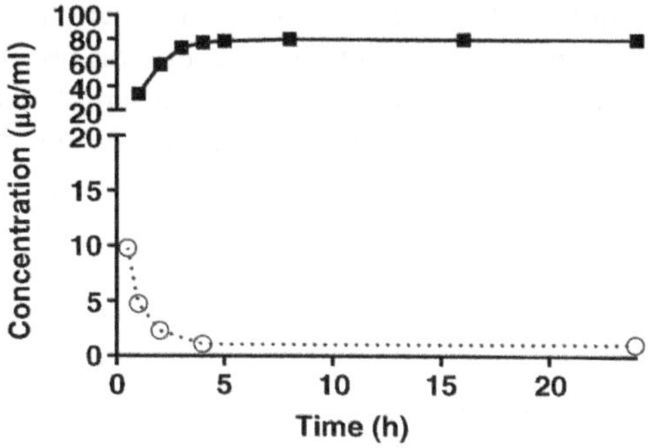

Figure 7.6. Concentration of suspensions of amorphous SDDs in simulated fluids show maintenance of supersaturation (filled squares) or not (open circles) according to the SDD physical stability under the conditions tested.

simulated gastric fluid (SGF) and simulated intestinal fluid (SIF). Figure 7.6 shows the differing behavior in simulated fluids of two hypothetical compounds; while one (solid squares) remains in suspension over a significant timescale, the other (open circles) precipitates in its crystalline form after less than 5 h. In general, given the residence time in the stomach of typical laboratory animals (rats, mice, and dogs), a suspension is at unacceptably high risk of recrystallizing if it nucleates in under 3 h. The exception could be rapid nucleation combined with slow growth of the resultant crystals; while risky, this situation could be manageable. The rate of crystal growth, given nucleation, can be evaluated with "seeding" experiments, in which amorphous dispersions are contaminated with small amounts of crystalline drug. In practice, this seeding simulates the possibility of small-scale crystalline regions occurring in the dispersion when it is produced at scale, and helps determine specifications for acceptable levels of residual crystallinity during manufacture. Frequently, a higher risk than a quick nucleation, slow growth SDD would be an SDD in which crystalline material nucleates slowly but then grows rapidly. Because nucleation is stochastic, results could be uneven.

The danger of solvent-mediated phase transitions was borne out in our own experience. During a Phase I clinical study of telaprevir, lower than expected exposures were obtained in one arm of the study. Amorphous suspensions crystallized and gave lower exposure [16].

In response, we developed an alternative, more robust suspension with a different polymer. We were able to arrive at a different choice of polymer without disrupting the timeline of telaprevir's Phase I trial by devising a novel assay to study the underlying process, a solvent-mediated phase transition. (An SMPT is a two-step process in which a metastable suspended solid such as an amorphous dispersion dissolves in the medium and then crystallizes to a stable solid form.) Previously, we had investigated the physical form of suspensions by pulling samples at discrete time points. To measure crystallinity, the samples were filtered and the solid subjected to X-ray powder diffraction (XRPD), which was able to detect a crystallinity level of about 5%. For dissolution, the samples were run on an HPLC to measure the level of dissolved solid. This approach was able to determine both dissolution and crystallization as a function of time, but only at discontinuous time points; thus, its resolution depended on how closely spaced these time points were. Moreover, the process of drawing the sample and filtering it, both at ambient temperature, may have accelerated the (stochastic) beginnings of crystallization. The new method solved both problems: it gave a continuous readout without disturbing the sample. The suspension was monitored in a MicroCal VP isothermal titration calorimeter (ITC), an extraordinarily sensitive device typically used to measure the thermodynamics of protein–ligand binding. The ITC is a two-cell calorimeter calibrated to maintain a sample cell at exactly the same temperature as a separate reference cell. As the sample cell undergoes endothermic or exothermic events, the calorimeter applies correspondingly more or less heat relative to the reference cell; the heat output of the cell is what is directly measured, at a resolution on the order of μcal. Ultimately, the ITC was able to detect and quantify thermal events, like crystallization, in real time and at specified temperatures.

Ever since, the ITC has been an essential part of our suspension formulation development. We begin each ITC study by adding SDD to the vehicle of choice (more on solvent selection below) as what we term the pseudo-equilibrium concentration. This concentration is the highest that an amorphous material is able to reach (i.e., it reflects the material's intrinsic solubility) at a given pH and temperature. The distinction between the dispersion's pseudo-equilibrium concentration and its solubility is subtle, but important. The "solubility" of a material applies, by implication, to the material's most thermodynamically stable physical form. When the most stable (i.e., lowest enthalpy) form dissolves, its concentration in solution eventually reaches equilibrium and forms a plateau. The amorphous dispersions, however, were metastable. Hence, even at what appears to be an equilibrium concentration, metastable amorphous material is still present and the solution is supersaturated with respect to the crystalline material. Only after phase transition, and sufficient time, does the API reach its true equilibrium concentration, or solubility.

When selecting a solvent to prepare the suspension for ITC, we begin by attempting to use pure water. The water is then pH buffered to reflect the important dissolution conditions that the compound will encounter—for example, pH ~2.5 represents gastric pH in the fasted state, pH ~4.5 represents fed-state gastric pH, and intestinal pH is ~7.

Ultimately, we collect the material at the end of an experiment—and filtering it—to subject it to HPLC and XRPD; this is to confirm that thermal events we witness on the ITC are what we take them to be (e.g., crystallization). Hence, we initially prepare a suspension that is ~10 mg/ml of API to ensure that we collect enough material at the end to analyze it.

The raw data collected by the ITC take the form of total power output from the system over time. An example is shown in Figure 7.7a.

Next, a baseline is drawn within the feature or features of interest and the peak is integrated. A progressive power curve is generated by integrating the peak in equal slices and graphed as mcal/g of API evolved over time, as shown in Figure 7.7b. Finally, this power curve is analyzed within a model system of four equations (Figure 7.7c). From the shape and directionality of the curve, one can model amorphous content of the

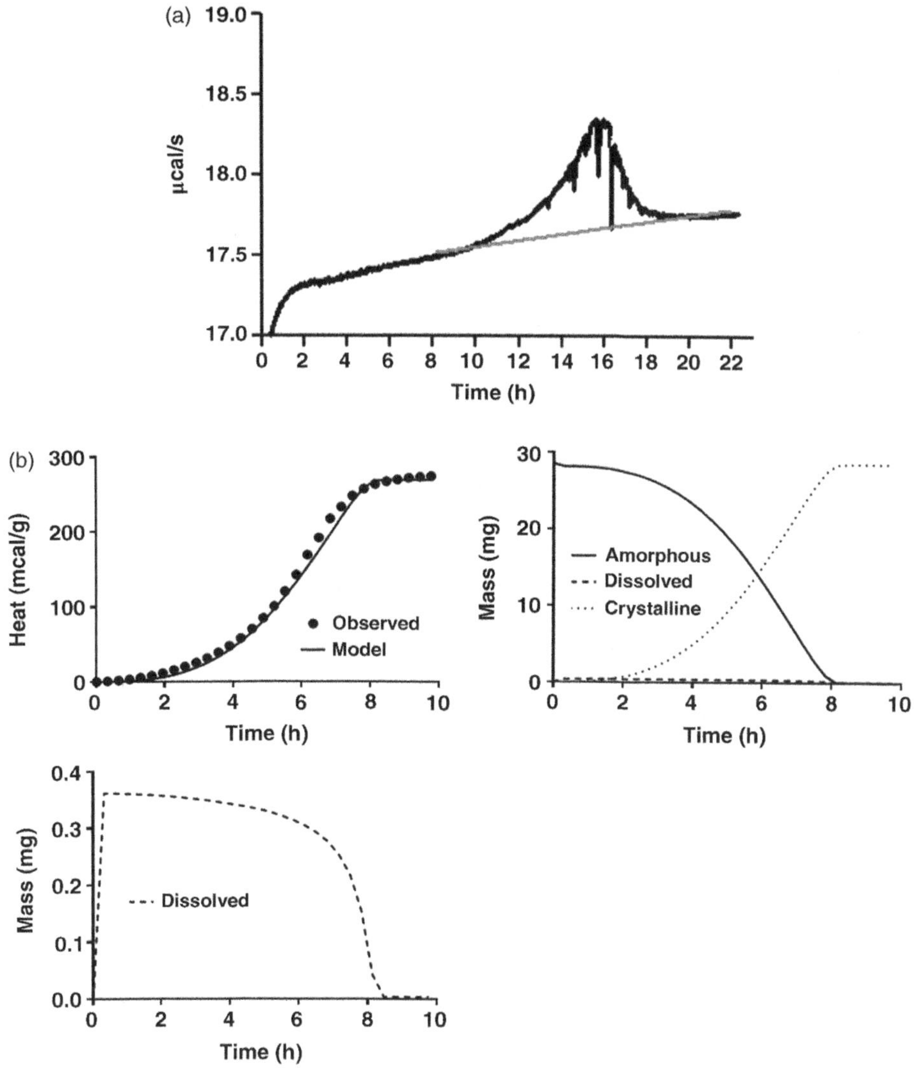

Figure 7.7. (a) Power output from the ITC as a function of time showing the continuous monitoring of the amorphous to crystalline transition in suspension at a fixed temperature. (b) Integrated calorimetric heat output for amorphous to crystalline solvent-mediated phase transition and deconvoluted progress curves for amounts of amorphous, crystalline, and dissolved compound. (c) System of differential equations describing a solvent-mediated amorphous to crystalline transition as monitored by isothermal calorimetry.

(c)

Amorphous $$\frac{dM_A}{dt} = -\chi h\left(\frac{M_A}{\rho}\right)^{2/3}(c_s - C_D)$$

Crystalline $$\frac{dM_C}{dt} = Vk_n(C_D - c_P)^n$$

Dissolved $$V\frac{dC_D}{dt} = \chi h\left(\frac{M_A}{\rho}\right)^{2/3}(c_s - C_D) - Vk_n(C_D - c_P)^n$$

Heat $$\frac{dQ}{dt} = \chi h\left(\frac{M_A}{\rho}\right)^{2/3}(c_s - C_D)\Delta_A H + Vk_n(C_D - c_P)^n\Delta_C H$$

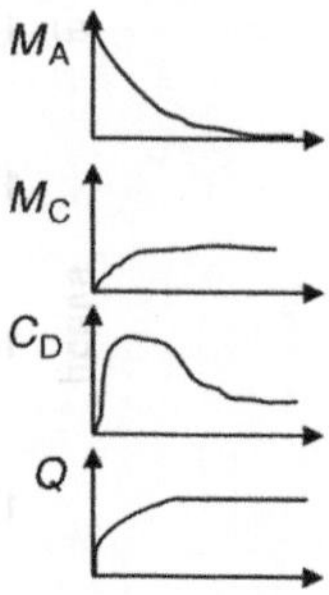

Figure 7.7. (*Continued*)

suspension, which decreases over time as amorphous material dissolves; crystalline content, which increases over time as the dissolved API nucleates and recrystallizes; the total amount dissolved, which peaks in the interval between amorphous dissolution and crystalline nucleation and then decreases with recrystallization; and heat, which increases in the example shown monotonically over time and finally plateaus.

In addition to a continuous readout of heat output, this ITC approach provides a derived measurement of concentration over time. Concentration as a function of time can be modeled with the data collected during the experiment. The total enthalpy, which is measured directly, actually represents the sum of two processes, that is, the two halves of a solvent-mediated phase transition: dissolution of the amorphous material, and its recrystallization and precipitation as a less soluble crystalline solid. The one caveat of ITC is that it is not always possible, in one experiment, to disaggregate these two processes. Hence, we are unable to determine the rate-limiting step in a single run, and assess whether a given SMPT is dissolution controlled or crystallization controlled. In some cases, we can get a crystallization signature from a sample that has been stirred to pseudo-equilibrium concentration, and then filtered and run in the ITC. In this way, we decouple the dissolution heat from the crystallization heat.In addition, we can use our model to simulate the characteristic shapes of dissolution- versus crystallization-controlled SMPTs.

With ITC, though, we can model the aqueous stability of a given SDD. Added to the absorption model described above (in Section 7.2.1), these data may give us an understanding of the parameters that are essential to choosing a suspension formulation for preclinical and even early clinical use.

Developing a suspension formulation, no matter how successful, does not end the formulation aspects of TDD, of course. Oral dosage forms are rarely commercialized as solutions or suspensions; the desired embodiment is typically a solid form, be it a capsule or a tablet. Hence, development scientists at this stage are already beginning to consider how the SDD formulation used for preclinical (and possibly FIH) studies can be translated into a solid dosage form. These considerations form the subject of the next section.

7.3 TRANSLATIONAL DEVELOPMENT AFTER DISCOVERY

7.3.1 Solid Form Development: Preparing Stable Dispersions

By the time a molecule formally transitions from the discovery to the development phase, the compound in question has passed animal toxicology studies and shown exposure *in vivo*. The next step is to shape it into at least a provisional solid formulation that will be not only soluble—a parameter that has already been optimized for—but also manufacturable, dosable, and stably stored downstream.

As noted above, at the end of Section 7.2.2, the recrystallization kinetics of most SDDs tend to require that some additional steps be taken to stabilize them. Analogously to the use of polymers in the suspension formulation, certain polymers, when added to the spray drying solution and thereby incorporated into the dispersion, stabilize it and markedly slow the transition to the crystalline state in the dispersion and, eventually, in the solid dosage form. Thus, the next step in this stage of SDD formulation development is to define a design space for these polymers and other excipients (such as surfactants to improve the wetting of the material). Since API is often in short supply this early in development, it is infeasible to empirically test each polymer with the drug. Instead, we consider which polymers are theoretically expected to mix favorably with the API in question, such that we can test promising polymers at the bench. Our polymer mixing calculations are grounded in statistical mechanics, specifically the Flory–Huggins theory developed in the 1940s and often used since that time to estimate the free energy of mixing between solids. Per this theory,

$$\Delta G_{\text{mixing}} = RT(n_1 \ln \Phi_1 + n_2 \ln \Phi_2 + n_1 \Phi_2 \chi_{12}),$$

where χ_{12} is the dimensionless interaction parameter, which is a guide to which polymers are miscible with the desired API at the desired ratio (e.g., 1:1 API:polymer). To determine the interaction parameter, we use an *a priori* method in which we compare the solubility parameter of the API with that of various polymers and choose the polymers that are most closely matched.

Once these polymers are selected, we prepare blends of amorphous drug and polymer at slightly larger scale. To that end, we again use small-scale spray dryers—a "mini" model that requires about 1 g of the material and a "nano" model that requires a few hundred mg (but tends to give poor yield, around 50%). Solvent film casting is another option. This approach has the benefit of being automated. However, it entails a greater risk of recrystallizing the dispersion by mistake, and for a given sample it is slower than spray drying.

To assess how readily the blended dispersions prepared here will lend themselves, in practice, to a solid dosage form, focus naturally returns to their solid-state properties. These properties, which had been temporarily set on the back burner during suspension formulation development, may affect an SDD's manufacturability, dosability, or storability.

7.3.2 Manufacturability

Manufacturability, as evaluated at this stage, depends on two parameters. The first is the solubility of the crystalline form in various organic media. This is a property that scientists developing an SDD will most likely already be familiar with, since the crystalline form will need to have been solubilized in organic media to be spray dried in the first place. Nonetheless, it may pose problems downstream if the crystalline solid is *only* soluble in, for instance, a class III solvent (e.g., dichloromethane). Vaporizing these solvents may pose health hazards at larger clinical or commercial manufacturing scales and, therefore, require more extensive (and costly) safety precautions. The maximum acceptable amount of residual solvent will also be lower, requiring more thorough secondary drying.

In addition to organic solubility, manufacturability also depends on the size of the potential dose, which will have been estimated by pharmacokinetics groups during preclinical dosing. If the dose is large, then one of two things will need to happen. One possibility is that the drug load of a given tablet will have to be large; that is, the tablet will have to contain relatively more drug compound and relatively less excipient. However, a lower level of excipients puts greater demands on the physical properties of the drug itself: less polymer means the drug will need to be relatively more stable on its own, while less surfactant means that the drug alone will need to be relatively more soluble. If neither is the case, then one possibility remains: a higher pill burden, meaning either a larger number of tablets or capsules or simply larger tablets or capsules. Patients find high pill burdens inconvenient or uncomfortable, which ultimately puts the drug at a commercial disadvantage and reduces patient compliance as well.

7.3.3 Dosability

At this stage, with a focus on large-scale clinical and commercial production downstream, the exposure that the formulation is able to give is no longer the overwhelming priority that it was in the preclinical phase. The translational approach calls for thinking ahead and setting a compound up for success, not passing it down the field and hoping for the best. For this reason, stability comes back into the picture, and—for a metastable form such as an SDD—remains the center of attention for the rest of development. It is the principal concern for both dosability and storage stability.

For an orally administered drug to be "dosable," it needs to resist recrystallization in the gastrointestinal tract until it can be absorbed into the systemic circulation. No matter how soluble the API is, if it converts to another, lower energy form before it is absorbed, it is of no use. This is, of course, a consideration as early as the preclinical (animal) tox phase, where generally a suspension is selected that has a minimum of 3 h stability. This, again, is just a guideline based on the typical gastric residence time of laboratory animals—that is, how much time is likely to elapse before the compound reaches the duodenum (the first portion of the small intestine) with the rest of the gastric chyme and can be absorbed. However, a more rigorous definition of acceptable stability is needed for formulations that have progressed into development.

At this later stage, the crucial point to bear in mind is that no level of stability is always acceptable, in absolute terms, in every case. Once a formulated amorphous dispersion is ingested, what happens next represents a race of sorts between two processes: the solvent-mediated phase transition by which the metastable form recrystallizes, and the absorption of the API into the systemic circulation. Hence, the speed of absorption—which, as mentioned in Section 7.1.3, is inherent to a given scaffold and is difficult to alter by formulation approaches—determines the length of time a formulation needs to resist recrystallizing in suspension. Some degree of precipitation *in vivo* is even acceptable—as long as the process is slower than absorption.

As we discussed in Section 7.2.3, the stability of the compound in aqueous suspensions of differing pH is studied during suspension formulation development, as preparation for preclinical *in vivo* work. For these studies, our group uses ITC. To compare this figure with absorption, we first derive the rate constant for the drug's permeation of the small intestinal wall. Given the surface area of the small intestinal mucosa (a known quantity for a given species), as well as the jejunal permeability (estimated for correlation with *in vitro* permeability), the rate constant is the product of the surface area and permeability divided by the volume of fluid in the stomach (generally 250 ml for humans). If the ratio of the permeation rate constant to the *in vivo* precipitation rate—determined by ITC—is much greater or much less than 1, the stability of that formulation is clearly sufficient or insufficient, respectively. If, on the other hand, the two parameters differ by an order of magnitude or less, a more detailed model will be required—one that takes account of, for instance, the likelihood that a given drug molecule will be a substrate of active transporters that take up substances from the intestinal lumen.

7.3.4 Storage Stability

Finally, the development phase is when our group assesses how readily a dispersion can be stored at time frames relevant to commercial manufacture, distribution, and dosing. This assessment consists of a mixture of modeling and empirical studies.

First, to predict behavior upon storage, we return our attention to the glass transition temperature, or T_g, of the dispersion. Since the T_g of a multicomponent amorphous system is generally somewhere between the T_g of each component, the addition of a polymer *antiplasticizes* the dispersion: antithetically to the plasticizing effects of a residual solvent, the polymer reduces molecular mobility and thereby slows the initiation of crystallization. To a first approximation, the T_g of the blended dispersion generally follows linear mixing rules:

$$T_{g,mix} = \Phi_1 T_{g,1} + \Phi_2 T_{g,2}.$$

However, other, harder to predict interactions between drug API and polymer may alter the actual T_g. The Gordon–Taylor or Fox equation (see Section 7.2.2) can be used here to more thoroughly describe the mixing of these components. Regardless, it is still the case that the combined T_g of drug and polymer should ideally be 40–50 °C above the expected storage temperature. This is not a hard and fast rule, but simply a check based on the behavior of amorphous materials that have been well investigated.

Second, the desired dispersion, blended with polymers and surfactants as needed, is "put up"—as in, put up on the shelf, sometimes literally—for stability studies under formal (i.e., ICH) conditions. The same applies to tablets that were made from the dispersion. (Our colleagues discuss tableting in detail in Chapter 10.) In these studies, we assess the propensity of the dispersion and the tablet to crystallize both directly—by checking samples for crystallinity with Raman spectroscopy, SSNMR, or XRPD—and by modeling (see below). The ultimate objective of these studies, which can last 24 months or more, is to unambiguously inform the technical operations aspect of pharmaceutical development—for example, the packaging that the drug will require, as well as any special handling (such as cold chaining or controlled humidity).

In physical stability studies, we explore the effect on crystallization of two parameters: residual crystalline content and residual water. Generally, these parameters are considered critical quality attributes, or CQAs, in the QbD framework. Our colleagues offer a fuller discussion of QbD (Chapter 10); here it suffices to note that a CQA is a parameter that is expected to influence the clinical efficacy or safety (or both) of the drug product. If a CQA deviates from its stipulated design space, efficacy or safety may become unacceptable. To define this design space—in other words, to determine what levels of crystalline and water content are not significantly likely to have deleterious effects on efficacy or safety and are, therefore, acceptable—is the second purpose of these stability studies.

In addition to a direct measurement of physical stability, there is one way to understand the temperature-dependent kinetics of recrystallization below the T_g. As an indirect measurement, we study the material's enthalpic relaxation rate. The relaxation of a metastable material, such as an amorphous dispersion, that is at a nonequilibrium state is the process of molecular rearrangement by which the material returns to a (still metastable) equilibrium. During this process, the system loses enthalpy, which can be used to calculate the maximum extent of relaxation possible:

$$\Delta H_{\infty} \Delta C_p^{T_g}(T_g - T),$$

where T is the storage temperature and $\Delta C_p^{T_g}$ is the heat capacity change at the T_g. The fraction of amorphous material that remains unrelaxed ϕ is a function of this enthalpy and the time-dependent enthalpy of relaxation (ΔH_{relax}):

$$\phi = 1 - \frac{\Delta H_{\text{relax}}}{\Delta H_{\infty}}.$$

When an amorphous material that has undergone some relaxation is heated to its T_g, it enters an equilibrium state and recovers the enthalpy it lost during relaxation. (The heats of relaxation and recovery are generally assumed to be equal.) This is the heat that is measured by thermal analysis—specifically, DSC—and allows a development team to measure, albeit indirectly, the extent of relaxation.

Relaxation, in turn, serves as an index of molecular mobility, which may correlate in turn with propensity to recrystallize. Since the process is a molecular rearrangement, it is kinetically controlled. In some cases, it has been our experience that the structural relaxation of an amorphous SDD is too slow for it to be measured reliably on an experimental timescale. To circumvent this problem, we accelerated the study by inducing relaxation at higher temperatures, and then extrapolating back to the temperature of interest. First, the observed relaxation data were fit to the Kohlrausch–Williams–Watts [25] stretched exponential function:

$$\phi(t) = \exp\left(-\left[\frac{t}{\tau}\right]^{\beta}\right),$$

where $\phi(t)$ is the extent of relaxation at time t, τ is the relaxation time constant, and β is known as the stretch function, a nonlinearity function that ranges between $0 < \beta < 1$ and represents the heterogeneity of the system. With this, the maximum relaxation is defined as

$$\Delta H_{\infty} \Delta C_p^{T_g}(T_g - T),$$

where T is the "aging" temperature, or temperature at which the material was held. The resulting values for τ and β were extrapolated to lower temperatures on an Arrhenius plot ($\beta \ln \tau$ versus $1/T$).

The commercially approved SDD of telaprevir is one example of a material that relaxes too slowly to be observed in an experimental time frame. Hence, the development team measured enthalpic relaxation by DSC at temperatures close to T_g ($T_g - 15\,°C$, $T_g - 20°C$, and $T_g - 35\,°C$), as shown in Figure 7.8, and then fit the data to the KWW equation by nonlinear least squares analysis to determine the degree of molecular mobility at lower temperature. With Arrhenius analysis, relaxation of these materials was extrapolated to room temperature (Figure 7.9). The time constant for relaxation at 25 °C proved to be 10^7 years—which we took to be, on a clinical timescale, clearly sufficient.

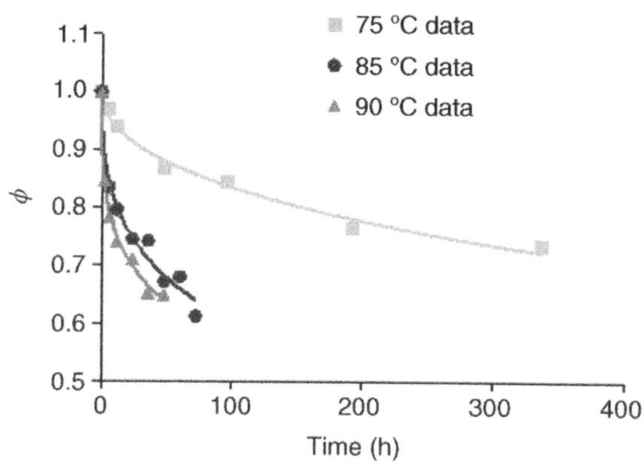

Figure 7.8. Extent of relaxation as a function of time at fixed temperatures.

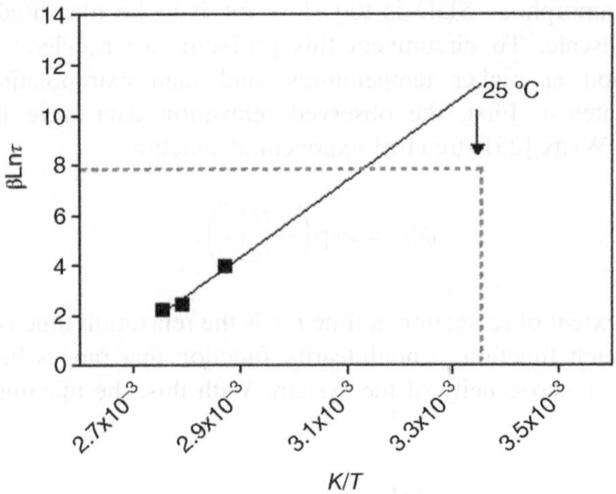

Figure 7.9. van't Hoff analysis of relaxation extrapolated to storage temperatures.

7.4 CONCLUSIONS

This chapter has described how our organization carried out early development of amorphous dispersions using certain aspects of telaprevir as an example. On the strategic level, we implement a strategy we call *translational drug development*, or TDD. TDD represents a departure from both of the previous paradigms of drug development, the simple handoff from discovery to development we classified as Development 1.0 and the somewhat more integrated, high-throughput, screening-based approach we think of as Development 2.0.

Compared with Development 1.0, TDD charges the pharmaceutical development division with understanding a broad range of physicochemical properties as early as possible in a compound's development. The department is responsible for ensuring that physicochemical properties present no surprises or obstacles downstream. In other words, instead of passing the compound onward and hoping for the best, TDD requires that this group *make sure* that the drug candidate can be translated from the *in vitro* laboratory into animal studies, into clinical trials, into the manufacturing plant, and into shelf-stable packaging. To do this requires conducting the most rigorous science possible—an approach that sounds *prima facie* unimpeachable but is, nonetheless, still not universal in the drug industry. For instance, in developing a suspension formulation of an amorphous dispersion, a mathematical, biopharmaceutical model that predicts absorption is created to ensure that a preclinical study will not result in inadequate exposure *in vivo*. In addition, quantitative examination of the stability of the dispersion in the desired suspension formulation and in simulated fluids with specialized assays (such as ITC) is conducted, in an effort to develop the optimum vehicle and ensure that solvent-mediated phase transitions do not cause the API to crystallize and render it useless. These extra steps at the discovery phase require some investment at an earlier time, but in our experience they effectively mitigate the risk of a failed animal study and wasted time.

This rigorous approach also distinguishes TDD *vis-à-vis* Development 2.0. This latter involves increased integration of development stage scientists into the discovery phase—a crucial advance. However, under the usual 2.0 paradigm, the studies conducted tend to be high-throughput, database-building work, in the hope of generating predictive guidelines such as Lipinski's Rule of 5. While guidelines such as these have been useful to a generation of development scientists, the added rigor of modeling and of derived measurements offers a higher resolution view of compound performance and greater predictive power. For instance, by executing extensive stability studies, which are semi-empirical in nature, in the best-case scenario, one can determine whether the API in dispersions and tablets remains amorphous (within the limit of detection) at expected storage and manufacturing conditions. One can also measure and calculate the extent of a material's structural relaxation as an index of molecular mobility. The input to these models is the amount of relaxation at high-stress conditions of increased temperature and humidity; by generating nonzero data point this way, we can estimate by extrapolation how long an API is likely to remain amorphous at room temperature. In practice (and with the proper stabilizing polymers), the timescale of crystallization is often less clinical than cosmological—sometimes in the millions of years.

In describing our approach and the positive results it has garnered to date, we hope that our recommendations prove useful to our colleagues across the industry. However, we do not mean to presume that TDD is the last word in pharmaceutical development. While TDD, or "Development 3.0," is, we believe, a salutary and important step beyond the *status quo*, much room for growth remains.

For instance, it will be crucial, in the near future, to build on recent advances to more fully apprehend the structure–function relationships of amorphous materials. One noteworthy 2012 paper described the paracrystalline structure of amorphous silicon [26]. Attaining the same level of understanding in pharmaceutical amorphous dispersions would enable researchers to better predict a dispersion's properties, such as its solubility and crystallization kinetics, *in silico*. It would not be an exaggeration to suggest that this degree of sophistication in modeling would help bring drug development into a new era. The aerospace industry, like the pharmaceutical industry, operates with extremely long product cycle times, but at the beginning of every decade-long development process is a period of extensive modeling and simulation. A similar advance in the pharmaceutical industry would move us away from much of our current *a posteriori* trial-and-error empiricism and into a level of rigor we have come to expect in other technically intensive sectors of society.

To reach this point, however, a great deal of work remains to be done. We hope that it is always so. A field that is constantly developing, growing more sophisticated and advancing more powerful tools, is one that will attract strong scientists and continually push past what were once taken to be its limitations. A stagnant field will do neither. And stagnation is particularly important to avoid, and constant growth is particularly important to strive for, in the pharmaceutical sciences—a field that produces not only knowledge but also proximate gains to the public good.

REFERENCES

1. Saxena, V., Panicucci, R., Joshi, Y., and Garad, S. (2009) Developability assessment in pharmaceutical industry: an integrated group approach for selecting developable candidates. *J. Pharm. Sci.*, 98(6):1962–1979.
2. Venkatesh, S. and Lipper, R. (2000) Role of the development scientist in compound lead selection and optimization. *J. Pharm. Sci.*, 89(2):145–154.
3. Sasseville, V.G., Lane, J.H., Kadambi, V.J., Bouchard, P., Lee, F.W., Balani, S.K., Miwa, G.T., Smith, P.F., and Alden, C.L. (2004) Testing paradigm for prediction of development-limiting barriers and human drug toxicity. *Chem.-Biol. Interact.*, 150(1):9–25.
4. Sun, D., Yu, L.X., Hussain, M.A., Wall, D.A., Smith, R.L., and Amidon, G.L. (2004) *In vitro* testing of drug absorption for drug developability assessment. *Curr. Opin. Drug Discov. Dev.*, 7(1):75–85.
5. Balani, S.K., Miwa, G.T., Gan, L.S., Wu, J.T., and Lee, F.W. (2005) Strategy of utilizing *in vitro* and *in vivo* ADME tools for lead optimization and drug candidate selection. *Curr. Top. Med. Chem.*, 5(11):1033–1038.
6. Balbach, S. and Korn, C. (2004) Pharmaceutical evaluation of early development candidates: the "100-mg approach." *Int. J. Pharm.*, 275(1–2):1–12.
7. Huang, L.-F. and Tong, W.-Q. (2004) Impact of solid state properties on developability assessment of drug candidates. *Adv. Drug Deliv. Rev.*, 56(3):321–332.
8. Kerns, E.H. and Di, L. (2002) Multivariate pharmaceutical profiling for drug discovery. *Curr. Top. Med. Chem.*, 2(1):87–98.
9. Lipinski, C.A., Lombardo, F., Dominy, B.W., and Feeney, P.J. (2001) Experimental and computational approaches to estimate solubility and permeability in drug discovery and development settings. *Adv. Drug Deliv. Rev.*, 46(1–3):3–26.
10. Amidon, G.L., Lennernäs, H., Shah, V.P., and Crison, J.R. (1995) A theoretical basis for a biopharmaceutic drug classification: the correlation of *in vitro* drug product dissolution and *in vivo* bioavailability. *Pharm. Res.*, 12(3):413–420.
11. Connelly, P.R., Vuong, T.M., and Murcko, M.A. (2011) Getting physical to fix pharma. *Nat. Chem.*, 3(9):692–695.
12. Accenture (2007) *In Pursuit of High Performance: Understanding Pharmaceutical Research and Development Cost Drivers*. Available at http://www.accenture.com/Microsites/rdtransformation/Documents/PDFs/Accenture_In_Pursuit_of_High_Performance.pdf. Accessed 17 December 2014.
13. Caldwell, G.W., Ritchie, D.M., Masucci, J.A., Hageman, W., and Yan, Z. (2001) The new pre-preclinical paradigm: compound optimization in early and late phase drug discovery. *Curr. Top. Med. Chem.*, 1(5):353–366.
14. Thomson Reuters (2009) *CMR International 2009 Pharmaceutical R&D Factbook.*
15. Gardner, C.R., Walsh, C.T., and Almarsson, O. (2004) Drugs as materials: valuing physical form in drug discovery. *Nat. Rev. Drug Discov.*, 3(11):926–934.
16. Kwong, A.D., Kauffman, R.S., Hurter, P., and Mueller, P. (2011) Discovery and development of telaprevir: an NS3-4a protease inhibitor for treating genotype 1 chronic hepatitis C virus. *Nat. Biotechnol.*, 29(11):993–1003.

17. Kolykhalov, A.A., Mihalik, K., Feinstone, S.M., and Rice, C.M. (2000) Hepatitis C virus-encoded enzymatic activities and conserved RNA elements in the 3′ nontranslated region are essential for virus replication *in vivo*. *J. Virol.*, 74:2046–2051.

18. Perni, P.B., Almquist, S.J., Byrn, R.A., Chandorkar, G., Chaturvedi, P.R., Courtney, L.F., Decker, C.J., Dinehart, K., Gates, C.A., Harbeson, S.L., Heiser, A., Kalkeri, G., Kolaczkowski, E., Lin, K., Luong, Y., Rao, B.G., Taylor, W.P., Thomson, J.A., Tung, R.D., Wei, Y., Kwong, A.D., and Lin, C. (2006) Preclinical profile of VX-950, a potent, selective and orally bioavailable inhibitor of hepatitis C NS3-4a serine protease. *Antimicrob. Agents Chemother.*, 50(3):899–909.

19. Connelly, P.R. and Thomson, J.A. (1992) Heat capacity changes and hydrophobic interactions in the binding of FK506 and rapamycin to the FK506 binding protein. *Proc. Natl. Acad. Sci. USA*, 89(11):4781–4785.

20. Connelly, P.R., Aldape, R.A., Bruzzese, F.J., Chambers, S.P., Fitzgibbon, M.J., Fleming, M.A., Itoh, S., Livingston, D.J., Navia, M.A., and Thomson, J.A. (1994) Enthalpy of hydrogen bond formation in a protein–ligand binding reaction. *Proc. Natl. Acad. Sci. USA*, 91(5):1964–1968.

21. Connelly, P.R. (1994) Acquisition and use of calorimetric data for prediction of the thermodynamics of ligand-binding and folding reactions of proteins. *Curr. Opin. Biotechnol.*, 5(4):381–388.

22. Connelly, P.R., Snyder, P.W., Zhang, Y., McClain, B., Quinn, B., Johnston, S., Medek, A., Tanoury, J., Waters, W.P., Dokou, E., Knezic, D., and Bransford, P. (2015) The potency–insolubility conundrum in pharmaceuticals: mechanism and solution for hepatitis C protease inhibitors. *Biophys. Chem.* 196:100.

23. Bauer, J., Spanton, S., Henry, R., Quick, J., Dziki, W., Porter, W., and Morris, J. (2001) Ritonavir: an extraordinary example of conformational polymorphism. *Pharm. Res.*, 18(6):859–866.

24. Hancock, B.C., Shamblin, S.L., and Zografi, G. (1995) Molecular mobility of amorphous pharmaceutical solids below their glass transition temperatures. *Pharm. Res.*, 12(6):799–806.

25. Williams, G. and Watts, D.C. (1970) Non-symmetrical dielectric relaxation behavior arising from a simple empirical decay function. *Trans. Farad. Soc.*, 66:80–85.

26. Treacy, M.M. and Borisenko, K.B. (2012) The local structure of amorphous silicon. *Science*, 335(6071):950–953.

8

PRECLINICAL AND CLINICAL STUDIES

Marcus E. Brewster,[1] Geert Verreck,[1] Jan Bevernage,[1] Joachim Brouwers,[2] Guy Van den Mooter,[2] and Patrick Augustijns[2]

[1]*Drug Evaluation—Pharmaceutical Sciences, Johnson & Johnson Pharmaceutical Research and Development, Janssen Pharmaceutica, Beerse, Belgium*

[2]*Laboratory for Drug Delivery and Disposition, KU Leuven, Leuven, Belgium*

8.1 INTRODUCTION

Contemporary drug pipelines are increasingly populated with difficult-to-formulate drug substances [1–3]. The genesis of this heightened complexity is rooted in three confluent trends: (i) the reliance of the drug discovery process on high-throughput screening [4,5], (ii) growing issues with drug form [6,7], and (iii) the nature of current drug targets that is often associated with structure–activity relationships (SAR) that deviate or disconnect from the chemical space associated with good oral bioavailability and acceptable biopharmaceutical fitness [8,9]. High-throughput screening is thought to subselect for compounds with high lipophilicities and low water solubilities, for compounds with higher molecular weights and entopic flexibility, and for related properties thought not to be conducive for good absorption [4]. Drug form concerns are similarly related to the

Pharmaceutical Amorphous Solid Dispersions, First Edition. Edited by Ann Newman.

rapid prototyping needs of modern drug candidate assessment that complicate both early formulate ability and robustness [10]. The drug target dilemma is especially vexing and can further diverge drug candidate properties from those required for a useful biopharmaceutical profile [8]. The trend toward receptor modulator development is a useful example with negative and positive allosteric modulators (NAMs and PAMs) attracting attention. For a variety of classes of these modulators, the SAR does not tolerate basic nitrogen in their structure removing one molecular tool for impacting drug solubility and permeability. Taken in aggregate, this evolution in dosage form development challenges has forced formulators to innovate with a variety of novel approaches promulgated in the last few years [11–14]. Generally, these approaches are associated with either increasing the chemical potential (μ) of the drug substance in the solid state through a change in the drug form or reducing the μ of the drug in its dissolved state. Technologies in the latter category include the use of complexation as in the case of cyclodextrin-based formulations [15–19], the use of cosolvent systems [20,21], and certain lipid-associated strategies [22–25]. Altering the drug form can be executed through the use of different salts, cocrystals, and less stable polymorphic forms or through the use of amorphous, smectic, or nematic mesophases of the drug or drug candidate [26–30]. The selection of an appropriate candidate formulation is generally tied to assessment of the active pharmaceutical ingredient (API) aspects of its formulation as well as biopharmaceutical factors. This entails assessing design space elements and drug ability information associated with preformulation, dosage form, and screening evaluations in animals. Brachau et al. suggested a three-part strategy in this context with step 1 associated with deciding whether a conventional or enabling formulation strategy is needed, if enablement is needed, which is the best strategy (step 2), and finally what process or excipient design space features should be considered for the selected trajectory (step 3) [31]. A biopharmaceutical dimension is also added usually by an indirect assessment of the *in vivo* dissolution rate in the rat or dog (Figure 8.1) [32,33].

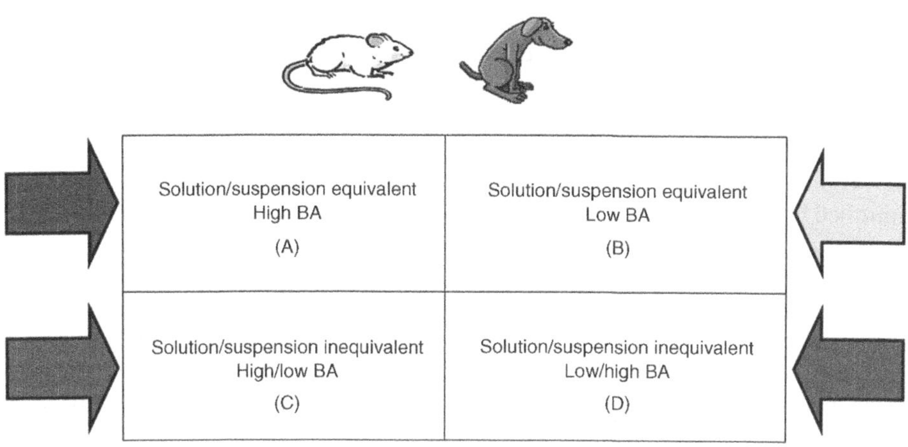

Figure 8.1. Solution–suspension comparison in the rat and dog to assess *in vivo* dissolution rate or solubility limitations.

Thus, a drug solution and suspension are sequentially dosed orally along with an i.v. comparison. If the solution and suspension give comparable blood levels, this suggests little or no limitation of bioavailability by *in vivo* dissolution and a conventional formulation route may be indicated. If by contrast the solution provides for a significantly higher exposure than does the suspension, this may suggest that drug enablement is required to optimize development [33–36]. Additional *in silico* or downscaled (96-well plate) endpoints might then be applied to generate data-driven decisions related to which technology is best suited for the specific task at hand. If a noncrystalline concept is suggested, the pure amorphous phase is possible only in a limited number of cases since this physical form is thermodynamically unstable requiring manipulation to render it pharmaceutically useful [37–39]. These pharmaceutical interventions usually include stabilizing the noncrystalline phase by vitrifying/dispersing it in a glassy carrier, by increasing the aggregate/mixing glass transition temperature (T_g) of the composite through the use of a high T_g carrier, and by encouraging specific drug–carrier interactions through selecting matrices with appropriate functionality and chemistry [37].

The solid dispersion thus generated has a number of advantages relative to a traditional pharmaceutical form. Based on the Nernst–Brunner/Noyes–Whitney relationship that is given by

$$\frac{\mathrm{d}C}{\mathrm{d}t} = \frac{DA}{Vh}(C_s - C_t),$$

where the rate of drug dissolution is a function of the saturation solubility of the drug (C_s), A is the surface area of the solid drug particles, h is the thickness of the unstirred water layer separating the solid drug particles from the bulk media, and D is the diffusion coefficient of the drug away from the drug solid [40]. In this context, amorphous solid dispersions can increase drug dissolution rate by (i) increasing A by reducing the effective drug particle size through its dispersion in the matrix. In theory, the drug could be reduced to the smallest size possible, that is, the isolated (dissolved) drug molecules in the glassy carrier. (ii) The apparent solubility of the drug can increase as a function of it becoming amorphous. That is, as the crystal lattice forces are eliminated, the apparent drug solubility increases. (iii) Finally, the dispersion of the drug in the carrier can impact drug wettability (h), thereby augmenting dissolution rate [41]. In principle, the rate of dissolution of the amorphous solid dispersion is controlled by the rate of dissolution of the carrier that can be orders of magnitude faster than that of a poorly soluble API [42].

Selecting a prototype solid dispersion generally involves posing key questions and then executing appropriate experiments to assess whether the questions can be answered as well as what risks/risk mitigations are associated with the system. As already suggested, key questions may include the following: (i) Can a polymer or polymers be found into which the drug can be dispersed with high-enough miscibility to be pharmaceutical useful at an appropriate excipient bulk? (2) Is the resulting dispersion stable as a function of phase separation and recrystallization? (iii) Does the system

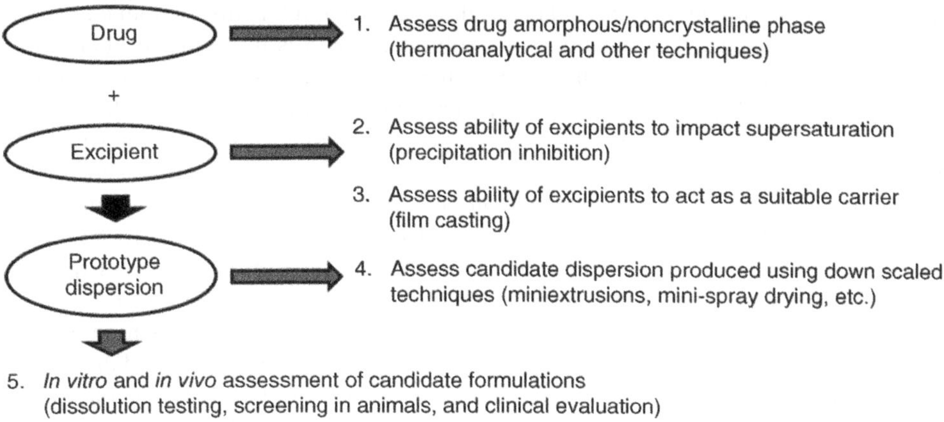

Figure 8.2. Flowchart for assessing and testing candidate amorphous solid dispersions.

release the drug at high concentrations and does it provide for supersaturation in a pharmaceutically useful manner? (iv) Does the dispersion lend itself to downstream dosage form fabrication and useful processability? (v) Does the formulation increase drug exposure in animals and man in an appropriate manner aligned with the efficacy and safety of its intended use. While many of these points are covered in other chapters of this book, a few words of explanation may be useful in assessing dispersion selection with regard to their preclinical and clinical attributes.

The selection of a dispersion can be assessed in several steps: (i) understanding the nature of the challenge by assessing the appropriate noncrystalline phase, (ii) selecting appropriate carrier(s) that provide for useful supersaturation extent and stability, (iii) screening these carriers in terms of miscibility and matrix stability, (iv) assessing which processing technique is likely to be best positioned to generate a dosage form for the specific purpose under consideration, (v) generating and pharmaceutically evaluating prototypes followed by their testing in appropriate *in vitro* and *in vivo* paradigms prior to human clinical testing (Figure 8.2) [43,44].

The nature of the physicochemical challenge can be addressed by extracting information related to the API, including those associated with its solution properties (pK_a, $\log P$, and solubility) as well as with its solid state (melting point and heat of fusion, polymorphic properties, and susceptibility to hydration or solvation). Other parameters of interest may include its tendency to form a glass (i.e., is it a good glass former, does it form a stable glass?) [45,46]. If the amorphous phase is tractable, information related to its fragility, configurational heat capacity, relaxation (β or τ) parameters, and its energy of activation for structural relaxation at the T_g should be collected [47]. In addition, the possible solubility advantage of the amorphous phase relative to the crystalline API can be suggestive of possible utility [48–51]. These data can give insight into the overall feasibility as well as into the nature of the stability issue to be overcome and how well the amorphous phase should be dispersed in the glassy carrier as a function of domain size. A critical success factor in the operation of a solid dispersion is its ability to induce and maintain supersaturation [52]. In this respect, the glassy carrier should preferably serve the dual role as a dispersant as well as a precipitation inhibitor [53]. High-throughput assays can be incorporated into the early dispersion finding protocols to aid in such assessments. Glassy elements that may serve as precipitation inhibitors are considered by assessing the effect of these materials on the tendency to generate and maintain supersaturation. Thus, supersaturation can be induced by various methods as reviewed by Bevernage et al. [54], including the use of the solvent-shift or cosolvent quench method wherein the drug of interest is solubilized in a water-miscible organic solvent such as DMSO, DMF, DMA, PEG, or propylene glycol [53,55]. This solution is then added to a solution containing the excipient to be tested or to blank media and the ability of the excipient to maintain supersaturation is assessed using UPLC/HPLC or nephelometry. These systems are simple and can be automated using robotics and 96-well plate technology and have the advantage of reducing the API budget as well as increasing the accessible excipient space [56]. Other supersaturation approaches include the pH shift approach where a base-solubilized acid or acid-solubilized base is pH adjusted such that supersaturation ensues subsequent to the loss of solubilizing capacity of the solution [52,54]. This is often done for weak bases through the use of a two-phase or transfer dissolution model where the test article is first assessed at pH values associated with the stomach and then at pH values associated with the proximal intestine. Finally, a potentiometric method is available as incorporated in Sirius CheqSol® system [57–59]. In this system, compounds are characterized as either chasers (i.e., they form supersaturated systems upon neutralization) or nonchasers (i.e., they do not supersaturate under these conditions). The effect of excipients on precipitation kinetics can also be assessed. Once efficient precipitation inhibitors are selected, their ability to act as a carrier is judged by assessing cast films [44]. In this screen, the drug and selected carrier (s) are dissolved in a common solvent, dispensed to specially designed 96-well plate, and the solvent is then removed. The resulting films are assessed for homogeneity (via white light and birefringence microscopy), crystallinity (through *p*XRD), and phase separation (via RAMAN and IR imaging). The ability of these systems to generate a rapidly dissolving, supersaturating system can then be assessed by adding dissolution media to the 96-well plates (either in a traditional or two-phase protocol) and determining concentration by the UPLC or by a UV plate reader. Importantly, cast films may or may not faithfully reflect the properties of the ultimate dispersion depending on which processing technique is eventually selected. To this point, downscaled processing approaches are often applied to the excipient(s) selected to confirm whether they are likely to be useful. Downscaled equipment available for this purpose include those for assessing spray drying (e.g., microspray drying and ProCepT), melt extrusion (minimelt extrusion, Haake, and melt compression), milling/cryomilling, and supercritical fluid-based processing. These techniques also allow a better assessment of stability to be garnered as well as downstream processability in addition to items such as allowing the assessment of Gordon–Taylor/Fox behavior [60–62]. These downscaled assessments can then be translated to prototype formulations that can be evaluated in appropriate test animals and eventually in man.

8.2 *IN VITRO* AND PHARMACEUTICAL CHARACTERIZATION

Based on the theorized way of working of amorphous solid dispersions, one of the most important design elements is their ability to dissolve and supersaturate. This is initially assessed *in vitro* through the use of dissolution protocols especially set up for that purpose. However, a number of issues have been raised that complicate the appropriate selection of a useful dissolution protocol. These concerns are associated with both the nature of the solid dispersion themselves and the biorelevance of the selected apparatus and media. While the use of traditional USP apparatus (i.e., USP I and II) as well as compendial buffers or media for assessing drug release from solid dispersions is possible, they are often suboptimal [63]. To address artifacts associated with rapid changes in solubility of the compound as a function of time, expedited sample handling has been suggested to optimize phase separation and dilution and include the use of rotating syringes and microcentrifuge tube-based systems [64,65]. Likewise, systems designed to produce a pH shift mimicking gastrointestinal passage are available using both single and multivessel apparatuses [66–68], including the artificial stomach duodenum (ASD) systems [69,70] and the TIM approach offered by TNO [71–74]. In many cases, these approaches still fail to appropriately characterize solid dispersions both as a function of their release properties and the ability to predict *in vivo* exposure. As reviewed by Augustijns and Brewster, one reason that traditional methods are particularly unsuited for solid dispersion testing and screening is that they are designed to maintain sink conditions that may mask or confuse dissolution, supersaturation, and precipitation dynamics [63]. Thus, one suggested element for dissolution methods for solid dispersions is that they are developed using a nonsink protocol. A second issue is related to the hydrodynamics and mechanical stress associated with the dissolution process [63]. Hydrodynamics can affect the induction period associated with crystallization and precipitation of a supersaturated solution by either increasing the drug diffusivity or lowering the barrier to nucleation. Unfortunately, test systems reflecting *in vivo* hydrodynamics are not yet available [75]. A third consideration is the impact of media and especially the fact that biorelevant fluids such as FaSSIF and FeSSIF may better simulate dissolution in the gastrointestinal tract. Bevernage et al. compared the tendency of drugs to supersaturate and precipitate in simple compendia buffers, simulated media, and aspirated human gastric and intestinal fluids [76–78]. In these studies, simple buffers poorly correlated with human fluids in that they significantly overpredicted supersaturation stability and underestimated precipitation. While FaSSIF was useful compared with aspirated human samples, FeSSIF underestimated precipitation relative to human aspirates. Similar trends were observed for gastric precipitation with simulated fluid (FaSSGF) more useful than simple SGF. Finally, the temperature of these assays has been considered [54]. Generally, media are thermostated to 37 °C; however, some studies suggested that the temperature can influence the tendency of the system to supersaturate with amorphous felodine forming supersaturated solutions at 25 °C but not at 37 °C [79].

To these points, drug release from a solid dispersion in which itraconazole was loaded into mesoporous silica was completed in both sink and nonsink methodologies [80]. In sink conditions, rapid dissolution occurred without an indication of precipitation and added precipitation inhibitors had no measureable effects. By contrast, nonsink condition revealed a dissolution–precipitation profile that was amenable to improvement by the addition of a precipitation inhibitor (i.e., HPMC). *In vivo* assessment of these systems in the rat demonstrated good concordance with the nonsink model with a similar increase in oral bioavailability when HPMC was added. A study with fenofibrate loaded into mesoporous silica of different pore dimensions suggested that, under sink conditions, the largest pore size was associated with the fastest release [81]. Under nonsink conditions and using biorelevant media, the small pore size was associated with the best profile (i.e., the faster release from large pores rapidly increased the supersaturation ratio and induced precipitation to a greater extent than did the more moderate release from smaller pores). Again, *in vivo* data were aligned with the finding of the nonsink method. As will be discussed later, the observation that supersaturating systems that more moderately release drug can improve biopharmaceutical performance has been observed for several solid dispersions in man suggesting that optimal profiles for these systems are different from those of traditional formulations. This counterintuitive aspect of solid dispersion optimization is therefore important, with respect to not only how dissolution profiles are generated but also how they are interpreted. Depending on the nature of the solid dispersion, experimental factors, including the apparatus, media, and analytical endpoint, can confound the interpretation of data generated from dispersions based on nonbiorelevance as well as on specific system-based artifacts.

A second area of investigation for the dissolution of solid dispersions is the impact of drug absorption [52,54]. In a traditional dissolution bath, a dissolving drug associated with a supersaturating drug delivery system has only two options over the measurement time course: it may dissolve, supersaturate, and remain in the metastable state or precipitate. Interestingly, the driving force for precipitation, that is, the supersaturation ratio, is also the thermodynamic driving force for increased uptake across a permeable membrane. That is, in systems without an absorptive component, precipitation may dominate; while in an absorptive system, effects may differ based on altered precipitation kinetics secondary to a lowering of the Gibb's free energy of the system [54]. This was assessed by Bevernage et al. using loviride as a probe compound [82]. In their system, a dissolution apparatus was generated using a solvent shift approach in the donor compartment of Transwell® inserts as such (i.e., with the membrane blocked) or with a Caco-2 monolayer in place using FaSSIF as the donor compartment media. These studies found that supersaturation increased the flux of drug across the Caco-2 membrane relative to a saturated drug solution and that induced precipitation was less in the presence of an absorptive sink [82]. In addition, the impact of the precipitation inhibitor, HPMC, was less pronounced in the absorptive setup. Taken in aggregate, the data suggest that simple dissolution methods that ignore absorption may poorly reflect the dissolution–supersaturation–precipitation behavior of a solid dispersion that occurs *in vivo*. Similarly, Kataoka et al. (2012) assessed a dissolution/permeability test for supersaturable self-emulsifying systems of both well-permeable (danazol) and poorly permeable (pranlukast) probe compounds [83]. While the poorly permeable probe did not demonstrate an increase in flux when formulated in a solubilized or supersaturable dosage form, danazol exerted improved uptake in a supersaturable formulation. These data were well corrected with *in vivo* studies in the rat. The increase in flux as a function of supersaturation results from the increase in free drug concentration and the associated increase in thermodynamic activity [21,84]. This is in contrast to other approaches that may increase apparent solubility but not the free drug concentration as in cases of micelle formation (i.e., the effect of surfactants), complexation (i.e., cyclodextrin interactions), or the use of cosolvents.

8.3 *IN VIVO* EVALUATION AND MODELS

A number of *in vivo* assessments of the potential utility of amorphous solid dispersions have been completed using various animal models. Most of these approaches assessed supersaturating solid dispersions indirectly in that they compared exposure and pharmacokinetics associated with an amorphous solid dispersion to a nonenabled formulation. Newman et al. have outlined the general use of animal models in this context along with accompanying *in vitro* characterization, including dissolution testing, and surveyed 40 published studies [85]. Bevernage et al. described indirect and direct approaches for assessing the *in vivo* performance of dispersions [54]. Traditional pharmacokinetic and bioavailability studies can be considered indirect assessments as the behavior of the solid dispersion, that is, the nature and extent of supersaturation at the absorption site, is unknown and not assessed. These studies are, however, critical to proof-of-concept testing to encourage development of potential formulations and further narrow the excipient space of concepts based on safety and related concerns. Newman et al. found that in published studies, dispersions increased oral bioavailability in more than 80% of cases with exposure increases of 2–80-fold relative to reference dosage forms or API [85]. In the review, dogs and rats were most commonly used in the bioavailability comparisons, although other species were also referenced including the rabbit and monkey. These data are useful in positioning the biopharmaceutical usefulness of dispersions and the effect of rendering the drug amorphous. In selecting a useful model, testing efficiency and translation to man based on factors such as intestinal pH, transit time, and enzymatic characteristics should be considered. The dog is arguably the easiest to test with regard to dosage form dimensions. The rat requires some modification in terms of amorphous dispersion testing, including administration of an amorphous dispersion powder by gavage or formulation of an amorphous solid dispersion in appropriate capsules (e.g., size No. 9). Other important criteria include the effect of food in the selected animal model as this impacts a large number of physiological factors such as GI pH, motility, regional water content, hydrodynamics, mechanical force, and various other parameters [54]. Taken in aggregate, it is clear that API, pharmaceutical, and biopharmaceutical considerations will all play a role in guiding the selection of the most useful animal model for testing a specific dispersion [32]. Blood or plasma levels of the drug administered as an amorphous solid dispersion can be analyzed using physiologically based pharmacokinetic (PBPK) model to estimate *in vivo* precipitation time, transit time, and related factors [86–88]. Alternatively, with appropriate input, PBPK modeling can attempt to predict blood levels from a solid dispersion with appropriately measured *in vitro* properties such as kinetic solubility, precipitation time, and dissolution rate. In this way, the weak base nelfinavir and the farnesyltransferase inhibitor FTI2600 have been assessed [89]. While traditional modeling relies on

PBPK assessments with very general parameter set inputs (e.g., first-order precipitation), a number of authors have specifically attempted to model supersaturation–precipitation profile to assess biopharmaceutical properties, including the use of classical nucleation theory [90–93].

Direct methods for assessing supersaturation in animal models rely on simultaneous blood sampling and sampling of gastric and intestinal fluids to assess drug state, degree of supersaturation, and form [54]. While generally applied in the human situation, animal-based approaches in this context are gaining popularity.

8.4 CLINICAL ASSESSMENTS

A number of amorphous solid dispersions have reached the market (Table 8.1). These include systems in which the drug is rendered amorphous and mixed with various excipients/carriers through the use of a number of processing techniques such as melt extrusion, melt blending, bead coating, solvent processing, spray drying, and antisolvent precipitation (i.e., the microprecipitated bulk powder approach). In all cases, postprocess manipulation is needed to generate the final dosage form be it a tablet or a capsule. Based on the traditional uncertainty of translatability of animal models, dosage form testing in man is often completed to ensure that the best concept is developed and forwarded to the market. In these studies, indirect and increasingly direct methods are being applied to better understand the applied solid dispersion and its mechanism for augmenting oral bioavailability. As has already been suggested, presenting the drug in a physically stable amorphous format meets only part of the challenges associated with the development of these systems. An appropriate processing approach also needs to be identified, confirmed, and implemented. Verreck and coworkers assessed the benefit of solid dispersion technologies to increase the bioavailability of a poorly soluble microsomal triglyceride

TABLE 8.1. Amorphous Drug Dispersions Approved by Regulatory Agencies

Product	API	Carrier	Processing Technology	Company
Norvir®	Ritonavir	PVP–VA	Melt extrusion	Abbott
Kaletra®	Ritonavir/ lopinavir	PVP–VA	Melt extrusion	Abbott
Isoptin SR-E 240®	Verapamil	HPC/HPMC	Melt extrusion	Abbott
Onmel®	Itraconazole	PVP-VA	Melt extrusion	GSK/Stiefel
Gris-PEG®	Griseofulvin	PEG	Melt blending	Pedinol Pharm
Cesamet®	Nabilone	PVP	Wet granulation	Lilly
Prograf®	Tacrolimus	HPMC	Solvent process	Fujisawa
Certican®	Everolimus	HPMC	Solvent process	Novartis
Sporanox	Itraconazole	HPMC	Bead coating	Janssen
Zelboraf	Vemurafenib	HPMCAS	Coprecipitation	Roche
Intelence	Etravirine	HPMC	Spray drying	Janssen
Incevik/Incivo®	Telaprevir	HPMCAS	Spray drying	Vertex/Janssen

transfer protein (MTP) inhibitor R103757 [94]. Dispersions were generated using three processing techniques, including melt extrusion and bead coating, both of which used HPMC as the glassy carrier, as well as through the use of a glass thermoplastic system (GTS) method. The GTS is a multicomponent dispersion containing HPMC, HP-β-CD, and citric acid, which are dissolved with the drug in a common solvent, the solvent is removed, and the residue formed is placed into dosage forms through encapsulation [95,96]. A dog model was applied after which clinical testing was completed in both the fed and fasted states. While no absorption was observed in the dog for the crystalline drug incorporated in a standard tablet, solubilizing the API using a 10% w/v aq. HP-β-CD solution resulted in good exposure. The coated bead concept was also assessed in this dog study that suggested an exposure lower than that of the cyclodextrin solution, but significantly higher than that of the crystalline drug formulated in a tablet. In the human oral bioavailability studies, three solid dispersion concepts, administered either in gelatin capsules (GTS and beads) or as pressed tablets (melt extrudate), were assessed and compared with a reference cyclodextrin (25% w/v aq. HP-β-CD) solution (Table 8.2). Results found that while all dosage forms contained amorphous API,

TABLE 8.2. Pharmacokinetic Parameters ($\pm$SD) for R103757 after Human Dosing of Various Prototype Formulations in Fed and Fasted Conditions

	AUC_{∞} (ng h/ml)	
Formulation	Fasted State	Fed State
Film-coated beads	148 ± 72	252 ± 13
ME tablet	225 ± 80	253 ± 46
GTS	381 ± 103*	590 ± 98*
HP-β-CD solution	402 ± 99*	
	C_{max} (ng/ml)	
	Fasted State	Fed State
Film-coated beads	17.2 ± 12.6	45.3 ± 24.0
ME tablet	37.9 ± 14.9	38.9 ± 15.1
GTS	62.1 ± 21.9	101 ± 17*
HP-β-CD solution	71.8 ± 21.1*	
	t_{max} (h)	
	Fasted State	Fed State
Film-coated beads	1.9 ± 0.3	2.3 ± 0.6
ME tablet	3.0 ± 0.6*	3.3 ± 1.7*
GTS	2.5 ± 1.0	1.8 ± 0.7
HP-β-CD solution	1.6 ± 0.6	

Asterisks denote significantly different means ($p < 0.05$) calculated using ANOVA and a *post hoc* multiple range test.

exposures significantly varied in the following order (in fasted individuals): GTS > melt extrusion > coated bead. The GTS was 2.6-fold more bioavailable than the film-coated beads ($p < 0.05$) and 1.7-fold more bioavailable than the melt-extruded tablets. There was a positive food effect with all three formulations with the GTS exposure increasing 2.3-fold. In the fasted state, the relative oral bioavailabilities for the three formations relative to the cyclodextrin solution were 27% (coated beads), 75% (melt-extruded tablets), and 97% (GTS).

Clinical assessments to investigate itraconazole supersaturation (indirectly) have also been completed. Itraconazole is a widely studied compound in this context with numerous *in vitro*, *in vivo*, and clinical evaluations completed [85]. The compound is an triazolon broad-spectrum antifungal agent and was the first orally bioavailable pharmaceutical with efficacy against both *Candida* and *Aspergillus*, the most commonly occurring human fungal pathogens. The development of dosage forms for itraconazole is complicated by its challenging physicochemical properties: log P >5, pK_a ~4, and aqueous solubility ~1 ng/ml at neutral pH and ~3 μg/ml in 0.1 N HCl [97]. One beneficial property of the compound is its ability to form a stable chiral nematic mesophase [98]. When developing potential solid dispersions, several formulation trajectories were assessed, including a film-coated bead (using HPMC as the carrier) filled into a gelatin capsule (and marketed as Sporanox® oral capsule) [99], a melt-extruded dispersion (also using HPMC) wherein the extrudate is milled to various particle sizes and compressed into tablets with other excipients [100], and a spray-dried dispersion in which lactose is used as a carrier and the powder is then compressed into a tablet with other excipients [101]. The assessment of the various possible dispersion-based formulations proceeded in several phases, including assessing dissolution in an appropriate model, testing dispersions in human pharmacokinetic studies, and evaluation of the *in vitro* and *in vivo* relationships of these dosage forms.

Dissolution methods that may be useful in constructing IVIVCs were considered. The solubility of itraconazole was determined in media at various pH values and in the presence of other solubilizing additives. The effect of paddle speed and surfactant or cyclodextrin concentration on dissolution rate of the innovator formulation (Sporanox 100 mg oral capsules) was assessed and correlated with a number of pharmacokinetic parameters derived from human clinical studies. These *in vitro–in vivo* relationships were optimized as a function of Tween20 concentrations and stirring speed. The optimized systems used a USP type II apparatus operating at 100 rpm and using 900 ml of 0.01 N HCl ($SGF_{w/oPepsin}$) containing 1% Tween20 maintained at 37 °C. Under these conditions, the following correlation could be generated: %dissolution (1 h) = 20.0 + 63.3·AUC ratio ($r = 0.96$) and %dissolution (1 h) = 21.9 + 61.7·C_{max} ratio ($r = 0.96$) (Figure 8.3). The solid dispersion formulations as well as crystalline itraconazole were compared. The formulations assessed included the coated bead (filled into a capsule, Sporanox 100 mg oral capsule), a spray-dried powder and a melt extrudate milled to two different particle sizes and included in a tablet. Crystalline itraconazole undergoes incomplete and slow dissolution (~5%). All other systems generated supersaturated levels of itraconazole that were sustained for at least 1 h. All of the dispersed systems were amorphous as assessed by DSC, modulated DSC, and powder X-ray diffraction (XRD) and released 100% of the formulated drug within 1 h. Having said that, the rate of dissolution was different with the

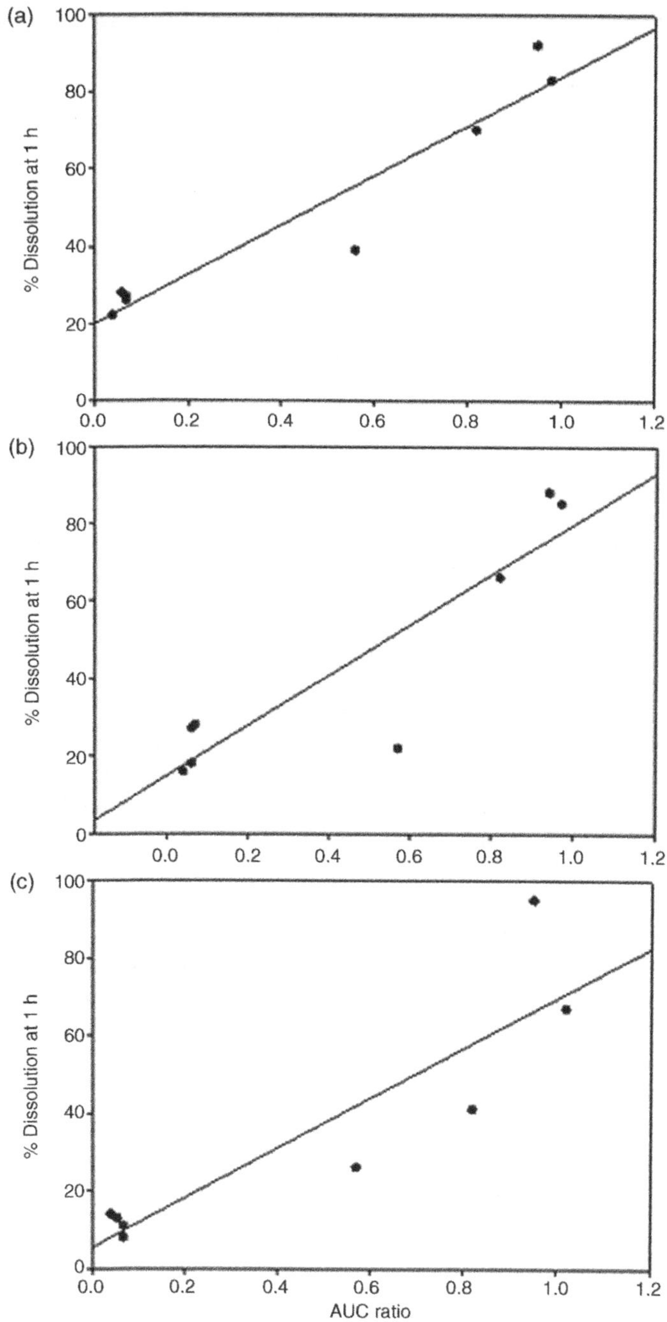

Figure 8.3. IVIVC models for itraconazole based on the following dissolution conditions: (a) $SGF_{w/o\ pepsin}$ + 1% Tween 20 at 100 rpm, $r = 0.96$, (b) $SGF_{w/o\ pepsin}$ + 1% Tween 20 at 75 rpm, $r = 0.89$, and (c) $SGF_{w/o\ pepsin}$ at 100 rpm, $r = 0.88$.

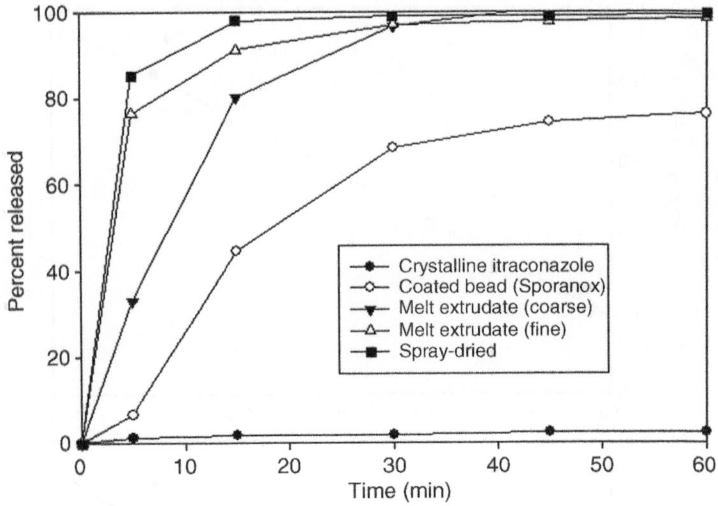

Figure 8.4. Dissolution profile for itraconazole containing formulations including the crystalline drug in capsule, an HPMC-based dispersion coated on a sugar sphere (Sporanox), an HPMC-based melt extrudate coarsely milled (>400 mm) or finely milled (<200 mm) and pressed into a tablet, and a tablet based on itraconazole spray dried on lactose.

bead-based (Sporanox) system providing relatively slow release (~40% at 15 min), the melt extrusion-based tablet with larger particles somewhat faster release (80% at 15 min), and the smaller particle extrudate and spray-dried formulation essentially dissolved by 15 min (Figure 8.4) in these sink-based models.

The human oral bioavailability of the formulations were pooled from a number of clinical studies and are provided relative to the marketed beads that were used as a reference. All comparisons were completed using $AUC_{24\,h}$ generated under fed conditions. In all cases, the highest oral bioavailability in this series of studies was generated by the marketed bead formulation. The melt-extruded tablet with the coarser milled particles was associated with a relative exposure of 91.5% and the tablet with the finer milled extrudate manifested a relative oral bioavailability of 85%. The tablet based on the spray-dried drug substance gave a relative value of 74% (Figure 8.5). The data suggest that an optimal dissolution rate is required in the sink-based dissolution apparatus for this formulation whose mechanism of increased oral bioavailability is generating supersaturated drug solutions. That is, the systems that provide for the most rapid drug dissolution *in vitro* have a lower oral bioavailability *in vivo*. Consistent with the observations of Augustijns and Brewster, this may be associated with rapid increase in the supersaturation ratio that biases the system toward precipitation [62]. Similarly, Six et al. assessed dissolution rate and oral bioavailability for three solid dispersions of itraconazole based on melt extrudates of HPMC, EUDRAGIT® E100, or a mixture of EUDRAGIT E100 and PVP–VA all filled into capsules and compared these with the marketed bead formulation [102]. While dissolution data indicated that the EUDRAGIT

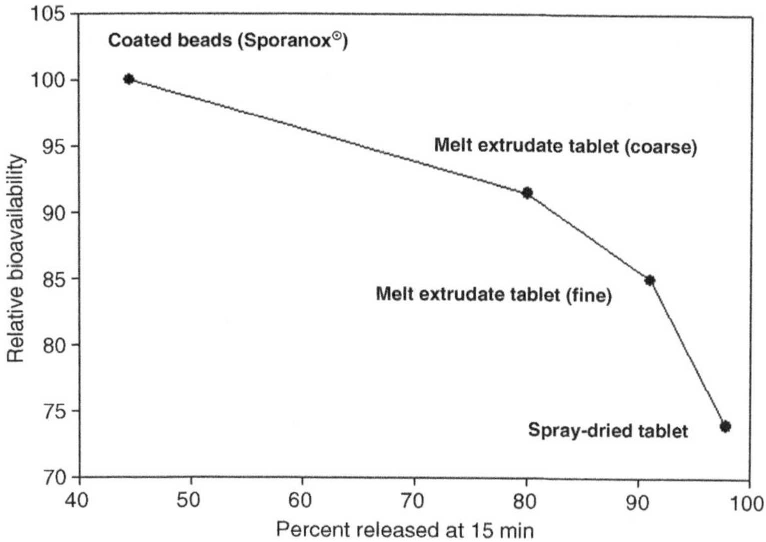

Figure 8.5. Relative normalized oral bioavailability as a function of percent dissolved at 15 min for a number of itraconazole-containing formulations.

E100 and EUDRAGIT E100–PVP–VA formulation were superior in terms of the rate of dissolution, they had the poorest oral bioavailability. The use of sink-based dissolution methods can therefore obfuscate formulation selection if traditional paradigms are followed.

The development of the marketed dispersion-based formulation for etravirine (Intelence®), an anti-HIV therapy, also demonstrated the importance of dispersion processing with regard to a useful biopharmaceutical effect and the essential role of clinical testing in arriving at an optimal dosage form [43,44,103]. Etravirine is a poorly water-soluble drug with a number of challenging physicochemical properties that impact its ability to be formulated [44]. Initial screening studies found that the drug was not orally bioavailable even when the crystalline API was milled into the Nanocrystal® formulation. This meant that early formulation assessments were completed using either a PEG400 solution or a PEG4000 dispersion. These systems rapidly proved themselves to be of limited value based on pill burden and low oral bioavailability. Based on the possible utility of solid dispersions, additional screening was conducted on various processing approaches [44]. Etravirine is a poor glass former and forms an unstable glass. Attempts to render the drug amorphous were complicated by its temperature sensitivity that precluded melt-based strategies. Furthermore, other techniques including spray drying did not provide for useful amorphous material. Eventually, cryomilling appeared to be the most suitable method with the amorphous phase generated by milling the API for 3 h at −196 °C [104]. While the T_g of the cryomilled material may have been suggestive of good stability ($T_g \approx 100\,°C$), the T_g/T_m suggested that a fragile glass was generated [47]. In addition, when the material was stored in open conditions at 30 °C below the T_g, it rapidly converted to the crystalline material [44]. The next screening component for the development of a dispersion was assessing the behavior of cast films. Screening experiments suggested that HPMC would be a useful glassy matrix and was used at various ratios with the API. Data for the API:HPMC (1:3) films were generated to assess the effect of temperature (below the T_g) on the extent of enthalpic relaxation. At 45 °C below T_g, the amorphous material relaxes only a few percent over the time course of the experiment, while the most activated conditions (25 °C below T_g) provide for ~20% relaxation. The companion data give the occurrence and increase of crystalline content over time at three conditions: at the T_g and at 10 and 25 °C above the T_g. By assessing temperatures at or above the T_g, crystallization can theoretically be disconnected from relaxation. The least aggressive condition shows a significant induction time and a small increase in crystalline content such that 4–5% could be detected by the end of the experiment. The induction time decreases and the final crystalline content increases as the temperature increases as would be expected. Importantly, the pure, unformulated amorphous phase underwent significant crystallization within 4 h at 30 °C below the T_g emphasizing the stabilizing effect of the polymeric dispersion [44].

Clinical trial formulations were generated for etravirine using three concepts, all of which were predicated on the use of the amorphous phase [43]. The formulations included a microcrystalline cellulose (MCC) bead onto which was coated a dispersion of the API and HPMC, a tablet derived from an amorphous granulo-layered powder in which API:HPMC was processed in a fluid bed with lactose serving as the carrier and a tablet based on a spray-dried dispersion containing HMPC, API, and other excipients. The spray-dried powder was densified by compaction, subjected to postdispersion drying, blended with other tableting excipients, and compressed into tablets. In all cases, the API was rendered amorphous and the dispersions were stable on storage. While the granulo-layered system was slightly better than the bead-based formulation in terms of oral bioavailability, the spray-dried system was almost ninefold more bioavailable relative to the coated beads (Figure 8.6). This increase in oral bioavailability meant that the pill burden could be dramatically reduced and was enabling the product from a compliance and commercial point of view [43]. Intelence was approved as a 100 mg tablet in 2008 with a 200 mg tablet, also based on a spray-dried dispersion, being introduced to the market in 2010.

New technologies for the configuration of dispersions are emerging and these often require optimized or improved *in vitro* or *in vivo* testing protocols to assess their possible usefulness. An important recent example is associated with the marketed solid dispersion for vemurafenib (Zelboraf®), a useful B-RAF inhibitor used in the treatment of melanoma [105–107]. Formulation of vemurafenib is complicated by its API properties: high melting point (272 °C), moderate $\log P$ (~3), and extremely poor water solubility [108]. In addition, the solvents that provide for useful solubility were high boiling, including DMA or DMSO. These properties eliminated spray drying and melt-based approaches for the preparation of a dispersion. The concept that was ultimately selected was a polymer-stabilized solid amorphous dispersion using solvent-controlled coprecipitation [108]. In this approach, the drug and an enteric (ionic) polymer are dissolved in a common (high-boiling) solvent and then introduced into a cooled acidic or basic aqueous medium generating a microprecipitated bulk powder (MBP) [109,110]. The solubility

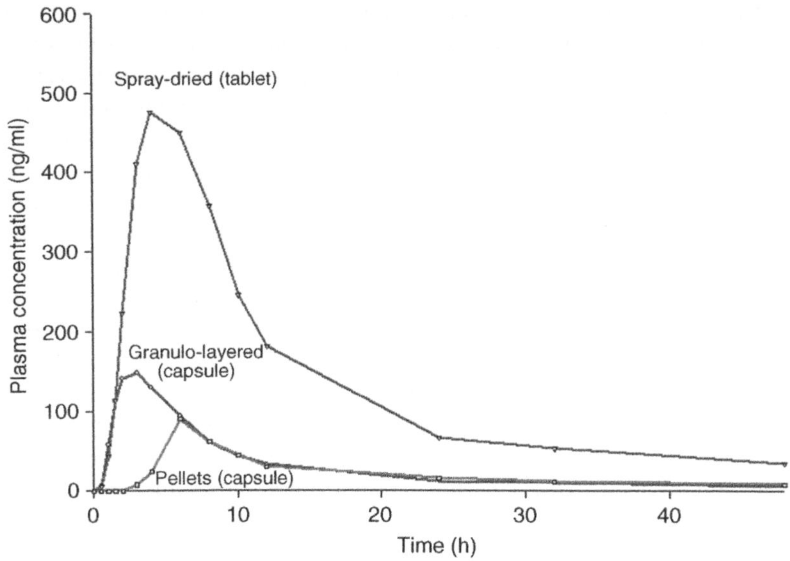

Figure 8.6. Human oral bioavailability of three amorphous dispersions of etravirine including coated beads, tablets prepared using granulo-layered powder, and tablets configured from a spray-dried dispersion.

change as well as mixing and precipitation dynamics generates particles containing a dispersion of the drug in the polymeric matrix. The fact that enteric polymers are used and that an acidic medium serves as the antisolvent means that the pH can be adjusted as a function of the polymer of interest to optimize dispersion formation and properties. In the screening assessments, three polymers were evaluated: HPMC-AS, EUDRAGIT L100-55, and HPMC-P. HPMC-P was eliminated based on the poor physical stability of the generated dispersion with crystallization detected in open conditions at 2 weeks of storage. Assessing the ability of the dispersion to maintain supersaturation was completed using a one-phase dissolution approach (USP II) and showed that the HPMC-AS polymer generated a significantly higher level of supersaturation than did the EUDRAGIT polymer consistent with a greater polymer solubility as a function of pH [108]. These and other factors suggested that HPMC-AS was the most optimal polymer for this dispersion generated using the MBP processing approach. Screening studies suggested that a 30:70 drug:polymer ratio was optimal and these systems were further characterized. Based on both the pH sensitivity of the MBP format and its particulate nature, dissolution of the powder was assessed using a USP IV flow-through powder cell. Dissolution media included pH 6.8 buffer containing hexadecyltrimethyl ammonium bromide. These assays were useful in showing not only the drug release but also the impact of MBP particle size on dissolution and supersaturation properties. In addition, a nonsink USP II apparatus containing FaSSIF was used to discriminate among amorphous, crystalline, and partially crystalline materials in addition to freshly prepared and stressed MBP samples [108]. The decrease in dissolution upon stressing the samples was attributed to partial crystalline conversion of the amorphous phase as well as to sample sintering. Based on these design criteria, solid dosage forms containing 40 mg vemurafenib were generated using the MBP with either dry or wet mixing of additional excipients and filling the resulting blends into capsules. These systems were compared with the micronized crystalline drug in capsules containing poloxamer F127. Human bioavailability assessments were completed using the two MBP-based formulations (dry blending or wet mixing) compared with the crystalline formulation. Exposure data suggested that a 160 mg dose provided for AUC values of 86 and 80 μMh for the dry-blended and wet-mixed formulations, respectively. This suggests an increase in AUC of 4.7 and 4.4-fold, respectively, compared with that of the crystalline drug formulation (Figure 8.7). In addition, C_{max} values increased 3.6- and 3.1-folds for the two MBP-based formulations relative to the crystalline system.

In addition to these studies, direct assessments of supersaturation in the human GI tract have been completed in a few cases. In these studies, drug dosage forms are administered traditionally or via gastric administration of solubilized forms of the API [54]. As can be appreciated, direct assessments of supersaturating solid dispersion in man require modalities to either sample gastric and/or intestinal contents or the development of probes that can assess drug levels remotely. The latter toolset is still in its

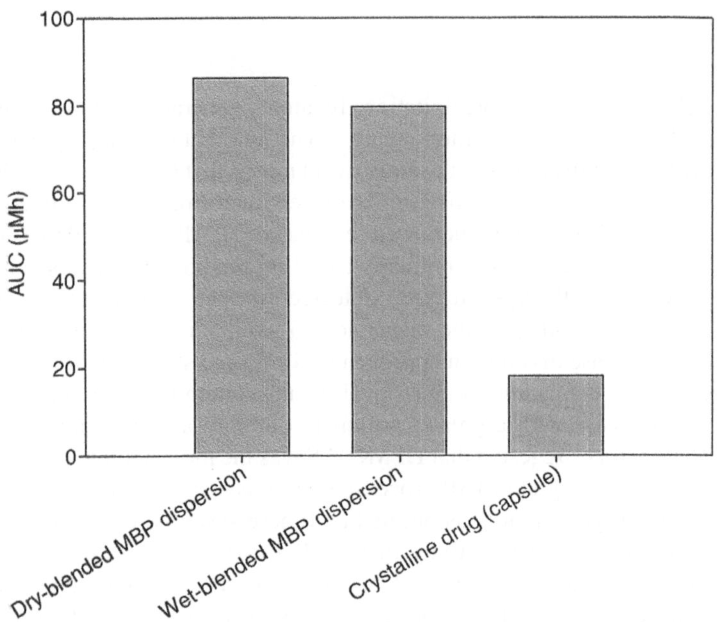

Figure 8.7. Exposure (AUC) in man after administration of vemurafenib as a microprecipitated bulk powder (MBP) dry or wet blended with other excipients and filled in a capsule or micronized as the crystalline API and filled into capsules with poloxamer F127. AUC values are normalized to a 160 mg dose.

infancy; however, sampling systems are available using a variety of different platforms. GI fluids are sampled by means of a double-lumen catheter, positioned either via the mouth or the nose into the upper GI tract [54]. Intubation with two catheters allows the simultaneous assessment of drug concentrations at two intraluminal positions: the stomach and duodenum or duodenum and upper jejunum. Positioning of the catheters is usually monitored by means of fluoroscopy. GI fluids are sampled as a function of time, typically for a period of 2–4 h following formulation administration. To minimize the effect of sampling drug on the absorption process, aspirated volumes should be kept to a minimum, especially when multiple GI sites are sampled and/or plasma concentrations are determined in parallel [54]. At the time of sampling, enzymatic processes should be inhibited, nondissolved and/or precipitated drug particles should be separated (centrifugation or filtration), and possible further precipitation should be avoided by diluting the particle-free sample. The observed drug concentrations can be related to characteristics of the GI fluids such as pH, osmolality, viscosity, surface tension, buffer capacity and concentration of bile salts, phospholipids, and dietary (digestion) products. Using these tools, Psachoulias et al. delivered acidic solutions of two weak bases, ketoconazole and dipyridamole, to the stomach and then assessed the compounds in aspirated drug sample [111]. Both the total drug and the soluble fraction of the drug in the aspirates were assessed. While the concentration of administered drug sampled from the duodenum was supersaturated, only limited precipitation was observed. Furthermore, the degree of precipitation observed *in vivo* seemed to have been overpredicted when using a transfer model. Importantly, these data were used to further optimize the transfer model [111]. Walravens et al. assessed the behavior of a commercially available suspension of posaconazole (Noxafil®) administered as such in the fasted state, with a cola-carbonated beverage, with a proton pump inhibitor (PPI), or with a cola-carbonated beverage and a PPI [112]. In this study both blood levels and GI samples (gastric and duodenal) were taken to assess *in vivo* behavior of the API and formulation. Blood levels suggested that coadministration of a cola beverage with 400 mg posaconazole significantly increased drug absorption (AUC increased twofold), while coadministration of the drug with a PPI (esomeprazole 40 mg) with or without a cola drink decreased exposure (with a decrease in AUC of 37 and 19%, respectively). Gastric and intestinal pH were invariant to dosing of the cola beverage, but both increased in response to treatment with the PPI. Drug concentrations were increased in the stomach upon dosing with a cola beverage and were lower in the presence of a PPI. Based on the variability, intestinal concentrations of posaconazole were not significantly higher after the cola beverage but were lower after treatment with the PPI. These data may suggest that the cola beverage has a number of effects. One of these may be a delay in gastric emptying as a function of the caloric content and other components that increase the extent of solubilization of posaconazole in the stomach [113]. Intestinal concentrations of the drug were dramatically lower than those measured in the stomach. This may be due to several factors including precipitation based on the accompanying pH shift or drug absorption that removed drugs from the intestine as a function of blood flow and permeability. The higher blood levels and lower intestinal levels associated with this treatment cohort may suggest that supersaturated levels of the drug were generated and that these were rapidly transferred to the systemic circulation by the applied sink. When the pH increased through the action of a PPI, exposure decreased as did the gastric and intestinal drug levels. It is apparent, therefore, that the assessment of the intraluminal behavior of solid dispersions using the GI sampling technique may be critical to unravel the precise mechanisms of absorption enhancement and to guide optimization of *in vitro* evaluation tools.

While data are limited to this point, there is a general feeling that current *in vitro* systems overestimate precipitation and excipient effects compared with *in vivo* observations. While there may be a variety of reasons for this, poor biorelevance may be a contributing factor. The nature of drug nucleation and crystal growth *in vitro* may poorly resemble that *in vivo* based on factors like secondary nucleation, hydrodynamics, absorption, and related features. Attempts to generate appropriate biorelevant models for solid dispersions are therefore an important goal [75].

8.5 CONCLUSIONS

Useful preclinical and clinical models for assessing solid dispersion fitness will contribute to the drug development process in many important ways, from both the drug enablement viewpoint and the increasing efficiency and probability of success by reducing cycle time and rework. While the ultimate goal may be to generate predictive *in silico* paradigms, a number of factors need to be elucidated prior to a true model-based drug development approach is at hand [114]. Specifically, *in vitro* methods with improved biorelevance need to be considered. In addition, those API, formulation, and biopharmaceutical features that impact the choice of an animal or human model need to be better articulated and systematized. All of these efforts will rely on a better understanding of human and animal gastrointestinal physiology and systems biology so that important critical success factors can be deconstructed and included into the relevant models. These processes include, but are not limited to, gastric motility, hydrodynamic and mechanical forces, volume effects and changes as a function of GI transit, the interaction of not only the API with the GI environment but also the possible excipient interactions [32,76,115–117]. As data accrue, improved formulation design and assessment will result.

REFERENCES

1. Kola, I. and Landis, J. (2004) Can pharmaceutical industry reduce attrition rates? *Nat. Rev. Drug Discov.*, 3:711–715.
2. Empfield, J. and Leeson, R. (2010) Lessons learned from candidate drug attrition. *IDrugs*, 13:869–873.
3. Lipinski, C. (2001) Avoiding investment in doomed drugs. *Curr. Drug Discov.*, 4:17–19.
4. Lipinski, C., Lombardo, F., Dominy, B., and Feeney, P. (2001) Experimental and computational approaches to estimate solubility and permeability in drug discovery and development settings. *Adv. Drug Deliv. Rev.*, 46:3–26.
5. Lipinski, C. (2000) Drug-like properties and the causes of poor solubility and poor permeability. *J. Pharmacol. Toxicol. Methods*, 44:235–249.

6. Gardner, C., Walsh, C., and Almarsson, Ö. (2004) Drugs as materials: valuing physical form in drug discovery. *Nat. Rev. Drug Discov.*, 3:926–934.
7. Huang, L. and Tong, W. (2004) Impact of solid state properties on druggability assessment of drug candidates. *Adv. Drug Deliv. Rev.*, 56:321–334.
8. Lipinski, C. and Hopkins, A. (2004) Navigating chemical space for biology and medicine. *Nature*, 432:855–862.
9. Gashaw, I., Ellinghaus, P., Sommer, A., and Asadullah, K. (2011) What makes a good drug target? *Drug Discov. Today*, 16:1037–1043.
10. Blagden, N., de Matas, M., Gavan, P., and York, P. (2007) Crystal engineering of active pharmaceutical ingredients to improve solubility and dissolution rate. *Adv. Drug Deliv. Rev.*, 59:617–630.
11. Liu, R. (Ed.), (2008) *Water-Insoluble Drug Formulations*, CRC Press, Boca Raton, FL.
12. Pinnamaneni, S., Das, N., and Das, S. (2002) Formulation approaches for orally administered poorly soluble drugs. *Die Pharm.*, 57:291–300.
13. Kawabata, Y., Wada, K., Nakatani, M., Yamada, S., and Onoue, S. (2011) Formulation design of poorly water-soluble drugs based on biopharmaceutics classification system: basic approaches and practical applications. *Int. J. Pharm.*, 420:1–10.
14. Williams, R., Watts, A., and Miller, D. (Eds.), (2012) *Formulating Poorly Water Soluble Drugs*, Springer, Arlington, VA.
15. Loftsson, T. and Brewster, M. (1996) Pharmaceutical applications of cyclodextrins: I. Drug solubilization and stabilization. *J. Pharm. Sci.*, 85:1017–1025.
16. Davis, M. E. and Brewster, M. E. (2004) Cyclodextrin-based pharmaceutics: past, present, future. *Nat. Rev. Drug Discov.*, 3:1023–1035.
17. Brewster, M. and Loftsson, T. (2007) Cyclodextrins as pharmaceutical solubilizers. *Adv. Drug Deliv. Rev.*, 59:645–666.
18. Loftsson, T. and Brewster, M. (2010) Pharmaceutical applications of cyclodextrins: basics science and product development. *J. Pharm. Pharmacol.*, 62:1607–1621.
19. Loftsson, T. and Brewster, M. (2011) Pharmaceutical applications of cyclodextrins: effects on drug permeation through biological membranes. *J. Pharm. Pharmacol.*, 63:1119–1135.
20. Strickley, R. (2004) Solubilizing excipients in oral and injectable formulations. *Pharm. Res.*, 21:201–230.
21. Miller, J., Beig, A., Carr, A., Webster, G., and Dahan, A. (2012) The solubility–permeability interplay when using cosolvents for solubilization: revising the way we use solubility-enabling formulations. *Mol. Pharm.*, 9:581–590.
22. Porter, C., Trevaskis, N., and Charman, W. (2007) Lipid and lipid-based formulations: optimizing the oral delivery of lipophilic drugs. *Nat. Rev. Drug Discov.*, 6:231–248.
23. Hauss, D. (2007) Oral lipid-based formulations. *Adv. Drug Deliv. Rev.*, 59:667–676.
24. O'Driscoll, C. and Griffin, B. (2008) Biopharmaceutical challenges associated with drugs with low aqueous solubilities: the potential impact of lipid-based formulations. *Adv. Drug Deliv. Rev.*, 60:617–624.
25. Dahan, A. and Hoffman, A. (2007) The effect of different lipid-based formulations on the oral absorption of lipophilic drugs: the ability of *in vitro* lipolysis and consecutive *ex vivo* intestinal permeability to predict *in vivo* bioavailability in rats. *Eur. J. Pharm. Biopharm.*, 67:96–105.

26. Stahl, P. and Wermuth, C. (Eds.), (2002) *Handbook of Pharmaceutical Salts: Properties, Selection and Use*, Wiley-VCH Verlag GmbH, Zürich, Switzerland.

27. Qiao, N., Li, M., Schlindwein, W., Malek, N., Davies, A., and Trappitt, G. (2011) Pharmaceutical co-crystals: an overview. *Int. J. Pharm.*, 419:1–11.

28. Yu, L. (2001) Amorphous pharmaceutical solids: preparation, characterization and stabilization. *Adv. Drug Deliv. Rev.*, 48:27–42.

29. Hancock, B. and Zografi, G. (2000) Characterization and significance of the amorphous state in pharmaceutical systems. *J. Pharm. Sci.*, 86:1–12.

30. Stevenson, C., Bennett, D., and Lechuga-Ballesteros, D. (2005) Pharmaceutical liquid crystals: the relevance of partially ordered systems. *J. Pharm. Sci.*, 94:1861–1880.

31. Branchu, S., Rogueda, P., Plumb, A., and Cook, W. (2007) A decision-support tool for the formulation of orally active, poorly soluble compounds. *Eur. J. Pharm. Sci.*, 22:128–139.

32. Sjögren, E., Bolger, M., Lukacova, V., Weitschies, W., Wilson, C., Augustijns, P., Reppas, C., Tannergren, C., Becker, D., Holm, R., Juretschke, H., Zane, P., Brewster, M., Van Peer, A., Münster, U., Abrahamsson, B., Flanagan, T., Lindahl, A., Kubbinga, M., Heinen, C., Harwood, M., Rostami, A., Neuhoff, S., Nguyen, M., Lennernäs, H., and Langguth, P. (2014) *In vivo* methods for drug absorption and API/formulation studies. *Eur. J. Pharm. Sci.*

33. Mackie, C., Mortishire-Smith, R., Wuyts, K., Lampo, A., and Brewster, M. (2008) Assessing solid dosage form feasibility and solubility/dissolution rate limitations for drug candidates: early solution–suspension comparisons in the rat. *American Association of Pharmaceutical Scientists (AAPS) Annual Meeting and Exposition, Atlanta, GA*, poster T2100.

34. Saxena, V., Panicucci, R., Joshi, Y., and Garad, S. (2009) Developability assessment in pharmaceutical industry: an integrated group approach for selecting developable candidates. *J. Pharm. Sci.*, 98:1962–1979.

35. Maas, J., Kamm, W., and Hauck, G. (2007) An integrated early formulation strategy: from hit evaluation to preclinical candidate profiling. *Eur. J. Pharm. Biopharm.*, 66:1–10.

36. Muenster, U., Pelzetter, C., Backensfeld, T., Ohm, A., Kuhlmann, T., Mueller, H., Lustig, K., Keldenich, J., Greschat, S., Göller, A., and Gnoth, M. (2011) Volume to dissolve applied dose (VDAD) and apparent dissolution rate (ADR): tools to predict *in vivo* bioavailability from orally applied drug suspensions. *Eur. J. Pharm. Biopharm.*, 78:522–530.

37. Leuner, C. and Dressman, J. (2000) Improving drug solubility for oral delivery using solid dispersions. *Eur. J. Pharm. Biopharm.*, 50:47–60.

38. Serajuddin, A. (1999) Solid dispersion of poorly water-soluble drugs: early promises, subsequent problems and recent breakthroughs. *J. Pharm. Sci.*, 88:1058–1066.

39. Janssens, S. and Van den Mooter, G. (2009) Physical chemistry of solid dispersions. *J. Pharm. Pharmacol.*, 61:1571–1586.

40. Dokoumetzidis, A. and Macheras, P. (2006) A century of dissolution research: from Noyes–Whitney to the biopharmaceutical classification system. *Int. J. Pharm.*, 321:1–11.

41. Verreck, G. and Brewster, M. (2004) Melt extrusion-based dosage forms: excipients and processing conditions for pharmaceutical formulations. *Sustained Release Formulations with Lipid Based Excipients*, 97th Bulletin Technique Gattefosse, Craig, D. (Ed.), Gattefosse, Saint-Priest, France, pp. 85–95.

42. Craig, D. (2002) The mechanism of drug release from solid dispersions in water-soluble polymers. *Int. J. Pharm.*, 231:131–144.

43. Brewster, M., Swinney, K., Stokbroekx, S., and Verreck, G. (2011) Solid dispersions as supersaturating drug delivery systems. *Supersaturating Drug Delivery Systems*, 104th Bulletin Technique Gattefosse, Augustijns, P. (Ed.), Gattefosse, Saint-Priest, France, pp. 33–48.

44. Weuts, I., Van Dycke, F., De Cort, S., Stokbroekx, S., Leemans, R., Brewster, M., Xu, D., Segmuller, B., Turner, Y., Roberts, C., Davies, M., Qi, S., Craig, D., and Reading, M. (2011) Physicochemical properties of the amorphous drug, cast films and spray dried powders to predict formulation probability of success for solid dispersions: etravirine. *J. Pharm. Sci.*, 100:260–274.

45. Kawakami, K., Usui, T., and Hattori, M. (2012) Understanding the glass-forming ability of active pharmaceutical ingredients for designing supersaturating dosage forms. *J. Pharm. Sci.*, 101:3239–3248.

46. Mahlin, D. and Bergström, C. (2013) Early drug development predictions of glass-forming ability and physical stability of drugs. *Eur. J. Pharm. Sci.*, 49:323–332.

47. Crowley, K. and Zografi, G. (2001) The use of thermal methods for predicting glass-former fragility. *Thermochim. Acta*, 280:79–93.

48. Hancock, B. and Parks, M. (2000) What is the true solubility advantage for amorphous pharmaceuticals? *Pharm. Res.*, 17:397–404.

49. Murdande, S., Pikal, M., Shanker, R., and Bogner, R. (2010) Solubility advantage of amorphous pharmaceuticals: 1. A thermodynamic analysis. *J. Pharm. Sci.*, 99:1254–1264.

50. Murdande, S., Pikal, M., Shanker, R., and Bogner, R. (2010) Solubility advantage of amorphous pharmaceuticals: 2. Application of quantitative thermodynamic relationships for prediction of solubility in structurally diverse insoluble pharmaceuticals. *Pharm. Res.*, 27:2704–2714.

51. Murdande, S., Pikal, M., Shanker, R., and Bogner, R. (2011) Solubility advantage of amorphous pharmaceuticals: 3. Is maximum solubility advantage experimentally attainable and sustainable? *J. Pharm. Sci.*, 100:4349–4356.

52. Brouwers, J., Brewster, M., and Augustijns, P. (2009) Supersaturating drug delivery systems: the answer to solubility-limited oral bioavailability? *J. Pharm. Sci.*, 98:2549–2572.

53. Vandecruys, R., Peeters, J., Verreck, G., and Brewster, M. (2007) Use of a screening method to determine excipients which optimize the extent and stability of supersaturated drug solutions and application of this system to solid formulation design. *Int. J. Pharm.*, 342:168–175.

54. Bevernage, J., Brouwers, J., Brewster, M., and Augustijns, P. (2013) Evaluation of gastrointestinal supersaturation and precipitation: strategies and issues. *Int. J. Pharm.*, 453:25–35.

55. Yamashita, T., Ozaki, S., and Kushida, I. (2011) Solvent shift method for anti-precipitant screening of poorly soluble drugs using biorelevant medium and dimethyl sulfoxide. *Int. J. Pharm.*, 419:170–174.

56. Wyttenback, N., Janas, C., Siam, M., Lauer, M., Jacob, L., Scheubel, E., and Page, S. (2013) Miniaturized screening of polymers for amorphous drug stabilization (SPADS): rapid assessment of solid dispersion systems. *Eur. J. Pharm. Biopharm.*, 84:583–598.

57. Box, K., Volgyi, G., Baka, E., Stuart, M., Takacs-Novak, K., and Comer, J. (2006) Equilibrium versus kinetic measurements of aqueous solubility and the ability of compounds to supersaturate in solution: a validation study. *J. Pharm. Sci.*, 95:1298–1307.

58. Box, K. and Comer, J. (2008) Using measured pK_a, log P, and solubility to investigate supersaturation and predict BCS class. *Curr. Drug Metab.*, 9:869–878.

59. Box, K., Comer, J., Gravestock, T., and Stuart, M. (2009) New ideas about the solubility of drugs. *Chem. Biodivers.*, 6:1767–1788.

60. Six, K., Berghmans, H., Leuner, C., Dressman, J., Van Werde, K., Mullens, J., Benoist, L., Thimon, M., Meublat, L., Verreck, G., Peeters, J., Brewster, M., and Van den Mooter, G. (2003) Characterization of solid dispersions of itraconazole and hydroxypropylmethylcellulose prepared by melt extrusion: Part 2. *Pharm. Res.*, 20:1047–1054.

61. Weuts, I., Kempen, D., Decorte, A., Verreck, G., Peeters, J., Brewster, M., and Van den Mooter, G. (2004) Phase behavior analysis of solid state dispersions of loperamide and two structurally related compounds with the polymers PVP-K30 and PVP-VA64. *Eur. J. Pharm. Sci.*, 22:375–385.

62. Van den Mooter, G., Wuyts, M., Blaton, N., Busson, R., Grobet, P., Augustijns, P., and Kinget, R. (2001) Physical stabilization of amorphous ketoconazole in solid dispersions with polyvinylpyrrolidone K25. *Eur. J. Pharm. Sci.*, 12:261–269.

63. Augustijns, P. and Brewster, M. (2012) Supersaturating drug delivery systems: fast is not necessarily good enough. *J. Pharm. Sci.*, 101:7–9.

64. Curatolo, W., Nightingale, J., and Herbig, S. (2009) Utility of hydroxypropylmethylcellulose acetate succinate (HPMCAS) for initiation and maintenance of drug supersaturation in the GI milieu. *Pharm. Res.*, 26:1419–1431.

65. Friesen, D., Shanker, R., Crew, M., Smithey, D., Curatolo, W., and Nightingale, J. (2008) Hydroxypropyl methylcellulose acetate succinate-based spray dried dispersions: an overview. *Mol. Pharm.*, 5:1003–1019.

66. Carlert, S., Palsson, A., Hanisch, G., von Corswant, C., Nilsson, C., Lindfors, L., Lennernäs, H., and Abrahamsson, B. (2010) Predicting intestinal precipitation: a case example for a basic BCS class II drug. *Pharm. Res.*, 27:2119–2130.

67. Van Speybroeck, M., Mols, R., Mellaerts, R., Thi, T., Martens, J., Humbeeck, J., Annaert, P., Van den Mooter, G., and Augustijns, P. (2010) Combined use of ordered mesoporous silica and precipitation inhibitors for improved oral absorption of the poorly soluble weak base itraconazole. *Eur. J. Pharm. Biopharm.*, 75:354–365.

68. McAllister, M. (2010) Dynamic dissolution: a step closer to predictive dissolution testing? *Mol. Pharm.*, 7:1374–1387.

69. Bjattachar, S., Perkins, E., Tan, J., and Burns, L. (2011) Effect of gastric pH on the pharmacokinetics of a BCS class II compound in dogs: utilization of an artificial stomach and duodenum dissolution model and GastroPlus^{TM} simulations to predict absorption. *J. Pharm. Sci.*, 100:4756–4765.

70. Polster, C., Atassi, F., Wu, S., and Sperry, D. (2010) Use of artificial stomach–duodenum model for investigation of dosing fluid effects on clinical trial variability. *Mol. Pharm.*, 7:1533–1538.

71. Minekus, M., Marteau, P., Havenaar, R., and Huis in 't Veld, J. (1995) A multicompartmental dynamic computer-controlled model simulating the stomach and small intestine. *Altern. Lab. Anim.*, 23:197–209.

72. Minekus, M., Smeets-Peeters, M., Bernalier, A., Marol-Bonnin, S., Havenaar, R., Marteau, P., Alric, M., Fonty, G., and Huis in 't Veld, J. (1999) A computer-controlled system to simulate conditions of the large intestine with peristaltic mixing, water absorption and absorption of fermentation products. *Appl. Microbiol. Biotechnol.*, 53:108–114.

73. Verwei, M., Freidig, A., Havenaar, R., and Groten, J. (2006) Predicted serum folate concentrations based on *in vitro* studies and kinetic modeling are consistent with measured folate concentrations in humans. *J. Nutr.*, 136:3074–3078.

74. Brouwers, J., Anneveld, B., Goudappel, G., Duchateau, G., Annaert, P., Augustijns, P., and Zeijdner, E. (2011) Food-dependent disintegration of immediate release fosamprenavir tablets: *in vitro* evaluation using magnetic resonance imaging and a dynamic gastrointestinal system. *Eur. J. Pharm. Biopharm.*, 77:313–319.

75. Lennernäs, H., Aarons, L., Augustijns, P., Beato, S., Bolger, M., Box, K., Brewster, M., Butler, J., Dressman, J., Holm, R., Julia Frank, K., Kendall, R., Langguth, P., Laplanche, L., Lindahl, A., McAllister, M., Münster, U., Müllertz, A., Ojala, K., Pepin, X., Reppas, C., Rostami, A., Verwei, M., Weitschies, W., Wilson, C., Karlsson, C., and Abrahamsson, B. (2014) Oral biopharmaceutical tools: time for a new initiative—an introduction to the IMI project, OrBiTo. *Eur. J. Pharm. Sci.*, 57:292–299.

76. Bevernage, J., Brouwers, J., Clarysse, S., Vertzoni, M., Tack, J., Annaert, P., and Augustijns, P. (2010) Drug supersaturation in simulated and human intestinal fluids representing different nutritional states. *J. Pharm. Sci.*, 99:4525–4534.

77. Bevernage, J., Forier, T., Brouwers, J., Tack, J., Annaert, P., and Augustijns, P. (2011) Excipient-mediated supersaturation stabilization in human intestinal fluids. *Mol. Pharm.*, 8:564–570.

78. Bevernage, J., Hens, B., Brouwers, J., Tack, J., Annaert, P., and Augustijns, P. (2012) Supersaturation in human gastric fluids. *Eur. J. Pharm. Biopharm.*, 81:184–189.

79. Alonzo, D., Zhang, G., Zhou, D., Goa, Y., and Taylor, L. (2010) Understanding the behavior of amorphous pharmaceutical systems during dissolution. *Pharm. Res.*, 27:608–618.

80. Mellaerts, R., Aerts, C., Van Humbeeck, J., Augustijns, P., Van den Mooter, G., and Martens, J. (2007) Enhanced release of itraconazole from ordered mesoporous SBA-15 silica materials. *Chem. Commun.*, 13:1375–1377.

81. Van Speybroeck, M., Mellaerts, R., Mols, R., Thi, T., Martens, J., Van Humbeeck, J., Annaert, P., Van den Mooter, G., and Augustijns, P. (2010) Enhanced absorption of the poorly soluble drug fenofibrate by tuning its release rate from ordered mesoporous silica. *Eur. J. Pharm. Sci.*, 41:623–630.

82. Bevernage, J., Brouwers, J., Annaert, P., and Augustijns, P. (2012) Drug precipitation–permeation interplay: supersaturation in an absorptive environment. *Eur. J. Pharm. Biopharm.*, 82:424–428.

83. Kataoka, M., Sugano, K., Mathews, C., Wong, J., Jones, K., Masaoka, Y., Sakuma, S., and Yamashita, S. (2012) Application of dissolution/permeation system for evaluation of formulation effect on oral absorption of poorly water-soluble drugs in drug development. *Pharm. Res.*, 29:1485–1494.

84. Takano, R., Takata, N., Saito, R., Furumoto, K., Higo, S., Hayashi, Y., Machida, M., Aso, Y., and Yamashita, S. (2010) Quantitative analysis of the effect of supersaturation on *in vivo* drug absorption. *Mol. Pharm.*, 7:1431–1440.

85. Newman, A., Knipp, G., and Zografi, G. (2012) Assessing the performance of amorphous solid dispersions. *J. Pharm. Sci.*, 101:1355–1377.

86. Gao, Y., Carr, A., Spence, J., Wang, W., Turner, T., Lipari, J., and Miller, J. (2010) A pH-dilution method for estimation of biorelevant drug solubility along the gastrointestinal tract: application to physiologically based pharmacokinetic modelling. *Mol. Pharm.*, 7:1515–1526.

87. Kuentz, M., Nick, S., Parrott, N., and Röthisberger, D. (2006) A strategy for preclinical formulation development using GastroPlus™ as pharmacokinetic simulation tool and a statistical screening design applied to a dog study. *Eur. J. Pharm. Sci.*, 27:91–99.

88. De Buck, S., Singa, V., Fenu, L., Nijsen, M., Mackie, C., and Gilissen, R. (2007) Prediction of human pharmacokinetics using physiologically based modeling: a retrospective analysis of 26 clinically tested compounds. *Drug Metab. Dispos.*, 35:1766–1780.
89. Shono, Y., Jantratid, E., and Dressman, J. (2011) Precipitation in the small intestine may play a more important role in the *in vivo* performance of poorly soluble weak bases in the fasted state: case example nelfinavir. *Eur. J. Pharm. Biopharm.*, 79:349–356.
90. Sugano, K. (2009) A simulation tool of oral absorption using classical nucleation theory. *Int. J. Pharm.*, 378:142–145.
91. Sugano, K. (2009) Introduction to computational oral drug simulation. *Expert Opin. Drug Metab. Toxicol.*, 5:259–293.
92. Ozaki, S., Kushida, I., Yamashita, T., Hasebe, T., Shirai, O., and Kano, K. (2012) Evaluation of drug supersaturation by thermodynamic and kinetic approaches for the prediction of oral absorbability in amorphous pharmaceuticals. *J. Pharm. Sci.*, 101:4220–4230.
93. Greco, K. and Bogner, R. (2012) Solution-mediated phase transformations: significance during dissolution and implications for bioavailability. *J. Pharm. Sci.*, 101:2996–3018.
94. Verreck, G., Vandecruys, R., De Conde, V., Baert, L., Peeters, J., and Brewster, M. (2004) The use of three different solid dispersion formulations – melt extrusion, film-coated beads, and a glass thermoplastic system – to improve the bioavailability of a novel microsomal triglyceride transfer protein inhibitor. *J. Pharm. Sci.*, 93:1217–1228.
95. Brewster, M., Vandecruys, R., Verreck, G., Noppe, M., and Peeters, J. (2003) A novel cyclodextrin-containing glass thermoplastic systems (GTS) for formulating poorly water soluble drug candidates: preclinical and clinical results. *J. Incl. Phenom. Macrocycl. Chem.*, 44:38–39.
96. Vandecruys, R., Verreck, G., Peeters, J., and Brewster, M. (2007) A surfactant-based glass thermoplastic system (GTS) for formulating poorly water-soluble weak base drug candidates. Abstracts, *34th Annual Meeting and Exposition of the Controlled Release Society, Long Beach, CA, July 7–11, 2007.*
97. Peeters, J., Neeskens, P., Tollenaere, J. P., Van Remoortere, P., and Brewster, M. (2002) Characterization of the interaction of 2-hydroxypropyl-β-cyclodextrin with itraconazole at pH 2, 4 and 7. *J. Pharm. Sci.*, 91:1414–1422.
98. Six, K., Verreck, G., Peeters, J., Binnemans, K., Berghmans, H., Augustijns, P., Kinget, R., and Van den Mooter, G. (2001) Investigation of thermal properties of glassy itraconazole: identification of a monotropic mesophase. *Thermochim. Acta*, 376:175–181.
99. Gilis, P., De Conde, V., and Vandecruys, R. (1997) Beads having a core coated with an antifungal and a polymer. U.S. Patent 5,633,015.
100. Verreck, G., Six, K., Van den Mooter, G., Baert, L., Peeters, J., and Brewster, M. (2003) Characterization of solid dispersions of itraconazole and hydroxypropylmethylcellulose prepared by melt extrusion: Part 1. *Int. J. Pharm.*, 251:165–174.
101. Engers, D., Teng, J., Jimenez-Novoa, J., Gent, P., Hossack, S., Cambell, C., Thomson, J., Ivanisevic, I., Templeton, A., Byrn, S., and Newman, A. (2010) A solid-state approach to enable early development compounds: selection and animal bioavailability studies of an itraconazole amorphous solid dispersion. *J. Pharm. Sci.*, 99:3901–3922.
102. Six, K., Daems, T., de Hoon, J., Van Hecken, A., Depre, M., Bouche, M., Verreck, G., Peeters, J., Brewster, M., and Van den Mooter, G. (2005) Clinical studies of solid dispersions of itraconazole prepared by hot-stage extrusion. *Eur. J. Pharm. Sci.*, 24:179–186.
103. Kakuda, T., Schöller-Gyüre, M., Workman, C., Arasteh, K., Pozniak, A., De Smedt, G., Beets, G., Peeters, M., Vandermeulen, K., Woodfall, B., and Hoetelmans, R. (2008) Single

and multiple dose pharmacokinetics of etravirine administered as two different formulation in HIV-1-infected patients. *Antivir. Ther.*, 13:655–661.

104. Qi, S., Weuts, I., De Cort, S., Stokbroekx, S., Leemans, R., Reading, M., Belton, P., and Craig, D. (2009) An investigation into the crystallization behavior of an amorphous cryomilled pharmaceutical material above and below the glass transition. *J. Pharm. Sci.*, 99:196–208.

105. Bollag, G., Hirth, P., Tsai, J., Zhang, J., Ibrahim, P., Cho, H., Spevak, W., Zhang, C., Zhang, Y., Habets, G., et al. (2010) Clinical efficacy of a RAF inhibitor needs broad target blockage in BRAF-mutant melanoma. *Nature*, 467:596–599.

106. Fisher, R. and Larkin, J. (2012) Vemurafenib: a new treatment for BRAF-V600 mutated advanced melanoma. *Cancer Manag. Res.*, 4:243–252.

107. Desai, D., Diodone, R., Go, Z., Ibrahim, P., Iyer, R., Mair, H., Sandhu, H., Shah, N., Visor, G., Wyttenbach, N., Lauper, S., Pudewell, J., and Weirschem, F. (2010) Compositions and uses thereof. World Patent WO2010114928.

108. Shah, N., Iyer, R., Mair, H., Choi, D., Tian, H., Diodone, R., Fähnrich, K., Pabst-Ravot, A., Tang, K., Scheubel, E., Grippo, J., Moreira, S., Go, Z., Mouskountakis, J., Louie, T., Ibrahim, P., Sandhu, H., Rubia, L., Chokshi, H., Singhal, D., and Malick, W. (2013) Improved human bioavailability of vemurafenib, a practically insoluble drug, using amorphous polymer-stabilized solid dispersions prepared by a solvent-controlled coprecipitation process. *J. Pharm. Sci.*, 102:967–981.

109. Dong, Z., Chatterji, A., Sandhu, H., Choi, D., Chokshi, H., and Shah, N. (2008) Evaluation of solid state properties of solid dispersions prepared by hot-melt extrusion and solvent co-precipitation. *Int. J. Pharm.*, 355:141–149.

110. Shah, N., Sandhu, H., Phuapradit, W., Pinal, R., Iyer, R., Albano, A., Chatterji, A., Anand, S., Choi, D., Tang, K., Tian, H., Chokshi, H., Singhal, D., and Malick, W. (2012) Development of a novel microprecipitated bulk powder (MBP) technology for manufacturing stable amorphous formulations of poorly soluble drugs. *Int. J. Pharm.*, 438:53–60.

111. Psachoulias, D., Vertzoni, M., Goumas, K., Kalioras, V., Beato, S., Butler, J., and Reppas, C. (2011) Precipitation in and supersaturation of contents of the upper small intestine after administration of two weak bases to fasted adults. *Pharm. Res.*, 28:3145–3158.

112. Walravens, J., Brouwers, J., Spriet, I., Tack, J., Annaert, P., and Augustijns, P. (2011) Effect of pH and comedication on gastrointestinal absorption of posaconazole: monitoring intraluminal and plasma drug concentration. *Clin. Pharm.*, 50:725–732.

113. Houston, J. and Levy, G. (1975) Effect of carbonated beverages and of an antiemetic containing carbohydrate and phosphoric acid on riboflavin bioavailability and salicylamide biotransformation in humans. *J. Pharm. Sci.*, 64:1504–1507.

114. Zhao, P., Zhang, L., Grillo, J., Liu, Q., Bullock, J., Moon, Y., Song, P., Brar, S., Madabushi, R., Wu, C., Booth, B., Rahman, N., Reynolds, K., Berglund, E., Lesko, L., and Huang, S. (2011) Application of physiologically based pharmacokinetic modeling (PBPK) and simulation during regulatory review. *Clin. Pharmacol. Ther.*, 89:259–261.

115. Bergström, C., Holm, R., Jørgensen, S., Andersson, S., Artursson, P., Beato, S., Borde, A., Box, K., Brewster, M., Dressman, J., Feng, K., Halbert, G., Kostewicz, E., McAllister, M., Muenster, U., Thinnes, J., Taylor, R., and Müllertz, A. Early pharmaceutical profiling to predict oral drug absorption: current status and unmet needs. *Eur. J. Pharm. Sci.*, 57, 173–199.

116. Kostewicz, E., Abrahamsson, B., Brewster, M., Brouwers, J., Butler, J., Carlert, S., Dickinson, P., Dressman, J., Holm, R., Klein, S., Mann, J., McAllister, M., Minekus, M.,

Muenster, U., Müllertz, A., Verwei, M., Vertzoni, M., Weitschies, W., and Augustijns, P. (2014) *In vitro* models for the prediction of *in vivo* performance or oral dosage forms. *Eur. J. Pharm. Sci.*, 57, 342–366.

117. Kostewicz, E., Aarons, L., Bergstrand, M., Bolger, M., Galetin, A., Hatley, O., Jamei, M., Lloyd, R., Pepin, X., Rostami, A., Sjögren, E., Tannergren, C., Turner, D., Wagner, C., Weitschies, W., and Dressman, J. (2014) PBPK modeling for the prediction of *in vivo* performance of oral dosage forms. *Eur. J. Pharm. Sci.*, 57, 300–321.

9

SPRAY DRYING AND SCALE-UP

Daniel E. Dobry, Dana M. Settell, and John M. Baumann

Bend Research, Inc., Bend, OR, USA

9.1 INTRODUCTION

Spray drying is a versatile, continuous process that is commonly used in the food, chemical, and pharmaceutical industries. The spray drying process converts liquid feedstocks including solutions, slurries, and suspensions into dry powders. Spray drying is applicable to a wide range of materials and across a wide range of scales. Furthermore, spray drying offers fast drying kinetics, flexibility in achieving the desired product properties, mild processing temperatures, and short residence times, making it an ideal process for manufacturing amorphous dispersions.

Spray drying has demonstrated scalability from milligram quantities of active pharmaceutical ingredient (API) during formulation screening to metric ton quantities at commercial scale. Minimal use of API during development and scale-up and demonstrated commercial viability make spray drying an efficient and effective manufacturing option for amorphous dispersions [1–3]. This chapter discusses amorphous dispersions manufactured using the spray drying process, which are referred to as spray-dried dispersions (SDDs). Key aspects of spray drying equipment, process development, and scale-up are covered.

Pharmaceutical Amorphous Solid Dispersions, First Edition. Edited by Ann Newman.

9.2 PROCESS BACKGROUND AND PHYSICAL SITUATION

Spray drying offers the opportunity to produce amorphous dispersions with superior physical stability using a well-established unit operation. During spray drying, the API is dissolved in a spray solution that may include a dispersion polymer, a spray solvent, and other excipients. The spray solvent is evaporated from the spray solution droplets within milliseconds [4–6]. This rapid drying from solution to solid limits the mobility of the API in the dispersion polymer and makes it possible to produce a wide range of physically stable amorphous dispersions.

A wide variety of polymers can be used in this process. Thermodynamic solubility of the API in the dispersion polymer is not required because the API can be kinetically trapped during spray drying [1]. Due to the rapid drying kinetics, it is often possible to achieve higher drug loadings in SDDs than in dispersions produced using other manufacturing methods (e.g., hot melt extrusion). It is also possible to use polymers with high glass transition temperatures (T_g) that can produce amorphous dispersions with superior physical stability.

Critical quality attributes (CQAs) for SDDs, such as particle size and density, can be tuned by controlling the spray solution composition, droplet formation, and drying rates. Particle size can be controlled by selection of atomizers and atomization parameters (e.g., nozzle geometry, nozzle pressure, atomizing gas pressure, and solution concentration) to produce the target droplet size. The density and morphology of SDDs, which are often important to downstream tablet or capsule manufacture, are controlled by the droplet drying rate through adjustment of the thermodynamic parameters (e.g., dryer outlet temperature (T_{out})).

Thus, by controlling atomization and drying parameters, SDD particles can be engineered to enable a wide range of downstream formulation options, including liquid presentations (e.g., suspensions or sachets), immediate-release solid oral dosage forms (e.g., tablets or capsules), and modified-release solid oral dosage forms (e.g., osmotic tablets).

Since droplet drying, or solidification, occurs very quickly, SDD particle attributes are defined within a key control volume around the exit of the atomizer, where the liquid droplets are formed and initially come into contact with the drying gas [4]. Figure 9.1 shows the physical situation and key control volume for a typical spray dryer. As the figure shows, the liquid feed is atomized into discrete droplets. Solvent immediately begins to evaporate from the droplet surface upon contact with the drying gas, leading to an initial free evaporation drying phase. Most SDDs are formulated with somewhat viscous (depending on the spray solution concentration) film-forming polymers, which will skin quickly (on the order of 10 ms) after the initial free evaporation stage [5]. After skin formation or viscous gelation of the polymer, the droplet drying rate is limited by internal mass transfer due to the added diffusional resistance provided by the solidified skin of the droplet. During this phase, the droplet drying rate is significantly slower than the rate of free solvent evaporation. Modeling and experimental results have shown that the solidified droplet or SDD particle subsequently experiences heating to temperatures equivalent to T_{out}. If the droplet experiences a temperature near or above the solvent boiling point at this stage of drying, the vapor pressure inside the particle will keep the particle inflated, leading to a final particle with a "hollow sphere" morphology. This morphology often results in low bulk density (e.g., <0.2 g/cm^3) with good compactability. If the droplet experiences a temperature significantly

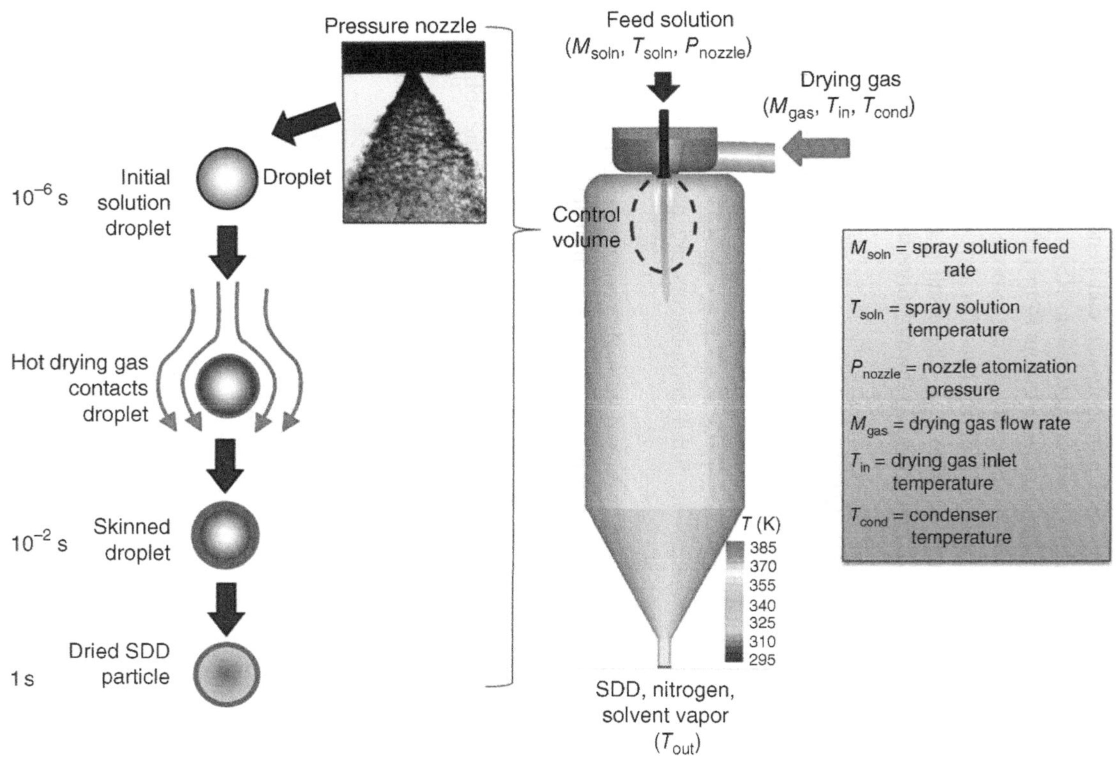

Figure 9.1. Physical situation and key control volume during spray drying.

below the solvent boiling point after skinning, the particle will maintain its surface area, but collapse to a "shriveled raisin" morphology. This morphology often results in higher bulk density (e.g., 0.2–0.4 g/cm^3) and lower compactability than the hollow spheres dried at higher temperatures. The "hollow sphere" and "shriveled raisin" particles represent two extremes on the particle morphology continuum that result from the spray drying process. As discussed in this chapter, SDD bulk densities and morphologies can generally be correlated to and modified by controlling T_{out}, relative solvent saturation (RS), or other process parameters.

Figure 9.1 shows the temperature contour output from a computational fluid dynamics (CFD) model of a typical nozzle-based spray drying process. As the figure shows, the evaporation plume is the coolest area in the drying chamber and the rest of the drying chamber is at a temperature approximately equal to T_{out}. The result is mild temperature exposure, which is especially important for compounds or formulations that may be thermally labile.

9.3 SPRAY DRYING EQUIPMENT

The spray drying process can be broken down into the following five key steps or focus areas, which are described below:

1. *Spray Solution Preparation*: Solid components are dissolved into a mutual solvent.
2. *Atomization*: The solution is passed through a nozzle, creating discrete droplets with a high surface area for rapid heat and mass transfer.
3. *Drying*: Atomized droplets are contacted with heated drying gas to remove solvent and produce a dried particle.
4. *Product Collection*: A cyclone separator is used to remove the solids from the drying gas.
5. *Secondary Drying*: A secondary drying process is used to remove residual solvent from the SDD to a target residual solvent level.

An overview of these key areas for a spray drying equipment train to support SDD manufacture is shown in Figure 9.2.

9.3.1 Spray Solution Preparation

The spray solution is prepared by dissolving the API with the dispersion polymer and other desired excipients in an organic solvent system that provides mutual solubility of each component. Solvents typically used include acetone, methanol, tetrahydrofuran (THF), and dichloromethane (DCM). In addition, binary blends of these solvents or solvent systems containing small fractions of water (e.g., 1–30% by mass) can be advantageous for maximizing solubility of the API and polymer.

Spray solution tanks are sized based on the throughput and desired batch size of the SDD at a given spray drying scale. Total solids concentration is defined by either the maximum solubility of the API or the maximum spray solution viscosity that can be atomized while maintaining target particle properties. Typically, API concentration is limited by the equilibrium solubility of the API in the spray solution over process-

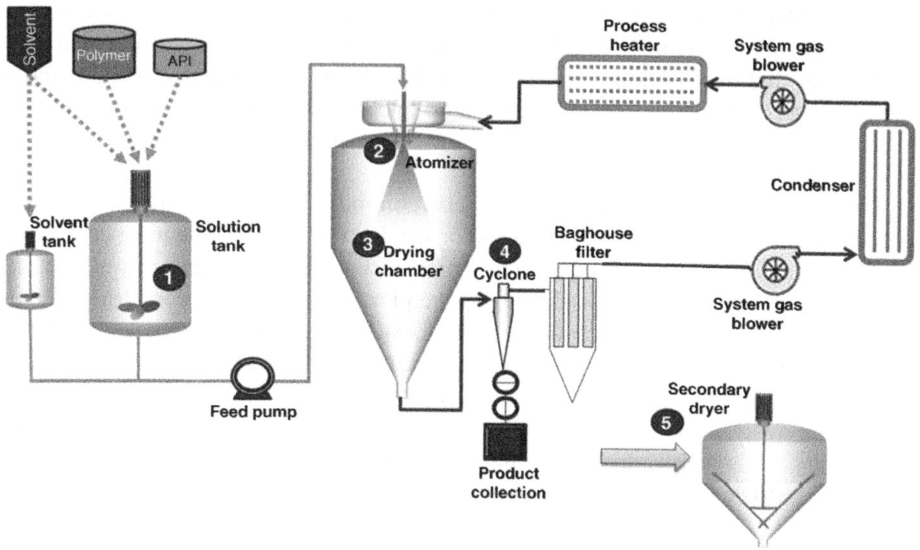

Figure 9.2. Spray drying process equipment.

relevant temperature ranges. A good rule of thumb is that <80% of equilibrium solubility is chosen to minimize the risk of undissolved API in the solution or API precipitation during droplet drying, which could lead to phase separation.

During spray solution preparation, powder components (e.g., API, polymer, or other excipients) are usually added in one of the two ways. The first method uses gravity feed through chutes attached to the top of the processing tanks, with devices such as metering valves to control the rate of powder addition. The second method uses a high-shear induction mixer (e.g., a Quadro YTron®) to provide good initial dispersion of the powders into the solvent prior to dissolution. It is particularly critical to avoid wetting large masses of polymer without proper mixing to avoid "fisheyes" (hydrated outer layer of gel around a dry core of powder) that can take a long time to dissolve. Design of the solution tank geometry and associated equipment should consider the wide range of solution properties that may be encountered during SDD manufacture and ensure adequate shear and solution turnover during mixing. Key process parameters that are optimized for solution preparation include agitation rate, total mixing time, and temperature.

The spray solution may be heated to increase solubility or the dissolution rate of the drug or polymer. Conversely, the spray solution may be cooled to enhance chemical stability or reduce the formation of degradants. To allow precise control of the spray solution temperature, jacketed tanks are commonly used. Nitrogen inertion is also employed to reduce the hazards associated with organic solvents and dust-producing excipients.

9.3.2 Atomization

The atomization process allows for the creation of high surface area droplets of a controlled size that can be rapidly dried. This is important to achieve a homogeneous, amorphous SDD with the desired particle size. A wide variety of atomizers are available, allowing the flexibility to address formulation or solution property constraints. Droplet sizes and distributions can be modified based on input energy or atomization technique (e.g., hydraulic, pneumatic, or electric), nozzle geometry, or solution property. By controlling atomization parameters, a range of particle sizes can be precisely engineered.

Table 9.1 shows the atomizers that are commonly used to spray dry SDDs, highlighting the advantages and drawbacks for specific nozzles [7]. The most commonly used atomizer types are two-fluid and pressure-swirl atomizers. They are well characterized and widely available from commercial vendors in a variety of geometries, so they can be selected to meet the needs for specific formulations or equipment scales. These nozzles also allow for the most straightforward scale-up of the atomization process and are robust across the range of SDD formulations.

Pressure-swirl nozzles are typically the preferred option for SDD manufacture because they (1) reliably and reproducibly produce a wide range of droplet sizes with narrow particle size distributions and (2) offer maintenance-free operation, including

TABLE 9.1. Atomizer Options for SDD Manufacture

Atomizer	Advantages	Drawbacks
Two-fluid	• Large liquid orifice prevents clogging • Capable of atomizing high-viscosity solutions • Broadly applicable • May be preferred if targeting small particle sizes	• Broad particle size span (e.g., >2) • Requires gas compressor or high usage of high-pressure nitrogen • Can be prone to bearding/ product buildup • Not preferred for larger particle sizes
Pressure-swirl	• Narrow particle size span (e.g., 1.5–2) • Robust and scalable (self-cleaning designs available for spray drying) • Tunable over a range of particle size targets • No atomizing gas is required	• Requires high pressure (up to 250 bar at large scale) • Limited ability to make small particles at large scale
Rotary disk	• Narrow particle size span (e.g., 1.5–2) • Independent control of liquid feed rate and atomization energy	• Requires spray dryer with short/wide aspect ratio • Not amenable to viscous film-forming feedstocks (makes floss and filaments)
Ultrasonic or electrostatic	• Capable of narrow span • Low shear/velocity of droplet ejection	• Limited flow range (not scalable) • Challenging with viscous feedstock • Particle charge must be dissipated (for electrostatic)

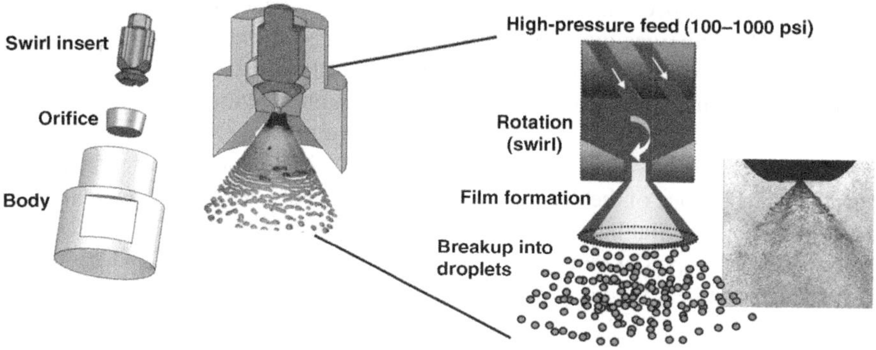

Figure 9.3. Key components for pressure-swirl atomizers.

self-cleaning designs to minimize product buildup, or "bearding" during long manufacturing campaigns. The key components of pressure-swirl atomizers are shown in Figure 9.3.

Pressure-swirl nozzle designs vary among different manufacturers. Generally, a swirl insert is placed into the chamber of the nozzle above an orifice of fixed geometry and diameter. Spray solution passes through the channels, which impart angular momentum to the fluid, producing a rotating (or "swirling") film that exits the orifice with a hollow air core in the center. After exiting the orifice, the film expands radially, due to the angular momentum imparted by the swirl channels, and subsequently breaks down into sheets, ligaments, and finally into droplets. The geometry and size of the swirl insert and orifice affect key attributes of the atomized droplets, including velocity, spray angle, droplet size, and droplet size distribution.

In addition to nozzle geometry, solution properties, including viscosity, density, and surface tension, are all important to the atomization process. Lefebvre and others have published extensive work correlating spray solution properties to droplet size [7]. Although surface tension and density can be important factors in defining droplet size, over the range of solutions typically used in the manufacture of SDDs, viscosity has the largest impact for solutions of somewhat viscous polymers in organic solvents of relatively low surface tension. Often the solids concentration in the spray solution is limited by the maximum viscosity that can be successfully atomized to create the target droplet size. Since the dispersion polymers are typically the main contributor to spray solution viscosity, placebo solutions containing just the dispersion polymer dissolved in the spray solvent can be used to evaluate atomizer performance, minimizing the use of valuable API during early stages of process development.

9.3.3 Drying

Drying gas is introduced into the drying chamber via the gas disperser. The gas disperser should be designed to adequately mix the drying gas with the atomized droplets while minimizing turbulence near the nozzle. If turbulence occurs near the nozzle, wet or skinned particles can become entrained and stick to the surface of the nozzle (creating "bearding"). The DPH gas disperser, which is integrated in the Niro PHARMASD^{TM} spray dryer, is an example of an effective design that is easily scaled.

To allow adequate drying time, spray drying chambers with high aspect ratios are typically preferred for SDD manufacture with pressure nozzles. A drying gas residence time of approximately 10–30 s ensures that the droplets dry to a solid particle before they contact the walls or cone of the drying chamber [9]. Typically, this can be achieved with a spray drying chamber that has a height:diameter ratio of greater than 2:1 (including the conical section). Droplet skinning and drying can be modeled with CFD packages to design drying chambers with appropriate dimensions and to evaluate drying conditions.

9.3.4 Product Collection

Typically, cyclone separators are used to remove the SDD particles from the drying gas. Cyclone geometry and pressure drop can be used to optimize collection efficiency [8]. Baghouse filters can also be used to collect the SDD from the drying gas, but they are generally considered as a fines collection filter in series with a cyclone (see Figure 9.2) and are not the preferred primary collection point except in cases of extremely fine powders.

9.3.5 Secondary Drying

Due to the short residence times in the spray dryer, a small amount of solvent, typically 2–10% by weight, remains in the SDD particles that are collected from the cyclone. This "residual solvent" is removed during a secondary drying step. The most commonly used unit operations are convection tray dryers, agitated vacuum dryers (e.g., Ekato VPT dryers), or biconical rotating vacuum dryers (e.g., Bachiller and Pfaudler).

9.4 PROCESS DEFINITION

Definition of a robust, scalable spray drying process for SDDs is based on integrated consideration of (1) formulation variables and (2) process parameters that control atomization and drying thermodynamics.

9.4.1 Formulation Selection

Formulation selection is covered in Chapter 7, but it is important to consider formulation selection as an activity that is coupled with process definition during SDD development. Use of compound guidance maps based on physicochemical properties of the API to identify the formulation approach and process-related risks can help increase development efficiency and decrease risk. Such approaches have been previously discussed in the literature [9].

Below is a list of some of the key formulation selection considerations that have a significant impact on spray drying process development.

- *Dissolution Mechanism*: SDDs should allow for rapid dissolution under biorelevant conditions; however, it is important to identify formulations that may have particle size-dependent dissolution (see Figure 9.4). For example, highly

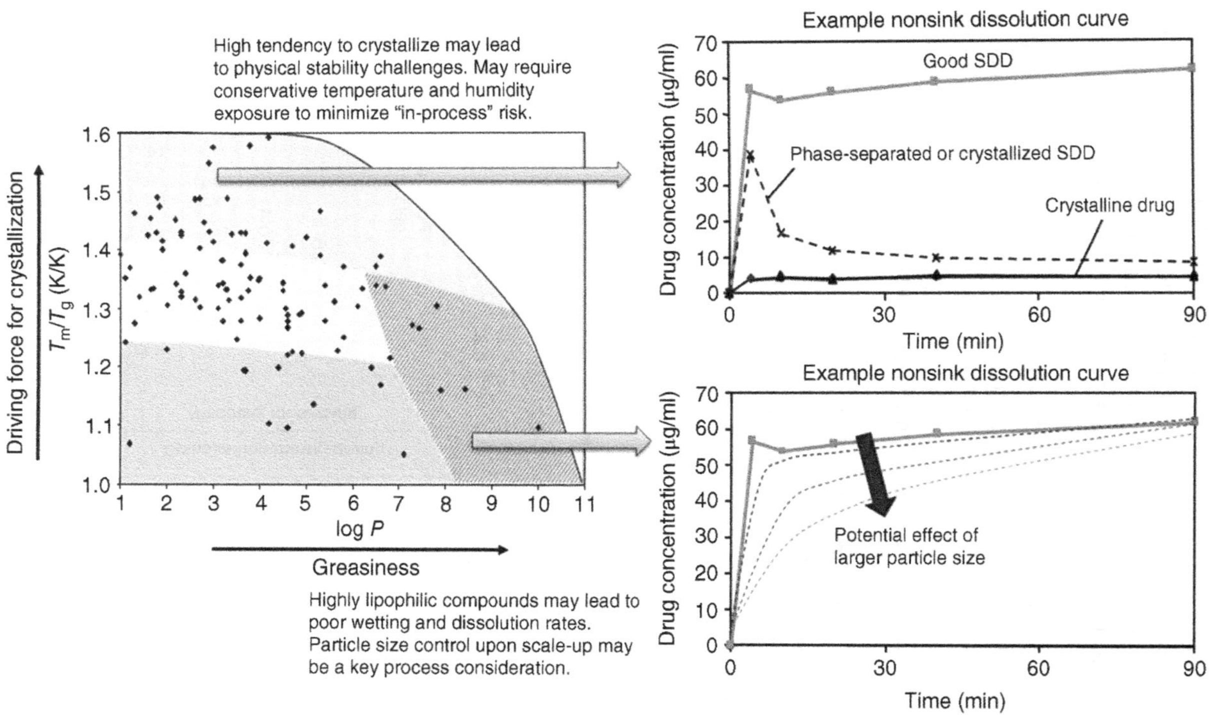

Figure 9.4. Guidance map of API physicochemical properties and examples of process-relevant formulation risks—high tendency to crystallize and high lipophilicity (T_m = melt temperature).

lipophilic compounds have a tendency to have particle size-dependent dissolution. Particle size is often a CQA that must be optimized around the dissolution performance of the SDD and maintained upon scale-up.

- *Physical Stability Risks*: SDDs are formulated with high-T_g polymers in either the stable or metastable region. Selection of the drug loading and the polymer should enable formulation robustness for process-relevant times and temperature exposures, as well as in-process hold times before and during secondary drying (see Figure 9.4).
- *Solubility in Common Solvents*: Selection of a common solvent for the drug, polymer, and other excipients is critical to forming a homogeneous dispersion. It is also important to identify a system that not only allows rapid drying, but also reduces the potential for physical instability due to potential concentration gradients as the droplet undergoes composition changes during drying (see Figure 9.5)

9.4.1.1 Spray Solution Definition Spray solution definition is a critical component during formulation and process selection. The following considerations should be accounted for to ensure both homogeneity of the SDD formulation and an efficient process.

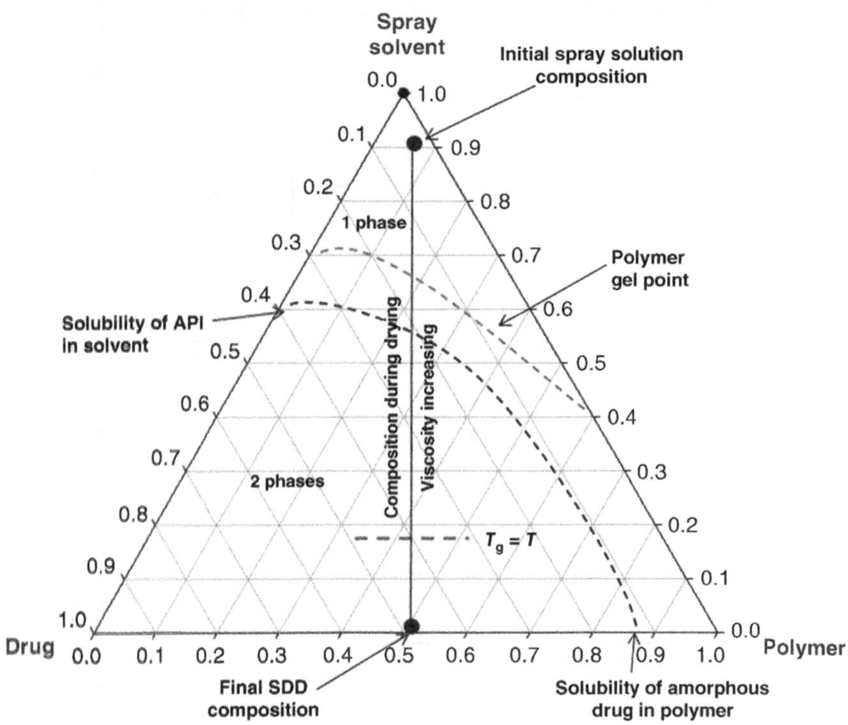

Figure 9.5. Theoretical ternary phase diagram with example path from droplet-to-particle composition.

Assessment of API solubility in solvent systems is critical. Solvents typically used during the spray drying process are limited by precedence and fall within the FDA Class 2 or 3 classification [10]. The most commonly used solvents in the spray drying process include acetone, methanol, THF, and DCM, although many others can be used.

Ideally, a single solvent can be found to provide adequate solubility of the drug, polymer, and excipients. However, it is also common to use mixed solvent systems to further enhance drug solubility or provide mutual solubility of the drug and excipients. Water is also commonly added to mixed solvent systems (typically less than 30 wt% of total solvent) due to its miscibility with a wide range of solvents and its ability to increase the solubility of the API or the excipients. The solubility and physical state of functional excipients also should be evaluated carefully to maximize formulation and process robustness. For example, use of cosolvents that allow for complete dissolution of a polymer versus a colloidal solution may be critical for achieving the desired physical state of the SDD.

Additional consideration is given to the overall volatility of the spray solvent. Key properties related to volatility are the latent heat of vaporization and the boiling point. Volatility defines the overall throughput of the spray drying process and impacts the cycle time of the secondary drying process.

9.4.2 Process Variables

An overview of process parameter definition is shown in Table 9.2 and Figure 9.6 and discussed in detail throughout this section. A rationale strategy (discussed previously) [4] can be used to streamline this approach during the development stage.

9.4.2.1 Thermodynamics The key process variables that are critical to understanding the spray drying process are summarized in Table 9.2. Each variable is important to the attributes of the SDD and has a specific relationship or input into the overall process definition. The goal is to rationally define each process parameter based on a known relationship to the SDD and eliminate the need for empirical approaches to process definition.

The thermodynamic operating space of the spray drying process can be defined by conducting a mass and energy balance for the process using the drying chamber as a

TABLE 9.2. Summary of Key Thermodynamic Spray Drying Process Parameters

Parameter	Variable Type	Defined by
T_{in}	Input/independent	Formulation physical or chemical stability constraints
M_{gas}	Input/independent	Scale of equipment
M_{soln}	Input/independent	Target evaporation rate of spray dryer
T_{cond}	Input/independent	Equipment constraints at large scale, drying rate of product
T_{out}	Output/dependent on inputs	Desired particle attributes, physical stability
RS_{out}	Output/dependent on inputs	Desired particle attributes, product recovery constraints

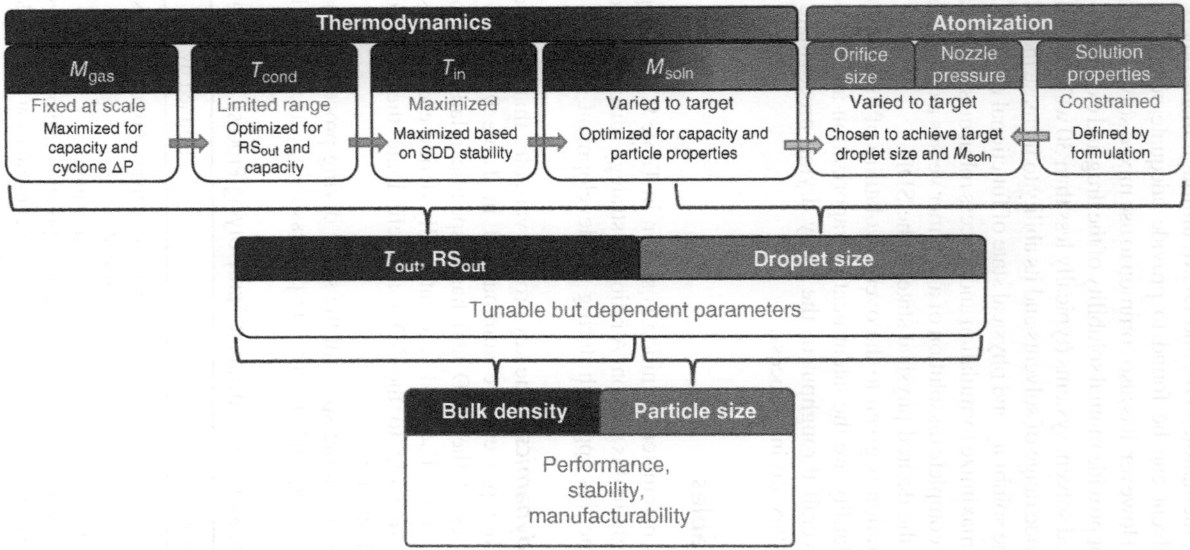

Figure 9.6. Overview of the process parameter selection process including consideration of thermodynamics and atomization (RS_{out} = outlet relative saturation).

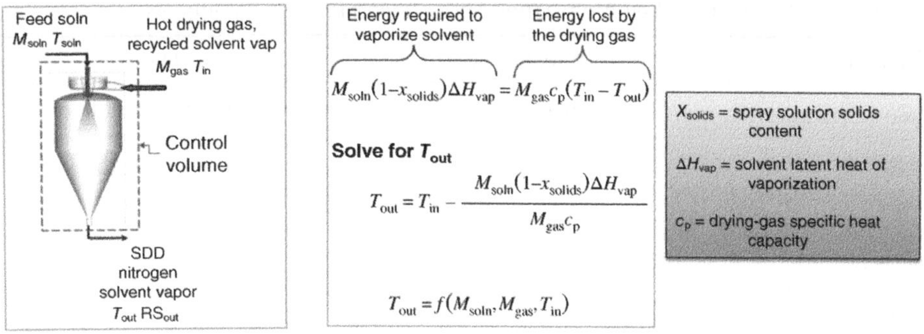

Figure 9.7. Simplified spray dryer mass and energy balance.

control volume. The energy required for evaporation is defined from the spray solution inputs into the system (i.e., mass flow rate, solids fraction, and latent heat of vaporization). Conversely, the energy lost to the evaporation process is defined by the drying gas inputs into the system (i.e., mass flow rate, specific heat, and temperature differential). This approach and the key constitutive relationships have been described previously and are summarized in Figure 9.7 [4]. Using the simplified equations shown, T_{out} and RS_{out} can be calculated for any combination of T_{in} and M_{soln}. Further, given that M_{gas} is fixed at a given drying scale, the ratio of M_{soln} and M_{gas}—the specific drying ratio—can be used to compare various scales of spray dryers on the same plot of key thermodynamic spray drying parameters. The resulting graphical representation of the mass and energy balance is shown in Figure 9.8, including five of the key parameters described in Table 9.3. T_{cond} is not discussed here and is dealt with later in Section 9.5.

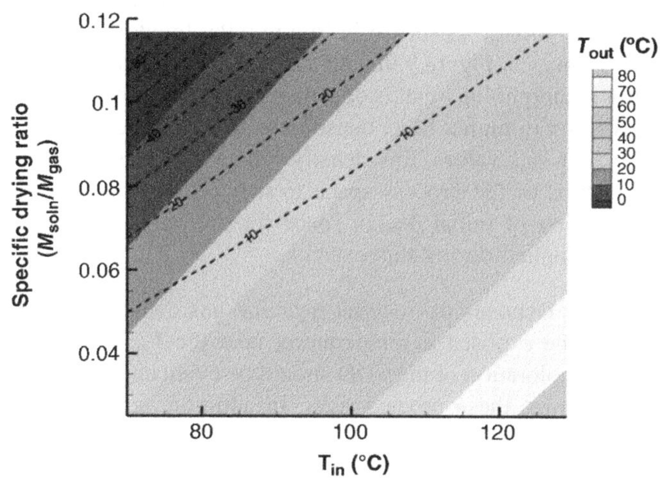

Figure 9.8. Plot of the spray drying process thermodynamic operating space for acetone as a spray solvent (T_{out} denoted by gradations in key; dotted lines indicate RS_{out}).

TABLE 9.3. Summary of Key Atomization Spray Drying Process Parameters

Parameter	Variable Type	Defined by
M_{soln}	Input/ independent	• Evaporation capacity of dryer • Desired physical properties of SDD • Target atomization pressure
P_{nozzle}	Input/ independent	• Equipment constraints • Target droplet size
Nozzle geometry (e.g., design, swirl channel open area, orifice diameter)	Input/ independent	• Nozzle manufacturer • Target droplet or SDD particle size
Solids content	Input/ independent	• Solubility limit of drug or polymer in the spray solvent • Can be limited by viscosity of spray solution (e.g., incomplete atomization or buildup) • Target droplet or SDD particle size
Droplet size	Output/ dependent	• Combination of all input variables

To expand upon the thermodynamic model output and begin to use quality-by-design (QbD) principles, a product-specific operating space can be defined for the spray drying process based on the thermodynamic constraints. An example follows.

The thermodynamic operating space for the spray drying process can be defined in a rational manner by understanding the impact of the key spray drying process variables on the process and SDD product. Although T_{out} is a dependent variable, it is known to have the greatest impact on drying rate and thus residual solvent content and SDD morphology. This is demonstrated by a typical response curve for an SDD formulation of bulk specific volume (BSV) to T_{out}, as shown in Figure 9.9a. In addition, scanning electron micrography (SEM) images (Figure 9.9b and c) for the extremes of the T_{out} range show the change in particle morphology that is observed across this range (Figure 9.9d and e). Lower T_{out} values result in higher SDD densities with a shriveled or buckled morphology, whereas at higher T_{out} values, low-density inflated spheres are observed [5,11].

Using relationships of the process space to the formulation attributes allows for straightforward selection of initial design space for the SDD process. Four different criteria are typically applied during this exercise.

1. *Maximum T_{in}*: Deposits of material near the gas disperser inlet to the drying chamber may be exposed to temperatures near the T_{in}. Chemical and physical stability or discoloration of the SDD should be evaluated to select a temperature that will minimize these process risks. In addition, active cooling of the gas disperser inlet can be implemented to reduce this risk (Niro DPH design).
2. *Process Efficiency*: M_{soln}, or specific drying ratio, is generally maximized within the operating space to achieve throughput goals for the product.

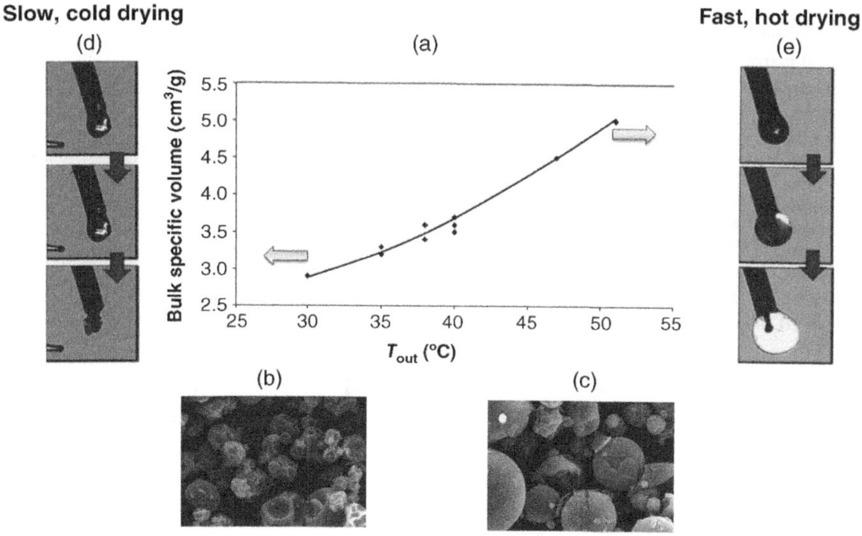

Figure 9.9. Particle property dependence on drying conditions. (a) Example SDD bulk specific volume (1/density) versus T_{out}. (b and c) SEM images of SDDs made by slow, cold drying and hot, fast drying, respectively. (d and e) High-speed camera image sequences of spray solution droplet suspended on a thermocouple, illustrating drying for slow, cold (25 °C) and hot, fast (65 °C) conditions, respectively.

3. *RS_{out}*: At drying conditions that result in high RS_{out}, residual solvent content in the SDD can be high. A constraint on RS_{out} must be set to avoid deposition of wet droplets in the drying chamber. RS_{out} can also be correlated to drying rate and thus SDD attributes such as SDD bulk density.
4. *Formulation*: Stability can be a key consideration for defining the maximum T_{out} of the spray drying process. Mapping T_g as a function of residual solvent can be used to determine if the formulation is at risk for physical instability. SDD bulk density may also define the upper limit of T_{out}.

Based on these criteria, a product-specific operating space can be determined, as seen in the example shown in Figure 9.10. Experimental runs can then be rationally selected to further refine the operating space. In addition, these constraints can be extended to scale-up methodology, which are discussed below in this chapter.

9.4.2.2 Atomization The atomization process is critical during the spray drying process as it provides particles with the high surface area necessary for rapid evaporation of the spray solvent during the drying process. In addition, the droplet size provides the process control point for the particle size of the SDD. Similar to how the thermodynamic parameters are defined during the process, the atomization process can be broken down into key process parameters. These are shown in Table 9.3.

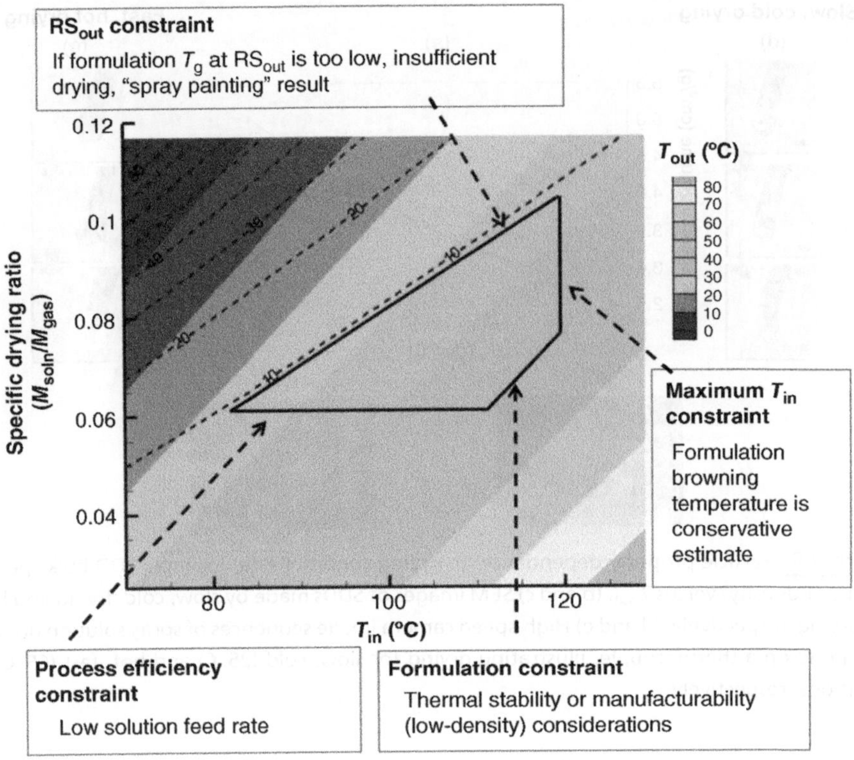

Figure 9.10. Example spray drying process thermodynamic operating space created for a specific SDD using key process definition constraints.

The droplet size of the atomized spray solution is generally proportional to the dried particle size of the SDD. The ideal relationship is based on a mass balance of the droplet drying to a perfect sphere of known density. This relationship is

$$d_{\text{powder}} = d_{\text{droplet}} \times \sqrt[3]{\text{solids fraction} \times (\rho_{\text{droplet}}/\rho_{\text{powder}})},$$

where d_{powder} is the diameter of the dried particle, d_{droplet} is the diameter of the initial droplet, ρ_{droplet} is the density of the spray solution, and ρ_{powder} is the density of an individual SDD particle. The relationship of the droplet size to particle size is generally more complex than this in practice. Formulation and drying history both have a significant impact on morphology of the particle, which in turn impacts the particle size distribution [5]. SDD particles typically have a hollow core that is formed during the drying process after solidification or skinning of the droplet. Depending on the final drying conditions, the solidified droplet can further expand, producing a hollow sphere or shriveled raisin morphology, as described above.

However, developing an understanding of the atomization process is routinely done through correlation of the key atomization process parameters to the final droplet size or final dried particle size. Khavkin and others have extensively researched and written about this for pressure-swirl atomizers [12]. These relationships are often derived experimentally, but all account for the fundamental aspects of solution properties, nozzle geometry, and atomization parameters (e.g., P_{nozzle} and M_{soln}).

Given the complexity of the atomization process and the wide range of formulations that are encountered during process development, experimental tools are commonly used to measure droplet size distributions as a function of atomizer geometries and operating conditions during process development. This can be particularly useful during scale-up where it is desired to maintain a similar droplet size distribution to achieve a similar particle size distribution. Solution properties for SDD formulations are typically dominated by the properties of the dispersion polymer (e.g., viscosity, density, and surface tension), so placebo (polymer-only) solutions can be used to avoid API consumption.

A phase Doppler particle analyzer (PDPA) (TSI Inc., Shoreview, MN) is commonly used to characterize sprays. The PDPA can be used as an off-line tool to understand the operating space and robustness of selected atomizers and to select the appropriate droplet size to meet the SDD particle size targets. An overview of the equipment configuration used for off-line nozzle characterization is shown in Figure 9.11. The PDPA also generates additional information that is useful during process development, including droplet velocity, volume flux, and overall droplet concentration.

Additional techniques that are commonly used during characterization of atomization performance during SDD manufacture include laser light scattering (e.g., Spraytec laser diffraction system, Malvern Instruments Ltd., Malvern, UK) and particle image velocimetry (e.g., Micro PIV laser systems, TSI Inc.) [13]. Each technique is well suited to generate key droplet data used during development or scale-up.

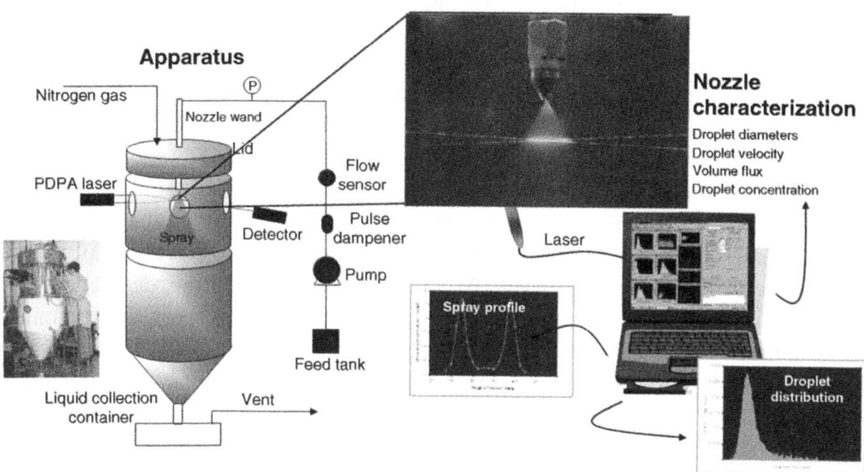

Figure 9.11. PDPA nozzle characterization set-up.

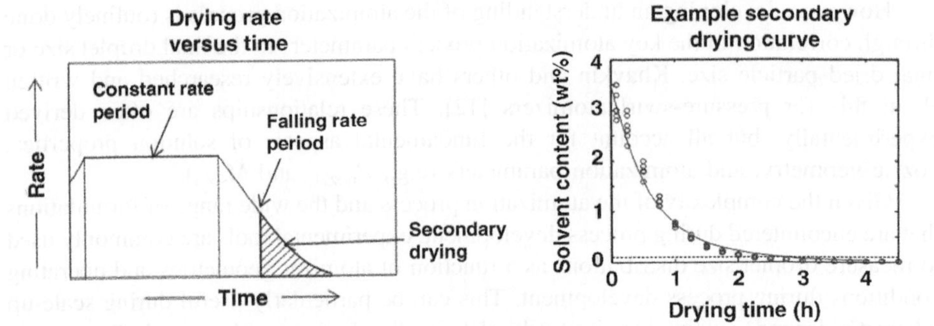

Figure 9.12. SDD secondary drying occurs in the falling rate drying period defined by internal diffusion limitations of solvent through the amorphous matrix.

9.4.2.3 Secondary Drying After spray drying, a secondary drying process is necessary to remove residual solvent from the spray-dried powder to meet the final residual solvent specification [14]. In addition, removal of residual solvent also helps maintain the physical and chemical stability of the SDD during storage.

In consideration of the classically defined solids drying periods [15], secondary drying of amorphous spray-dried powders can be classified as occurring wholly in the "falling rate" drying regime, as shown in Figure 9.12. Common unit operations used in secondary drying are tray dryers (vacuum or convection), agitated vacuum dryers such as the Ekato VPT (Ekato System GmbH, Schopfheim, Germany), and biconical rotary drum dryers such as the Pfaudler CDB (Pfaudler Inc., Rochester, NY). Other pharmaceutical unit operations that achieve driving force for heat and mass transfer leading to solvent removal may also be used. The pros and cons of each unit operation are summarized in Table 9.4.

Various mechanisms are used to achieve the driving force for solvent removal in secondary drying. These include heat, agitation, and reduction in partial pressure with vacuum and/or purge gas. The temperature of the process is selected to maximize the drying rate while maintaining the physical stability of the SDD. Early in the drying curve, the T_g is depressed by the residual solvent that is present; as the drying proceeds, the T_g increases to that of the dry powder. Vacuum level and purge rate are also key variables for vacuum dryers. Table 9.5 describes the common variables encountered in SDD secondary drying.

Removal of residual solvents from amorphous dispersions is generally considered a diffusion-limited process. Residual spray solvent, which is distributed throughout the amorphous matrix, must diffuse to the surface of the particle before it is transferred into the headspace of the secondary dryer. At small scale and when powder bed depth is small, the diffusion through the bed of powder can be ignored. However, during scale-up to larger equipment, the diffusion through the bed of powder may need to be considered.

As shown in Figure 9.13, agitation techniques are most commonly used to decrease the external mass transfer limitations during secondary drying. However, during use of tray dryers where the bed is fixed, this can be accomplished by limiting the overall bed depth to a height where external mass transfer resistance becomes negligible relative to single-particle diffusion kinetics.

TABLE 9.4. Common SDD Secondary Drying Unit Operations

Unit Operation	Advantages	Disadvantages
Tray drying	• Effective, broadly applicable approach at small/moderate scale • Control of temperature and humidity with gas stream • High surface area, control of bed depth	• Personnel exposure during charging/discharging • Limitations on use at large scale due to tray handling
Agitated vacuum dryer	• Contained charging/discharging • Mild fluidization minimizes powder bed mass transfer resistance, leading to short drying times • Easily scalable to large volumes	• Potential for particle attrition or compaction with friable particles
Biconical tumble dryer	• Contained charging/discharging • Easily scalable to large volumes	• Powder bed turnover not as efficient as agitated dryer, leading to longer drying times • Localized heating of wet powder at dryer wall due to lack of active mixing
Fluid bed dryer	• High heat and mass transfer rate	• Challenges with fluidization and product recovery of fine powders with broad particle size distribution

An additional technique that can be employed during secondary drying includes humidification of the drying gas stream. This serves two purposes: (1) to decrease the T_g, thus increasing molecular mobility and effectively increasing the diffusion coefficient; and (2) molecular replacement of solvents by water through hydrogen bonding interactions, typically with the drug molecule.

TABLE 9.5. Common Variables Defined in SDD Secondary Drying

Parameter	Variable Type	Defined by
Product temperature	Input/independent	• Desired cycle time • Physical stability constraints of SDD
Dryer RH	Input/independent	• Equipment limitations • SDD physical stability constraints
Agitation rate	Input/independent	• SDD attributes • Equipment limitations
Pressure	Input/independent	• Equipment limitations
Drying time	Output/dependent	• Input variables • SDD formulation

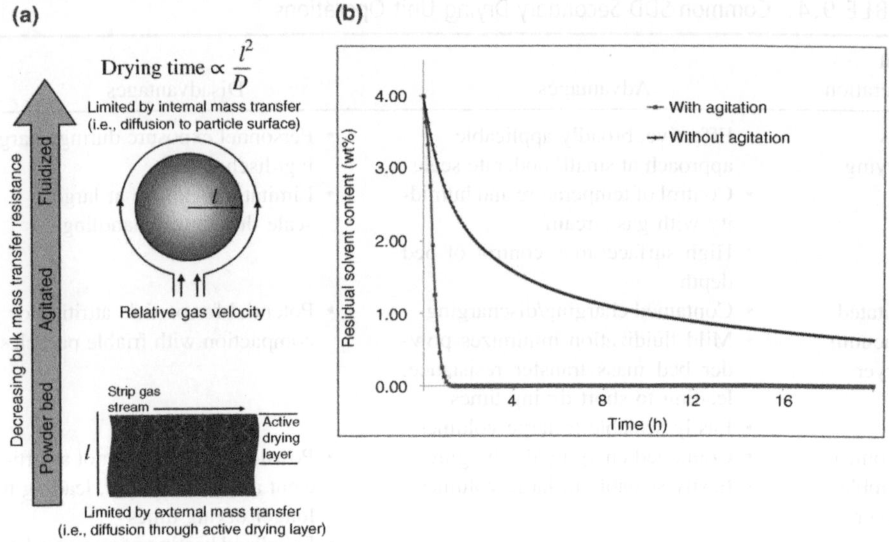

Figure 9.13. (a) Physical situation of SDD particle drying and limiting factors depending on mode ($l \approx$ particle radius; D = diffusivity of solvent vapor through amorphous SDD matrix). (b) Example SDD secondary drying curves with and without bed agitation (1000 l scale).

9.5 SPRAY DRYING SCALE-UP

The spray drying process is a continuous unit operation that allows flexibility during manufacturing to accommodate the desired throughput or batch size. It is readily scalable, accommodating formulation screening at milligram quantities to commercial-scale manufacture at metric ton quantities. Equivalent particle attributes are achievable across a wide range of spray dryer scales. Spray drying processes have been successfully scaled up in the food, pharmaceutical, and other industries with the precision to meet specific powder property targets. Masters cites an example spray dryer that produces nearly 10 tons of milk powder an hour with stringent requirements on the spray-dried powder [16].

Figure 9.14 shows typical SDD spray drying scales and qualitatively describes the overlap of the achievable product properties. It is important to keep scale-based differences in mind during process development at the laboratory scale in preparation for future scale-up or commercialization. However, by using a methodical approach to process definition and understanding the key changes that occur between scales, process parameters can be efficiently selected during scale-up while maintaining the CQAs of the SDD formulation across scales.

It is important to maintain consistency in equipment during scale-up for items such as the gas disperser, atomization technique (e.g., pressure-swirl nozzle, two-fluid nozzle), chamber aspect ratio, and cyclone efficiency. If the equipment is selected carefully and well understood, the SDD spray drying process can be efficiently scaled up by focusing on the differences across scales that impact the key physical situations of (1) droplet formation and (2) drying rate.

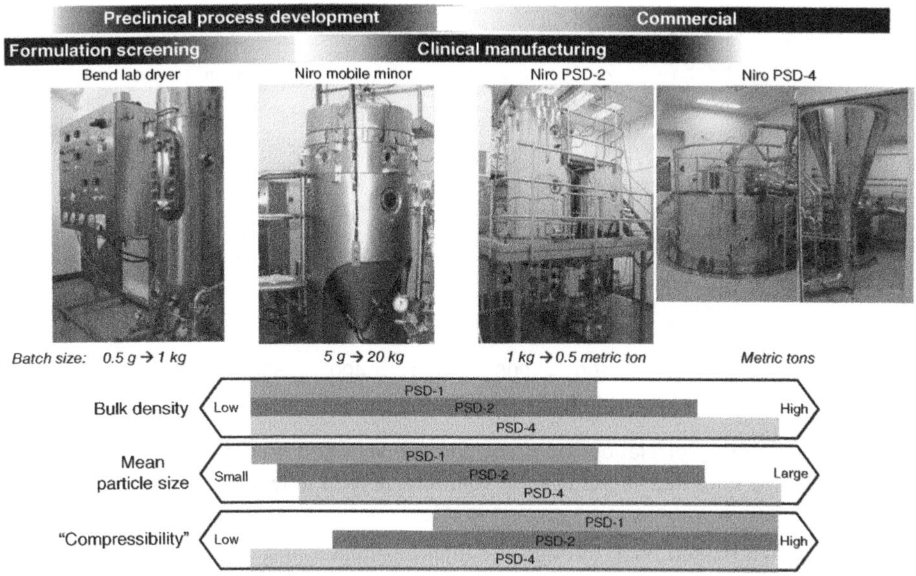

Figure 9.14. Overview of spray drying equipment scales and expected particle property control across scales. (PSD-4 picture courtesy of Hovione Inc.; other photos courtesy of Bend Research Inc.)

9.5.1 Scale-Up Considerations for Droplet Formation

As M_{soln} increases upon scale-up, the atomization process requires increasing energy input to obtain the same droplet size and subsequent particle size of the SDD. For atomization using a pressure-swirl nozzle, the increased energy is provided from the hydraulic pressure used for atomization. As the nozzle diameter increases to allow a higher M_{soln}, the applied pressure must increase to provide a similar droplet size.

Figure 9.15 shows an example correlation indicating the pressure required to maintain droplet size across a wider range of M_{soln} values. It is important to keep this in mind during early process development to ensure that reasonable particle size targets are selected that can be achieved at commercial scales. This is particularly important for SDDs that have particle size-dependent dissolution performance, which is common for compounds with high lipophilicity.

9.5.2 Scale-Up Considerations for Droplet Drying Rate

There are two important considerations that impact the droplet drying rate upon scale-up: (1) the introduction of solvent vapor in the drying gas and (2) increased droplet density or flux in the spray plume.

To minimize the volume requirements for inert drying gas (most commonly, nitrogen), it is typical to recycle the drying gas in large-scale spray dryers. As shown in Figure 9.2, after passing through the cyclone and/or baghouse, the drying gas is conveyed by fans through a condenser to remove the majority of the solvent vapor. The

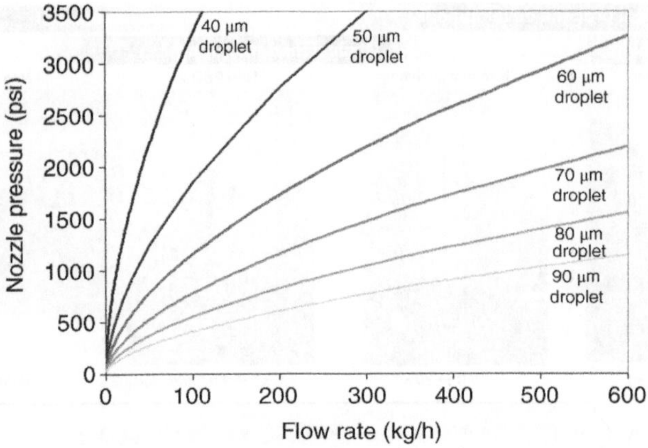

Figure 9.15. Example correlation for predicting the required nozzle pressure to maintain target droplet size for a high-viscosity spray solution (12 wt% polymer in acetone).

drying gas is then passed back through the inlet gas heater and into the drying chamber through the gas disperser. Recycling the drying gas is referred to as "closed cycle" operation, whereas smaller scale spray dryers without recycle operate in "single-pass" operation (i.e., without the condenser loop). Although the purpose of the condenser is to remove solvent from the exhaust vapor in closed-cycle operation, the recycled drying gas will contain significant solvent vapor. The amount will be the concentration in equilibrium at T_{cond}. This recycled solvent vapor can result in slower droplet drying rates and higher residual solvent contents relative to smaller open-cycle spray dryers. Slower drying rates typically result in denser SDD particles with raisin-like morphology. If the formulation is sensitive to drying rate and solvent content, this could also impact the physical state of the SDD since the API will have higher mobility in the SDD and thus increase the potential for amorphous phase separation or inhomogeneity.

The effect of this solvent vapor can easily be mitigated by adjusting drying parameters (e.g., T_{in}, M_{gas}, and M_{soln}) and T_{cond}, which controls the quantity of recycled solvent vapor in the recycled gas stream. The recycled solvent vapor can be accounted for in heat and mass balance equations, and thermodynamic models can be developed to predict target drying parameters at larger scales.

A second scale-up consideration that impacts droplet drying rate is the droplet and drying gas mixing efficiency. Since typical nozzles have spray angles of less than 90° and are essentially a "point source" of droplet formation, the flux of droplets in the key control volume (near the exit of the atomizer) increases proportionally to the spray rate. Higher M_{soln} in similarly sized control volumes leads to denser spray plumes upon scale-up. The denser plumes are an added resistance to mass and energy transfer and slow down droplet drying. The effects of the denser plume can also be easily mitigated by adjusting drying parameters (e.g., T_{in}, M_{gas}, M_{soln}, and T_{cond}). The effect of denser droplet plumes can be modeled with CFD, and scale correlations can be developed to predict target drying parameters at larger scales. It is important to consider the recycled solvent vapor and denser droplet plumes early in development to evaluate the impact of slower droplet drying on the physical stability of the SDD formulation. Slower drying rates can lead to SDDs with higher densities, which can be desirable from a powder handling perspective. For these reasons, it can be advantageous to conduct early development work on small-scale closed-cycle spray dryers rather than the more commonly available small-scale open-cycle spray dryers.

9.5.3 Scale-Up Methodology

The goal of scale-up is often to ensure that the SDD particle attributes (e.g., particle size, density, physical state, and morphology) remain consistent while achieving the necessary throughput to meet clinical and/or commercial demands. This can be achieved by matching droplet size and drying rate during scale-up. A methodology similar to the one described in Section 9.4 can be employed.

Correlations can be developed to predict droplet size based on spray solution attributes (e.g., viscosity), nozzle geometry, and atomization parameters (e.g., P_{nozzle}) at the larger scale spray solution throughputs. As described above, the PDPA can be used to characterize the droplet size distributions produced by the larger scale atomizers. By using atomization correlations and/or measuring the droplet size distribution, an atomizer can be selected that produces the same droplet size distribution as that produced by the smaller scale nozzles.

A thermodynamic mass and energy balance model can be developed, taking into account T_{cond} and the recycled solvent vapor. The thermodynamic plot can be extended from two dimensions to three and can be used to show all six key process parameters described in Table 9.2. Example three-dimensional thermodynamic plots are shown in Figure 9.16. The thermodynamic operating space can be defined using the same methodology described in Section 9.4. In addition to the T_{in}, process efficiency, RS_{out}, and formulation constraints described in the previous section, there may be equipment

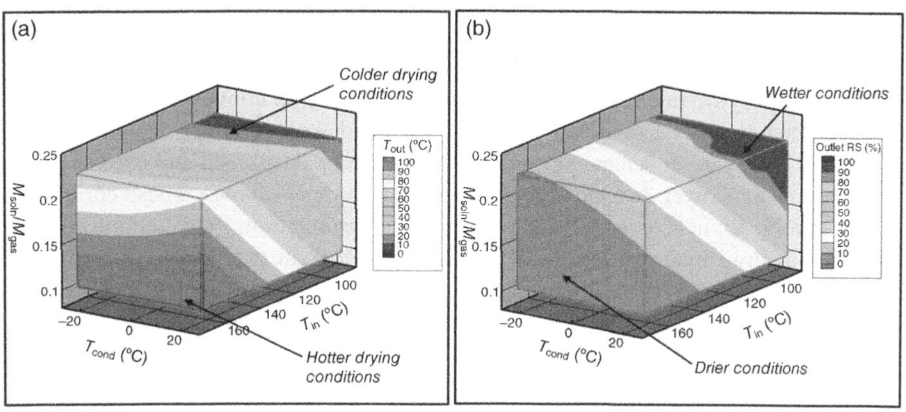

Figure 9.16. Three-dimensional thermodynamic plots showing the six key parameters including T_{cond}, which is important for larger scale dryers. (a) Contours of constant T_{out} and (b) contours of constant RS_{out}.

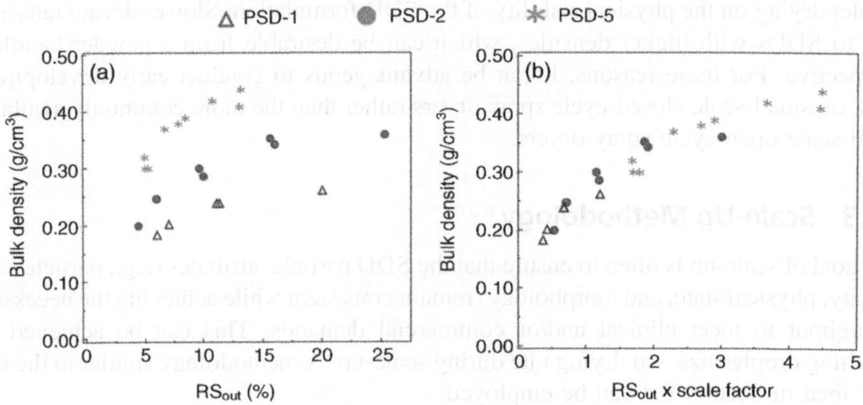

Figure 9.17. Use of scale factor to describe effect of higher droplet in the spray plume for larger scale spray dryers, resulting in a multiscale correlation. (a) Correlation of bulk density versus RS_{out} for three spray dryer scales and (b) multiscale correlation of bulk density versus RS_{out} using scale factor.

limitations dictated by plant utilities (e.g., maximum achievable T_{in} or minimum achievable T_{cond}). These constraints can be depicted as planes in the three-dimensional plots similar to the constraints shown as lines in Figure 9.16 to guide the identification of target spray drying conditions and to illustrate the operating space.

As described previously, correlations with T_{out}, RS_{out}, or other combinations of key process parameters can be used to predict the important SDD attributes. An example correlation between RS_{out} and SDD bulk density is shown in Figure 9.17a for three spray dryer scales. All of these spray dryers were operated in closed-cycle mode and the target median droplet size was the same for all of the conditions shown. The three different spray dryer scales have separate correlations with SDD density. SDD manufactured with the same RS_{out} results in denser SDD on the larger PSD-5 spray dryer than on the smaller PSD-1 spray dryer. This illustrates the scale effect of the denser droplet plume and thus slower droplet drying rate resulting in denser SDD described above. The droplet-to-gas mixing efficiency can be modeled by CFD or studied with carefully designed experiments across spray dryer scales. Scale correlations can be developed and used to predict the impact of the slower droplet drying due to denser spray plumes. These correlations can be widely applicable to SDDs with similar formulations (e.g., cellulose-based polymer with drug loadings less than 50%). The use of an example scale correlation is shown in Figure 9.17b. A dimensionless Damköhler number, which is the ratio of reaction rate to convective mixing rate, can be adopted [17]. In this case, the Damköhler number is defined as the ratio of evaporation rate to convective droplet and gas mixing and a simplified equation is shown below.

$$Da = \frac{L_{\text{diffusion}}^2/D}{L_{\text{mixing}}/u} \propto \frac{L_{\text{mixing}}}{u}.$$

The characteristic droplet and drying gas mixing length (L_{mixing}) can be determined experimentally or by evaluating the output of CFD models. Different spray dryer scales and conditions can be modeled and the length of the spray plume can be used as the characteristic mixing length for that scale and spray drying condition. Since evaporation rate is dominated by diffusion (D) of the solvent through the skin and in this example the formulation and droplet size ($L_{diffusion}$) is held constant, the Damköhler number simplifies to the L_{mixing} divided by the drying gas velocity (u). The modified Damköhler number can be used to derive the scale factor shown in Figure 9.17b. In this way, the scale factor can be used to predict target spray drying parameters at each scale and ensure that droplet drying rate is maintained upon scale-up.

9.6 CONCLUSIONS

This chapter has described the physical situation, equipment considerations, process development considerations and approaches, and methodology for scale-up of the process for manufacture of SDDs. In each of these sections, common challenges and science-based mitigation approaches such as modeling and experimental strategies have been suggested. While spray drying processes can be rationally developed to avoid most common pitfalls, the importance of coupled consideration of process- and formulation-related parameters cannot be overstated. Formulation performance mechanism (e.g., disintegration or erosion), physical and chemical stability risks (e.g., crystallization and degradation), and dosage form requirements (e.g., drug loading and dose) should be considered as integral parts of process design.

When considering the evolution of process definition across the development cycle, it is possible to rationally reduce the numerous degrees of freedom that seem to exist early on down to a well-defined operating space based on combined risk assessment of performance, stability, manufacturability, and patient considerations. This approach is at the heart of the current ICH Q8(R2) guidance [18]. Figure 9.18 graphically describes an

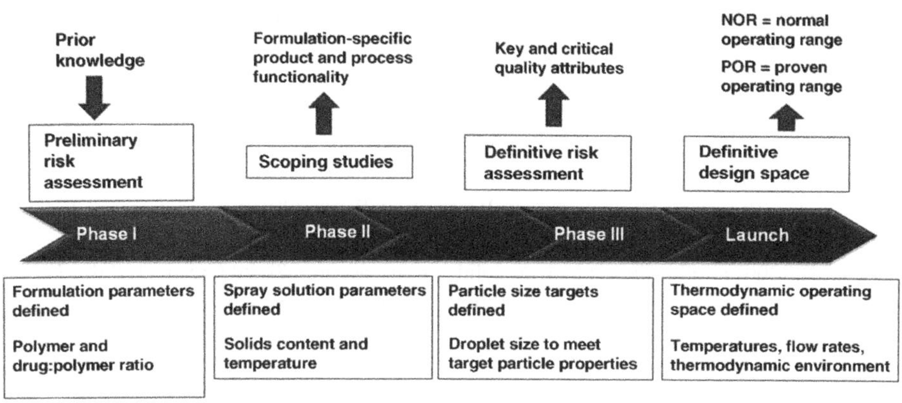

Figure 9.18. Example QbD design space evolution throughout the development cycle, leading to a design space for filing of a spray drying process.

example evolution of the spray drying process considerations, ultimately leading to a rationally constrained, but appropriately flexible process that can stand up to contemporary regulatory scrutiny and play a key role in enabling delivery of a quality product to the patient.

REFERENCES

1. Friesen, D., Shanker, R., Crew, M., Smithey, D., Curatolo, W., and Nightingale, J. (2008) Hydroxypropyl methylcellulose acetate succinate-based spray-dried dispersions: an overview. *Mol. Pharm.*, 5(6):1003–1019.
2. Paudel, A., Worku, Z., Meeus, J., Guns, S., and Van den Mooter, G. (2013) Manufacturing of solid dispersion of poorly water soluble drugs by spray drying: formulation and process considerations. *Int. J. Pharm.*, 453:253–284.
3. Smithey, D., Gao, P., and Taylor, L. (2013) Amorphous solid dispersions: an enabling formulation technology for oral delivery of poorly water soluble drugs. *AAPS Newsmagazine*, January.
4. Dobry, D., Settell, D., Baumann, J., Ray, R., Graham, L., and Beyerinck, R. (2009) A model-based methodology for spray-drying process development. *J. Pharm. Innov.*, 4(3):133–142.
5. Vehring, R. (2008) Pharmaceutical particle engineering via spray drying. *Pharm. Res.*, 25(5):999–1022.
6. Vehring, R., Foss, W., and Lechuga-Ballesteros, D. (2007) Particle formation in spray drying. *J. Aerosol Sci.*, 38(7):728–746.
7. Lefebvre, A. (1989) *Atomization and Sprays*, Taylor & Francis, Bristol, PA.
8. Masters, K. (1991) *Spray-Drying Handbook*, 5th edition, Longman–Wiley, New York.
9. Graham, L., Taillon, R., Mullin, J., and Wigle, T. (2010) Pharmaceutical process/equipment design methodology case study: cyclone design to optimize spray-dried-particle collection efficiency. *Comput. Chem. Eng.*, 34(7):1041–1048.
10. U.S. Food and Drug Administration, Center for Drug Evaluation and Research (CDER) (2009) *Guidance for Industry. Residual Solvents in Drug Products Marketed in the United States.* U.S. FDA, Rockville, MD.
11. Walton, D. (2002) Spray-dried particle morphologies. *Dev. Chem. Eng. Miner. Process.*, 10(3/4):323–348.
12. Khavkin, Y. (2004) *The Theory and Practice of Swirl Atomizers*, Taylor & Francis.
13. Thybo, P. and Hovgaard, L. (2008) Droplet size measurements for spray dryer scale-up. *Pharm. Dev. Technol.*, 13:93–104.
14. U.S. Food and Drug Administration, Center for Drug Evaluation and Research (CDER), Center for Biologics Evaluation and Research (CBER) (2012) *Guidance for Industry. Q3C—Tables and List, Revision 2.* U.S. FDA, Rockville, MD.
15. Perry, R. (2007) Solids drying and gas solid system, Section 20. In: Green, D. (Ed.), *Perry's Chemical Engineers' Handbook*, 8th edition, McGraw-Hill, New York.
16. Masters, K. (1994) Scale up of spray dryers. *Drying Technol.*, 12(1–2):235–257.
17. Fogler, S.H. (2006) *Element of Chemical Reaction Engineering*, 4th edition, Prentice Hall.
18. International Conference on Harmonisation of Technical Requirements for Registration of Pharmaceuticals for Human Use (2009) *ICH Harmonised Tripartite Guideline: Pharmaceutical Development Q8(R2), Current Step 4 version.* U.S. FDA, Rockville, MD.

10

HOT MELT EXTRUSION OF AMORPHOUS SOLID DISPERSIONS

Kieran Crowley[1] and Andreas Gryczke[2]

[1]*Quotient Clinical Ltd., Nottingham, U.K.*
[2]*Global Development and Technical Marketing Solubilisation, BASF SE, Ludwigshafen, Germany*

10.1 INTRODUCTION

The term extrusion describes a process of force feeding a material through an opening of well-defined dimension to form an extrudate with modified physical and possibly chemical properties. Hot melt extrusion (HME) is a process that uses rotating screws to mix and feed materials at elevated temperatures through a die opening(s) followed by downstream cooling to produce solidified phase. The typical requirement for HME is that at least one component, usually polymer, is heated to significantly above its thermodynamic glass transition temperature so that the necessary softening of materials for extrusion takes place. The opportunity to form drug in polymer solid dispersions with HME is obvious, as much theory defining pharmaceutical solid dispersions centers on physical and chemical properties above and below the glass transition temperature [1,2].

During the hot melt extrusion process, a series of well-controlled events must occur to ensure that solid dispersion phase is generated with consistent phase structure and high chemical purity. These process events include some or all of the following:

Pharmaceutical Amorphous Solid Dispersions, First Edition. Edited by Ann Newman.

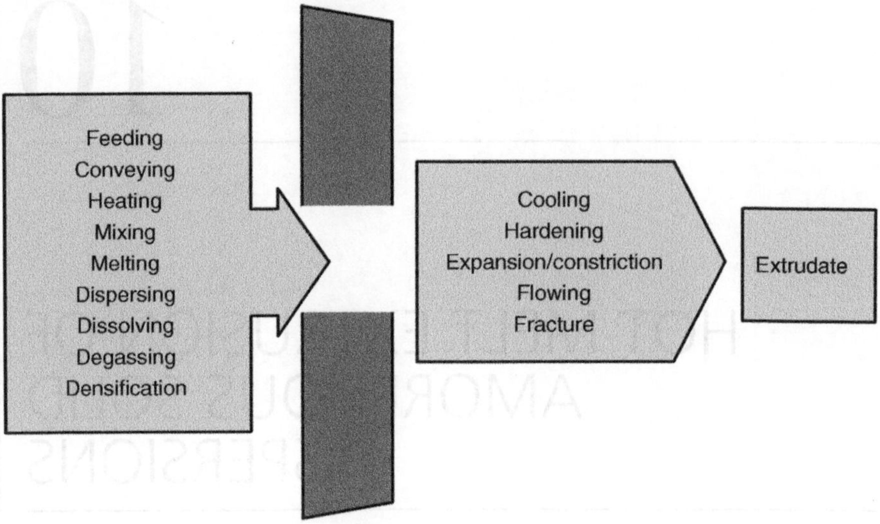

Figure 10.1. Physical, chemical, and mechanical events that may occur during the hot melt extrusion process, displayed as before and after extrudate passes through a die opening.

feeding, conveying, heating, mixing, dispersing, dissolving, homogenizing, degassing densification, and discharging (see Figure 10.1). The main material and equipment factors contributing to these events are detailed later in this chapter.

HME is proven as a commercially viable technology for solid dispersion dosage forms. From bench to commercial scale, the equipment uses essentially the same operating principles and so in-process data collected at small scale allows the development scientist to approach scale-up in a data-driven mode. Sections 10.4 and 10.5 will provide details on these topics.

Prior to an HME product development project, consider the following three questions:

- What is the target unit dose and dosage form?
- What phase structure is required (or preferred) in the solid dispersion?
- What preformulation data do I have to support selection of materials and process conditions for hot melt extrusion feasibility trials?

Analysis of these questions should allow some development targets to be specified and a series of feasibility experiments to be designed. A key consideration is that HME always generates a solid dispersion intermediate (extrudate) that must undergo downstream processing to form the finished dosage form. The topic of formulating the solid dispersion into a finished dosage form is addressed in a separate chapter, but it is implicit that the design of an HME formulation and process occurs "with the end in mind."

10.2 MATERIALS SELECTION

All materials used in a hot melt extrusion process have high potential to be intimately mixed, often forming new molecular interactions between active pharmaceutical ingredient (API)–excipient and excipient–excipient and eliminating the molecular interactions that existed in the single-component states of each material. Each component may significantly impact the properties of an HME finished product, even if present at low concentration. This feature creates both challenges and opportunities for the formulator using HME. Irrespective of the drug delivery application of an HME formula or the excipients used, robust characterization of the API–excipient and possibly excipient–excipient interactions in the final product is important.

10.2.1 Polymers

It is typical to initiate HME formulation development in a stepwise manner, beginning with drug/polymer binary mixtures before advancing to multicomponent mixtures. Many polymeric excipients are available for inclusion in hot melt extrusion processes. For solid dispersions of poorly soluble drugs, a polymer with some hydrophilic properties is desired along with the capability to form intermolecular interactions with the drug. Ideally, preformulation data will be available to screen candidate polymers and identify candidates that exhibit some miscibility with the drug molecule and have the opportunity to maintain a physically stable dispersed phase.

The most important physical property of polymers used in HME is melt viscosity, so that the heat-induced (i.e., thermoplastic behavior) and possibly shear-induced (i.e., thixotropic behavior) liquid flow is sufficient to provide controlled movement of material through the extruder. Many polymers meet these requirements, either alone or with plasticizer. Crowley et al. [3] provide extensive overview of polymers used for hot melt extrusion applications. If seen from the industrial viewpoint for products formulated using hot melt extrusion or solid dispersions in general, it can be concluded that most prominent polymers are as follows:

1. Copovidone (Kollidon® VA 64 or Plasdone® S-630)
2. Povidone (Kollidon 12/17 or Plasdone K-12/K-17)
3. HPMC-AS
4. Polyethylene oxide (e.g., Polyox®)
5. Polyethylene glycol
6. HPMC-E5
7. Methacrylates (e.g., EUDRAGIT® E)

Besides these polymers used in industrial applications, a large variety of other polymers are suitable for hot melt extrusion applications. BASF launched in 2009 under the brand Soluplus® the first polymer developed specifically for hot melt extrusion applications. Soluplus is a polyvinyl caprolactam–polyvinyl acetate–polyethylene glycol graft copolymer. Due to its amphiphilic nature, it is a polymeric emulsifier that is suitable for obtaining glassy solid solutions with APIs.

Once lead polymer(s) are selected, consideration must be given to processing temperatures. It is important to pay particular attention to the product (i.e., dispersion) melt temperature, and that there can be a significant delta between the extruder's barrel temperature and the product temperature in the extruder. The following are the other key process considerations:

1. Selecting process temperatures at which polymer has reduced viscosity but avoids chemical degradation.
2. Selecting process temperatures at which API can melt and/or solubilize in polymer, but avoid chemical degradation.
3. Selecting process temperatures at which the drug–polymer dispersion can be extruded to yield a processable solid upon cooling.

The melting temperatures and glass transition temperatures of drug and polymer support the selection of process temperature. It is common to target process temperatures 20–50 K above polymer T_g. Depending on drug loading in the polymer and its solubility rate in molten polymer, the drug does not need to exceed its melting temperature for successful solid dispersion formation by HME. If a dispersion forms in the extruder, the miscible drug may have a large effect on dispersion T_g. Predictions of dispersion T_g should be made using approaches such as in Ref. [4] to support selection of suitable process temperatures (see Chapter 9 for description of T_g prediction theory and methods). Often the T_g of the amorphous active ingredient is lower than the polymer T_g, producing a plasticizing effect that creates the risk of extrudate lacking the targeted room-temperature physical properties for simple downstream processing. For example, a lack of brittle behavior at room temperature may prevent preparation of a granule or pellet-size extrudate as the material is too soft for milling or cutting operations.

Polymer selection for HME formulation feasibility represents the initial step toward designing a viable solid dispersion formulation. It provides the basis for a viable HME process and a carrier excipient that can stabilize drug in the targeted dispersed state. Additional components will likely be introduced into the extrudate formula to optimize the manufacturing conditions and the physical and chemical properties of the extruded material.

10.2.2 Plasticizers

Plasticizers may be used to aid the HME process by lowering thermal transition temperatures of individual components and the extruded product. Polymers with high T_g values or a complex temperature–viscosity relationship, for example, HPMC polymers, are common candidates for plasticization. To function, the plasticizer must melt at the target process temperatures and be miscible with at least one component of the formulation. Common plasticizers include low and medium molecular weight polyethylene glycols, poloxamers, and triethyl citrate.

Plasticizer may be directly fed as part of the preblend or injected through liquid feed port. Alternatively, plasticizer may be preprocessed with the polymer excipient to lower processing temperatures for extrusion with the active component [5].

10.2.3 Dissolution Adjuvants

The general term dissolution adjuvant is used to define an additive that alters dissolution via a mechanism other than the thermodynamic solubility advantage produced by the amorphous dispersion. A common approach is inclusion of a surfactant component within the dispersion (i.e., intragranular) to enhance the wetting, disintegration, and dissolution of solid dispersed drug into solution. Surfactants commonly used include vitamin E polyethylene glycol succinate, poloxamer, polysorbate 80, and polyoxyl castor oil. Other reasons for dissolution adjuvants are to extend the duration that supersaturation is sustained or to delay drug release, typically using a polymeric third component.

10.2.4 Other Additives

Chemical stability enhancement of active ingredient and/or polymer matrix may be achieved with inclusion of antioxidants or other chemical stabilizers in the HME process. The process may generate molecularly dispersed stabilizer within the matrix that maximizes the stability enhancement effect of the additive.

10.2.5 General

The grade of excipient is a critical detail for both processing and formulation reasons. The first consideration is ensuring reliable flow properties of materials fed to the extruder. If feeding blends of materials, it is common to avoid major differences in particle size distribution of each blend component so that good blend homogeneity is obtained. Waxy or semisolid excipients may be used that need pre-extrusion processing to generate a flowable powder form. Liquid feeding may be considered if the liquid excipient or mixture has viscosity that permits pumping with simple apparatus. This configuration is common in food industry applications, for instance, introducing water into carbohydrate and protein-based systems to form extruded product with specific texture (e.g., chewable materials or puffed high-porosity solids) and color attributes caused by controlled in-process chemical reaction [6–8].

Another excipient consideration is content of volatile component(s). Most pharmaceutical excipients contain residual water and possibly solvent. Depending on equipment configuration and process conditions, volatiles may be reduced in the extrudate product during processing. Such change will require investigation if in-process stability is a concern, or if the volatile content in extrudate is a critical material property. While the authors do not consider this common practice, a preprocessing drying step can be used to reduce the amount of volatiles in a feed material. Such action should not be necessary in most cases, as the correct HME design should provide opportunity for in-process volatile removal.

10.3 EQUIPMENT SELECTION

HME is a well-established technology in the plastics, metals, and food industries. Pharmaceutical application of HME has lagged behind these industries, until the past 15 years that have seen intense pharmaceutical industry activity. The opportunity to form solid dispersions by HME is a primary reason for the increased interest in the technology, but another factor in driving pharmaceutical industry growth relates to HME processing equipment. Historically, some barriers to HME use included inadequate laboratory-scale extruders for performing small-scale R&D formulation and "perceived" manufacturing and quality risks from the high-temperature/high-wearing equipment. These concerns are no longer significant barriers, with recent approvals of commercial oral and parenteral products using HME processing supporting this conclusion.

A general schematic of a hot melt extruder is provided in Figure 10.2, accompanied by the commonly targeted before- and after-state of the materials processed by HME. As outlined below, there are three separate items of processing equipment functioning during hot melt extrusion: feeder, extruder, downstream process. Equipment is readily available that meets CGMP standards, with an example of a CGMP extruder shown in Figure 10.3.

10.3.1 Feeder

The feeding operation delivers raw material or a preblend into the extruder at a controlled rate and with controlled composition. Feeders are often manufactured and supported by specialist equipment vendors (e.g., K-Tron, Switzerland; Brabender Technologies, Germany; and Schenck Process, Germany) and either integrated with the extruder control panel or operated independently, as preferred. Exceptions are some small-scale extruders, including Thermo Fishers 11 mm model or the Leistritz Nano16 that have highly integrated feed mechanisms. For powdered material, the simplest option is a volumetric feeder, designed with a variable-speed auger system to transfer material from the hopper via a feed chute to an open port on the extruder. The feed rate may be correlated with the auger speed (rpm) setting before batch manufacture, with the assumption that the feed rate–rpm relationship remains constant thereafter. The feed

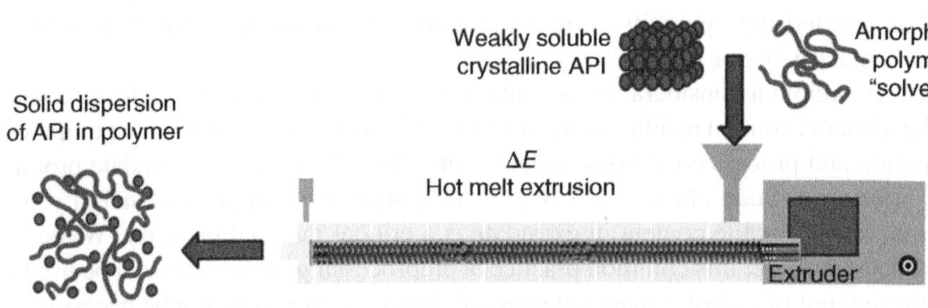

Figure 10.2. Hot melt extrusion process schematic accompanied by before- and after-physical properties for amorphous solid dispersion applications.

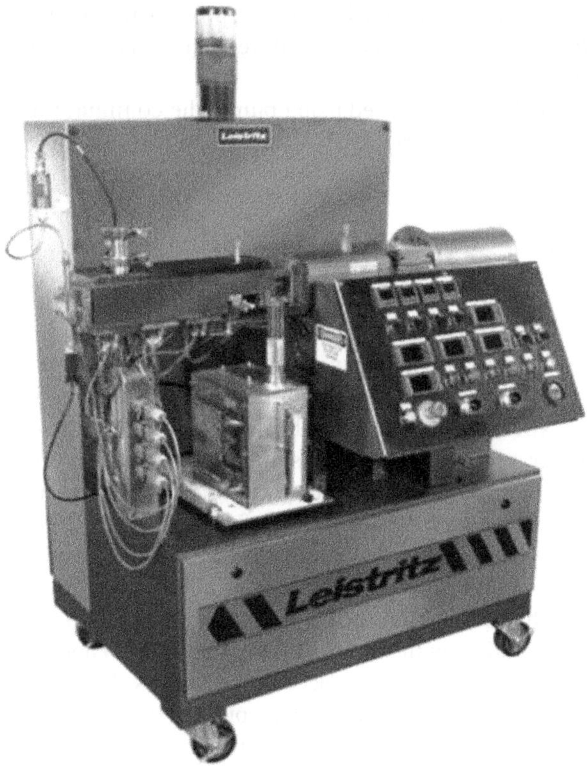

Figure 10.3. Image of a 16 mm corotating twin-screw extruder (model Nano-16). Reproduced with permission from American Leistritz Extruder Corp., USA.

rate–rpm relationship may not be constant, for instance, when feed material densifies causing higher bulk density in the lower section of the feeding vessel with lower bulk density in higher regions of the feeding hopper. Variable auger screw designs allow process optimization to be performed for a volumetric feeder, the requirement being that consistent flow must be achieved. The changing vertical height of powder in the hopper can affect the output of the volumetric feeder.

A loss-in-weight feeder is set to deliver material at a measured "mass per unit time" rate with real-time feed rate control. The hopper and feeding module are mounted on a load cell that measures weight at frequent intervals. The measured mass flow data are used by the controller to adjust the motor speed (or vibratory tray for some units) as necessary to maintain mass flow within the control limits.

When performing extended processing, recharging of the feed hopper is required. The action of recharging the hopper creates simultaneous weight loss and weight gain that will confound the controller input data. Loss-in-weight feeders manage this need with a function to enter a temporary feeding mode that uses the historical mass flow data to continue feeding at a predicted rate. Once charging is complete, real-time control can be reactivated.

Multiple feeders may be employed to deliver powder into one or more feed ports, which provides flexibility for the process if feed materials of different particulate properties or quantities are used.

Liquid feeding in HME is achieved using pump–die configurations on the extruder, and similar choices of volumetric or loss-in-weight equipment are available. For handling waxy materials, heatable liquid feeding systems are available.

10.3.2 Extruder

A hot melt extruder consists of a temperature-controlled barrel within which rotating screw(s) convey feed material to a die. The size of an extruder is defined by screw diameter. Three-Tec (Germany) offers very small-scale twin-screw extruders starting from a screw diameter of 5 mm. In early formulation development and screening, a new class of 10–12 mm extruders have been established with the pharmaceutical industry in mind. These extruder provide geometric similarity to their larger counterparts for properties such as outer-to-inner screw diameter ratio. These extruders are usually operated with feed rates between 100 g/h and 2 kg/h. The next equipment sizes for formulation development are the 16–18 mm extruders that are operated with feed rates usually up to 12 kg/h. Supporting pilot-scale or smaller production manufacturing are extruders with screw diameter between 24 and 27 mm, handling up to 40 kg/h depending on the formulation. Extruders with screw diameter between 30 and 50 mm are common for commercial manufacturing of pharmaceutical products. For reference, images are provided of two extruders commonly used in pharma for formulation development (Figure 10.3) and process development (Figure 10.4).

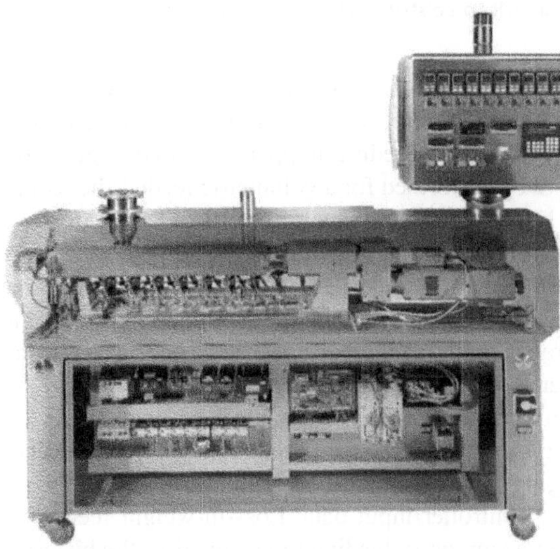

Figure 10.4. Image of an 18 mm CGMP corotating twin-screw extruder (model ZSE-18). Reproduced with permission from American Leistritz Extruder Corp., USA.

Screw length, expressed as the screw length:diameter ratio, typically ranges from 15:1 up to 40:1. Usually the extruder is operated not at the maximum possible feed rate (volume limited) but at an adequate feed rate and process temperature that results in both moderate torque (which is proportional to the applied shear stress) and moderate mean residence time. The mean residence time strongly correlates with the feed rate and hence a too high feed rate results in a too low mean residence time and poor axial mixing.

The equipment is built and operated in a modular manner, with multiple temperature-controlled blocks assembled along the length of the screw shaft to form the barrel of the extruder. Each modular block has heating and cooling control that allows different temperature set points, referred to as heating zones, to be maintained along the barrel. In an optimized extrusion process, the majority of process heat generated in the heating zones is from frictional forces between feed materials, the rotating screw surfaces, and the barrel walls [9].

Durability of extrusion equipment depends upon the process conditions applied as well as on the level of owner care during operation, cleaning, and storage of product contact parts. Equipment vendors provide barrel and screw components in different grades of stainless steel that meet high specifications against abrasion and corrosion [10].

Single- or twin-screw extruders are available, with twin-screw designs offering counterrotating or corotating operation. The conveying and mixing dynamics within these different designs differ profoundly, as well as material output. Corotating twin-screw extruders are most commonly used for pharmaceutical HME, providing a high degree of material mixing within the extruder that aligns with the typical process goal of material melting, dissolution, and extensive molecular interaction of the components. An additional advantage of the corotating design is high output rate coupled with relatively low residence time distribution within the extruder.

The rotating screw consists of multiple screw elements with geometric variables such as screw pitch and flight (Figure 10.5), the sequence and proportions of which are defined as the *screw configuration*. Different screw geometries provide separate regions for material conveying in the forward direction, material conveying in the reverse direction, material kneading and mixing of varying intensity, and material compacting. The best practice is to design the screw configuration based on the material properties (both feed materials and extrudate) being processed during formulation development. As such, it is rare for the scientific literature or an equipment vendor to specify default screw configurations. Information is provided in Table 10.1 and in the following text to support screw configuration definition and selection:

First Region (Feeding and Transport): Conveying screw elements are required at the feeding zone and beyond to provide opportunity for ambient temperature feed material to become distributed onto the screw surfaces in powder form and begin heating toward softening temperatures. Without correct conditions for material conveying along the extruder, risks include partial fluidizing and backflow of powder or sticking of materials in the feeding zone. These scenarios could desegregate the feed material, or increase the torque requirement for the screw shafts.

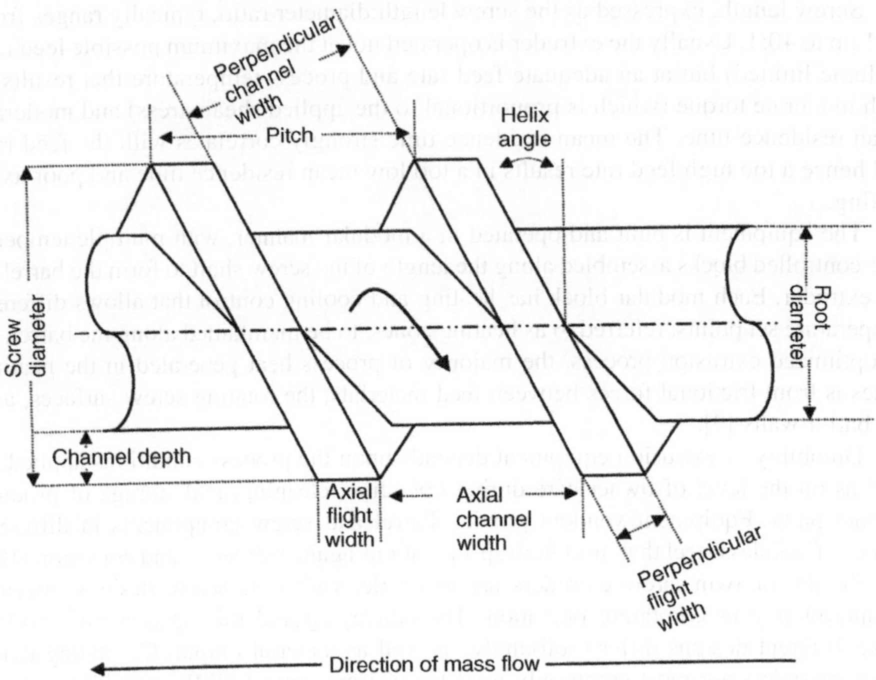

Figure 10.5. Design and geometry descriptions of a conveying screw. Reproduced with permission from Breitenbach [9].

Second Region (High-Shear Mixing and Transport): At sufficient barrel length after which powder feeding and liquid feeding (if applicable) are complete and partial melting is anticipated, one or more kneading zones may be introduced. A kneading zone will generate localized torque on the screw shaft, high-shear forces between feed material and equipment surfaces, causing intense melting and mixing of materials. Different kneading screw elements are available that affect material dynamics, from forwarding (30 or 60° offset) to neutral (90° offset) to reversing (reverse 30° offset). The kneading element offset also affects the degree of shear, with larger offset angles (forwarding or reversing) providing greater shear.

More than one kneading zone may be included in the second barrel region, separated by some conveying screw elements. Such a design may be selected for multiple reasons, for example, to ensure that the maximum torque for the extruder equipment is not reached, to reduce the material residence time in the extruder, or to minimize the risk of localized overheating of materials in the kneading zone due to too high local shear. The contribution of kneading elements in producing drug–polymer solid dispersion was demonstrated by Nakamichi et al. by comparing solid-state and dissolution properties of materials made with and without kneading elements in the screw design [11]. Using a clamshell design

TABLE 10.1. Tabulated Guide to the Relative Importance of Different Process Parameters and Material Parameters versus Different Screw Element Designs

Process/Product Property	Solid feeding	Melting	Melt Conveying	Dispersing	Melt Conveying (Optimal Degassing)	Homogenizing	Degassing	Discharging
Screw speed	X	X	X	X	X	X	X	X
Throughput	X	X	X	X	X	X	X	X
Free screw volume	X						X	
Pitch	X		X	X	X			X
Bulk Density	X							
Friction between product and barrel wall	X							
Viscosity				X				
Screw geometry		X				X		
Heat flux		X						
Specific heat capacity		X						
Heat conduction		X						
Melting enthalpy		X						
Particle size		X						
Vacuum							X	
Gap between screw and barrel								X

X indicates that the process/material parameter is critical in relation to the corresponding screw element. Coperion GmbH, Stuttgart, is acknowledged for contributing information to this table.

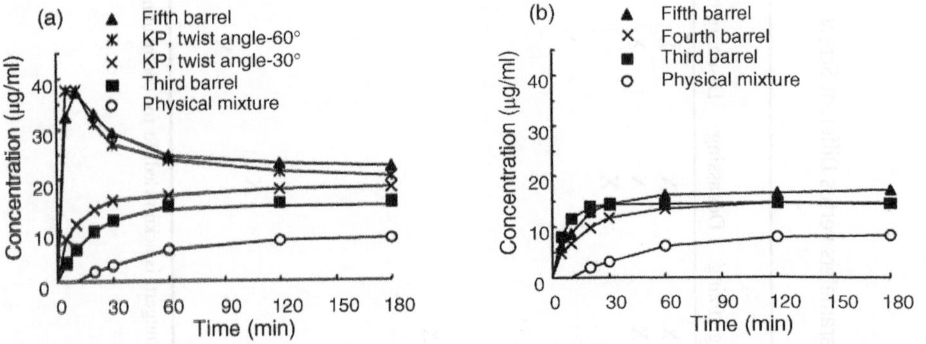

Figure 10.6. Effect of kneading paddle on dissolution profiles of NP. (a) With kneading paddle elements. (b) Without kneading paddle elements. Reproduced with permission from Ref. [11].

extruder with no end plate, extruder screws were stopped and exposed mid-process for sampling of in-process material at multiple regions along the screw. Kneading elements of specific geometry were required to generate a glassy amorphous phase with the required dissolution characteristics, unlike a reference configuration with only feeding elements that generated a dispersion with residual crystallinity (Figure 10.6). The effect of screw design on the solid dispersion may directly relate to the additional shear force in a kneading zone, but other changes in residence time or mixing efficiency may also be a factor. For reasons including scientist health and safety, the difficulty in representative sampling, and ease of data interpretation, isolating materials from the extruder in this way is not recommended as a first-choice approach.

Third Region (Pre-extrusion): The final barrel sections are designed to prepare the extruder contents for extrusion at the die plate. The section must build up material and associated pressure to enable mixture passage through the die in a controlled manner. These functions are typically achieved with screws of wider flight and shorter pitch. Low temperature in zones at or close to the die may be used to ensure that extrudate exiting the extruder is of a viscosity suited to the postextrusion processing step. A venting port may also be included in this region to allow controlled removal of air and/or volatile components from the barrel mixture.

Extruder Die: The final component of the extruder barrel is the die plate that contains the die opening(s) of specified shape and size. The die configuration is selected based on the postextrusion process being used. The simplest configuration is a single round die producing a single extrudate strand, which is compatible with chill roll cooling, conveyor cooling, cutting, or pelletization downstream processes. Multiple die openings in the die plate allow larger quantities of extrudate to simultaneously exit the extruder, a design that is well suited to die face pelletization or use with a chill roll. Up to 12 die orifices are common, as described in Ref. [12]. Dies with a thin slit opening allow formation of extruded sheets that may be collected via torque winders as a cooled film.

10.3.3 Downstream Process

The final HME processing step provides controlled cooling and collection of extrudate. This step, referred to as downstream processing, will determine the physical attributes of the collected extrudate and has the potential to effect the amorphous energy state formed. Referred to as downstream processing, a wide selection of downstream processing methods are available, providing the advantage of tailoring the physical form of extrudate over the properties of the finished dosage form.

Conveyor cooling transports extrudate from the die along a slow moving surface (typically a conveyor belt with durable product contact properties) of sufficient length for cooling. A conveyor is the most basic process option, compatible with a circular die but less suited for ribbon extrudate. The speed of the conveyor can affect the extrudate thickness via a pulling effect of material from the die. Strand extrudate cooling occurs primarily at the extrudate–air interface, so strand thickness will affect cooling rate. One or more air knives may be configured along the conveyor for assisted cooling. Partially cooled extrudate strands may have sufficient brittleness to be fragmented and collected in bulk manually or with a rotary cutting device (sometimes referred to as a strand granulator). Additional processing is usually required to generate a multiparticulate form with size range and distribution for finished product manufacture.

Pelletizers allow in-line cutting of extrudate at high frequency to generate pellets of controlled size distribution. A die-face cutting assembly is typically used that cuts heated extrudate to particles that are then transported and cooled in a vacuum collection system. Removal of fine or large particles may be a component of the pelletizer system, producing a final size distribution that can be directly filled into hard gelatin capsules. Die face cutting has been applied to materials that swell upon exiting the die to form spherical and more porous particles. Alternatively, pelletizers that collect and cut the semicooled extruded strand are available. Both strand cutting and die-face cutting configurations are displayed as schematics in Figure 10.7.

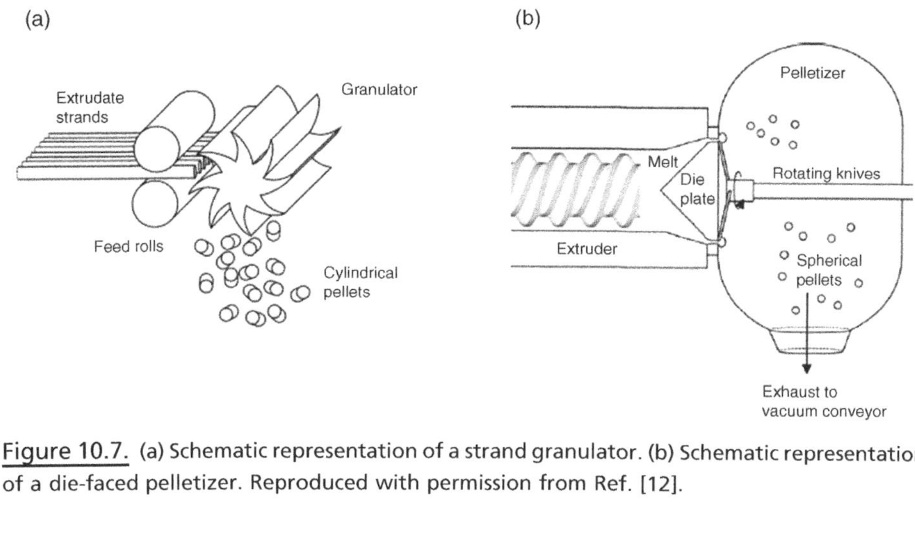

Figure 10.7. (a) Schematic representation of a strand granulator. (b) Schematic representation of a die-faced pelletizer. Reproduced with permission from Ref. [12].

Chill rolls can be used to deliver cooling to particulates or intact ribbon quickly after exiting the die. Examples of vendors providing this equipment are Thermo Electron, Germany, and the CCR units from BBA Innova, Switzerland. The roller has active cooling, which ensures that low surface temperature is maintained through repeated surface contact with extrudate. This technology is standard for collection of ribbon extrudate as the material can pass through subsequent rollers controlling torque and thickness, and ultimately wound onto a collecting roll. A rapid cooling cycle delivered by a chill roll may be advantageous for amorphous solid dispersions, delivering consistency of the glassy state formed. Chill rolls can be perfectly combined with online connected mills to grind the cooled extrudate immediately into a powder [13]. Calendering equipment feeds heated extrudate ribbon through counterrotating rollers, each with aligned cavities that produce molded tablets in-line [14]. This technology has been commercialized for solid dispersion products Kaletra® and Norvir® [15].

When milling of cooled extrudate intermediate is needed, conventional granule milling equipment can be used. Due to the physical properties of many extrudates, excessive temperature increase during the milling operation must be avoided. This precaution may be achieved by careful selection of equipment variables and feed rate, otherwise milling equipment with chilling functionality may be required. Homogenous, free-flowing granules are often generated from extruded drug–polymer dispersions that are suited to final blending and compression in a conventional tableting process.

10.4 PROCESS DESIGN AND CONTROL

Solid dispersion development using hot melt extrusion typically includes *in vitro* and/or *in vivo* assessment of early formulation prototypes before any process development has occurred. Screening of early prototypes is common because demonstration of a solubility or bioavailability increase will be a critical selection criterion for a solid dispersion. Once a lead prototype is identified, process development experiments are commonly initiated in parallel to formulation optimization.

There are two main reasons why HME process development activities begin early in a project life cycle. First, HME must generate multicomponent dispersions for which both the physical and chemical phase homogeneities are extremely critical. The "phase uniformity" quality attribute for HME may represent a tighter control requirement than, for example, the "blend uniformity" requirement for a dry powder blend used for direct compression of a tablet. Second, the HME process has a large number of process variables with a strong interrelationship with formulation variables. An example is a thixotropic polymer for which dynamic viscosity at a certain process temperature will vary depending on shear rate, which is in turn a function of screw design and rotation speed. Adjustment of critical process variables may have a significant effect on the drug delivery or stability properties of a solid dispersion prototype, possibly affecting the decision to select an HME prototype or not.

The most important equipment settings are powder feed rate, temperature, and screw speed (screw configuration is not listed but can be considered as another equipment variable). These equipment parameters are dependent variables, which together affect the mixing, heating, and rheological properties of materials inside the various zones of the hot melt extruder. Extruder system parameters such as specific mechanical energy consumption and residence time distribution are employed to provide a more useful process measurements.

Specific mechanical energy consumption (SMEC) represents the amount of mechanical energy or work input into the extruding materials from the drive motor, as described in the following equation:

$$\text{SMEC}(\text{kJ/kg}) = \frac{\text{screw speed (rpm)} \times \text{power (kW)} \times \text{torque (\%)}}{\text{max. screw speed (rpm)} \times \text{throughput}\left(\frac{\text{kg}}{\text{sec}}\right) \times 100}.$$

SMEC provides a good measure of materials response to the equipment settings with a single parameter. It may be used to support equipment parameter selection during DOE design, or for equipment parameter selection when changing equipment manufacturer or upon scale-up. Application of SMEC using this or related equations are described in the literature [7].

Residence time distribution may be measured using a tracer die method [16]. During steady-state operation of the extruder, a single addition of die may be made directly into the extruder feed port. Visual, spectroscopic, or photometric measurement of color change at the extruder die allows parameters such as residence time onset, median, or offset to be determined. The mean residence time ($\bar{t}$) can be calculated, as described in the following equation:

$$\bar{t} = \frac{\int_0^\infty tc\,\mathrm{d}t}{\int_0^\infty c\,\mathrm{d}t} = \frac{\sum_0^\infty tc\Delta t}{\sum_0^\infty c\Delta t}.$$

The experiment is valuable when comparing process performance between different extruders, as matching process parameters is less effective due to subtle equipment design variances between different equipment vendors and scales of equipment. If a quantitative tool is used to generate a residence time profile, more advanced analysis may be performed. For example, DiNunzio et al. showed direct correlation between residence time and chemical degradation of hydrocortisone in HME-processed material [17]. To optimize the extrusion process in terms of chemical degradation, mixing, and other critical quality attributes, Kolter et al. [18] describe the combined approach of reading specific mechanical energy consumption and residence time distribution is effective in finding the optimal balance between sufficient (molecular) mixing and controlling degradation of, for example, an active component.

10.5 AMORPHOUS SOLID DISPERSION APPLICATIONS

A growing body of original research is published describing the use of hot melt extrusion to generate amorphous solid dispersion systems. Selected research from this field is described in this chapter, with a goal of providing the reader practical examples spanning preclinical formulation selection work to development and scale-up of commercial-ready dosage forms.

10.5.1 HME Formulation Feasibility

Many research groups describe HME feasibility that integrate preformulation screening, formulation ranging, and physical characterization activities. The availability of small-scale extruders is an important tool to enable such research. An example of such work is presented by Albers et al. investigating a celecoxib/EUDRAGIT EPO dispersion [19]. Preformulation experimental and theoretical tools (including use of solubility parameter and ideal mixing theory) were applied to identify target drug/polymer compositions for generating amorphous solid dispersions by extrusion and to predict mixture T_g. A 1:1 (w/w) solid dispersion could be prepared by HME. The amorphous dispersion produced a 58-fold increase in API solution concentration under a nonsink condition. The generation of a dispersion that is amorphous by X-ray powder diffraction and the measurement of kinetic solubility enhancement are the typical outcomes needed to judge that HME has potential feasibility for developing an amorphous solid dispersion. An important observation with the celecoxib/EUDRAGIT EPO system was that creation of an X-ray amorphous dispersion was critical to obtaining the large solubility advantage. For reference, a dispersion extruded at lower process temperature formed an opaque extrudate containing partially crystalline drug dispersed in polymer that did not deliver as large solubility increases (Figure 10.8). Two notes of caution are required when reviewing feasibility experiments that use very high drug loading dispersions, such as described by Albers et al. drug:polymer ratio of 1:1 often presents challenges from the processing perspective at larger scales, and is rarely seen in advanced product development. In addition, the level of phase characterization performed during feasibility may

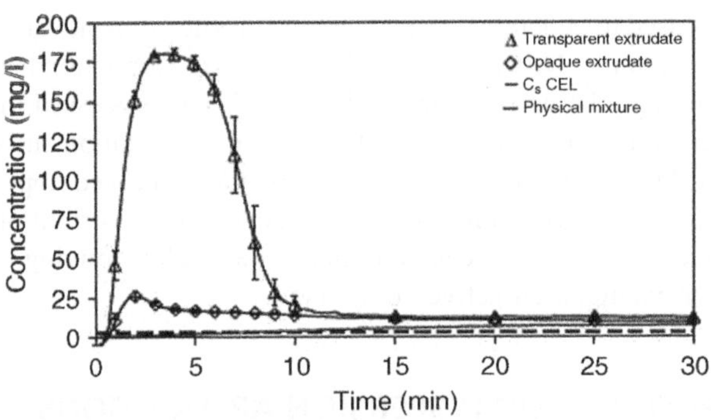

Figure 10.8. Drug release of CEL, CEL/aPMMA 1:1 (w/w) physical mixture, CEL/aPMMA 1:1 (w/w) transparent extrudate, and CEL/aPMMA 1:1 (w/w) opaque extrudate; 200 mg drug; 900 ml 0.1 N HCl, nonsink conditions, 37 °C, paddle, 50 rpm, 250 nm; mean ±SD, $n = 3$. Reproduced with permission from Ref. [19].

not be thorough because many samples are being screened and the characterization tools used are not yet optimized.

Another example of a well-integrated HME feasibility study reports polymer screening experiments for two active ingredients: indomethacin and lacidipine [20]. The authors successfully used solubility parameters to predict miscible mixtures in different molecular weight PEG and PVP polymers, plus copovidone. The detection of a single T_g measured by DSC was described as an important criterion for selecting miscible mixtures, and these mixtures were successfully prepared as amorphous dispersions by HME. While Forster et al. [20] show a single T_g may be a good predictor of amorphous state miscibility, some drug and polymer systems are harder to characterize by DSC due to complex phase behavior and complex thermal properties. Screening for a single T_g should not be a prerequisite for investigating the feasibility of an amorphous dispersion by HME.

The screening experiments and feasibility criteria must be designed around the nature of the dispersion formers under investigation. An example is a study investigating dispersion formation of drug in cyclodextrin using HME [21]. Solution-phase solubility experiments were first performed to confirm that the drug, ketoprofen, was capable of molecular interaction with the cyclodextrins. These experiments were followed by extrusion of 1:1 (w:w) blends of drug:cyclodextrin that yielded solid phase with reduced drug crystallinity. The process conditions were in the region of drug melting temperature to enable excipient dispersion in the molten drug phase. The degree of molecular interaction between drug and cyclodextrin was greater for HME-prepared dispersions relative to freeze-drying or cogrinding methods.

Formulation screening will typically include *in vivo* assessment, partly because the poor solubility of the API means that *in vitro* dissolution data have limited value. The *in vivo* assessment is also important because changes in absorption have been reported for amorphous dispersions. For example, the polymer Soluplus was shown to increase permeability of poorly soluble drugs in a Caco-2 model, an effect thought to contribute to the increased bioavailability seen in a dog model [22]. Understanding how different carrier molecules may interact with drugs in dispersed form and how the carrier molecules may affect factors other than solubility *in vivo* is essential when comparing and selecting prototypes at the feasibility stage of development.

10.5.2 HME Formulation Optimization

It is common for initial HME formulation feasibility work to focus on binary mixtures of drug/polymer. One benefit of this approach is that a direct assessment of the supersaturation provided by an amorphous dispersion can be measured. It is recommended that formulation optimization experiments be performed with the goal of enhancing the dissolution properties of the dispersion so that the solubility and dissolution advantage of the system is maximized. Examples of the mechanism for such enhancement include improving the wetting and disintegration of the dispersion, as well as prolonging the length of time a supersaturated solution of active is maintained.

Inclusion of surfactant as a third component in drug/polymer dispersions manufactured by HME was investigated by Ghebremeskel et al. [23]. Of the four ternary systems to undergo *in vitro* dissolution testing, two systems (API/copovidone/Tween 80 and API/HPMC E5/Tween 80) showed that surfactant had a positive effect on dissolution rate relative to the corresponding drug/polymer mixture. The same surfactant did not change the dissolution profile for an API/PVP K30/Tween 80 composition. Interestingly, an API/HPMC E5/docusate sodium HME dispersion exhibited slower dissolution relative to the drug/polymer reference, an effect that is opposite to that of Tween 80 added to the same matrix. Surfactant was included at 10% w/w content in all test samples, and particle size of milled extrudate was kept constant to remove these variables from the experiment. This study demonstrates that drug release into solution from a multicomponent amorphous dispersion is sensitive to low levels of surfactant, but that the effect on dissolution may not be favorable. The mechanism of surfactant effects was not investigated, but the probable role of different micellar structures formed by surfactant and polymer was discussed. Just as supersaturation profile of an amorphous solid is primarily kinetically controlled, the kinetics of surfactant dissolution and aggregation (both in solution and at interfaces) can be anticipated as contributing to the dissolution behavior of API from an amorphous dispersion.

Recognizing the important bioavailability enhancement opportunity with a third component, Shanbhag et al. report a high-throughput screening approach for ternary mixtures that uses single-point dissolution measurements at the microscale to select lead systems [24]. The screen successfully identified API/polymer/surfactant compositions for pharmacokinetic assessment in rats by administering ground HME dispersions in a miniature capsule. An amorphous solid dispersion prepared by HME produced similar oral bioavailability to a reference oral solution for the poorly soluble compound.

Dissolution behavior of three-component extrudates containing poorly soluble drug fenofibrate with two polymers was reported by Kalivoda et al. [25]. Hot melt extrusion was used to prepare binary mixtures of fenofibrate with copovidone and with HPMC, plus different ratio ternary mixtures of fenofibrate/copovidone/HPMC. All mixtures generated amorphous dispersions suitable for downstream pelletization to generate controlled pellet size. Dissolution testing was performed using USP II apparatus in test media containing 0.1% polysorbate 80 surfactant, and the profiles are provided in Figure 10.9. For binary mixtures, supersaturation was greatest for the fenofibrate/copovidone dispersion (1/3 weight ratio), with a 12-fold solubility increase seen at the 12 min timepoint, followed by a rapid decrease in dissolved fenofibrate. A fenofibrate/copovidone/HPMC dispersion (1/3/1 weight ratios) produced a similar supersaturation effect to the copovidone reference, but with a prolonged period of supersaturation. The authors postulate that HPMC introduced a gelling property to the dispersion that complements the supersaturation performance of the drug/copovidone dispersion. Such gelling may have limited the formation of crystal nuclei or growth in a high supersaturated microenvironment, and slowed down the release of water-soluble polymer away from drug and into the disperse phase. The example given by Kalivoda et al. helps teach that controlling the local environment of the dissolving amorphous solids is an important aspect of formulation optimization so that the supersaturation

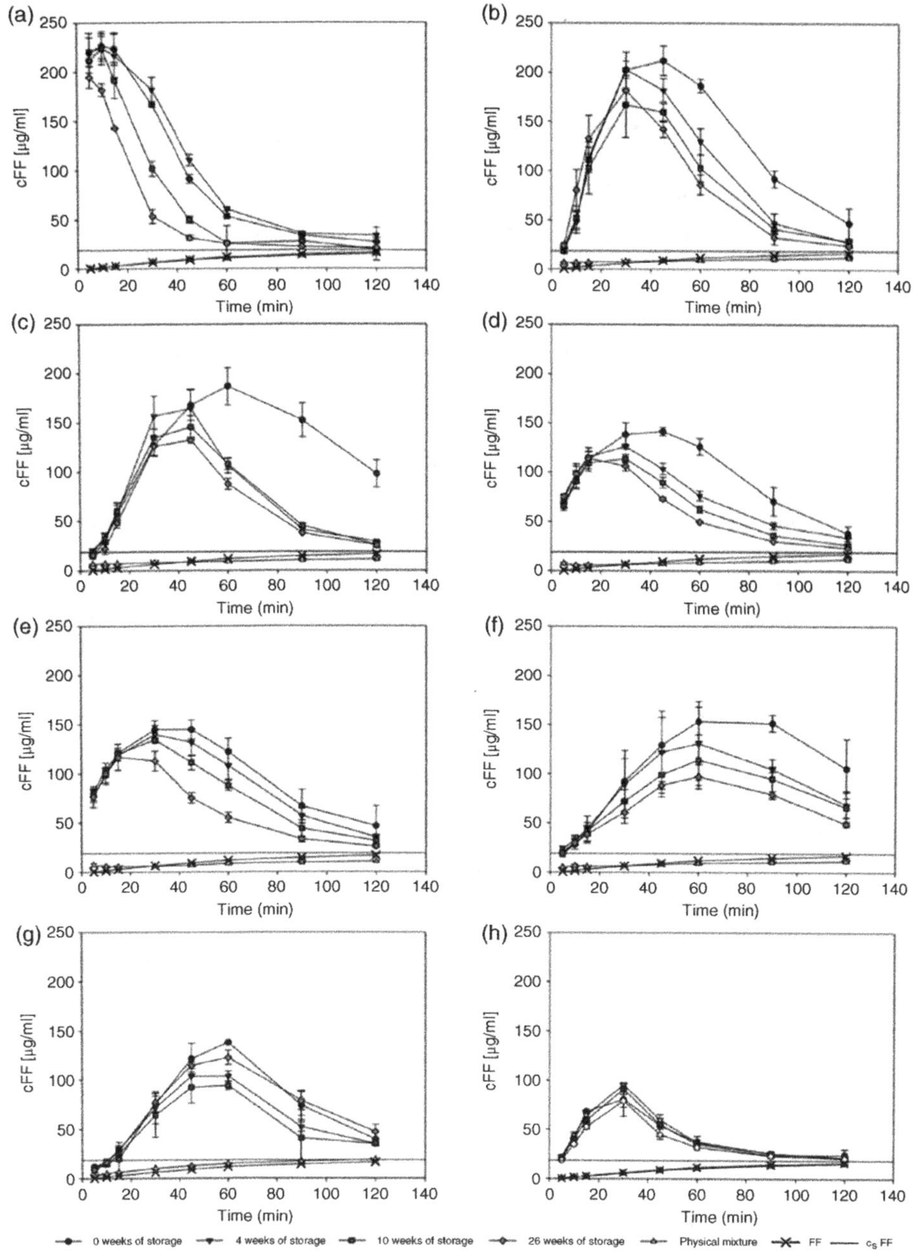

Figure 10.9. Dissolution profiles of formulations (a)–(h) in comparison with pure fenofibrate and physical mixtures of the corresponding components. Reproduced with permission from Ref. [25].

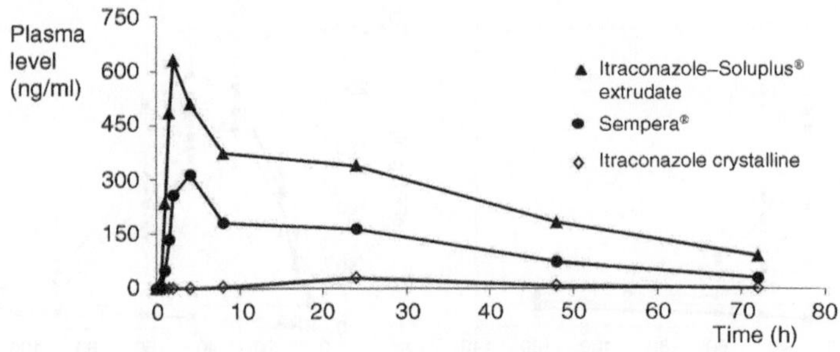

Figure 10.10. *In vivo* blood plasma concentration versus time data for various forms of itraconazole orally dosed to beagle dogs. HME-processed Soluplus dispersion exhibits 2.3-fold higher area under the curve (AUC) than the marketed product Sempera and 23-fold higher AUC than the crystalline itraconazole. Reproduced with permission from BASF SE.

effect is not only large but is also sustained. Various substituted HPMC polymers were dispersed with a model drug using HME that again shows how the duration of supersaturation can vary between related dispersions [26]. All dispersions exhibited gelling properties around the dissolving particle, but dissolution was not consistent, showing that specific molecular interactions with drug in the supersaturated environment are also significant.

Guth et al. [27] showed improved bioavailability using Soluplus in a solid dispersion with itraconazole. The formulation formed by hot melt extrusion showed a 2.3-fold increase in bioavailability compared with a commercialized formulation using HPMC-based drug-layered pellets. Beagle dog pharmacokinetic data for the solid dispersion are shown in Figure 10.10 compared with the reference pellet product (Sempera®) and crystalline API all dosed at 100 mg itraconazole.

More detailed characterization of dissolution processes have been reported for the commercialized Meltrex® HME technology [28]. Photon correlation spectroscopy measurements identified 50–60 nm mean particle size for the disperse phase that was formed from a ritonavir/copovidone/Span 20/silica solid dispersion. Analysis of placebo controls indicated that the dispersed particles were neither pure drug nor pure silica. The colloidal phase likely contains drug/polymer/surfactant, forming a relatively stable structure that maintains this mean size over 24 h. Changes in turbidity and the existence of larger particles indicate the system is heterogeneous and not thermodynamically stable, but the temporary formation of this dispersion is important for understanding dissolution behavior from the multicomponent solid dispersion. This colloidal system did not yield large concentrations of dissolved drug in a traditional dissolution test condition, but the addition of bile salt to the dissolution media produced a large change in dissolution behavior, with significant supersaturation observed.

It is evident that the main focus of formulation optimization for amorphous dispersions of poorly soluble drugs relates to dissolution performance. The topic of designing *in vitro* dissolution test conditions is a critical aspect of the development effort for these systems, but is beyond the scope of this chapter.

10.5.3 Maximizing Performance in the Finished Dosage Form

Converting a solid dispersion intermediate material into a final dosage form is required with the goal of maximizing the solubility enhancement potential of the dispersion from the dosage form. In the case of a granular solid dispersion blended with other excipients, the extragranular phase may perform critical dissolution functions similar to that of a conventional immediate-release tablet or capsule. Considerations such as the ratio of extragranular to granule components, the selection and quantity of extragranular disintegrant, lubricant, or other flow aid are all part of identifying a tablet or capsule finished dosage formulation.

Lakshman et al. investigated the oral bioavailability of a poorly soluble drug dispersed in a PVP-based extrudate [29]. Extrudate was milled and screened to generate a size-controlled granular form and then blended with poloxamer 188 surfactant with microcrystalline cellulose (Avicel PH102) before filling into hard gelatin capsules. The amount of extragranular phase ranged from 38 to 48% w/w. Capsule formulations containing solid dispersion blends produced a sevenfold increase in oral bioavailability relative to a reference capsule of crystalline API with poloxamer 188. It was hypothesized that the solubility advantage of the amorphous dispersion benefited from the wetting properties of extragranular surfactant to produce enhanced oral bioavailability. The dissolution benefits produced by the extragranular excipients may be specific to the chemistry of the polymer used in the solid dispersion. For example, use of specific salts in tablets containing carbamazepine–Soluplus solid dispersions were found to have disintegration and dissolution properties via rapid hydration and possibly effervescent actions [30]. This approach was identified to negate the gelling tendency of a hydrophilic polymer used at high loading in a dosage form.

HME processing methods are emerging for pharmaceutical solid dispersions that generate porous dispersions. Introduction of supercritical liquid into an extrudate is viable, an approach that can modify the rheological properties of materials inside the extruder as well as adjust the behavior of extrudate exiting the die as the supercritical phase converts from liquid back to gas. Two groups report the supercritical CO_2 plasticizes model drug–polymer dispersions allowing lower processing temperatures and forming more porous extrudate product [31,32]. Enhanced dissolution was reported in both systems.

10.6 SUMMARY

Hot melt extrusion is a frontline manufacturing solution for pharmaceutical amorphous solid dispersions. The status of HME technology is validated by the presence of successful commercial solid dispersion products, and is seeing widespread use in early-stage R&D laboratories that are formulating poorly soluble compounds.

This chapter has described a number of important factors that justify HME suitability for solid dispersion applications, ranging from materials performance and quality, equipment performance and versatility, and the proven ability to generate robust finished dosage forms that provide enhanced and controlled solubility behavior leading to improved bioavailability.

REFERENCES

1. Hancock, B. and Zografi, G. (1997) Characteristics and significance of the amorphous state in pharmaceutical systems. *J. Pharm. Sci.*, 86(1):1–12.
2. Baird, J. and Taylor, L. (2012) Evaluation of amorphous solid dispersion properties using thermal analysis techniques. *Adv. Drug Deliv. Rev.*, 64:396–421.
3. Crowley, M., Zang, F., Repka, M., Thumma, S., Upadhye, S., Kumar, S., McGinity, J., and Martin, C. (2007) Pharmaceutical applications of hot-melt extrusion: Part 1. *Drug Dev. Ind. Pharm.*, 33:909–926.
4. Gordon, M. and Taylor, J. (1952) Ideal copolymers and the second-order transitions of synthetic rubbers: I. Non-crystalline copolymers. *J. Appl. Chem.*, 2:493–498.
5. Bruce, L.D., Shah, N., Malick, A., Infeld, M., and McGinity, J. (2005) Properties of hot-melt extruded tablet formulations for the colonic delivery of 5-aminosalicylic acid. *Eur. J. Pharm. Biopharm.*, 59:85–97.
6. Carvalho, C., Takeiti, C., Onwulata, C., and Pordesimo, L. (2010) Relative effect of particle size on the physical properties of corn meal extrudates: effect of particle size on the extrusion of corn meal. *J. Food Eng.*, 98:103–109.
7. Guerrero, P., Beatty, E., Kerry, J., and de la Caba, K. (2012) Extrusion of soy protein with gelatin and sugars at low moisture content. *J. Food Eng.*, 110:53–59.
8. Zhu, L.-J., Shukri, R., Jhoe de Mesa-Stonestreet, N., Alavi, S., Dogan, H., and Shi, Y.-C. (2010) Mechanical and microstructural properties of soy protein: high amylose corn starch extrudates in relation to physiochemical changes of starch during extrusion. *J. Food Eng.*, 100:232–238.
9. Breitenbach, J. (2002) Melt extrusion: from process to drug delivery technology. *Eur. J. Pharm. Biopharm.*, 54:107–117.
10. Leistritz Corporation, Germany (2012) Leistritz extrusion technology brochure. http://www.leistritz.com/extrusion/de/00_pdf/EXTRUS_TECHNOLOGY_GB_2013.pdf.
11. Nakamichi, K., Nakano, T., Yasuura, H., Izumi, S., and Kawashima, Y. (2002) The role of the kneading paddle and the effects of screw revolution speed and water content on the preparation of solid dispersions using a twin-screw extruder. *Int. J. Pharm.*, 241:203–211.
12. Bialleck, S. and Rein, H. (2011) Preparation of starch-based pellets by hot-melt extrusion. *Eur. J. Pharm. Biopharm.*, 79:440–448.
13. Modern in-line hot melt extrusion process solutions. http://www.bba-innova.com/modern-in-line-hot-melt-extrusion-process-solutions.html. BBA Innova AG, Aarau, Switzerland.
14. Breitenbach, J., Shrof, W., and Neumann, J. (1999) Confocal Raman spectroscopy: analytical approach to solid dispersions and mapping of drugs. *Pharm. Res.*, 16(7):1109–1113.
15. Pharmaceutical solid dosage form. U.S. Patent 8,399,015. (2013).
16. Guns, S., Mathot, V., Martens, J., and Van den Mooter, G. (2012) Upscaling of the hot-melt extrusion process: comparison between laboratory scale and pilot scale production of solid dispersions with miconazole and Kollicoat IR. *Eur. J. Pharm. Biopharm.* 81(3):674–682.
17. DiNunzio, J., Brough, C., Hughey, J., Miller, D., Williams, R., and McGinity, J. (2010) Fusion production of solid dispersions containing a heat-sensitive active ingredient by hot melt extrusion and Kinetisol dispersing. *Eur. J. Pharm. Biopharm.*, 74:340–351.
18. Kolter, K., Karl, M., and Gryczke, A. (2012) Hot Melt Extrusion with BASF Pharma Polymers.

19. Albers, J., Alles, R., Matthee, K., Knopp, K., Schulze Nahrup, J., and Kleinebudde, P. (2009) Mechanism of drug release from polymethacrylate-based extrudates and milled strands prepared by hot melt extrusion. *Eur. J. Pharm. Sci.*, 71:387–394.

20. Forster, A., Hempenstall, J., Tucker, I., and Rades, T. (2001) Selection of excipients for melt extrusion with two poorly water-soluble drugs by solubility parameter calculation and thermal analysis. *Int. J. Pharm.*, 226(1–2):147–161.

21. Fukuda, M., Miller, D., Peppas, N., and McGinity, J. (2008) Influence of sulfobutyl ether-β-cyclodextrin (Captisol®) on the dissolution properties of a poorly soluble drug from extrudates prepared by hot melt extrusion. *Int. J. Pharm.*, 350:188–196.

22. Linn, M., Collnot, E.-M., Djuric, D., Hempel, K., Fabian, E., Kolter, K., and Lehr, C.-M. (2012) Soluplus® as an effective absorption enhancer of poorly soluble drugs *in vitro* and *in vivo*. *Eur. J. Pharm. Sci.*, 45(3):336–343.

23. Ghebremeskel, A., Vemavarapu, C., and Lodaya, M. (2007) Use of surfactants as plasticizers in preparing solid dispersions of poorly soluble API: Selection of polymer-surfactant combinations using solubility parameters and testing the processability. *Int. J. Pharm.*, 328(2):119–129.

24. Shanbhag, A., Rabel, S., Nauka, E., Casadevall, G., and Shivanand, P. (2008) Method for screening of solid dispersion formulations of low-solubility compounds: miniaturization and automation of solvent casting and dissolution testing. *Int. J. Pharm.*, 351, 209–218.

25. Kalivoda, A., Fischbach, M., and Kleinebudde, P. (2012) Application of mixtures of polymeric carriers for dissolution enhancement of fenofibrate using hot-melt extrusion. *Int. J. Pharm.*, 429:58–68.

26. Tajarobi, F., Larsson, A., Matic, H., and Abrahmsen-Alami, S. (2011) The influence of crystallization inhibition of HPMC and HPMCAS on model substance dissolution and release in swellable matrix tablets. *Eur. J. Pharm. Biopharm.*, 78:125–133.

27. Guth, F., Becker, M., Buesen, R., and Kolter, K. (2012) Bioavailability enhancement of itraconazole with solid solutions based on soluplus. Presented at *CRS Annual Meeting, July 15–18.*

28. Tho, I., Liepold, B., Rosenberg, J., Maegerlein, M., Brandl, M., and Fricker, G. (2010) Formation of nano/micro-dispersions with improved dissolution properties upon dispersion of ritonavir melt extrudate in aqueous media. *Eur. J. Pharm. Sci.*, 40(1):25–32.

29. Lakshman, J., Cao, Y., Kowalski, J., and Serajuddin, A. (2008) Application of melt extrusion in the development of a physically and chemically stable high-energy amorphous solid disperison of a poorly water-soluble drug. *Mol. Pharm.*, 5(6):994–1002.

30. Hughey, J., Keen, J., Miller, D., Kolter, K., Langley, N., and McGinity, J. (2013) The use of inorganic salts to improve the dissolution characteristics of tablets containing Soluplus-based solid dispersions. *Eur. J. Pharm. Sci.*, 48:758–766.

31. Verreck, G., Decorte, A., Heymansa, K., Adriaensen, J., Cleeren, D., Jacobs, A., Liu, D., Tomasko, D., Arien, A., Peeters, J., Rombautuc, P., Van den Mooter, G., and Brewster, M. (2005) The effect of pressurized carbon dioxide as a temporary plasticizer and foaming agent on the hot stage extrusion process and extrudate properties of solid dispersions of itraconazole with PVP-VA 64. *Eur. J. Pharm. Sci.*, 26:349–358.

32. Lyons, J., Hallinan, M., Kennedy, J., Devine, D., Geever, L., Blackie, P., and Higginbotham, C. (2007) Preparation of monolithic matrices for oral drug delivery using a supercritical fluid assisted hot melt extrusion process. *Int. J. Pharm.*, 329(1–2):62–71.

11

FORMULATION DEVELOPMENT OF AMORPHOUS DISPERSIONS

Tapan Sanghvi,[1] Jeff Katstra,[2] Brian Patrick Quinn,[3] Hayden Thomas,[1] and Patricia Hurter[4]

[1]*Formulation Development, Vertex Pharmaceuticals Incorporated, Boston, MA, USA*
[2]*Formulation Development, Vertex Pharmaceuticals Incorporated, Cambridge, MA, USA*
[3]*Parthenon - Ernst & Young, New York, NY, USA*
[4]*Global Pharmaceutical Development and Regulatory Affairs, Vertex Pharmaceuticals Incorporated, Boston, MA, USA*

11.1 PREPARING DISPERSIONS FOR DRUG PRODUCTS

11.1.1 Introduction

A pharmaceutical development group may decide to move forward with an amorphous dispersion for a number of reasons [1]. One of these reasons is solubility [2–4]. For a promising compound with a high affinity for its target that nonetheless has poor aqueous solubility, creating an amorphous dispersion is one means for keeping the molecule in play. Simply asking the medicinal chemistry group for more scaffolds may not always be an option; for some protein targets, most effective ligands are likely to have the same or similar characteristics. For example, as our colleagues discussed in Chapter 7, competitive inhibitors of proteases, which generally have to bind tightly to an apolar active site, are

Pharmaceutical Amorphous Solid Dispersions, First Edition. Edited by Ann Newman.

frequently apolar and hydrophobic. Identifying and validating an alternative target, meanwhile, can be both an uncertain and a resource- and time-intensive process. Hence, when a discovery group identifies an effective inhibitor of a valuable target, it is useful to try to get that molecule into a state that will give it the desired exposure *in vivo*. For certain compounds, amorphous dispersions are a possible option.

To date, only a handful of amorphous pharmaceutical dispersions have been brought to market (Table 11.1 [5]). The pace of new releases has increased in recent years. However, amorphous dispersions pose certain risks and demand certain considerations that crystalline drugs do not. They are not broadly used in commercial products, mainly because there is the possibility that during processing (mechanical stress) or storage (temperature and humidity stress) the amorphous state may undergo crystallization [6–9]. The effect of moisture on the storage stability of amorphous pharmaceuticals is also a significant concern, because it may increase drug mobility and promote crystallization [8,10]. Moreover, most of the polymers used in solid dispersions can absorb moisture, which may result in phase separation, crystal growth, or conversion from the amorphous to the crystalline state or from a metastable crystalline form to a more stable form during storage. This may result in decreased dissolution rate and solubility [6,11]. Therefore, exploitation of the full potential of solid dispersions requires their stabilization in solid state as well as during *in vivo* performance and a sound scientific understanding of scale-up for defining a robust manufacturing process. As these materials become more common in pharmaceutical development organizations—and pharmacy shelves—it is important that our collective knowledge keeps pace with our activity. Being thoroughly informed, while adopting a rigorous technical approach allowed by the state of the art, will solidify this knowledge and may mitigate these risks.

This chapter examines the development of clinical and commercial formulations of amorphous dispersions. We consider first, in this introduction, the various means available to render a crystalline drug amorphous in the first place. Next we consider the downstream ramifications of each technique, such as on the dosage forms available, and special processing and handling. In Section 11.2, we give the high-level strategic picture of amorphous formulation development, a process that best fits the framework of quality by design (QbD) advocated for by the FDA. Finally, in Section 11.3, we turn to a detailed case study of the first (of two) amorphous dispersion-based drugs we developed and commercialized in-house—the hepatitis C protease inhibitor VX-950 or telaprevir (Incivek®). In this case study, we focus, first, on developing a drug product manufacturing process, including defining parameters, developing the design space, modeling, and applying the QbD framework to manufacture. Next we turn to the analytical methods used to discriminate dispersion attributes and tablet properties that could affect performance of the product in the clinic. Finally, we turn to a scientific understanding of product stability, as it is used to define specifications for the raw materials, the finished product, and the product's packaging.

Throughout, we remain continually mindful that we are describing a QbD approach—one that uses statistical models and process controls to influence technical decision making, always mindful of upstream influences on final products. While moving forward with an amorphous dispersion with a limited scientific or technical understanding *is* risky, we emphasize (and, we hope, demonstrate) that with a rigorous

TABLE 11.1. Marketed Oral Drug Products Produced Using Solid Dispersions

Product	Compound	Company	Dispersion Polymer	Dispersion Process	Dosage Form	Approval
Cesamet	Nabilone	Valeant	PVP			
Sporanox	Itraconazole	Janssen/Ortho-McNeil	HPMC	Fluid bed bead layering	Capsule	1992
Prograf	Tacrolimus	Astellas Pharma	HPMC	Spray drying		1994
Rezulin	Troglitazone	Pfizer	PVP	HME	Tablet	1997 (withdrawn in 2000)
Kaletra	Lopinavir, ritonavir	Abbott	PVP–VA	HME	Tablet	2007
Intelence	Etravirine	Tibotec/J&J	HPMC	Spray drying	Tablet	2008
Zortress	Everolimus	Novartis	HPMC	Spray drying	Tablet	2010
Onmel	Itraconazole	Merz Pharma	HPMC	HME	Tablet	2010
Norvir	Ritonavir	AbbVie	PVP–VA	HME	Tablet	2010
Incivek	Telaprevir	Vertex	HPMCAS	Spray drying	Tablet	2011
Zelboraf	Vemurafenib	Roche	HPMCAS	Solvent/antisolvent precipitation	Tablet	2011
Kalydeco	Ivacaftor	Vertex	HPMCAS	Spray drying	Tablet	2012

Adapted from Ref. 5.

PVP: polyvinylpyrrolidone; PVP–VA: polyvinylpyrrolidone–vinyl acetate; HPMC: hydroxypropylmethylcellulose (hypromellose); HPMCAS: hydroxypropylmethylcellulose (hypromellose) acetate succinate; HME: hot melt extrusion.

framework such as QbD, and an adequate attention to every step, the risks become more than manageable.

11.1.2 Creating an Amorphous Drug Substance

The most common groups of techniques for amorphization (to use an admittedly awkward term of art) are melting techniques and solvent evaporation techniques. Each group has its strengths and weaknesses.

For melting techniques such as hot melt extrusion (HME) [12,13], spray congealing [14,15], and melt agglomeration [16–18], two considerations predominate. First, the active pharmaceutical ingredient (API) needs to be thermostable enough to survive the temperatures involved, and second, the API has to be miscible in the polymer matrix with which it is co-melted. (Choice of polymers is, of course, also limited by the same consideration for thermostability.) On the up side, melting techniques are less expensive than solvent evaporation approaches, the instruments involved have a smaller footprint, and the approach lends itself well to being one stage of a continuous process. As attention in the pharmaceutical industry increasingly turns to the speed and efficiency of continuous over batch processing for drug product manufacture [19,20], the latter may prove an advantage.

One common melting technique is hot melt extrusion. In this approach, the drug compound and a polymer are added to the HME device. (A surfactant may be added too, if needed, to improve the bioavailability downstream. Adding a surfactant also lowers the melting point, since it acts as an impurity in the crystalline drug; this may be useful for high-melting APIs.) Heat is applied as the mixture is agitated by twin screws that shear it, mix it, convey it, and ultimately extrude it from the device. The resulting material is often quite dense, which may make the later process of compressing it into tablets relatively more difficult. To reduce the density, two additional steps are often followed: the extrudate may be milled, mixed with a more compressible substance (e.g., lactose and methylcellulose), or both.

Vis-à-vis solvent evaporation methods (see below), HME affords more flexibility in choosing a downstream dosage form; on the other hand, the need to add a filler such as lactose to increase compressibility represents an extra step of production. With respect to other melting methods—the most common being melt agglomeration—HME has the advantage of being the only technique used in formulation development that is also scalable to clinical and commercial manufacture.

One solvent evaporation method is spray drying [21–23]. In spray drying, the API is solubilized in organic solvents (also aqueous or a mixture) that have relatively low boiling temperatures. At a temperature above the solvent's (or solvents') boiling point (or points), the material is sprayed through a high-pressure or two-fluid nozzle and collected. Because the solvents used are generally organic (particularly to take advantage of low boiling points), the process is geared toward lipophilic APIs, which are also, of course, those most in need of being rendered water soluble. The resulting dispersion not only is soluble, but also can be stabilized—that is, made more resistant to recrystallization—by the addition of polymers (e.g., methylcellulose, polyvinylpyrrolidone, etc.).

Solid dispersion produced by either process (melting or solvent evaporation) can yield similar bioavailability and physical stability as long as both processes yield a homogeneous amorphous dispersion with appropriate final powder particle size. However, HME can often yield nonhomogeneous amorphous dispersion that will likely underperform due to incomplete dissolution of the compound. Either the high melting temperature of the compound or the low solubility of the compound in the molten polymer during the HME process can result in crystallization or phase separation when the melt cools. Also, the creation of the amorphous state in spray drying happens very quickly and at lower temperature, so the API and polymers are exposed to less heat for less time. Finally, the process can be repeated with minimal loss: if a sample is only incompletely converted to an amorphous material and some crystalline content remains, that sample can readily be redissolved and spray dried again. Spray drying also has several advantages over other evaporation methods (coprecipitation [24–26], supercritical fluids [27,28], etc.); most notably, it is highly scalable. During discovery, small-scale spray dryers can handle samples on the order of a few hundred milligrams; later in development, material can be spray dried at a rate of tens of kilograms per hour.

The spray drying process does have several drawbacks, though. As scale changes, so does particle size (PS)—so some adaptation will still be necessary. Second, and exactly the opposite of HME, the bulk density (BD) of a spray-dried dispersion (SDD) is generally low. As a result, powder handling is difficult; also, the material creates fine particles, which both pose a safety issue (a potential inhalation hazard) and lower the harvested yield. For the spray drying process, desired particle size and bulk density can be achieved to render flowable solid dispersion. Alternatively, another process, such as a dry or wet granulation step, may be necessary to increase density and flowability before encapsulating or compressing the dispersion into tablets. Also, the design space is somewhat limited by the requirement that the API and polymer both be sufficiently soluble in the same spray drying solvent system. And because of the amount of solvent vaporized during the process (in addition to the cost of large quantities of nitrogen), spray drying is somewhat more costly than other methods of amorphization and requires certain measures to ensure that the vaporized solvents do not escape significantly into the environment.

Finally, the freshly sprayed dispersion will contain some residual spray drying solvent, which poses two difficulties. First, since the solvents used are generally organic, they are acceptable only at very low levels in a material meant to be dosed in humans. Second—as our colleagues discuss in some detail in Chapter 7—any moisture in the dispersion plasticizes the material, increasing its molecular mobility and increasing the likelihood that the amorphous dispersion will nucleate and recrystallize. (This plasticization is reflected quantitatively in the material's glass transition temperature, or T_g, which represents the points at which sudden or "catastrophic" crystallization is likely to occur; residual moisture lowers the T_g.)

One final consideration in selecting a melting or solvent evaporation technique is the way in which that technique may constrain the eventual choice of delivery mechanism downstream. HME creates extrudate, which can be dosed in capsules or even simply by itself. However, while an extrudate can be formed into a tablet as well, this approach is not ideal: the material needs to be rendered less dense, as unadulterated extrudates are

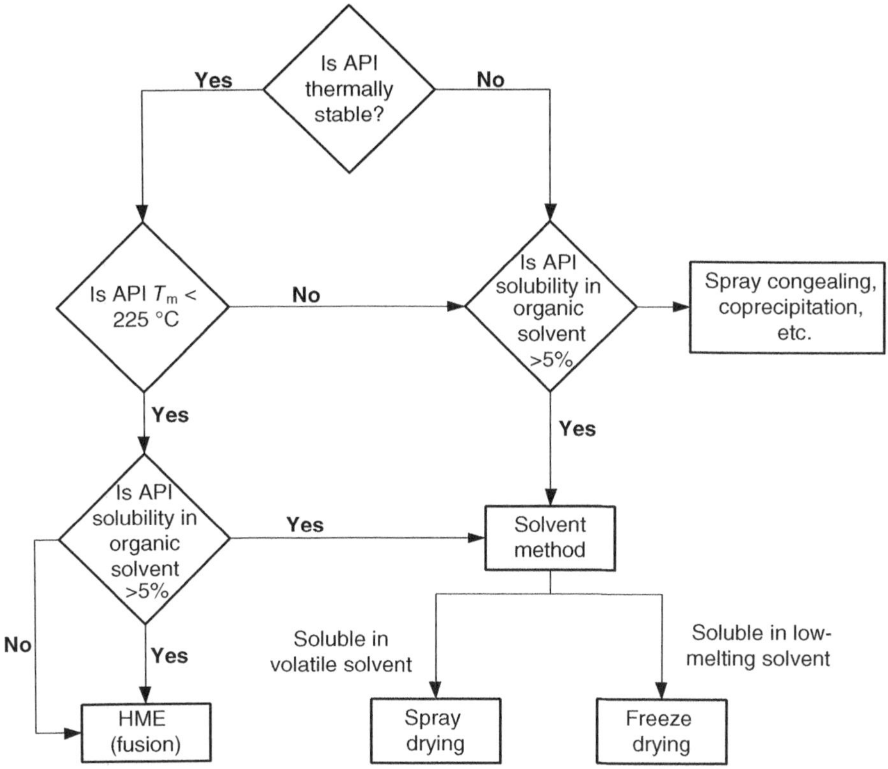

Figure 11.1. Decision tree for the selection of process for solid dispersion.

generally not compressible enough for tableting. Spray drying, meanwhile, is versatile (although the spray-dried material may need to be rendered more flowable and compressible before tableting, as described above). However, SDDs often function poorly in capsule form; in fact, they might not be released at all. When a capsule is immersed in an aqueous environment (for instance, the stomach), a small amount of fluid penetrates the capsule and begins to dissolve it. With an SDD, this fluid can mingle with the polymer present in the dispersion, cross-linking individual polymer molecules and creating a thick gel that entraps the API and prevents it from being released [29,30].

The considerations described above are summarized below in a decision tree (Figure 11.1). Ultimately, as this figure reflects, the choice of amorphization technique is a multifactorial decision, and no one technique is right for every compound or every development program. Regardless, once a development group has chosen its approach, the project enters a distinct new phase—one that should be guided, at every step, by the paradigm of QbD. The next section briefly describes the strategic ramifications of running a development program under QbD. Subsequent sections specifically address the processing, methods, and stability considerations of an actual program, the development of our first drug commercialized in-house, telaprevir.

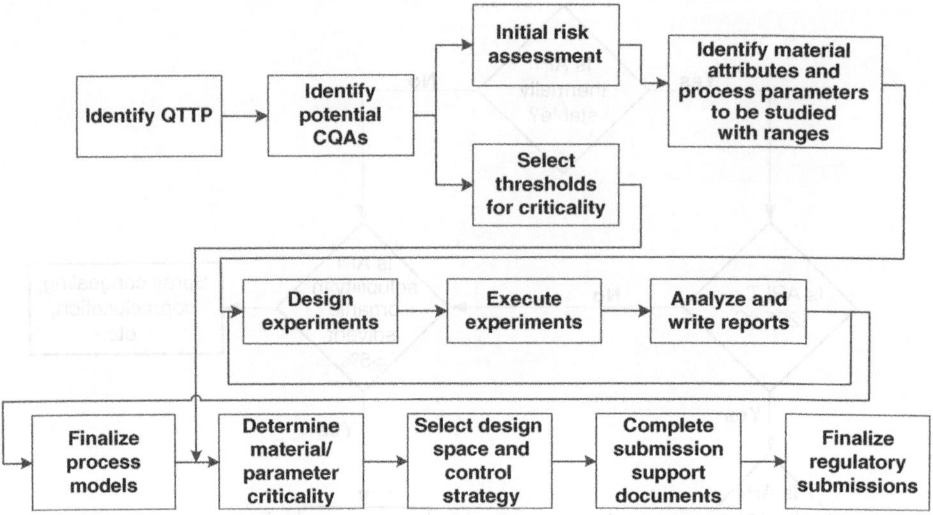

Figure 11.2. Quality by design process flow.

11.2 THE STRATEGY OF QUALITY BY DESIGN

No doubt many readers of this book will already be acquainted with the methodical, systematic approach to formulation development called quality by design [31–37]. However, QbD is so central to our strategy and our day-to-day operations that it is worthwhile to describe the approach here in some detail as a prelude to the case study that follows.

At the heart of QbD is scientific understanding of both the product and the process of development—understanding sufficiently rigorous to allow development teams to set predefined objectives and meet them with thorough risk management. The anticipated payoff is a better product, as well as a more reliable development process with fewer unforeseen failures, dead ends, and wasted time. This process, which was originally laid out in the ICH documents Q8, Q9, and Q10, is summarized by the flow diagram in Figure 11.2.

As the figure reflects, QbD begins with a quality target product profile, or QTPP, which is derived from a clinical target product profile that states the clinical goals of the product. The QTPP lays out the quality-related goals that the product will have to meet in order to reach the clinic. It could include the desired dose size, the limits on impurities and degradation products, and the shelf life in commercial packaging. Table 11.2 lists the definition of key terminologies used as the foundation for the material and process criticality analyses.

With the QTPP in hand, a development team can proceed to step 2 of QbD, which consists of defining critical quality attributes, or CQAs. The CQAs are the characteristics that need to be within a certain range or to display a certain distribution in order for the

TABLE 11.2. Key Terminologies in QbD

Terminology	Definition
Critical quality attribute	Attribute for which deviations from established limits may result in decreased assurance of quality, safety, and/or efficacy
Critical and key parameters	An aspect of the process whose variability has a meaningful impact on a critical quality attribute and therefore should be monitored or controlled to ensure that the process produces the desired quality. The difference between key and critical is determined by a risk assessment of the potential impact on product quality, which includes factors such as the magnitude of the effect, closeness of the design space to the edge of failure, location of the NOR within the design space, and robustness of the process. If the risk to the critical quality attribute is high, the aspect of the process is deemed critical; if the risk is moderate, the aspect is deemed key
Noncritical	A noncritical aspect of the process (i.e., material attribute, process parameter, and/or in-process control) has no significant potential of impacting a CQA
Design space	The multidimensional combination and interaction of input variables (e.g., material attributes) and process parameters that have been demonstrated to provide assurance of quality. Working within the design space is not considered as a change
Normal operating range	The range of material attributes and process parameters that are utilized during normal manufacturing
Process parameters	Procedures that can be adjusted to control the manufacturing process (e.g., stoichiometry, temperature, and time)
Proven acceptable range	A characterized range of a process parameter for which operation within this range, while keeping other parameters constant, will result in producing a material meeting relevant quality criteria. The characterized range is obtained using empirical models, experimental range examined, scientific fundamentals, and equipment limits
Design space limit	The maximum allowable range for each individual parameter that is part of a multivariate design space. The design space is described by DSLs and may include a limiting equation
Knowledge boundary	The combination of scientific fundamentals and experimental ranges for all parameters that represent the extremes of the variables used to develop the limits of process understanding and/or equipment limits

QTPP goals to be met. They could be physical, chemical, biological, or microbiological properties. What makes them critical are the detrimental effects—to safety or efficacy—that would be expected if some variation in the product or process causes the CQA to fall outside its target range. For this purpose, the potential for harm is assessed *before* taking into account any other measures that might mitigate the risk. Certain properties are generally considered to be CQAs. For instance, in the case of amorphous dispersions, the level of residual spray drying solvents is usually a CQA, especially in the likely event that the solvent is restricted to certain limits for human exposure. Amount of an observed degradant of the API will generally also be a CQA. Other properties may or may not be critical because they may or may not impact safety or efficacy, which needs to be evaluated on a case-by-case basis. For example, particle size would be a CQA if it affects dissolution to a degree that impacts the drug's exposure, but may not be critical otherwise.

Importantly, the risk assessment that determines criticality of an attribute is not a once-and-for-all process; instead, risk levels are re-evaluated during development as the material under consideration is better understood. In fact, the initial risk assessment aims primarily to identify areas of concern, where ongoing experimentation will be necessary. This determination may be based on prior familiarity with the process or product. To carry out this and any subsequent assessment, the development team may assign a risk priority number. This is a semiqualitative determination that multiplies the magnitude of potential harm, the likelihood of the harm occurring, and the likelihood that the harm would be detected. This is similar to failure modes and effects analysis (FMEA) commonly used in process development. Clearly, the process of assigning a risk priority number is not precise, but it does allow the team to classify a CQA broadly as low, medium, or high risk.

A number of parameters may have some effect on a CQA. The effect becomes an *impact*, in regulatory terms, when it causes the CQA to cross a *criticality threshold*. This threshold is crossed when the change to the CQA is likely to cause changes to safety or efficacy beyond accepted limits. To summarize, when a parameter is (statistically significantly) likely to change (or impact) a CQA beyond its acceptable limits (the criticality threshold), that parameter needs to be controlled during development and manufacture.

Regardless of the risk level, once a parameter is defined as a CQA, it is subject to quite different handling from other measurements of the drug. As far as regulators are concerned, an attribute cannot be *a little bit* critical—either it is critical or it is not. While the target range of a noncritical parameter could be changed without any pre-approval from regulatory authorities, such a change would require an update in an annual report. Conversely, changes to CQAs that expand the design space are major maneuvers and require a lengthy process of approval and review.

We propose utilizing a conceptual approach to design space development as illustrated in Figure 11.3. In the first stage (step 1), preliminary design of experiments (DoEs) should be done on individual unit operations (i.e., solution preparation, spray drying, secondary drying, and so on for the spray drying process; and blending, compression, and so on for the tablet process). These experiments will provide information on both viable operating ranges for the individual unit operations and the effect of material and process parameters for the given operation on output parameters from that unit operation (e.g., the effect of spray drying process parameters on SDD properties such as particle size and bulk density).

In the second stage (step 2), the interaction between outputs from one particular operation on downstream unit operations should be investigated by performing DoEs incorporating more than one unit operation. For example, the effect of spray drying on final tablet dissolution should be investigated, as a function of incoming SDD properties (such as particle size and bulk density), by performing DoEs incorporating the spray drying and tablet compression steps, in a discontinuous mode of operation.

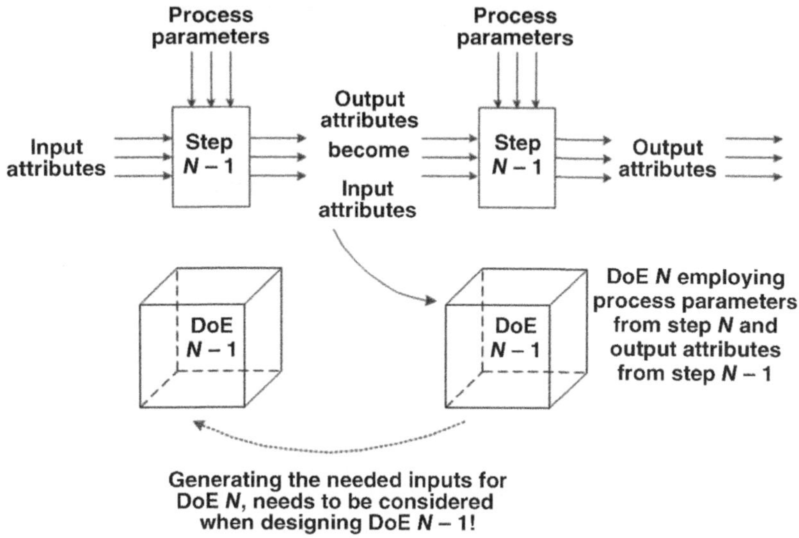

Figure 11.3. Connectivity of upstream and downstream experiments.

The final design space should incorporate results of experiments from both stages of development, and the process parameters for the individual unit operations will either be the same as established in step 1 or be further constrained to a subset of the ranges established by the step 2 experiments (e.g., if particle size of intermediate SDD from the spray drying step to tablet compression constrains acceptable tablet dissolution for acceptable bioavailability).

The acceptable limits of CQAs, in turn, are what inform the delineation of a design space—the combination of attributes, stipulated by the team designing the product that provides an assurance of quality. The crucial word here is *assurance*: in other words, if a material stays within its design space, it is understood that it *will* meet all CQAs and be safe and effective. The design space of any material is likely multidimensional (i.e., more than one process parameter will affect a CQA) and may involve interaction between parameters; hence, it cannot be defined by experiments that vary just one parameter at a time.

Figure 11.4 displays two examples of possible two-dimensional design spaces. The first, on the left, is simply rectangular. For a univariate or non-interacting design space, the NOR (normal operating range) appears in the step-by-step batch record instructions, while the broader DSL (design space limit) range can be consulted during operation outside the NOR to ensure that product quality is maintained. For the second example, on the right, the interaction between the two parameters limits the upper-right corner of the space, as described from a *post hoc* empirical limiting equation. In practice, it can be more challenging to implement an irregular design space, so the trade-off between shrinking the design space to be regular and maintaining a larger design space with more complexity needs to be made. As a consequence, when faced with a limiting equation for

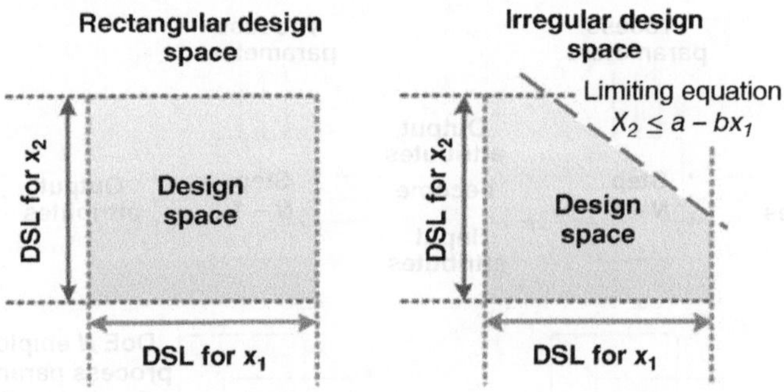

Figure 11.4. Illustration of design space definition.

one of our compounds, we shortened the design space along the x-axis to ensure that the NOR remained rectangular or univariate. We note, however, that equation-based design spaces like the second example may be easier to implement in the future as continuous (versus batch) manufacturing processes become more widespread; a continuous manufacturing rig can be programmed with control software that is able to perform complex calculations like the limiting equation on the fly.

To constrain a process to the predetermined acceptable design space, a development team would implement a *control strategy*. This strategy could consist of any of several elements. CQAs and operating conditions are constrained to within their acceptable limits. In-process controls (IPCs) are in-line or off-the-line tests that demonstrate that a process is operating as expected or producing the expected final or intermediate results. A statistically valid sampling strategy guides the testing that composes an in-process control. This control strategy is implemented continuously, and the trending of these results is monitored over time.

Our organization can report favorable experiences with the QbD framework sketched out in this section. Both of the pharmaceuticals we developed and commercialized in-house were designed under QbD—and both were amorphous dispersions. The rest of this chapter aims to give a practical example of how we used QbD to successfully develop a spray-dried dispersion—in this case, our first drug commercialized in-house, telaprevir. We hope that this case study, while only one example, will offer actionable insights to other groups seeking to derisk the development of a metastable API.

11.3 DESIGNING THE TELAPREVIR AMORPHOUS DISPERSION UNDER QbD

11.3.1 Telaprevir and Its Process Development

Telaprevir is a small molecule that targets the hepatitis C virus (HCV) [38]. The global prevalence of HCV infection is estimated to be 130–170 million people, approximately 2–3% of the world population [39], with more than 365,000 deaths per year resulting from long-term complications [40]. Approximately 55–85% of infections become chronic, and chronic hepatitis C (CHC) can lead to serious liver disease [41]. Cirrhosis develops within 20 years in 4–20% of patients with CHC [42]. Patients diagnosed with cirrhosis have an 18–29% risk of developing decompensated liver disease within 5–10 years, and a 10–30% risk of developing hepatocellular carcinoma after 20 years [43].

Telaprevir acts by binding competitively to the viral NS3:4a protease, which is essential for HCV replication [44]. As an active-site inhibitor, the molecule needs to be both sterically and electrostatically predisposed to attach to the site where NS3:4a normally binds and cleaves other viral proteins. From a pharmaceutical perspective, this poses a certain set of challenges. As our colleagues describe in Chapter 7, the NS3:4a active site, like the active sites of many proteases, is predominantly apolar; thus, telaprevir needs to be largely apolar as well in order to bind effectively. Telaprevir does in fact bind effectively, with nanomolar affinity, but because it is apolar, the compound is only sparingly soluble in water—less so, in fact, than marble [38]. Low aqueous solubility is a less than optimal trait for a drug candidate meant to be dosed orally; without an adequate concentration of dissolved drug in the aqueous environment of the stomach and (particularly) the intestine, the compound is unlikely to be absorbed and reach the systemic circulation.

After initial investigations on several options, the decision was made to advance telaprevir through development as an amorphous dispersion as discussed in Chapter 7. To render the compound amorphous, management decided to rely on spray drying. The advantages of spray drying are described in Section 11.1.2 and include the need for organic solvents (congenial to a lipophilic API like telaprevir), the low wastage of (at the time) scarce compound during spray drying method development, and the relative ease of scaling up the process. As our colleagues described in Chapter 7, excipients, namely, stabilizing polymers [45], would be a necessary addition to the spray drying mixture. The mixture would have to be spray dried [46], secondary dried to remove excess residual solvent, and then tableted—a procedure that involved blending the dried SDD with tablet-specific excipients, compressing the blend into tablets, and covering the tablets with a film coating. The overall process for the compound's formulation development is shown in Figure 11.5.

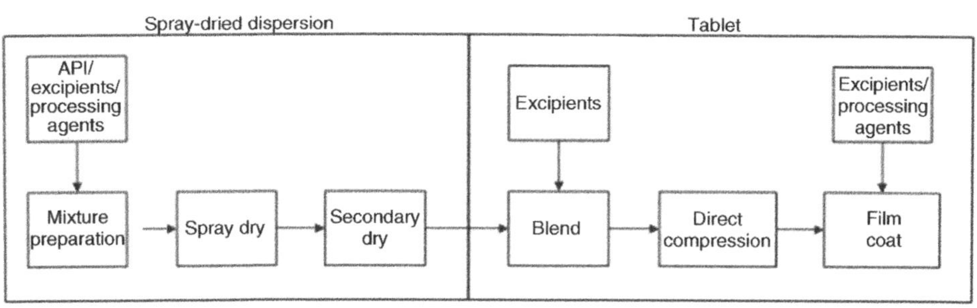

Figure 11.5. High-level manufacturing process flow diagram for telaprevir spray-dried dispersion and tablet.

As was stressed earlier in the chapter, our group developed the formulation for telaprevir within the QbD framework. As the opening step, the quality target product profile for telaprevir had specified that the commercialized drug should be

- an immediate-release tablet;
- stable during 24 months of shelf storage;
- orally administered;
- bioavailable;
- safe;
- efficacious;
- easily distinguishable from other medications.

With these requirements as the starting point for telaprevir's design, the project team moved on to define the CQAs for SDD and drug product (coated tablet). These attributes and their anticipated impacts on quality, safety, or efficacy are presented in Table 11.3.

11.3.2 Conceptual Approach to Telaprevir Process Design Space Development

Once telaprevir CQAs were identified (as shown in Table 11.3), the QbD process flow as depicted in Figure 11.2 was executed to establish the design space for telaprevir drug product. For the purpose of illustration, the process from initial risk assessment to the development of design space is described, focusing on the particular tablet (drug product) CQA of dissolution.

For tablet CQA of dissolution, an initial risk assessment was performed on tablet manufacturing process and materials to determine which process steps and materials potentially afford control of telaprevir tablet CQA of dissolution. If a material, process parameter, or in-process control in a particular step can potentially impact the CQA of dissolution, then that material and/or step is potentially critical, and is further investigated via a detailed criticality analysis. The initial materials risk assessment is performed to evaluate the potential of CQA of dissolution to be impacted by variability of the material within the incoming materials' specifications. For example, if lot-to-lot variability in the particle size of telaprevir SDD could potentially affect tablet dissolution, then telaprevir SDD material has the potential to impact the dissolution CQA of tablet. The risk assessment summarized in Table 11.4 is a preliminary assessment of the potential of particular materials or process steps to affect tablet CQA of dissolution. If a material or process step is deemed potentially critical, then further experiments and analysis should be performed before reaching a final conclusion on the criticality of the material or parameter.

The initial risk assessment of materials and process steps (Table 11.4) suggested that telaprevir SDD is potentially critical material. In addition, all the processing steps are determined to be potentially critical as each impacts the tablet CQA of dissolution. All potentially critical unit operations were studied to determine potential critical and key processing parameters by performing design of experiments. Table 11.5 lists parameters

TABLE 11.3. Summary of Critical Quality Attributes for Telaprevir

Quality Attributes	Spray-Dried Dispersion	Drug Product	Potential Impact on Quality, Safety, and/or Efficacy
Identification	X	X	Ensures correct active ingredient
Appearance	X	X	Visual indicator of product quality
Assay	X	X	Ensures potency of subsequent tablet
Degradation products	X	X	Ensure that levels of impurities do not affect assay or safety (by exceeding the levels deemed safe in toxicology or clinical studies)
Residual solvents	X		Affect safety, potency, and physical stability of the product
Water content	X	X	Ensures that water content is not so high as to lower chemical stability or impact physical stability during the product's shelf life
Physical form	X[a]	X[a]	Ensures that telaprevir remains amorphous and, therefore, bioavailable
Particle size	X		Affects flowability and compactability of the powder during tableting, which could lead to low potency or broken tablets (efficacy). Incorrect particle size affects downstream manufacturability through inability to meet content uniformity, tablet weight, and/or hardness. Particle size may impact dissolution
Powder bulk density	X		Affects flowability and compactability of the powder during tableting, which could lead to low potency or broken tablets (efficacy). Incorrect bulk density affects downstream manufacturability through inability to meet content uniformity, tablet weight, hardness, and/or appearance. These variations could have an impact on tablet dissolution
Tablet hardness		X	Ensures that the integrity of tablets is not compromised
Dose uniformity		X	Ensures dose is uniform across the batch
Dissolution		X	Impacts the rate of drug release
Microbial limits		X	Ensures safety of the drug product

[a]Physical form should be amorphous for dispersion and tablet.

and responses considered for each of the potentially critical process steps identified to affect tablet CQA of dissolution. Table 11.6 provides an overview of experiments that either need to be performed on individual unit operations or need to be linked across multiple operations to define process design space for tablet CQA of dissolution.

SDD material attributes that could affect tablet dissolution are SDD particle size and bulk density, as well as physical form. The SDD physical form must be amorphous in

TABLE 11.4. Initial Risk Assessment for Telaprevir Materials and Processing Steps for the CQA of Dissolution

	Potential for Lot-to-Lot Variability in Materials to Affect Tablet CQA of Dissolution				Potential for Process Step to Affect Tablet CQA of Dissolution		
Tablet CQA	Telaprevir SDD	Filler	Disintegrant	Lubricant	Blending	Compression	Coating
Dissolution	Yes	No	No	No	Yes	Yes	Yes
Conclusion	Potentially critical	Not critical	Not critical	Not critical	Potentially critical	Potentially critical	Potentially critical

TABLE 11.5. List of Parameters Identified to Affect the CQA of Dissolution

	SDD Process Steps		Tablet Process Steps		
	Spray Drying	Secondary Drying	Blending	Compression	Film Coating
Parameters	Equipment Solids load Nozzle diameter Feed pressure Outlet temperature Condenser temperature Drying gas flow rate	Drying temperature N_2 flow rate % Fill volume Wet SDD property Agitation rate	Lube blend revolution	SDD PS SDD BD Main compression force	Coating weight gain
Output to evaluate	PS BD Physical form	PS BD Physical form	% RSD	Hardness Dissolution Friability Thickness Appearance	Dissolution Water content

PS: particle size distribution; BD: bulk density.

order to ensure acceptable tablet dissolution. These SDD attributes are controlled as part of spray drying and secondary drying process and thus it is important to perform experiments to understand the functional relationships among the process parameters (of spray drying) and SDD CQAs of particle size and bulk density.

For telaprevir SDD, multiple factorial experimental designs were conducted evaluating nine factors that included both formulation and process parameters. Two types of spray dryers (FSD12.5 and PSD4) were investigated, along with eight parameters that have the potential to affect particle size and bulk density: solids load, nozzle orifice diameter, feed pressure, outlet temperature, condenser temperature, drying gas flow rate, sheath gas flow rate, and sheath gas temperature. A summary of factorial designs that were used to quantify the effect of the spray drying process parameters on particle size and bulk density are outlined in Table 11.7. The particle size and bulk density of the

TABLE 11.6. Overview of Experiments for Defining Telaprevir Design Space for the CQA of Dissolution

Experiment	Spray Drying	Secondary Drying	Blending	Compression	Film Coat	Predecessor	
1	P						Effect of spray drying on wet SDD properties and physical form
2		P				1	Effect of drying profile on SDD properties and physical form
3				L		2	Effect of SDD properties on compression and dissolution
4					L	3	Effect of core tablet properties and coating weight gain on dissolution
5			L				Effect of blend parameters on blend uniformity
6				L		5	Effect of PAR blend parameters on dissolution
7	C	C		C			Effect of SDD properties and compression on dissolution

L: laboratory scale; P: pilot scale; C: commercial scale.

TABLE 11.7. Summary of the Experiments Used to Determine Telaprevir Spray Drying Design Space

	Design	N	Spray Dryer Type	Solids Load (%, w/w)	Nozzle Diameter (mm)	Feed Pressure (bar)	Outlet Temperature (°C)	Condenser Temperature (°C)	Drying Gas Flow Rate (kg/h)	Sheath Gas Flow Rate (kg/h)	Sheath Gas Temperature (°C)
1	$2^{4-1}+2+2$	12	FSD12.5		✓	✓	✓			N/A	N/A
2	2^2+1	5	FSD12.5			✓	✓			N/A	N/A
3	3×2^2	12	FSD12.5	✓		✓	✓			N/A	N/A
4	2^2	4	FSD12.5		✓		✓	✓		N/A	N/A
5	2^2+1	5	FSD12.5			✓	✓			N/A	N/A
6	2^2+1	5	FSD12.5			✓	✓			N/A	N/A
7	2^3	8	FSD12.5			✓	✓	✓		N/A	N/A
8	$2^{3-1}+2^3+2^{3-1}+2$	18	FSD12.5			✓	✓	✓		N/A	N/A
9	$2^{5-1}-1+4+2^{4-1}$	27	FSD12.5	✓	✓	✓	✓	✓		N/A	N/A
10	2^3+3	11	PSD4			✓	✓	✓		✓	N/A
11	2^3+7	15	PSD4			✓		✓	✓	✓	✓
12	$2^{4-1}+6$	14	PSD4		✓	✓	✓	✓		✓	
13	$2^{4-1}+4$	12	PSD4	✓		✓	✓	✓			

The factorial experimental design nomenclature is given as $nL^{a-b}+c$, where L is the number of parameter levels evaluated (two levels were evaluated throughout these designs, that is, high and low levels); a is the number of parameters; b defines the degree of fractionation in the design, given as 1/2b, that is, $b=1$ generates a half factorial, $b=2$ generates a quarter factorial, and so on; c is the number of center points (or additional data points outside the design); and n defines the number of times the design was repeated. The "✓" symbol denotes process parameters that were varied in the study design.

spray-dried dispersion prior to secondary drying (hereinafter referred to as wet SDD) were the response factors. A total of 148 runs were completed in the 13 factorial designs. Another 14 runs were added from data attained during the production of the associated designs, creating a final data set of 162 runs. These large numbers of runs, when both API and spray drying time were available, provided opportunities to learn and establish a more thorough understanding of the spray drying process. This understanding will be extremely helpful for development of future molecules.

Particle size spray drying model statistics [47] (regression model, parameter estimates, and prediction profiler) are summarized in Figure 11.6a–d. The prediction profiler plots indicate the relative magnitude of the effect of each process parameter. For particle size, the significant process parameters are solids content of the spray drying mixture, nozzle orifice diameter, feed pressure, outlet temperature, and spray dryer equipment. Each of these parameters has a similar magnitude of effect on particle size in the range investigated, except for spray dryer type that has a smaller effect: on average the particle size shifts by 7 μm comparing the PSD4 with the FSD12.5. Also shown is that process variables can be adjusted when changing spray dryers to attain a similar particle size. Equation 11.1 summarizes a particle size spray drying model. In the similar manner, a bulk density spray drying model is generated.

$$\begin{aligned} \mathrm{PS}_{\mathrm{wet}} = a - bP_{\mathrm{feed}} - c\mathrm{SD}_{\mathrm{Type}} - dT_{\mathrm{out}} - e\mathrm{TS} - fD_{\mathrm{Nozzle}} \\ + gP_{\mathrm{feed}} \cdot \mathrm{SD}_{\mathrm{Type}} + hD_{\mathrm{Nozzle}} \cdot \mathrm{TS} \end{aligned} \tag{11.1}$$

where $\mathrm{PS}_{\mathrm{wet}}$ is the volume median diameter of wet SDD (μm), P_{feed} is the feed pressure (bar), $\mathrm{SD}_{\mathrm{Type}}$ is the spray dryer type (FSD12.5 = 1; PSD4 = −1), T_{out} is the outlet temperature (°C), TS is the total solids content (%, w/w), and D_{Nozzle} is the nozzle diameter (mm).

The spray drying design space for the CQAs of particle size and bulk density is defined by the relationships between (i) wet particle size, wet bulk density, and spray drying process parameters and (ii) the change in particle size and bulk density from wet SDD (post-spray drying) to dry SDD (post-secondary drying) by a series of design equations. The design equations are bounded by the control limits for particle size and bulk density.

Contours plots can be used to illustrate spray drying design space, although the effect of only two process parameters on the CQAs of PS and BD can be shown in each contour plot. The effects of outlet temperature and feed pressure on particle size and bulk density are illustrated in Figure 11.7. The contours in Figure 11.7a show predicted particle size as a function of feed pressure and T_{out}. The acceptable particle size range is 50–110 μm, and all combinations of feed pressure and outlet temperature studied produced particle size within acceptable limits, apart from a small region in the upper right-hand quadrant represented by the dotted black line. For the CQA of particle size, the acceptable range of feed pressure and outlet temperature is defined approximately by the rectangle that encompasses the knowledge space. Figure 11.7b shows the predicted bulk density as a function of these parameters. The acceptable range of bulk density is 0.33–0.52 g/cm^3, and the dotted black line defines the boundary of unacceptably low bulk density, which constrains the acceptable region for outlet temperature and feed pressure, as illustrated. The region in which both particle size and bulk density are within

(a) Particle size regression (left-hand plot) and residuals (right-hand plot). Open = FSD12.5; closed = PSD4 a

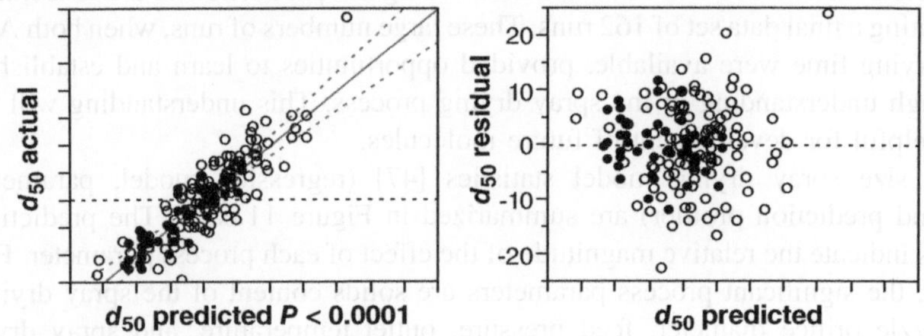

(b) Summary of fit

R^2	0.758
Adjusted R^2	0.747
Root mean square error	7.30
Mean of response	80.4
Observations	162

(c) Particle size sorted parameter estimates

Term	*t* Ratio	Prob > |*t*|
Spray dryer type	11.42	<0.0001*
Feed paressure (bar)	–11.26	<0.0001*
T_{out} (°C)	–10.93	<0.0001*
TS (%)	7.23	<0.0001*
Nozzle diameter (mm)	6.63	<0.0001*
TS (%) x nozzle diameter (mm)	3.63	00.0004*
Feed pressure (bar) x spray dryer type	3.56	00.0005*

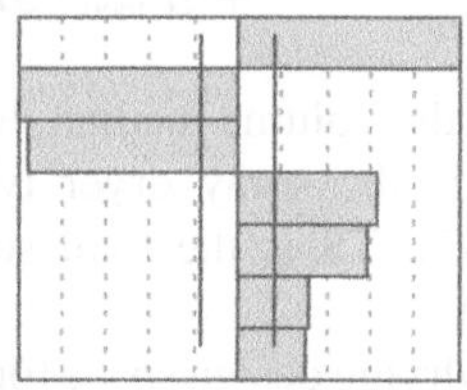

(d) Prediction profiler showing magnitude of the effect of each significant process variable on median particle size (d_{50}) for telaprevir SDD

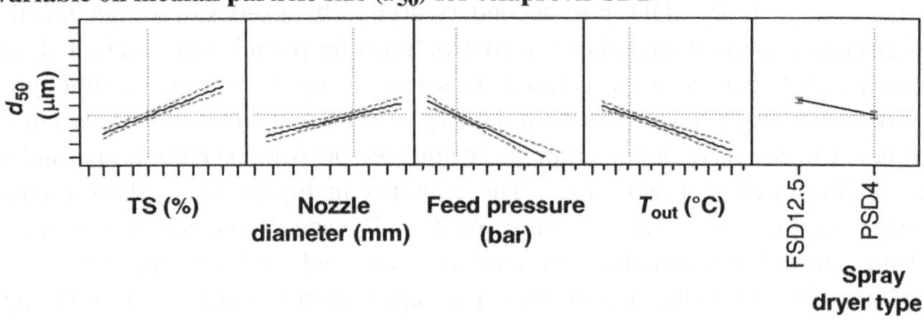

Figure 11.6. Particle size model for telaprevir SDD.

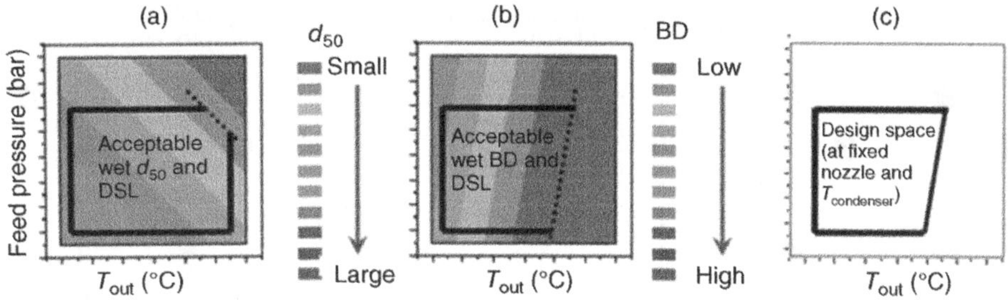

Figure 11.7. Example design space for feed pressure and outlet temperature for the telaprevir spray-dried dispersion manufacturing process.

range is shown in Figure 11.7c, which shows that the design space is an irregular shape, incorporating all possible combinations of feed pressure and outlet temperature that result in both particle size and bulk density being acceptable.

When design space complexity precludes definition of a straightforward NOR, alternative approaches are used to maintain simplicity during manufacturing. The SDD spray drying model relates the impact of various adjustable process parameters on PS and BD. The process parameter NORs (i.e., combinations of process parameters that will result in SDD within the particle size and bulk density NOR range) vary depending on other fixed and variable inputs. Therefore, model calculations are made in real time, and a continuously updated NOR is presented to the operator in a "dashboard" format. The dashboard identifies updated NOR and PAR (proven acceptable range) limits and the location of current operation, and suggests what changes can be made to adjustable parameters to maintain both PS and BD within their NOR ranges.

The intended use of the spray drying design equations is to (i) establish the initial spray drying conditions (set points) and (ii) estimate the adjustment in spray drying parameters required to recenter the response following in-process particle size and bulk density measurement.

SDD material attributes (particle size and bulk density) in combination with the tablet compression process parameters would be expected to affect tablet CQA of dissolution. For example, small PS SDD compressed to the same tablet hardness as large PS SDD would be expected to provide faster dissolution profile. Therefore, DoEs linking spray drying and compression processes have to be executed to study the interaction between these steps. This is illustrated in Figure 11.8. The spray drying DoE produces SDD with a range of particle size and bulk density, which are then used as inputs to a downstream DoE on tablet compression.

A regression model was developed to predict tablet dissolution as a function of SDD particle size and bulk density and average tablet hardness. A total of 241 tablet sublots were used to produce the dissolution models. The project team used mixed stepwise linear regression to determine the parameters affecting the mean cumulative release of telaprevir API from the tablet after 15 min, normalized by the 90 min time point. This early time point was selected as this part of the dissolution curve was considered to be more sensitive to processing and SDD particle size and bulk density effects on

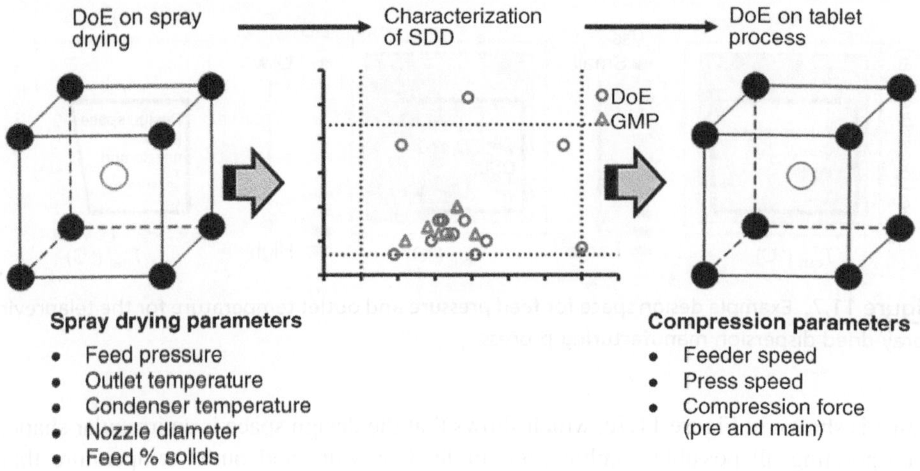

Figure 11.8. Interaction between outputs of spray drying and tablet compression.

dissolution. A criterion of mean dissolution ≥80% at 15 min was selected as a somewhat conservative criterion to define the design space for compression, to ensure that the dissolution specification (Q at 45 min = 80%) would be met, taking into account expected normal dissolution variability. The results of the normalized t_{15} dissolution model are summarized in Figure 11.9a–d.

The relationship between dissolution, tablet hardness, and SDD particle size and bulk density is shown in Equation 11.2. This is illustrated by the contour plots in Figure 11.10, showing the effect of particle size and bulk density on dissolution for tablets having low, target, and high hardness, respectively. The contour plots show that dissolution at a given hardness decreases with increasing SDD particle size and bulk density. As hardness increases, dissolution decreases and the range of SDD particle size and bulk density at which it is possible to achieve acceptable dissolution becomes smaller.

$$D_{15} = a + b \cdot \mathrm{PS} + c \cdot \mathrm{BD} + d\overline{H} - e \cdot \mathrm{PS} \cdot \mathrm{BD} - f \cdot \mathrm{BD} \cdot \overline{H} - g\overline{H}^2 \qquad (11.2)$$

where D_{15} is the average cumulative release after 15 min normalized to the 90 min infinity spin (%), PS is the SDD particle size (d_{50}, μm), BD is the bulk density (g/cm^3), and $\overline{H}$ is the average hardness of five tablets (kP).

This model was then reduced to a relationship between particle size and bulk density by assuming the minimum mean dissolution at 15 min to be 80%, and a mean average tablet hardness of 13 kP (which was the lower limit of the average tablet hardness DSL). The relationship is described by the following equation:

$$\mathrm{PS} \leq \frac{a + b \cdot \mathrm{BD}}{c \cdot \mathrm{BD} - d}$$

(a) Regression and residual plots of the normailized t_{15} dissolution model

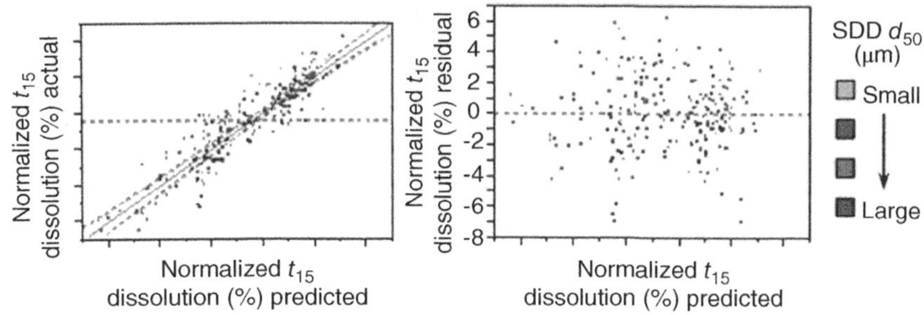

(b) Summary of fit

R^2	0.831
Adjusted R^2	0.826
Root mean square error	2.36
Mean of response	84.0
Observations (or sum weights)	241

(c) Parameter estimates of the normalized t_{15} dissolution model

Term	Prob>|*t*|
Intercept	<0.0001
d_{50} (μm)	<0.0001
SDD bulk density (g/cm³)	<0.0001
Average hardness of 5 tablets (kP)	<0.0001
(d_{50} (μm)-*x*) x (SDD bulk density (g/cm³)-*y*)	<0.0001
(SDD bulk density (g/cm³)-*y*) x (average hardness of 5 tablets (kP)-*z*)	0.0003
(Average hardness of 5 tablets (kP)-*z*) x (average hardness of 5 tablets (kP)-*z*)	<0.0001

(d) Prediction profiler for dissolution

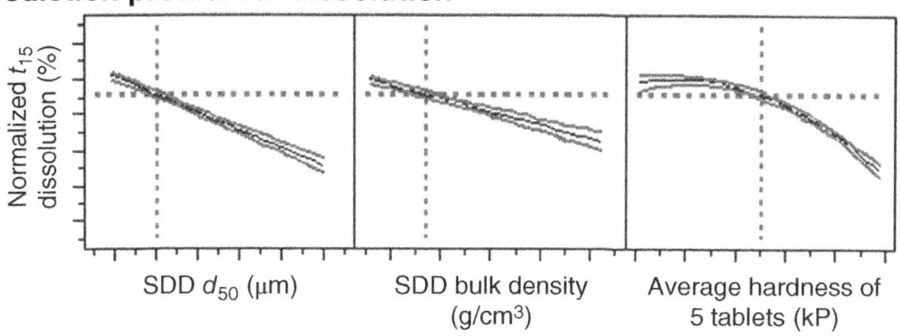

Figure 11.9. Normalized t_{15} dissolution model.

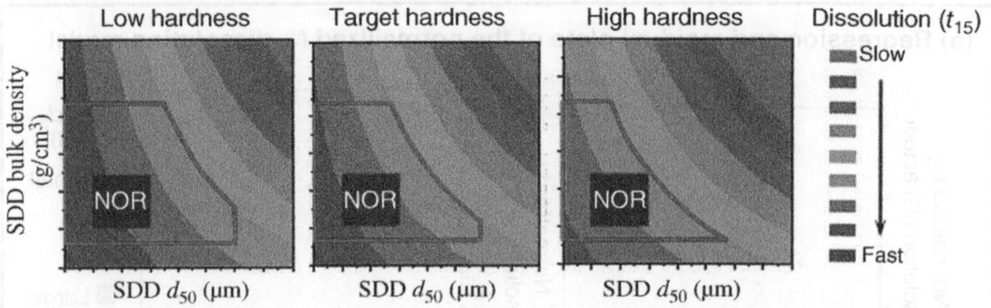

Figure 11.10. Design space for spray-dried dispersion particle size and bulk density at low, target, and high tablet hardness.

This equation represents the limiting design space equation for SDD particle size and bulk density that can be compressed to achieve acceptable dissolution; that is, the combined SDD particle size and bulk density must produce $D_{15} \geq 80\%$ for a minimum tablet hardness of 13 kP.

The NORs for particle size and bulk density ensure that the tablet dissolution CQA will be met across the acceptable range of tablet hardness. The DSLs for particle size and bulk density reflect the ranges over which acceptable dissolution could be achieved, with at least one acceptable combination of other parameters. For example, SDD with particle size of 110 μm can produce tablets with acceptable dissolution, provided that the bulk density is lower than 0.38 g/cm^3 at 13 kP average hardness. The design space equation for particle size and bulk density ensures that the dissolution CQA will be met with a minimum average tablet hardness of 13 kP.

The NORs for SDD particle size and bulk density are both well within the DSL ranges, at the target hardness, as shown in Figure 11.10. In addition, the spray drying process is under excellent control for the CQAs of SDD particle size and bulk density, as described in the spray drying model. Therefore, although SDD particle size and bulk density are critical quality attributes controlled via the SDD control strategy, with respect to the tablet CQA of dissolution they were defined as key since they have the potential to affect dissolution. However, the NOR is well within the DSL and they are well controlled. The NORs and DSLs for the SDD particle size and bulk density, average tablet hardness following compression, and compression processing parameters are summarized in Table 11.8.

11.3.3 Conceptual Approach to Defining Telaprevir Shelf Life

Although the solid dispersions may enhance aqueous solubility, they bring numerous stability issues such as recrystallization of the amorphous material and/or reduced chemical stability. Therefore, careful formulation design and process control is needed (during both spray drying and downstream processing) to prevent thermodynamic and chemical instability during processing and subsequent storage. Such instability can arise from interactions between excipients and the amorphous dispersion. In addition, compression, heat, and exposure to moisture, which can occur at different stages on

TABLE 11.8. Criticality Summary for Effect of Compression on the CQA of Dissolution

Step	Process Parameter/ In-Process Control	Normal Operating Range	Design Space Limit	Criticality
SDD critical material attributes	Particle size (μm)	60–80	NLT 47 and NMT 110 or $\frac{a + b \cdot \mathrm{BD}}{c \cdot \mathrm{BD} - d}$, whichever is lower	Key
	Bulk density (g/cm^3)	0.35–0.42	0.33–0.56	Key
	Average tablet hardness (kP)	NLT 13 and NMT 22	NLT 13 and NMT 22 or $a - b \cdot \mathrm{BD} + c\sqrt{d + e \cdot \mathrm{BD} + f \cdot \mathrm{BD}^2 + g \cdot \mathrm{PS} - h \cdot \mathrm{PS} \cdot \mathrm{BD}}$, whichever is lower	Critical
Compression	Precompression force (MT)	0.2–1.6	0.2–2.4	Noncritical
	Main compression force (MT)	0.8–5.0	0.8–7.0	Noncritical
	Press speed (rpm)	20–32	20–32	Noncritical

TABLE 11.9. Minimum Guideline from the ICH Q1A(R2) for Room-Temperature Product

Study	Storage Condition	Minimum Time Period Covered by Data at Submission
Long term[a]	25 ± 2 °C/60 ± 5% RH or 30 ± 2 °C/65 ± 5% RH	12 months
Intermediate[b]	30 ± 2 °C/65 ± 5% RH	6 months
Accelerated	40 ± 2 °C/75 ± 5% RH	6 months

[a]It is up to the applicant to decide whether long-term stability studies are performed at 25 ± 2 °C/60 ± 5% RH or 30 ± 2 °C/65 ± 5% RH.

[b]If 30 ± 2 °C/65 ± 5% RH is the long-term condition, there is no intermediate condition.

the way to a coated solid dosage form, are all known risk factors for promoting physical phase transformations and chemical instability.

Stability is examined in two different manners: first by meeting at least the minimum regulatory requirement needed to launch a drug and then by examining the root causes of degradation, if any. From a scientific/engineering perspective, it is important to determine what affects the stability of the drug in order to design a drug product process around these factors. Regulatory requirements for stability need to be done in the intended primary commercial package. ICH guidance [48] governs the expectations of pivotal stability studies. The minimum required for submission is shown in Table 11.9; real-time data are needed for shelf life dating over 2 years.

The QbD approach incorporates studies that increase our overall understanding of stability and allow identification of the critical parameters for stability. These studies (see Table 11.10 for details on design and testing) include the following:

1. The end-to-end (E2E) stability study justifies the independent shelf lives of the SDD and tablets.
2. The temperature and relative humidity (*T*/RH) stability study allows for mechanistic understanding of factors responsible for chemical instability of tablet.
3. Physical stability studies and solid-state experimental investigations assess physical stability of SDD and tablets under the proposed storage conditions and shelf life.

The purpose of the *T*/RH stability study was to develop a drug product chemical stability model that correlates the formation of epimer, the primary degradant of telaprevir, with the parameters that impact its formation, namely, surface pH of the excipient (traced to residual impurity from excipient), moisture content of the tablet, storage temperature, and time.

Tablets were equilibrated at varying RH (typically ranging from 10 to 60% RH, with corresponding tablet moisture contents of 0.4–3.4%), and then hand-sealed in foil pouches to ensure constant moisture content. These tablets were then placed on stability at storage conditions of 25, 30, and 40 °C for up to 12 months. Multiple studies were conducted, using excipient having varying levels of impurity. Figure 11.11 shows %

TABLE 11.10. Overview of Stability Study Designs for Defining Shelf Life for Solid Dispersion-Based Products

Study	Formulation	Purpose	Design	Tests
End-to-end study	SDD and tablet	Define separate shelf lives for SDD and tablet. Tablets manufactured from SDD at the end of its shelf life will still meet specifications at the end of the tablet shelf life	SDD: stored at 5 °C and 40 °C/75% RH for 3 and 6 months Tablets: prepared using initial SDD and aged SDD and stored at 5 °C, 25 °C/60% RH, and 30 °C/65% RH up to 36 months and at 40 °C/75% RH for up to 6 months	SDD: chemical stability, physical stability (crystallinity, relaxation), and moisture Tablets: physical stability (crystallinity), moisture, chemical stability, and dissolution
Temperature and relative humidity study	Tablet	Develop a drug product physical and chemical stability model	Include variables that are found to impact stability (e.g., moisture, excipient ROI, presence of oxygen, temperature, etc.) for 12 months	Tablets: physical stability (crystallinity), moisture, chemical stability, dissolution, and tablet color (if coated)
Solid-state studies	SDD	Characterization of the amorphous SDD with respect to T_g	SDD equilibrated to varying water amounts	T_g measurement (mDSC)
Solid-state studies	SDD	Characterization of the amorphous SDD with respect to structural relaxation	SDD: stored at $T_g - 15$ °C, $T_g - 25$ °C, and $T_g - 35$ °C	Relaxation using mDSC
Solid-state studies	SDD and tablet	Determining rate of crystallization using both fully amorphous SDD and SDD containing known amounts of crystalline seeds	SDD and tablet: stored open at 40 °C/100% RH and 60 °C/100% RH up to 8 weeks	% Crystallinity in SDD and tablets

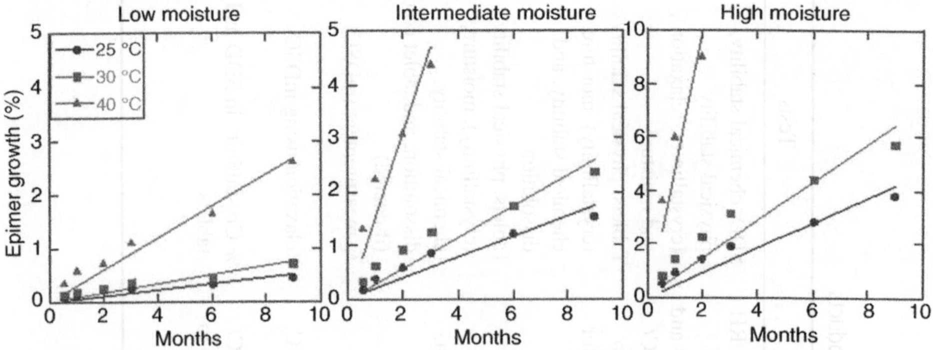

Figure 11.11. Effect of temperature and relative humidity on epimer growth in tablet.

epimer growth over time as a function of temperature and relative humidity at a constant impurity level.

A regression model was developed that predicts growth of epimer as a function of storage temperature, time, tablet moisture content, and excipient impurity. Stepwise linear regression was used to determine significant effects. The functional form of the regression equation was chosen based on the expected dependence of epimer growth on temperature and time according to the following equation:

$$k = A \exp(-E_a/RT),$$

where k is the degradation rate, A is a constant that depends on both tablet water content and excipient quality, E_a is the activation energy and is also dependent on the excipient quality, R is the universal gas constant, and T is temperature.

The actual and predicted increase in epimer using the model are in excellent agreement, as shown in Figure 11.12, and the model parameter estimates are tabulated in Table 11.11.

The regression model can be used to determine appropriate acceptance criteria for water content in tablets and impurity content in excipient based on the following:

- The predictions of the regression model are also in good agreement with traditional ICH stability data (Figure 11.13).
- The model appropriately predicts that formation of epimer increases as a function of impurity level of the excipient and storage temperature (Figure 11.13).

Finally, the model was used to project the increase in epimer as a function of tablet water content and excipient impurity after storage at 25 °C for 24 months and 30 °C for 12 months (assuming constant tablet moisture; thus, storage humidity was not considered) (Figure 11.14). The NOR for the proposed impurity level and moisture content combination is indicated in the shaded area on the 25 °C/60% RH contour plot in Figure 11.14. The DSL is chosen based on the assumption that the maximum epimer

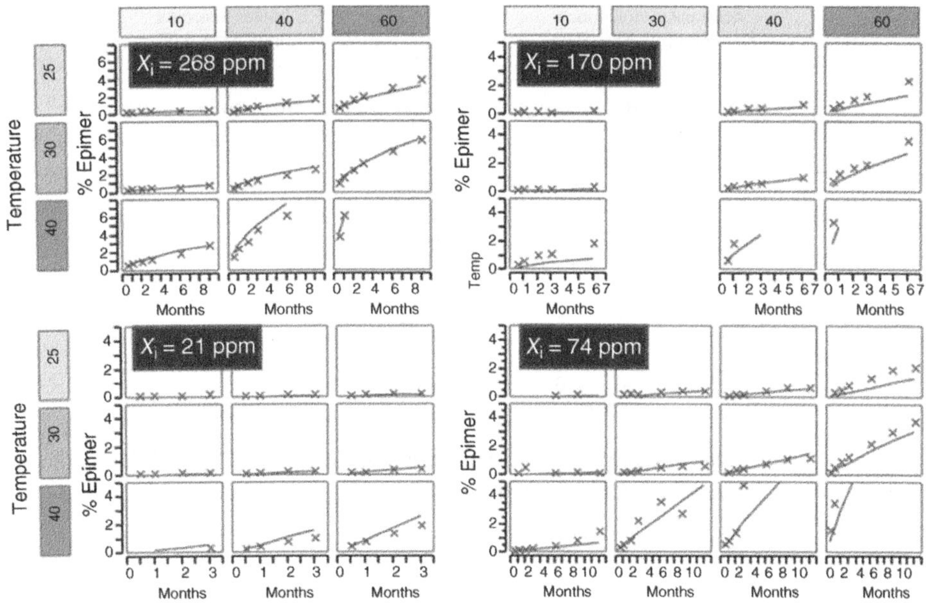

x: data; --: model; X_i = excipient impurity level

Figure 11.12. Effect of tablet water content, storage temperature, and excipient impurity level on the formation of epimer from the "*T*/RH" stability study.

content present in drug product at release is 2.5%, growing to a maximum of 4.5% during storage for 24 months at 25 °C/60% RH. As a result, the proposed DSL for impurity level and moisture content combination is indicated on the 25 °C/60% RH contour plot in Figure 11.14, which can be described by the following equation:

$$\mathrm{KF} = \left[\frac{\mathrm{e}^a}{b} \times t^{c-d\cdot\mathrm{Imp}}\ \mathrm{e}^{(-f+g\cdot\mathrm{Imp})/T}\ \mathrm{e}^{-h\cdot\mathrm{Imp}}\right]^i$$

TABLE 11.11. Parameter Estimates for Epimer Growth Regression Model

Term	Prob > $\lvert t \rvert$
Intercept	<0.0001
ln(time)	<0.0001
$1/T$ (K^{-1})	<0.0001
ln(KF)	<0.0001
Impurity (ppm)	<0.0001
(Impurity (ppm)-*a*) × (ln(time)-*b*)	<0.0001
(Impurity (ppm)-*a*) × (1/*T* (K)-*c*)	<0.0004

KF: Karl Fischer (water content titration).

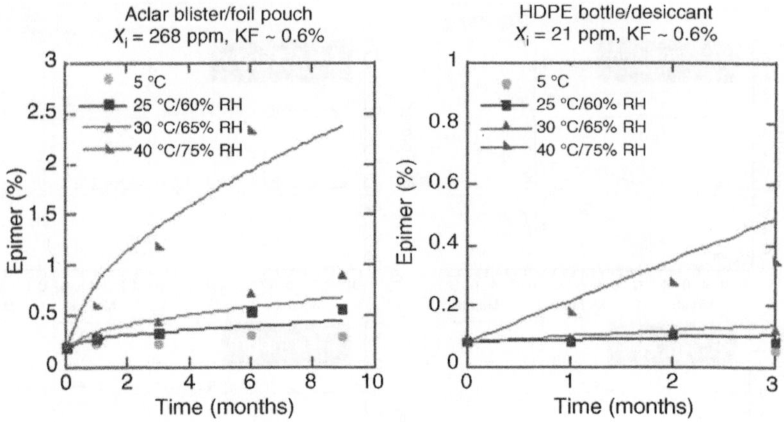

Figure 11.13. Predicted (regression line) versus observed (solid data points) formation of epimer in tablets in traditional ICH stability studies.

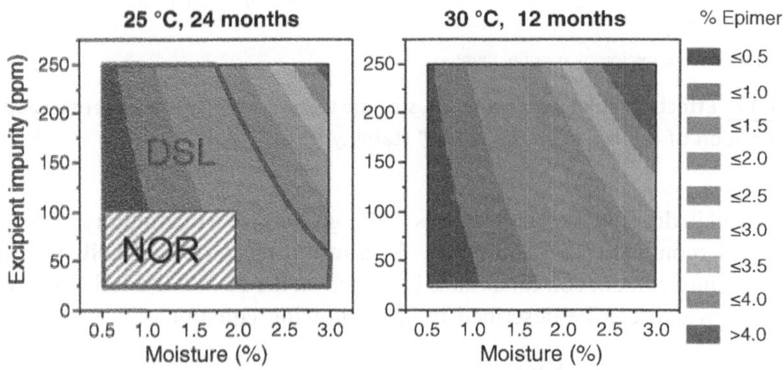

Figure 11.14. Predicted design space for epimer as a function of tablet water content and excipient impurity level with tablet storage at 25 °C for 24 months and 30 °C for 12 months from the "*T*/RH" stability study.

Based on the regression model described above, the allowable tablet water content level can be calculated for a target impurity level of the excipient (100 ppm). Two limiting cases were examined: (1) constant tablet moisture (corresponding to moisture-impermeable packaging), and (2) increasing tablet moisture (corresponding to moisture-permeable packaging). A 24-month shelf life at 25 °C/60% RH for drug product was assumed, as well as the backup accelerated condition of 12 months at 30 °C/65% RH.

For the first case (impermeable packaging), a maximum tablet water content of 2.4% must be maintained to keep epimer growth below 2.0% during a 24-month shelf life of drug product with storage at 25 °C/60% RH (Figure 11.14). To limit epimer growth to a maximum of 2.0% at the accelerated condition (30 °C/65% RH for 12 months), a lower tablet water content is required, about 2.1%. Based on manufacturing process capability,

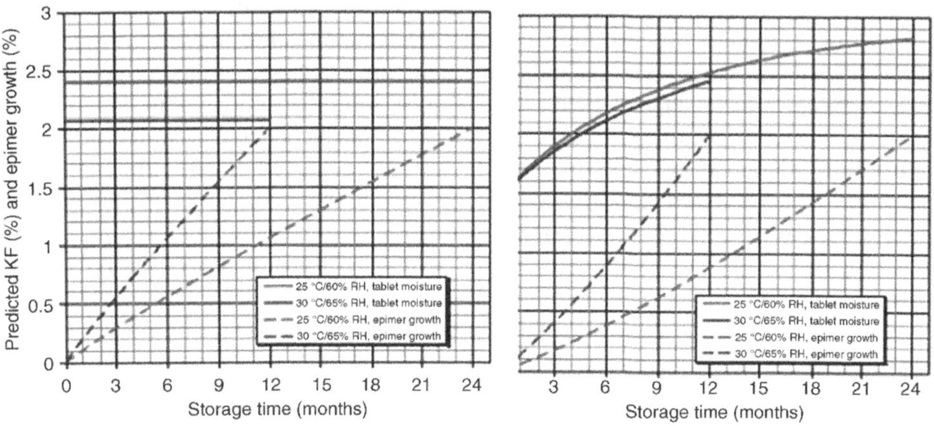

MVTR = 0 mg/day/tablet

MVTR = 0.149 mg/day/tablet for 24 months at 25 °C/60% RH
MVTR = 0.137 mg/day/tablet for 12 months at 30 °C/65% RH

Figure 11.15. Summary of model predicting the effect of storage condition on tablet water content and epimer growth for moisture-impermeable packaging and packaging with maximum moisture permeability.

a tablet water content of 1.5% at release is believed to be achievable, so for this limiting case (impermeable packaging) epimer growth would be limited to <2.0%.

For the second case (permeable packaging), moisture vapor transmission rate (MVTR) calculations were performed to determine the maximum package permeability that would maintain ≤2.0% epimer growth during 24 months storage of tablets at 25 °C/60% RH. In this scenario, the initial tablet water content is assumed to be 1.5%. For storage at 25 °C/60% RH for 24 months, the packaging MVTR must be below 0.149 mg/day/tablet (Figure 11.15). With this permeability, the tablet water content increases during storage from 1.5 to 2.8%. The accelerated condition is more stringent, and, at 30 °C/65% RH for 12 months, the maximum MVTR is 0.137 mg/day/tablet, corresponding to an increase in tablet water content during storage from 1.5% to a little less than 2.5%.

From this analysis, the maximum tablet water contents at release (1.5%) and on stability (2.5%) are defined. Also, the primary container closure system must have a maximum MVTR of 0.137 mg/day/tablet based on the more stringent requirement.

11.4 CONCLUDING REMARKS

Over the past two decades, the pharmaceutical industry has made significant progress on the drug discovery and design front that has yielded a number of potential drug candidates with poor or limited solubility and hence limited bioavailability. Solid dispersions are currently considered an effective method to solve the low bioavailability problem of poorly water-soluble drugs, but decades of research has only resulted in a handful of marketed drug products (Table 11.1). The primary challenges in developing solid dispersions as a drug delivery technology are technical challenges related to method of preparation, material processability, and dosage form variability. Physical stability and lack of reproducibility also limit the commercial applications of solid dispersions. Through proper material selection and formulation process based on first principles and state-of-the-art process characterization techniques, solid dispersion technology can be applicable to compounds with a broad range of physicochemical properties. The preparation of solid dispersions may be more easily achieved at both the laboratory and large scales through the advances in the development of novel manufacturing process technologies such as hot melt extrusion, spray drying, and freeze drying technologies that typically employ the poorly soluble API in admixture with polymeric materials as stabilizing agents.

The case study exemplified the application of QbD principles and tools to product and process development of telaprevir. The emphasis was on understanding the formulation and each unit operation at a mechanistic level and then integrating the outputs from one unit operation into the subsequent downstream processes. An experimental approach based upon DoE was outlined to identify CPPs and material attributes. The range of CPPs and material attributes that produce the desired product characteristics as outlined in CDAs formed the basis for establishing design space for telaprevir process. The knowledge gained was used to implement a flexible and robust manufacturing process that resulted in telaprevir product with consistent quality. To date, using defined design space we have manufactured tons of spray-dried dispersion and final coated tablets that are of high quality and low variability (Figure 11.16), further reinforcing the premise of QbD approach. This approach allowed us to derisk manufacturing, scale-up, and reproducibility challenges associated with commercialization of solid dispersion-based products.

With this case study, we show how QbD was successfully implemented for telaprevir. This approach is in alignment with the existing regulatory guidance documents. We hope the telaprevir case study, discussed in this chapter, will be very helpful in facilitating QbD adoption by the pharmaceutical industry and sharing best practices in research and development, manufacturing, quality, and regulatory areas.

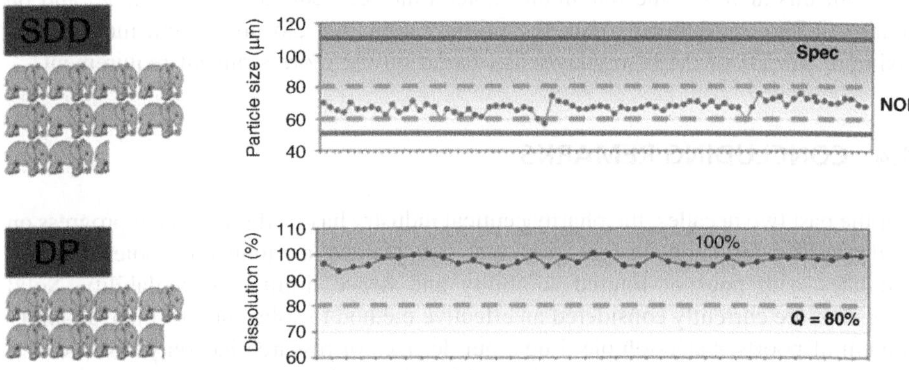

Figure 11.16. QbD benefit—tons of telaprevir product: high quality, low variability.

REFERENCES

1. Vasconcelos, T., Sarmento, B., and Costa, P. (2007) Solid dispersions as strategy to improve oral bioavailability of poor water soluble drugs. *Drug Discov. Today*, 12(23–24):1068–1074.
2. Chiou, W.L. and Riegelman, S. (1969) Preparation and dissolution characteristics of several fast-release solid dispersions of griseofulvin. *J. Pharm. Sci.*, 58(12):1505–1510.
3. Hancock, B.C. and Parks, M. (2000) What is the true solubility advantage for amorphous pharmaceuticals? *Pharm. Res.*, 17(4):397–404.
4. Six, K., Verreck, G., Peeters, J., Brewster, M., and Van den Mooter, G. (2004) Increased physical stability and improved dissolution properties of itraconazole, a class II drug, by solid dispersions that combine fast- and slow-dissolving polymers. *J. Pharm. Sci.*, 93(1):124–131.
5. Kawabata, Y., Wada, K., Nakatani, M., Yamada, S., and Pnoue, S. (2011) Formulation design for poorly water-soluble drugs based on biopharmaceutics classification system: basic approaches and practical applications. *Int. J. Pharm.*, 420(1):1–10.
6. Pokharkar, V.B., Mandpe, L.P., Padamwar, M.N., Ambike, A.A., Mahadik, K.R., and Paradkar, A. (2006) Development, characterization and stabilization of amorphous form of a low T_g drug. *Powder Technol.*, 167(1):20–25.
7. Van den Mooter, G., Weuts, I., De Ridder, T., and Blaton, N. (2006) Evaluation of Inutec SP1 as a new carrier in the formulation of solid dispersions of poorly soluble drugs. *Int. J. Pharm.*, 316(1–2):1–6.
8. Chauhan, B., Shimpi, S., and Paradkar, A. (2005) Preparation and evaluation of glibenclamide-polyglycolized glycerides solid dispersions with silicon dioxide by spray drying technique. *Eur. J. Pharm. Sci.*, 26(2):219–230.
9. Vasanthavada, M., Tong, W.-Q., Joshi, Y., and Kislalioglu, S. (2004) Phase behavior of amorphous molecular dispersions I: determination of the degree and mechanism of solid solubility. *Pharm. Res.*, 21(9):1598–1606.
10. Johari, G.P., Kim, S., and Shanker, R.M. (2005) Dielectric studies of molecular motions in amorphous solid and ultraviscous acetaminophen. *J. Pharm. Sci.*, 94(10):2207–2233.
11. Wang, X., Michoel, A., and Van der Mooter, G. (2005) Solid state characteristics of ternary solid dispersions composed of PVP VA64, Myrj 52 and itraconazole. *Int. J. Pharm.*, 303(1–2):54–61.
12. Breitenbach, J. (2002) Melt extrusion: from process to drug delivery technology. *Eur. J. Pharm. Biopharm.*, 54(2):107–117.
13. DiNunzio, J.C., Zhang, F., Martin, C., and McGinity, J.W. (2012) Melt extrusion. In: Williams, R.O., III, Watts, R.O, and Miller, D.A. (Eds.), *Formulating Poorly Water Soluble Drugs*, AAPS Press/Springer, New York, pp. 311–362.
14. Miller D.A. and Gil, M. (2012) Spray drying technology. In: Williams, R.O., III, Watts, A.B., and Miller, D.A. (Eds.), *Formulating Poorly Water Soluble Drugs*, AAPS Press/Springer, New York, pp. 363–442.
15. Passerini N., Albertini, B., Perissutti, B., and Rodriguez, L. (2006) Evaluation of melt granulation and ultrasonic spray congealing as techniques to enhance the dissolution of praziquantel. *Int. J. Pharm.*, 318(1–2):92–102.
16. Vilhelmsen, T., Eliasen, H., and Schaefer, T. (2005) Effect of a melt agglomeration process on agglomerates containing solid dispersions. *Int. J. Pharm.*, 303(1–2):132–142.

17. Seo, A., Holm, P., Kristensen, H., and Schaefer, T. (2003) The preparation of agglomerates containing solid dispersions of diazepam by melt agglomeration in a high shear mixer. *Int. J. Pharm.*, 259(1–2):161–171.
18. Gupta, M.K., Tseng, Y.-C., Goldman, D., and Bogner, R.H. (2002) Hydrogen bonding with adsorbent during storage governs drug dissolution from solid-dispersion granules. *Pharm. Res.*, 19(11):1663–1672.
19. Hurter, P., Thomas, H., Nadig, D., Embiata-Smith, D., and Paone, A. (2013) Implementing continuous manufacturing to streamline and accelerate drug development. *AAPS Newsmagazine*, August, pp. 14–19.
20. Schaber, S.D., Gerogiorgis, D.I., Ramachandran, R., Evans, J.M., Barton, P.I., and Trout, B.L. (2011) Economic analysis of integrated continuous and batch pharmaceutical manufacturing: a case study. *Ind. Eng. Chem. Res.*, 50(17):10083–10092.
21. Paudel, A., Worku, Z., Meeus, J., Guns, S., and Van den Mooter, G. (2013) Manufacturing of solid dispersions of poorly water soluble drugs by spray drying. *Int. J. Pharm.*, 453(1):253–284.
22. Srinarong, P., de Waard, H., Frijlink, H.W., and Hinrichs, W.L.J. (2011) Improved dissolution behavior of lipophilic drugs by solid dispersions: the production process as starting point for formulation considerations. *Expert Opin. Drug Deliv.*, 8(9):1121–1140.
23. Bikiaris, D.N. (2011) Solid dispersions, part I: recent evolutions and future opportunities in manufacturing methods for dissolution rate enhancement of poorly water-soluble drugs. *Expert Opin. Drug Deliv.*, 8(11):1501–1519.
24. Shah, N., Iyer, R.M., Mair, H.-J., Choi, D., Tian, H., Diodone, R. et al. (2013) Improved human bioavailability of vemurafenib, a practically insoluble drug, using an amorphous polymer-stabilized solid dispersion prepared by a solvent-controlled coprecipitation process. *J. Pharm. Sci.*, 102(3):967–981.
25. Sertsou, G., Butler, J., Scott, A., Hempenstall, J., and Rades, T. (2002) Factors affecting incorporation of drug into solid solution with HPMCP during solvent change co-precipitation. *Int. J. Pharm.*, 245(1–2):99–108.
26. Shah, N., Sandhu, H., Phuapradit, W., Pinal, R., Iyer, R., Albano, A. et al. (2012) Development of novel microprecipitated bulk powder (MBP) technology for manufacturing stable amorphous formulations of poorly soluble drugs. *Int. J. Pharm.*, 438(1–2):53–60.
27. Won, D.-H., Kim, M.-S., Lee, S., Park, J.-S., and Hwang, S.-J. (2005) Improved physicochemical characteristics of felodipine solid dispersion particles by supercritical anti-solvent precipitation process. *Int. J. Pharm.*, 301(1–2):199–208.
28. De Zordi, N., Moneghini, M., Kikic, I., Grassi, M., Del Rio Castillo, A., Solinas, D. et al. (2012) Applications of supercritical fluids to enhance the dissolution behaviors of furosemide by generation of microparticles and solid dispersions. *Eur. J. Pharm. Biopharm.*, 81(1):131–141.
29. Puri, V., Dantuluri, A.K., and Bansal, A.K. (2011) Investigation of atypical dissolution behavior of an encapsulated amorphous solid dispersion. *J. Pharm. Sci.*, 100(6):2460–2468.
30. Puri, V., Dantuluri, A.K., and Bansal, A.K. (2012) Barrier coated drug layered particles for enhanced performance of amorphous solid dispersion dosage form. *J. Pharm. Sci.*, 101(1):342–353.
31. Food and Drug Administration (2012) *Guidance for Industry PAT—A Framework for Innovative Pharmaceutical Development, Manufacturing, and Quality Assurance*. Available at http://www.fda.gov/CDER/GUIDANCE/6419fnl.htm.

32. Food and Drug Administration (2012) *Guidance for Industry. Q8(R2). Pharmaceutical Development.*
33. Food and Drug Administration, CDER (2012) *Draft Guidance for Industry and Review Staff: Target Product Profile-Strategic Development Tool.*
34. Food and Drug Administration, CDER (2012) *Guidance for Industry. Q8(R2). Pharmaceutical Development.*
35. ICH (2005) *Q9: Quality Risk Management.* Available at http://www.ich.org/fileadmin/Public_Web_Site/ICH_Products/Guidelines/Quality/Q9/Step4/Q9_Guideline.pdf.
36. ICH (2008) *Q10: Pharmaceutical Quality Systems.* Available at http://www.ich.org/fileadmin/Public_Web_Site/ICH_Products/Guidelines/Quality/Q10/Step4/Q10_Guideline.pdf.
37. ICH (2009) *Q8(R2): Pharmaceutical Development.* Available at http://www.ich.org/fileadmin/Public_Web_Site/ICH_Products/Guidelines/Quality/Q8_R1/Step4/Q8_R2_Guideline.pdf.
38. Kwong, A.D., Kauffman, R.S., Hurter, P., and Mueller, P. (2011) Discovery and development of telaprevir: an NS3-4a protease inhibitor for treating genotype 1 chronic hepatitis C virus. *Nat. Biotechnol.*, 29(11):993–1003.
39. Lavanchy, D. (2009) The global burden of hepatitis C. *Liver Int.*, 29(Suppl. 1):74–81.
40. Perz, J.F., Armstrong, G.L., Farrington, L.A., Hutin, Y.J., and Bell, B.P. (2006) The contributions of hepatitis B virus and hepatitis C virus infections to cirrhosis and primary liver cancer worldwide. *J. Hepatol.*, 45(4):529–538.
41. Hoofnagle, J.H. (2002) Course and outcome of hepatitis C. *J. Hepatol.*, 36(5 Suppl. 1): S21–S29.
42. Brass, V., Moradpour, D., and Blum, H.E. (2006) Molecular virology of hepatitis C virus (HCV): 2006 update. *Int. J. Med. Sci.*, 3(2):29–34.
43. Lawrence, S.P. (2000) Advances in the treatment of hepatitis C. *J. Adv. Int. Med.*, 45:65–105.
44. Kolykhalov, A.A., Mihalik, K., Feinstone, S.M., and Rice, C.M. (2000) Hepatitis C virus-encoded enzymatic activities and conserved RNA elements in the 3′ nontranslated region are essential for virus replication *in vivo*. *J. Virol.*, 74(4):2046–2051.
45. Cui, Y., Murphy, M., Dinehart, K., Hurter, P., and Connelly, P. (2006) Pharmaceutical compositions. U.S. Patent Application 20060089385.
46. Bittorf, K.J., Katstra, J.P., and Gaspar, F. (2010) Fluidized spray drying. U.S. Patent Application 20100011610.
47. SAS Publishing (2009) *JMP8: Statistics and Graphics Guide.*
48. ICH (2003) *Q1A(R2): Stability Testing of New Drug Substances and Products.* Available at http://www.fda.gov/downloads/drugs/guidancecomplianceregulatoryinformation/guidances/ucm073369.pdf.

12

SCIENTIFIC AND REGULATORY CONSIDERATIONS IN PRODUCT DEVELOPMENT*

Abhay Gupta, Ziyaur Rahman, and Mansoor A. Khan

Division of Product Quality Research, Office of Pharmaceutical Science, U.S. Food and Drug Administration, Silver Spring, MD, USA

12.1 INTRODUCTION

Ease of administration makes the oral route the most preferred route of drug delivery. Orally administered drugs must dissolve in the gastrointestinal tract to get absorbed and become bioavailable. Hence, the aqueous solubility of the drug substances becomes important. With a large number of new drug substances showing very low aqueous solubility, their development into commercially successful drug products is severely hampered by their limited bioavailability, unless appropriate pharmaceutical and formulation approaches are employed. Amorphous solid dispersions offer a practical and viable approach to enhance the oral bioavailability of these compounds, without the limitations of some other commonly used approaches such as particle size reduction, salt formation, and/or use of cosolvent systems. This chapter focuses on the scientific and regulatory considerations for developing amorphous solid dispersions of small-molecule drugs with poor aqueous solubility for oral administration.

* The findings and conclusions in this chapter have not been formally disseminated by the Food and Drug Administration and should not be construed to represent any Agency determination or policy.

Pharmaceutical Amorphous Solid Dispersions, First Edition. Edited by Ann Newman.

12.2 APPROVAL PROCESS IN THE UNITED STATES

Drug development is a multistep and multidisciplinary process involving discovery, development, Food and Drug Administration (FDA) review and approval, and post-approval, postmarketing safety surveillance to ensure continued safety and efficacy of the drug. The drug discovery process involves screening of a large number of novel compounds for pharmacological activity, followed by acute toxicity studies in animals of the promising compounds. For the new molecular entities that show promising pharmacological activity, the next logical step is to determine diagnostic and/or therapeutic potential in humans. The Federal Food, Drug, and Cosmetic Act (FD&C Act) considers these new molecular entities as new drugs that are subjected to specific requirements before they can be legally tested in human subjects.[1] The FDA is the agency of the U.S. government that is responsible for enforcing the provision of this FD&C Act. Drugs intended for human use are evaluated by FDA's Center for Drug Evaluation and Research (CDER) to ensure that they are safe and effective for their proposed use. Drugs that are approved by the FDA are said to be "safe and effective when used as directed." All drugs have side effects and no drug is absolutely safe. "Safe" in this sense means that the benefits of the drug appear to outweigh the known risks.

All new molecular entities undergo rigorous scrutiny by the FDA before they are approved for human use. Investigational new drug (IND) application is the vehicle that allows a sponsor to perform clinical trials in human subjects and also provides, to the sponsor, technical exemption for interstate shipment of the IND, without an approved market authorization.[2] IND submission is regulated by 21 CFR Part 312. There are three types of IND applications. An *Investigator IND* is submitted by a physician who both initiates and conducts an investigation and under whose immediate direction the investigational drug is administered or dispensed. A physician might submit a research IND to propose study on an unapproved drug, or on an approved product for a new indication or in a new patient population. An *Emergency Use IND* allows the FDA to authorize use of an experimental drug in an emergency situation that does not allow time for submission of an IND in accordance with 21CFR, Section 312.23 or Section 312.34. It is also used for patients who do not meet the criteria of an existing study protocol, or if an approved study protocol does not exist. A *Treatment IND* is submitted for experimental drugs showing preliminary evidence of drug efficacy for treating serious or immediately life-threatening conditions, or if no comparable alternative drug or therapy is available to treat that stage of the disease in the intended patient population. An immediately life-threatening disease is defined as a stage of a disease in which there is a reasonable likelihood of death occurring within a few months or in which premature death is likely without early treatment.

An IND application must contain information on three broad areas [1]: *Animal Pharmacology and Toxicology Studies* section includes all pertinent preclinical data that would permit an assessment as to whether the product is reasonably safe for initial testing in human subjects, including information on the pharmacological profile of the drug and

[1] Title 21, United States Code, Chapter 9: Federal Food, Drug, and Cosmetic Act, 2012 ed.

[2] Title 21, Code of Federal Regulations, Part 312. Investigational New Drug Application, 2012 ed.

the acute toxicity of the drug in at least two species of animals. If novel excipient is used to prepare the drug product, then the information on the pharmacological and toxicological profiles of the novel excipient is also included. Any previous experience with the drug in humans, for example, from another country whose population is relevant to the U.S. population, is also included as part of the information provided in this section. *Manufacturing Information* section includes information, pertaining to the composition, manufacturer, stability, and controls (used for manufacturing the drug substance and the drug product), that would ensure that the sponsor can adequately produce and supply consistent batches of the drug. *Clinical Protocols and Investigator Information* section includes detailed protocols for proposed clinical studies to assess whether the initial-phase trials will expose subjects to unnecessary risks. This section also includes information on the qualifications of clinical investigators, that is, professionals (generally physicians) who oversee the administration of the experimental compound, to allow an assessment of their qualification in fulfilling their clinical trial duties and commitments to obtain informed consent from the research subjects, to obtain review of the study by an institutional review board (IRB), and to adhere to the investigational new drug regulations.

Clinical trials are performed to evaluate the safety and effectiveness of an IND in treating, preventing, or diagnosing a specific disease or condition. The most important consideration in these trials is the safety of the human subjects. Hence, the FDA carefully monitors the design and conduct of these clinical trials to ensure that subjects in these trials are not exposed to unnecessary risks. The sponsors are therefore required to ensure the identity, quality, purity, and strength of the IND during the clinical trials and submit this information to the FDA. A sponsor must wait 30 days from the date of submitting an IND application before starting a clinical trial. This allows FDA to review the prospective study. If FDA finds problem(s) with the study, it may order a clinical hold to delay the start of the clinical trials. In some cases, FDA interrupts a clinical trial if problems are found during the study.

Clinical trials of an IND are generally divided into three phases, which are generally conducted sequentially. *Phase 1* is the initial introduction of the IND into human subjects. The subjects in these studies are generally healthy volunteers, although they may also be conducted in patients. The total number of subjects and patients included in phase 1 studies varies with the drug, numbering anywhere from 20 to 80 subjects. The studies in this phase are designed to determine the metabolism and pharmacological actions of the drug in humans, the side effects associated with increasing doses, and, if possible, to gain early evidence on effectiveness. These studies also evaluate drug metabolism, structure–activity relationships, and mechanism of action in humans, as well as explore the use of the IND in biological phenomena or disease processes. The information on the drug's pharmacokinetics and pharmacological effects collected in phase 1 clinical trials is used to design well-controlled, scientifically valid, phase 2 studies. *Phase 2* includes the early controlled clinical studies conducted in a relatively small number of patients, generally no more than several hundred subjects. The studies in this phase are designed to obtain preliminary data on the effectiveness of the drug for a particular indication or indications in patients with the disease or condition under study. These studies also help determine the common short-term side effects and risks associated with the drug. *Phase 3* includes expanded controlled and uncontrolled trials once enough evidence suggesting effectiveness of the drug has been obtained. The studies in this phase are intended to gather additional information about effectiveness and safety that is needed to evaluate the overall benefit–risk relationship of the drug. The studies in this phase, performed on several hundred to several thousand patients, also provide adequate basis for the information in the physician labeling.

New Drug Application (NDA) is the vehicle through which the sponsor formally proposes that the FDA approves the sale of a new drug in the United States.[3] It includes test data and analysis from the clinical trials in human subjects, along with full information on manufacturing specifications, stability and bioavailability data, method of analysis of each of the dosage forms the sponsor intends to market, packaging and labeling for both physician and consumer, and the results of any additional toxicological studies not already submitted in the IND application. The 1992 Prescription Drug User Fee Act (PDUFA) established a two-tiered system for the drug approval process. *Standard Review* is applied to a drug that offers, at most, only minor improvement over existing marketed therapies. The goal for completing a Standard Review is 10 months. Most NDAs undergo a Standard Review. *Priority Review* designation is given to drugs that offer major advances in treatment, or provide a treatment where none existed. The goal for completing a Priority Review is 6 months.

The results of the clinical trials comprise the single most important factor in the approval or disapproval of a new drug. In addition to the clinical and nonclinical data and analyses, an NDA must also provide sufficient information on chemistry, pharmacology, medical, biopharmaceutics, statistics, manufacturing and control of the drug to allow FDA to determine safety and effectiveness of the drug for the proposed use, risk–benefit of the drug, proposed labeling, and whether the manufacturing methods and controls are adequate to preserve the identity, strength, quality, and purity of the drug.

The FD&C Act that was enacted in 1938 only required NDAs to contain information pertaining to the safety of the new drug. The Kefauver–Harris Amendments to the FD&C Act, 1962, added the requirement of the NDAs to contain evidence of the effectiveness of the new drug for its intended use and that the established benefits of the drug outweighed its known risks. The requirements of the NDA were again modified in 1985, mainly to restructure the ways in which information and data are organized, thereby expediting the FDA review process. Although the exact requirements are a function of the nature of a specific drug, the NDA must provide all relevant data and information that a sponsor has collected during the product's research and development.

A review of an NDA involves FDA reviewers, from multiple scientific aspects, including, chemistry, pharmacology/toxicology, medical, biopharmaceutics, and statistics, reconfirming and revalidating the sponsor's conclusions on the safety and efficacy of the drug. If needed, additional or reanalysis of the submitted data is performed. The review team looks for possible issues with the application, such as weaknesses of the study design or analyses. Reviewers determine whether they agree with the sponsor's results and conclusions, or whether additional information is needed to make a decision. Each reviewer prepares a written evaluation containing conclusions

[3] Title 21, Code of Federal Regulations, Part 314. New Drug Application, 2012 ed.

and recommendations about the application. The technical review and recommendations from different review teams are combined in an action letter recommending approval, approvable, or nonapprovable decision on the application and a justification for that recommendation. If an application is not approved, FDA issues a Complete Response Letter to the sponsor listing all deficiencies in the application that the sponsor must address in order for the application to be reconsidered for approval. The deficiencies may be major (e.g., unexpected safety issues or failure to demonstrate effectiveness of the drug that may require additional clinical trials) or minor (e.g., labeling changes).

The final approval of an NDA, in some cases, is also dependent on a preapproval inspection of the sponsor's manufacturing facilities and clinical trial sites in which the FDA investigators audit and verifies manufacturing-related statements and commitments made in the NDA against the sponsor's manufacturing practices. Preapproval inspections are generally conducted for product (i) that is a new chemical or molecular entity, (ii) that has a narrow therapeutic range, (iii) that represents first approval for the applicant, or (iv) that is sponsored by a company with a history of CGMP problems or that has not been the subject of a CGMP inspection over a considerable period. Approval can be delayed if the facility is not ready for inspection. In addition, manufacturing deficiencies found during inspection must be corrected before an approval is granted by the FDA. Upon approval, all new drugs are available by prescription only.

Some new drugs, especially those for serious and life-threatening diseases that lack satisfactory treatments, may be granted an *accelerated approval* by the FDA before measures of effectiveness are available. For these drugs, FDA relies on less traditional measures called surrogate endpoints to evaluate effectiveness. These are laboratory findings or signs that may not be a direct measurement of how a patient feels, functions, or survives, but are considered likely to predict benefit. If studies do not confirm the initial results, FDA may withdraw an approval granted under the accelerated approval rules.

Most new drugs, like other new products, are frequently under patent protection giving their sponsors sole right to sell these drugs while the patent is in effect. When the patents or other periods of exclusivity expire, other drug product manufacturers may apply to the FDA to obtain market authorization to sell generic versions of these drug products in the United States. A generic drug product is one that contains the same active ingredient(s) as the innovator drug product, although the inactive ingredients may vary. It must also be identical in dosage form, strength, and route of administration and have the same use indications. They are required to meet the same requirements for identity, strength, purity, and quality, and be manufactured under the same strict standards of FDA's good manufacturing practice regulations required for the innovator drug products. The Waxman–Hatch Act of 1984 requires the generic drugs to be bioequivalent to the innovator drug, that is, they must deliver the same amount of active ingredients into a patient's bloodstream in the same amount of time as the innovator drug. For approval, generic drugs sponsors need to provide scientific evidence that the generic drug is interchangeable with or therapeutically equivalent to the originally approved drug. All approved products, both innovator and generic, are listed in FDA's publication titled *Approved Drug Products with Therapeutic Equivalence Evaluations*, commonly known as the Orange Book. The publication also identifies the Reference Listed Drug (RLD) with which new generic drug products need to be compared to show that they are bioequivalent, thus avoiding the possibility of significant variations among generic drugs and their brand name counterpart.

Approval for market authorization for the generic drugs is obtained upon successful review and approval of an *Abbreviated New Drug Application* (ANDA) filed with the FDA. Under the FD&C Act, a drug is considered "new" if it is made by a different manufacturer, uses different excipients or inactive ingredients, is used for a different purpose, or undergoes any substantial change.[4] Generic drug applications are called "abbreviated" because the sponsors are generally not required to include preclinical (animal) and clinical (human) data to establish safety and effectiveness. Instead, they are required to scientifically demonstrate that the generic drug is bioequivalent (i.e., performs in the same manner) to the innovator drug product. Bioequivalence may be demonstrated by comparing the rate of absorption, or bioavailability, of the generic drug with that of the innovator drug in 24–36 healthy volunteers [2]. An ANDA sponsor must refer to the RLD in its application when seeking approval to market a generic equivalent in the country. The ANDA review process focuses on the review of the chemistry, bioequivalence and microbiology data, labeling information, and inspection of the sponsor's manufacturing facilities. Upon approval, the generic drugs are considered to be therapeutically equivalent to the innovators product.

Over-the-counter (OTC) drugs are developed and marketed under the OTC Monograph process or through the already described NDA process. Each OTC drug monograph is a kind of "recipe book" covering acceptable ingredients, doses, formulations, labeling, and, in some cases, testing parameters. Although preapproval by FDA for drugs marketed under a drug monograph is not required, many companies seek assurance that the product they intend to market under the drug monograph complies with the regulations. FDA continually updates the OTC drug monographs to add new ingredients and labeling as needed. Products conforming to an OTC monograph may be marketed without FDA preapproval.[5] For products not confirming to OTC drug monograph, an approved NDA or ANDA is required before they could be marketed in the United States. A sponsor may also use the NDA process to introduce these products into the OTC marketplace. For example, OTC drug products previously available only by prescription are first approved through the NDA process and their "switch" to OTC status is approved via the NDA process. OTC drugs marketed overseas can be introduced into the U.S. market via a monograph under a Time and Extent Application (TEA) as described in 21 CFR 330.14.

Since all possible side effects of a drug cannot be identified or anticipated during the clinical trials, FDA maintains a system of postmarketing surveillance and risk assessment programs to identify adverse events that did not appear during the drug approval process [3]. FDA monitors adverse events such as adverse reactions and poisoning. The Agency uses this information to take preventive actions such as updating the drug labeling and, on rare occasions, to reevaluate the approval or marketing decision.

[4] Title 21, Code of Federal Regulations, Part 310. New Drugs, 2012 ed.

[5] Title 21, Code of Federal Regulations, Part 330. Over the-Counter (OTC) Human Drugs Which Are Generally Recognized as Safe and Effective and Not Misbranded, 2012 ed.

FDA also conducts periodic, unannounced inspections of drug production and control facilities to ensure that the drug is being manufactured in a consistent and controlled manner and the sponsor is adhering to the terms and conditions of approval described in the application.

In addition to the regulations of the FD&C Act, FDA routinely issues guidance documents to provide guidelines to the processing, content, and evaluation/approval of applications and also to the design, production, manufacturing, and testing of regulated products. These documents ensure consistency in the Agency's regulatory approach and establish inspection and enforcement procedures. Because guidance documents are not regulations or laws, they are not enforceable, either through administrative actions or through the courts.

12.3 THE ICH

Drug products are regulated in different parts of the world by the respective national or regional regulatory agencies, be it the European Medicines Agency (EMA) in the European Union (EU), Health Canada in Canada, or the Ministry of Health, Labour and Welfare (MHLW) in Japan. The approval process may include a comprehensive review of all data, similar to the FDA approval process, or a limited review if the drug has already been approved in one of the major foreign markets (the United States, Europe, and/or Japan). Different regulatory agencies may also use different approaches to the type and amount of data required as part of the submission process and may request additional studies/data that are not required by the FDA.

Differences in the technical requirements for drug product registration from country to country means unnecessary duplication of many time-consuming and expensive test procedures. This led to the realization of the need for greater harmonization in the interpretation and application of technical guidelines and requirements for product registration across different regions of the world. The International Conference on Harmonisation of Technical Requirements for Registration of Pharmaceuticals for Human Use (ICH) was initiated in 1990 in an effort to harmonize the scientific and technical aspects of drug registration and to ensure that safe, effective, and high-quality medicines are developed and registered in the most efficient and cost-effective manner across the European Union, Japan, and the United States (the ICH region). The harmonized aspects, that is, the ICH guidelines, have been adopted as law in several countries. However, they are used only as guidance documents by the U.S. FDA.

The ICH guidance documents were initially developed on safety, quality, and efficacy topics and later expanded to include multidisciplinary topics, including MedDRA (Medical Dictionary for Regulatory Activities) and the CTD (Common Technical Document). CTD is an agreed upon common format for the preparation of a well-structured harmonized application that will be submitted to regulatory authorities in the ICH region significantly reducing the time and resources needed to compile applications for registration of human pharmaceuticals. The CTD is organized into five modules [4,5]. Module 1 is region specific. Modules 2–5 are intended to be common for all regions. This ensures that the organization of the contents of a submission is the same for all

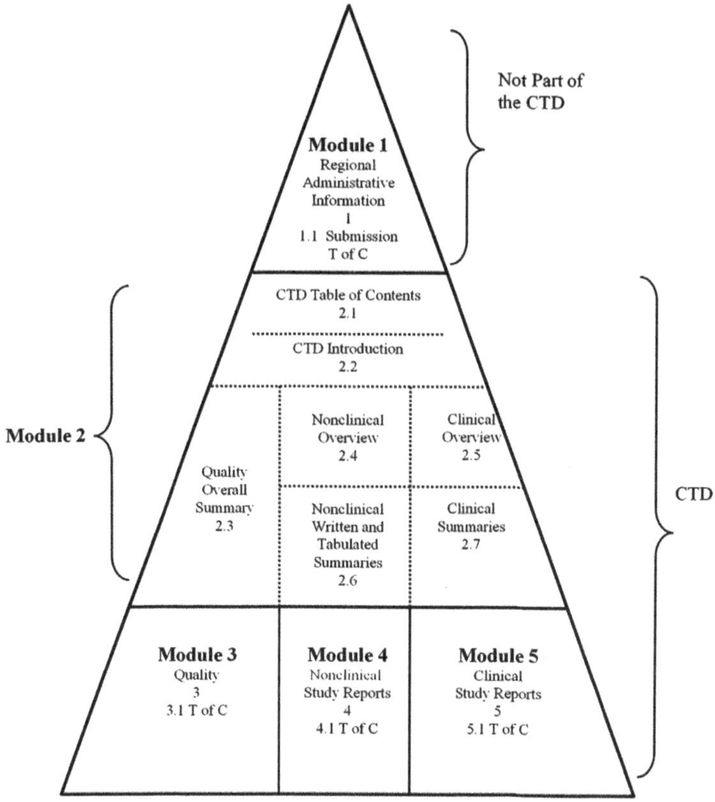

Figure 12.1. A diagrammatic representation of the CTD [4].

applications and related submissions. Conformance with the CTD guidance also ensures that these four modules are provided in a format acceptable to the regulatory authorities in the ICH region. A diagrammatic representation of the CTD is presented in Figure 12.1 and the location of regulatory requirements in relationship to various CTD modules is summarized in Table 12.1.

12.3.1 Module 1: Administrative Information and Prescribing Information

This module contains documents specific to each region; for example, application forms or the proposed label for use in the region. The content and format of this module are specified by the relevant regulatory authorities. In the United States, the documents in this section include FDA form 356h; comprehensive table of content; administrative documents such as patent information and certification, debarment certification, and user fee cover sheet; prescribing information; and the annotated labeling text for the NDAs, as described in 21 CFR 314.50(c)(2)(i) and 21 CFR 201.57, or the labeling comparison for the ANDAs, as described in 21 CFR 314.94(a)(8).

TABLE 12.1. Location of Regulatory Requirements under CTD Format [5]

CTD	NDA: 314.50	ANDA: 314.94 (Unless Otherwise Indicated)	BLA: 601.2 (Unless Otherwise Indicated)
Module 1	(a) Application form	(a)(1) Application form	(a) Application form
	(c)(2)(i) Annotated text of proposed labeling	(a)(2) Table of contents	
	(d)(1)(v) Statement of field copy	(a)(3) Basis for ANDA submission	
	(e) Samples and labeling	(a)(4) Conditions of use	(a) Labels, enclosures, and medication guides
	(h) Patent information	(a)(5) Active ingredients	Section 306(k)(1) and (2) debarment certification/list of convictions (FDC Act)
	(i) Patent certification	Section 306(k)(1) and (2) debarment certification/list of convictions (FDC Act)	
	(j) Claimed exclusivity	(a)(6) Route of administration, dosage form, and strength	
	(k) Financial certification or disclosure	(a)(8) Labeling requirements	
		(a)(12) Patent certification	
		(a)(13) Financial certifications or disclosure statement	(a) Financial certification or disclosure statement
		(d)(5) Certification of field copy	(a) Claim of categorical exclusion or environmental assessment
Module 2	(b) Comprehensive table of contents	N/A	
	(c) Summaries		(a) Summaries
	(d)(5)(vii) abuse potential		
Module 3	(d)(1) CMC	(a)(9) CMC	(a) Full description of manufacturing methods
		(a)(10) Samples	(a) Samples
Module 4	(d)(2) Nonclinical pharmtox	N/A	(a) Data from nonclinical studies
Module 5	(d)(3) Human pharmacokinetics	(a)(7) Bioequivalence/bioavailability information	
	(d)(4) Microbiology	Section 320.22(d)(2)(i) waiver of *in vivo* BA/BE	
	(d)(5) Clinical data		(a) Data from clinical studies
	(d)(6) Statistical section		
	(d)(7) Pediatric use		
	(f) Case report forms and tabulations		

12.3.2 Module 2: Common Technical Document Summaries

It begins with a general introduction to the pharmaceutical, including its pharmacological class, mode of action, and proposed clinical use, and contains seven sections in the following order: (i) CTD Table of Contents, (ii) CTD Introduction, (iii) Quality Overall Summary, (iv) Nonclinical Overview, (v) Clinical Overview, (vi) Nonclinical Written and Tabulated Summaries, and (vii) Clinical Summary.

12.3.3 Module 3: Quality

It contains information on *Quality* presented in the structured format described in the guidance document M4Q: The CTD—Quality [6]. For the drug substance, this includes general information such as name, physicochemical and other relevant properties, and name, address, and responsibility of each manufacturer, as well as description of manufacturing process and process controls, control of materials, critical steps and intermediates, process validation and/or evaluation, manufacturing process development, and information on the characterization and control of drug substance and impurities, materials used as reference standards, container closure system, and stability. For the drug product, this includes a description of the drug product and its composition, information on the development studies conducted to establish the dosage form, formulation, manufacturing process, container closure system, microbiological attributes, and usage instructions, the name, address, and responsibility of each manufacturer, including batch formula and description of the manufacturing process and process controls, especially of critical steps and intermediates, excipients, and the final drug product. Information on the materials used as reference standards, container closure system and stability of the drug product, and key literature referenced, if applicable, should also be provided.

12.3.4 Module 4: Nonclinical Study Reports

It contains information on *Safety* presented in the order described in the guidance document M4S: The CTD—Safety [7]. This includes study reports on the pharmacology (primary and secondary pharmacodynamics, safety, and drug interactions), pharmacokinetics (absorption, distribution, metabolism, and extraction), and toxicology (single and multiple dose toxicity, genotoxicity, carcinogenicity, reproductive and developmental toxicity, and toxicity in juvenile animals).

12.3.5 Module 5: Clinical Study Reports

It contains information on *Efficacy* presented in the order described in the guidance document M4E: The CTD—Efficacy [8]. This includes reports on biopharmaceutic studies, pharmacokinetics studies using human biomaterials, human pharmacokinetic and pharmacodynamic studies, and efficacy and safety studies. Copies of referenced documents, including important published articles, official meeting minutes, or other regulatory guidance or advice should be provided.

12.4 QUALITY-BY-DESIGN

A drug product should not only be safe and effective but also be of high quality to ensure consistent and reliable delivery of the clinical performance and other characteristics stated in the product label. A comprehensive understanding of the influence of the formulation components, environmental condition, and manufacturing process on the product quality over the life of the product is therefore needed to ensure that the product and the process consistently deliver the desired quality. Historically, the quality of pharmaceutical drug products was confirmed by end product testing to ensure that the product met the empirically derived product performance standards.

In an effort to improve the efficiency and effectiveness of both manufacturing and regulatory decision-making, FDA's Pharmaceutical CGMPs for the 21st Century: A Risk-Based Approach (the Pharmaceutical cGMP initiative) was launched in August 2002. The initiative aims to apply scientific and engineering principles toward improving the efficiency and effectiveness of both manufacturing and regulatory decision-making process. It encourages voluntary development and implementation of innovative approaches in pharmaceutical development, manufacturing, and quality assurance and the early adoption of new technological advances by the pharmaceutical industry. It also encourages implementation of modern quality management techniques and risk-based approaches to all aspects of pharmaceutical production and quality assurance, with focused attention on critical areas. *Quality- by-design* (QbD) is an important element of this initiative. It is a systematic approach toward product development that begins with predefined objectives and emphasizes use of sound science and quality risk management principles in product and process understanding and control [9].

A QbD approach to product development comprises of a systematic evaluation of product attributes, understanding and refining of the formulation and manufacturing processes that are critical, application of quality risk management principles to establish appropriate control strategies that ensure product quality, and use of knowledge management and formal design of experiments to generate, monitor, and refine the design space throughout the life cycle of the product. *Design space* is the multi-dimensional combination and interaction of input variables (e.g., material attributes) and process parameters that have been demonstrated to provide assurance of quality. It links the critical quality attributes (CQA) of the drug product to the critical material attributes and critical process parameters. CQAs are the physical, chemical, biological, or microbiological properties or characteristics that should be within an appropriate limit, range, or distribution to ensure the desired product quality. An input variable or process parameter need not be included in the design space if it does not influence CQAs when varied over the full potential range of operation. A movement within an approved design space is not considered to be a change. However, movement out of the design space is considered to be a change and normally initiates a regulatory postapproval change process, depending upon the nature of the change.

Design space is proposed by the applicant and is subject to regulatory assessment and approval. It may be based on an analysis of historical data or be defined in terms of ranges of input variables or parameters. It may also be defined in terms of simple or complex mathematical relationships linking CQAs to the input variables and process parameters. Individual design spaces, which are often simpler to develop, may be established for individual unit operations. Conversely, a single design space may be established spanning multiple operations based on a comprehensive evaluation, thus providing more operational flexibility. For the design space to be applicable to multiple operational scales, it should be described in terms of relevant scale-independent parameters. Use of scale-independent parameters also provides greater regulatory flexibility during technology transfer and/or site change.

12.5 DEVELOPMENT AND CHARACTERIZATION OF AMORPHOUS SOLID DISPERSIONS

A large number of new drug candidates are characterized by low oral bioavailability. Often the cause is poor aqueous solubility rather than the limited permeability in the gastrointestinal tract. The aqueous solubility of an orally administered drug governs its *in vivo* dissolution and thus its bioavailability. Several methods have been used to enhance their solubility and dissolution rate and thus improve their bioavailability. These include particle size reduction, complexation, solubilization, cosolvency, and use of prodrugs, salts, polymorphs, solvates, or hydrates with higher aqueous solubility [10–14]. However, every approach has its limitation. Size reduction may not be possible for all drugs or may not enhance the dissolution rate. Complexation with cyclodextrin is expensive and may not result in a suitable complex. Solubilization and cosolvency are typically used to prepare liquid formulations that may exhibit poor stability characteristics. Similarly, not all drugs are suitable candidates for prodrug, salt, polymorph, solvate, or hydrate formation. The aqueous solubility of a molecule is typically higher in its amorphous state because little or no energy is required to break up the crystal lattice during the dissolution process. Hence, amorphous solid dispersions (ASDs) have been used to successfully improve the solubility, dissolution, and bioavailability of a number of drugs with poorly aqueous solubility [15–18].

A number of drugs are currently marketed as ASD formulations, for example, tacrolimus, ritonavir, itraconazole, and entravirine. [19]. ASDs can be bi- or multicomponent systems. Bicomponent ASDs contain drug and carrier, while multicomponent ASDs contain drug with more than one carrier or with a combination of carrier(s) with other excipient(s). The carrier is responsible for ASD formation and for keeping the drug in the amorphous form. It can be a polymer such as HPMC, PEG, or an inorganic mineral like silicon dioxide [15,16,20]. The excipients are used as processing aids, for example, diluent and lubricant. The selection of suitable carrier is the key in obtaining a successful ASD formulation since the dissolution of one component in a multicomponent system is affected by the dissolution of the other component(s) present in that system. Carriers with high aqueous solubility are preferred for preparation of immediate release dosage forms, while carriers that are water insoluble, slowly soluble, or swell in aqueous environment are used to prepare controlled- or delayed-release dosage forms. Carriers with a high solubilization capacity are preferred because they can dissolve large quantities of drug. It is imperative to select a carrier that shows good physical and chemical compatibility with the drug and is stable at the processing temperature. In addition, the carrier should be nontoxic, pharmacologically inert, and have high solubility in aqueous/organic solvents. A high glass transition temperature is also considered to be an invaluable property for a good carrier as it helps in improving the physical stability of the drug and fastens the melting and miscibility of the drug in the carrier matrix [19,21]. The melting or the glass transition temperature of the carrier should be such that it allows a practical processing temperature and decreases the potential of drug degradation during processing. Amorphous stabilizers may be added to increase the miscibility of the drug in the carrier, increase the drug load, decrease the processing temperature, enhance the dissolution behavior of solid dispersions, or improve the stability upon storage.

Only excipients that are generally recognized as safe (GRAS), have an official compendial monograph, are listed in the FDA's inactive ingredients database, or have documented human use in the proposed level may be used to prepare drug products. Generic drug products, if not intended for parenteral, ophthalmic, or otic use, are not required to contain the same excipients as those present in the reference listed drug product. However, the sponsor of generic drug products must demonstrate that the inactive ingredients do not affect the safety or efficacy of the proposed drug product. An excipient is not considered new if it has appeared in an approved drug product for a particular route of administration. Otherwise all new excipients require extensive evaluation and review for pharmacological activity using a battery of standard test, including tests for chronic toxicity and carcinogenicity if needed, before being approved for use as an ingredient in a drug product [22]. This ensures that all potential new excipients are safe for use in human pharmaceuticals at the concentrations being used when administered via the proposed route of administration.

A number of methods have been used to prepare ASD [12,23,24]. Organic or hydroalcoholic solvent systems have been used to dissolve drug and carrier, followed by the removal of the solvent at room or slightly elevated temperature, or by spray/freeze-drying. These methods typically result in fine dispersion of the drug in the carrier polymer matrix at molecular levels. The limitations include the toxicity associated with the residual content of the organic solvents. Recently, use of inert, nontoxic, and nonflammable supercritical fluid as solvent, for example, carbon dioxide, has also been reported for the preparation of ASD. Other methods of preparation include complexing the drug with cyclodextrin and use of hot melt extruder (HME) to fuse the drug into the carrier matrix by application of heat and high shear forces. Use of HME has the advantage of being a fast, simple, continuous, and solvent-free process, but is limited by the high processing temperature.

The selection of the method for preparing the ASD is dependent on the thermal stability of the drug. If the drug is thermally stable, hot melt extrusion may be used. The process involves introducing a drug–carrier powder blend via a hopper into a heated barrel with co- or counterrotating twin screws, where the powder blend is first heated to a molten state in which it undergoes intensive mixing, before being pushed out through a die as a strand. The amount of heat and shear forces applied, as well as the rate of cooling after leaving the die determines the physical structure and stability of the ASD extruded material. The most important requirement with this method is that drug and carrier should be stable at the process temperature. The optimal process temperature is governed by the melting point of the drug, glass transition temperature of the amorphous carrier, and the thermoplastic properties of the carrier [24]. Additives may be added as plasticizers to lower the process temperature [25]. A porous extrudate may be obtained by using supercritical carbon dioxide as the plasticizer resulting in faster dissolution [26]. The selection of extrusion conditions such as feed rate, screw speed, screw type, and so on also plays a critical role in ensuring consistent product quality as they determine the shear stress and the dwell time. Screw speed, length, diameter, and the type should allow sufficient time for the material to melt and mix inside the barrel. Screening design may be used to determine variables that may be critical to the product quality.

If the drug is not thermally stable, then solvent evaporation methods, for example, spray drying, may be used for preparing the ASD formulation. Unlike HME process, which produces dense product with low moisture content, solvent evaporation methods generally produce bulky product with higher moisture content [19,27]. The solvent used should be of low toxicity with ability to dissolve the drug and carrier into a single phase. It is critical to reduce the solvent level in these formulations to the lowest possible level, as the presence of solvent can induce physical instability of the amorphous system due to their plasticizing effect. If spray drying is used to prepare the ASD, the feed rate of the solution, nozzle configuration, flow rate of the atomizing air, and the temperature and flow rate of the drying gas are critical process parameters that may affect the morphology, yield, physicochemical characteristics, and the amount of residual solvent in the ASD [28–30]. Colloidal silicon dioxide may be added to solutions or dispersions containing drug and carrier to be spray dried to provide very large specific surface area onto which the drug can deposit. This not only improves the flow and the yield of the resulting ASD but also results in a fine dispersion of the drug with excellent dissolution characteristics [31].

Since the drug delivery advantages of ASD result from the amorphous nature of the drug, preventing recrystallization of the drug during the shelf life is critical. To ensure consistent product quality, it is desirable to control the amorphous-to-crystalline ratio of the drug and monitor it during the product shelf life. Presence of water, due to its plasticizing effect, may also be detrimental to the physical and/or chemical stability of the ASD. Hence, the moisture content should also be monitored and controlled in the ASD formulations [32]. X-ray powder diffraction and the differential scanning calorimetry are the most common techniques used to detect the presence and extent of crystalline material, while Karl–Fisher titration may be used to detect and quantitate the moisture content of the ASD formulations. Fourier transform infrared spectroscopy, midinfrared spectroscopy, scanning electron microscopy, small-angle X-ray scattering, confocal laser scanning microscopy, proton nuclear magnetic resonance, nanothermal analysis technique, hot-stage microscopy, and atomic force microscopy have also been used to characterize ASD formulations. The formulation and process variables that may influence the amorphous-to-crystalline ratio and the moisture content of the final drug product should also be thoroughly understood during the product development process. Dissolution is another CQA as a change in the rate and extent of dissolution may indicate a change in the ASD. Thus, the dissolution medium should not only be biorelevant but also have discriminating ability. Surfactants and organic solvents-based dissolution media should be avoided, whenever possible, due to their negative impact on the discriminating ability of the dissolution method.

12.6 CONCLUSIONS

Over the last few years, ASDs have provided a useful tool in improving the solubility, dissolution, and bioavailability of poorly water-soluble drugs. The properties of ASDs are strongly dependent on the manufacturing process, processing conditions, and the properties of the drug substance and the carrier polymer(s). A thorough understanding of the critical material attributes and processing parameters would ensure consistent quality and physicochemical stability of these drug products. Approval of these drug products is on the basis of the criteria outlined in the FD&C Act and involves a comprehensive regulatory review of the quality, safety, and efficacy information by a team of reviewers that include physicians, statisticians, chemists, pharmacologists, and other scientists. FDA is also encouraging real-time monitoring and control of critical material attributes and process parameters through timely in-, on-, or at-line measurements. A number of FDA and ICH guidance documents have also been issued to provide consistency in the regulatory process and to establish inspection and enforcement procedures.

REFERENCES

1. U.S. Department of Health and Human Services (1995) *Guidance for Industry. Content and Format of Investigational New Drug Applications (INDs) for Phase 1 Studies of Drugs, Including Well-Characterized, Therapeutic, Biotechnology-Derived Products.*
2. U.S. Department of Health and Human Services (2003) *Guidance for Industry. Bioavailability and Bioequivalence Studies for Orally Administered Drug Products: General Considerations.*
3. FDA Adverse Event Reporting System. (2013) Available at http://www.fda.gov/Drugs/GuidanceComplianceRegulatoryInformation/Surveillance/AdverseDrugEffects/default.htm (January 12, 2013).
4. U.S. Department of Health and Human Services (2001) *Guidance for Industry. M4: Organization of the CTD.*
5. U.S. Department of Health and Human Services (2001) *Draft Guidance for Industry. Submitting Marketing Applications According to the ICH-CTD Format: General Considerations.*
6. U.S. Department of Health and Human Services (2001) *Guidance for Industry. M4Q: The CTD—Quality.*
7. U.S. Department of Health and Human Services (2001) *Guidance for Industry. M4S: The CTD—Safety.*
8. U.S. Department of Health and Human Services (2001) *Guidance for Industry. M4E: The CTD—Efficacy.*
9. U.S. Department of Health and Human Services (2009) *Guidance for Industry. Q8(R2): Pharmaceutical Development.*
10. Hammond, R.B., Pencheva, K., Roberts, K.J., and Auffret, T. (2007) Quantifying solubility enhancement due to particle size reduction and crystal habit modification: case study of acetyl salicylic acid. *J. Pharm. Sci.*, 96(8):1967–1973.
11. Serajuddin, A.T.M. (2007) Salt formation to improve drug solubility. *Adv. Drug Deliv. Rev.*, 59(7):603–616.

12. Rahman, Z., Zidan, A.S., and Khan, M.A. (2010) Risperidone solid dispersion for orally disintegrating tablet: its formulation design and non-destructive methods of evaluation. *Int. J. Pharm.*, 400(1–2):49–58.
13. Rahman, Z., Zidan, A.S., Samy, R., Sayeed, V.A., and Khan, M.A. (2012) Improvement of physicochemical properties of an antiepileptic drug by salt engineering. *AAPS PharmSciTech.*, 3(3):793–801.
14. Rahman, Z., Siddiqui, A., and Khan, M.A. (2013) Assessing the impact of nimodipine devitrification in the ternary cosolvent system through quality by design approach. *Int. J. Pharm.*, 455(1–2):113–123.
15. Rahman, Z., Zidan, A.S., and Khan, M.A. (2010) Formulation and evaluation of a protein-loaded solid dispersions by non-destructive methods. *AAPS J.*, 12(2):158–170.
16. Zidan, A.S., Rahman, Z., Sayeed, V., Raw, A., Yu, L., and Khan, M.A. (2012) Crystallinity evaluation of tacrolimus solid dispersions by chemometric analysis. *Int. J. Pharm.*, 423(2):341–350.
17. Rahman, Z., Siddiqui, A., and Khan, M.A. (2013) Characterization of a nonribosomal peptide antibiotic solid dispersion formulation by process analytical technologies sensors. *J. Pharm. Sci.*, 102(12):4337–4346.
18. Rahman, Z. and Khan, M.A. (2013) Hunter screening design to understand the product variability of solid dispersion formulation of a peptide antibiotic. *Int. J. Pharm.*, 456(2):572–82.
19. Van den Mooter, G. (2009) Solid dispersions as a formulation strategy for poorly soluble compounds. *20th Annual Symposium of the Finish Society of Physical Pharmacy, Vithi, Finland, January 28–29, 2009.*
20. Patel, V.I. and Dave R.H. (2013) Evaluation of colloidal solid dispersions: physiochemical considerations and *in vitro* release profile. *AAPS PharmSciTech.*, 14(2):620–628.
21. Janssens, S. and Van den Mooter, G. (2009) Review: physical chemistry of solid dispersions. *J. Pharm. Pharmacol.*, 61(12), 1571–1586.
22. U.S. Department of Health and Human Services (2005) *Guidance for Industry. Nonclinical Studies for the Safety Evaluation of Pharmaceutical Excipients.*
23. Liu, H., Pan, W.S., Zhou, L.L., and Zhang, Z.H. (2007) Preparation of budesonide-poly (ethylene oxide) solid dispersions using supercritical carbon dioxide and *in vitro* evaluation. *Yao Xue Xue Bao*, 42(2):206–210.
24. Repka, M.A. and McGinity, J.W. (2000) Influence of Vitamin E TPGS on the properties of hydrophilic films produced by hot-melt extrusion. *Int. J. Pharm.*, 202(1–2):63–70.
25. Newman, A., Engers, D., Bates, S., Ivanisevic, I., Kelly, R.C., and Zografi, G. (2008) Characterization of amorphous API: polymer mixtures using X-ray powder diffraction. *J. Pharm. Sci.*, 97(11):4840–4856.
26. Andrews, G.P., Abu-Diak, O., Kusmanto, F., Hornsby, P., Hui, Z., and Jones, D.S. (2010) Physicochemical characterization and drug-release properties of celecoxib hot-melt extruded glass solutions. *J. Pharm. Pharmacol.*, 62(11):1580–1590.
27. Sakurai, A., Sako, K., and Maitani, Y. (2012) Influence of manufacturing factors on physical stability and solubility of solid dispersions containing a low glass transition temperature drug. *Chem. Pharm. Bull.*, 60(11):1366–1371.
28. Patterson, J.E., James, M.B., Forster, A.H., Lancaster, R.W., Butler, J.M., and Rades, T. (2005) The influence of thermal and mechanical preparative techniques on the amorphous state of four poorly soluble compounds. *J. Pharm. Sci.*, 94(9):1998–2012.

29. Marsac, P.J., Rumondor, A.C.F., Nivens, D.E., Kestur, U.S., Stanciu, L., and Taylor, L.S. (2010) Effect of temperature and moisture on the miscibility of amorphous dispersions of felodipine and poly(vinyl pyrrolidone). *J. Pharm. Sci.*, 99(1):169–185.
30. Wu, J.X., Yang, M., Berg, F.V.D., Pajander, J., Rades, T., and Rantanen, J. (2011) Influence of solvent evaporation rate and formulation factors on solid dispersion physical stability. *Eur. J. Pharm. Sci.*, 44(5):610–620.
31. Chauhan, B., Shimpi, S., and Paradkar, A. (2005) Preparation and evaluation of glibenclamide-polyglycolized glycerides solid dispersions with silicon dioxide by spray drying technique. *Eur. J. Pharm. Sci.*, 26(2):219–230.
32. Hancock, B. and Zografi, G. (1994) The relationship between the glass transition temperature and the water content of amorphous pharmaceutical solids. *Pharm. Res.*, 11(4):471–477.

13

PATENTING AMORPHOUS SOLID DISPERSIONS OF PHARMACEUTICALS*

Jeffrey A. Lindeman

J.A. Lindeman & Co. PLLC, Falls Church, VA, USA

13.1 INTRODUCTION

Inventions come from solving problems. An invention is new technology that solves a problem. Patents are intellectual property rights in new technology, in inventions. While a patent is a legal document, it is also a scientific document. Within the legal framework of a patent, an inventor describes the invention and defines the technology property right. How one describes and defines the invention to be patented often determines the strength and worth of a patent. This is particularly true for patents directed to amorphous solid dispersions of pharmaceuticals.

* The views expressed in this chapter are those of the author and not those of his company, its clients, or its customers. Because this chapter surveys basic requirements and considerations for patenting amorphous solid dispersions of pharmaceuticals, the discussion by its own nature is general. Specific amorphous solid dispersions, patents, and the circumstances surrounding them may, and are likely to, require further and other considerations. The chapter also presents a legal discussion of patents and amorphous dispersion patents in general terms and with reference to U.S. patent law as an example for discussion purposes only. The discussion is for educational purposes only. It is not legal advice and, knowing that the legal landscape is an ever-changing one, should not be substituted for obtaining current legal counsel in regard to a specific patent, amorphous solid dispersion, or other fact situation.

Pharmaceutical Amorphous Solid Dispersions, First Edition. Edited by Ann Newman.

As with solid-state chemistry, a patent application has both form and substance through which it satisfies the legal requirements for patentability. There are essentially two fundamental parts to a patent: the specification and claims. Each has a different function. The specification describes the invention and includes a written description, drawings, and examples. The patent claims are particularly important because they delineate the patented invention and thereby define the intellectual property right. Whether or not something infringes the patent depends on whether or not it comes within the scope of the patent claims. This chapter considers some key aspects of patenting amorphous solid dispersions of pharmaceuticals by surveying the information to be included in a patent (or patent application) to support and present robust patent claims to amorphous solid dispersions. The chapter describes the patent examination process and three common types of patent applications.

13.2 AN AMORPHOUS SOLID DISPERSION AS A PATENTABLE INVENTION

Consider the following patent claim set for an amorphous solid dispersion of a pharmaceutical compound:[1]

The claimed invention is

1. A solid dispersion of amorphous "drug" in a "polymer matrix."[2]
2. A solid dispersion of claim 1, wherein the weight ratio of drug to polymer matrix ranges from about 20:80 to about 5:95.
3. A solid dispersion of claim 1, having a single glass transition temperature (T_g).
4. A solid dispersion of claim 1, being stable for at least 72 h at 60 °C and 75% relative humidity.
5. A pharmaceutical composition comprising a solid dispersion of claim 1 and at least one pharmaceutically acceptable excipient, wherein the drug is present in a therapeutically effective amount.
6. A method for treating a condition in a human patient comprising the step of administering to the human patient a therapeutically effective amount of drug in the form of a solid dispersion according to claim 1 or in the form of a pharmaceutical composition of claim 5.

These claims represent those found in patents directed to an amorphous solid dispersion of a pharmaceutical. Independent claim 1 recites the amorphous solid dispersion

[1] This claim set is illustrative and for discussion purposes only. It does not pertain to, nor is it developed from, any particular patent or patent application.

[2] Note that we are treating the words "drug" and "polymer matrix" as if they are not general terms but, as they would in a patent, they define a specific drug and a specific polymer matrix that would be done in a patent. Note also that while a "polymer matrix" is mentioned here, the matrix material dispersing the amorphous drug may not be a polymer or a mixture of polymers. It may be any number of materials that can disperse an amorphous drug. The terms used here are used for ease of discussion and their easy recognition in the art.

itself, while dependent claims 2, 3, and 4 recite further limitations directed to the composition of the dispersion, a solid-state characteristic of the dispersion, and a property of the dispersion, respectively. Claim 5 recites a pharmaceutical composition containing the amorphous sold dispersion and claim 6 recites a method of treatment using the amorphous dispersion or a pharmaceutical composition containing it. Other claims could also be (and generally are) written to flesh out additional aspects of the amorphous solid dispersion. For example, one or more claims could be added to further define the polymer(s) in the polymer matrix, to identify the specific glass transition temperature, to provide a limit on the amount of crystalline drug that could be present in the amorphous solid dispersion of the invention, or to claim the method of making the amorphous solid dispersion.

The claims in a patent define the intellectual property right in the patent grant. To consider how to patent an amorphous solid dispersion of a pharmaceutical is fundamentally considering how to claim the inventive amorphous solid dispersion. The patent claims will take various shapes and forms depending on the dispersion to be patented and the information on hand to define and characterize it. As with any patent, the goal is robust patent claims that provide commercially meaningful protection.

13.3 CONSIDERING AMORPHOUS SOLID DISPERSIONS AS PATENTABLE COMPOSITIONS OF MATTER

The patent laws of most countries define the types of inventions that may be patented in that country. The U.S. patent law[3] provides that

> Whoever invents or discovers any new and useful process, machine, manufacture, or composition of matter, or any new and useful improvement thereof, may obtain a patent therefor, . . . [4]

Thus, in the United States a patentable invention must fit within a statutory class of subject matter. The invention must be classifiable as a process, a machine, a composition of matter, or an improvement of one of those. An amorphous solid dispersion of a pharmaceutical (commonly known as a "drug," an "active pharmaceutical ingredient," or an "API") clearly finds patentability as a "composition of matter" or "an improvement thereof." An amorphous solid dispersion is a composition of matter that is combination of an API and another material, such as a polymer matrix, in a defined manner. There is,

[3] This chapter references U.S. patent law for the purposes of its discussion and as representative of general patent law principles common to many countries. Any judicial decisions discussed are from a U.S. perspective and may not find a parallel in the judicial decisions of other countries. On September 16, 2011, the U.S. patent law went through a substantial revision by the enactment of the America Invents Act (AIA). The AIA changed the U.S. patent system from a first-to-invent system to a first-to-file system. Patent applications filed after March 15, 2013, very generally speaking, are under the AIA. The statutory references are to the post-AIA statute. The U.S. judicial decisions are generally pre-AIA, although the principles discussed here are expected to be applied in the same way post-AIA.

[4] 35 U.S.C. § 101 (2011).

however, more to this than just fitting an amorphous solid dispersion within a statutory class of patentable subject matter. Instead, considering an amorphous solid dispersion as a composition of matter begins to explain why it is patentable, to establish how the patent claims will be structured, and to determine what patent protection may be obtained. And this begins with considering the chemical aspects and properties of the dispersion—what does the chemistry teach you and how it should be reflected in the claims and the specification.

For example, the dispersion of an amorphous API in a particular polymer matrix may simply be a specific example of a more general invention. The simple dispersion of the API within the polymer matrix, regardless of the solid-state form of the API (amorphous or crystalline), may itself represent a new and patentable composition of matter. The API being amorphous in that dispersion then represents a further embodiment of the invention. Or, the particular polymer matrix may be shown to stably disperse not just one API but a class of drugs. Along the same lines, the amorphous API could be dispersed by not just the exemplified polymer but by the class of polymers it represents. Alternatively, the amorphous API could be dispersed by a mixture of matrix materials in contrast to a single polymer. In other words, a particular amorphous solid dispersion may be a species within a broader composition of matter genus with variation in the API, in the polymer matrix, or both. All of this may be patented to provide broader patent protection than a single amorphous solid dispersion. Considering what the chemistry teaches then provides an opportunity for developing broader patent protection while at the same time protecting the dispersion of most commercial interest.

Broad, effective patent protection may also be obtained by considering and claiming the inventive amorphous solid dispersion in all its various embodiments. The claim set at the beginning of the chapter illustrates this. Claims 1–4 recite amorphous solid dispersions that are patentable compositions of matter. Claim 5 presents an additional but different composition of matter—"A pharmaceutical composition comprising a solid dispersion of claim 1 and at least one pharmaceutically acceptable excipient, wherein the drug is present in a therapeutically effective amount." Claim 6 recites a "method" of treatment, which is a separate class of patentable subject matter. Another patentable method might be the "method" of making the amorphous solid dispersion. If, as another example, the pharmaceutical dosage form of the amorphous solid dispersion is to be prepared in a blister pack, that blister pack with the dosage form may also be claimed—an apparatus. Such claims are directed to different statutory classes of subject matter, each representing a different embodiment of the invention. By considering the various embodiments of the invention, how the invention might be made, used, or sold, and claiming each using different statutory classes of subject matter, the patent protection may be expanded and extended beyond just the composition of matter, the amorphous solid dispersion itself, to other compositions, methods, and even possible to apparatus.

13.4 CLAIMING AN AMORPHOUS SOLID DISPERSION

Patent claims, the numbered paragraphs at the end of a patent, are, by far, the most important part of a patent. They are, as discussed above, the measure of the intellectual property right granted by the patent. Patent claims are much like a deed to land. They do not recite each and every aspect of the invention but, when they are well written, only those aspects that delineate the boundaries of the patented invention and that differentiate it from the "prior art."[5] For patents, the language of the patent claims controls the intellectual property right.

By statute in the United States, the patent claims must particularly point out and distinctly claim the subject matter that the applicant regards as his invention.[6] This is essentially a requirement for precision and definiteness of the claim language. The language of the claims must make clear what they encompass.[7] This standard has also been stated in somewhat different terms: whether "the claims at issue [are] sufficiently precise to permit a potential competitor to determine whether or not he is infringing."[8] Clarity in claim language does not necessarily preclude broad patent protection but clarity does give robust patent claims. Clarity creates clear lines and sets the boundaries of the intellectual property—defining the scope of the patent.

Consider again the sample claim set at the beginning of this chapter. Claim 1 recites "A solid dispersion of amorphous drug in a polymer matrix." Those 10 words set out several defining elements for the claimed invention: The invention is (1) a solid composition of matter. That solid is made up of (2) a dispersion of a certain drug in a certain polymer matrix. The drug itself is (3) in its amorphous form, not in its crystalline form. These defining elements show how the words of a patent claim pack a lot of meaning. In a patent claim every word tells, none are extraneous.

Although patent claims are initially presented in a patent application examined by a patent examiner, it is later in litigation that patent claims are fully tested. To determine the meaning of a particular word in a patent claim, courts have increasingly relied on the "ordinary and customary meaning" of that word.[9] "Ordinary and customary meaning" is "the meaning that term would have to a person of ordinary skill in the art in question at the time of invention."[10] It is presumed that the person of ordinary skill in the art read the claim in the context of the entire patent, including the specification, not confining his understanding to the claim at issue.[11] While being examined for patentability, patent claims are given their broadest reasonable interpretation, consistent with the specification. During examination, the inventor and the patent examiner often carry on a written

[5] Being different from the prior art is the touchstone of patentability. The prior art, generally speaking, is what is known from what has been done prior to the patent application filing date. To be patentable, an invention must be "novel" or new (35 U.S.C. § 102) and nonobvious (35 U.S.C. § 103) from the prior art. A particular amorphous dispersion may in fact be new. Whether or not it is nonobvious from prior dispersion (should similar dispersions exist) is often a question of the properties of the inventive amorphous solid dispersion.

[6] 35 U.S.C. § 112 (b). *Conclusion*: The specification shall conclude with one or more claims particularly pointing out and distinctly claiming the subject matter that the inventor or a joint inventor regards as the invention.

[7] *PPG Industries, Inc. v. Guardian Industries Corp.*, 75 F.3d 1558, 1562 (Fed. Cir. 1996) (citations omitted).

[8] *Exxon Research and Engineering Corp. v. U.S.*, 265 F.3d 1371, 1375 (Fed. Cir. 2001) citing *Miles Labs., Inc. v. Shandon, Inc.*, 997 F.2d 870, 875 (Fed. Cir. 1993).

[9] See *Vitronics Corp. v. Conceptronic, Inc.*, 90 F.3d 1576, 1582 (Fed. Cir. 1996).

[10] *Phillips v. AWH Corp.*, 415 F.3d 1303, 1313 (Fed. Cir. 2005) (*en banc*).

[11] *Phillips v. AWH Corp.*, 415 F.3d 1303, 1314 (Fed. Cir. 2005).

conversation of what the terms in a patent claim mean and the scope of the invention the claims define. From the U.S. Patent and Trademark Office (USPTO)'s perspective, a patent applicant always has the opportunity to amend the claims and broad interpretation by the examiner reduces the possibility that the claim, once issued in a patent, will be interpreted more broadly than is justified ([1], § 2111).

In litigation, analysis of patent infringement involves two steps: (1) interpreting the claims and (2) determining whether the accused object or process is covered by the claims as interpreted.[12] The interpretation of a patent claim, known as "claim construction," is a question of law for the court (the judge) to decide, not one of fact for the jury to decide.[13] In litigation, claim language is to be interpreted as one reasonably skilled in the art would have interpreted the claim at the time of invention.[14] The importance of carefully and wisely choosing the language to describe and claim the invention cannot be overemphasized. For the patentee, the time to do that really only exists when the patent application is being prepared. During claim construction, the actual words of a patent claim are to be given their ordinary and usual meaning unless it is apparent from the patent disclosure that the inventor intended to define those words differently from their ordinary meaning.[15] An inventor may be his/her own lexicographer, define his/her own terms, but if he/she does those terms will be construed according to that definition.[16]

When drafting a patent application and preparing claims to an amorphous solid dispersion, an inventor might not be satisfied with the ordinary and customary meanings of important claim terms such as "amorphous" or "amorphous solid dispersion." The inventor may want to define such terms in a particular way to particularly point out and distinctly claim the invention. For example, as discussed below, there is some debate on just what an amorphous solid is—when is there sufficient order in the solid such that it is no longer amorphous. It may be difficult to prove infringement where the definition of amorphous depends upon determining the degree of ordering within the solid. In contrast, where an analytical technique is used to show the amorphous nature of the solid, defining "amorphous" or "amorphous solid dispersion" with reference to that technique takes the definition of "amorphous" from the theoretical to the measurable.

[12] *Omega Eng'g, Inc. v. Raytek Corp.*, 334 F.3d 1314 (Fed. Cir. 2003); *Fromson v. Advance Offset Plate, Inc.*, 720 F.2d 1565 (Fed. Cir. 1983).

[13] *Markman v. Westview Instruments, Inc.*, 517 U.S. 370, 384 (1996); *Gechter v. Davidson*, 116 F.3d 1454, 1457 (Fed. Cir. 1997); *Vitronics Corp. v. Conceptronic, Inc.*, 90 F.3d 1576, 1582 (Fed. Cir. 1996).

[14] *Specialty Composites v. Cabot Corp.*, 845 F.2d 981, 986 (Fed. Cir. 1988) (quoting *Loctite Corp. v. Ultraseal Ltd.*, 781 F.2d 861, 867 (Fed. Cir. 1985)).

[15] *O2 Micro Int'l Ltd. v. Beyond Innovation Tech. Co., Ltd.*, 521 F3d 1351, 1360 (Fed. Cir. 2008) (ordinary and customary meaning is "the meaning a term would have to a person of ordinary skill in the art after reviewing the intrinsic record at the time of the invention").

[16] See, for example, *Phillips v. AWH Corp.*, 415 F.3d 1303, 1316 (Fed. Cir. 2005) (*en banc*) (where "the specification may reveal a special definition given to a claim term by the patentee that differs from the meaning it would otherwise possess. . . . the inventor's lexicography governs"); *Thorner v. Sony Computer Entertainment America LLC*, 669 F.3d 1362, 1365 (Fed. Cir. 2012) ("To act as its own lexicographer, a patentee must 'clearly set forth a definition of the disputed claim term' other than its plain and ordinary meaning") (quoting *CCS Fitness, Inc. v. Brunswick Corp.*, 288 F.3d 1359, 1366 (Fed. Cir. 2002)).

The purpose of a patent claim is to set out the boundaries or scope of the intellectual property right embodied in the patent. The patent claim must particularly point out and distinctly claim that which the inventor regards as the invention.[17] Patent claims then do not, and should not, delineate every feature of the claimed invention. If they did, the patent claims would be too narrow or too burdened by limitations and quite possibly be incapable of being infringed. The detailed discussion of the invention occurs in another part of the patent, not in the claims but in the patent specification. The patent claims are interpreted in light of the patent specification, that is, in light of the description of the inventive amorphous solid dispersion and all its embodiments.

13.4.1 Describing Amorphous Solid Dispersions in a Patent

Under U.S. law, the patent specification must provide (1) a written description of the invention, (2) the manner of making and using the invention, and (3) the best mode contemplated for carrying out the invention.[18] The written description requirement is intended to show that the inventor had possession of the invention as of the patent application's filing date.[19] The second requirement, how to make and use the invention, is known as the "enablement" requirement. A patent must contain an enabling disclosure. That is, the specification must contain sufficient technical information to enable one of ordinary skill in the art to make and use (or practice) the invention without "undue experimentation."[20] The third requirement is that the specification must disclose the best mode contemplated by the inventor(s) for making and using the invention. The best mode is subjective—for the inventor(s) only—and does not have to be separately called out in the specification, only presented somewhere in there. The best mode requirement is unique to the United States. Under the America Invents Act of 2011, failure to disclose the best mode is no longer a ground for patent invalidity.[21] The discussion here focuses

[17] 35 U.S.C. § 112(b) (2011).

[18] In general, the specification shall contain a written description of the invention, and of the manner and process of making and using it, in such full, clear, concise, and exact terms as to enable any person skilled in the art to which it pertains, or with which it is most nearly connected, to make and use the same, and shall set forth the best mode contemplated by the inventor or joint inventor of carrying out the invention. 35 U.S.C. § 112(a) (2011).

[19] "The 'written description' requirement implements the principle that a patent must describe the technology that is sought to be patented; the requirement serves both to satisfy the inventor's obligation to disclose the technologic knowledge upon which the patent is based and to demonstrate that the patentee was in possession of the invention that is claimed." *Capon v. Eshhar*, 418 F.3d 1349, 1357 (Fed. Cir. 2005).

[20] See, for example, *United States v. Telectronics, Inc.*, 857 F.2d 778, 785 (Fed. Cir. 1988) ("The test of enablement is whether one reasonably skilled in the art could make or use the invention from the disclosures in the patent coupled with information known in the art without undue experimentation."). A patent need not teach, and preferably omits, what is well known in the art. *In re Buchner*, 929 F.2d 660, 661 (Fed. Cir. 1991); *Hybritech, Inc. v. Monoclonal Antibodies, Inc.*, 802 F.2d 1367, 1384 (Fed. Cir. 1986), *cert. denied*, 480 U.S. 947 (1987); and *Lindemann Maschinenfabrik GMBH v. American Hoist & Derrick Co.*, 730 F.2d 1452, 1463 (Fed. Cir. 1984).

[21] The America Invents Act left the best mode requirement intact but eliminated the failure to comply with the best mode requirement as a basis for patent invalidity. 35 U.S.C. § 283(3).

on the written description and enablement requirements and on meeting them when preparing a patent application for an amorphous solid dispersion.

The specification must "describe the invention." Typically, the specification begins with a background section that provides the context for the invention. The background section can set out problem the invention solves and discuss prior attempts to address the same problem. The background of the invention may include the following parts: (1) the Field of the Invention (a statement of the field of art to which the invention pertains) and (2) a Description of the Related Art (the background of the invention). The background of the invention is a description to the extent practical of the state of the prior art or other information disclosed known to the applicant, including references to specific prior art or other information where appropriate. Where applicable, the problems involved in the prior art or other information disclosed that are solved by the applicant's invention should be indicated ([1], § 608.01(c)). For amorphous solid dispersions of pharmaceuticals, the background of the invention generally will describe the API as a compound, other known forms of the API (e.g., polymorphs), its therapeutic properties, and the deficiencies that give rise to the need for an amorphous solid dispersion of the API.

Following the background of the invention, the specification then turns to the description of the invention. This portion of the specification will contain a summary of the invention, a brief description of any figures, and a detailed description of the invention. These sections of the specification contain the technical description of the invention. The patent claims and an abstract summarizing the invention disclosed in the application follow and complete the specification.

The specification, to satisfy the written description requirement, must describe or "disclose" all aspects of the claimed invention. Patent drafters typically prepare a working set of claims and use them as a guide to construct the specification. This helps to ensure that the language and features of the claims appear in and correspond to the description in the specification. Using a working set of claims focuses the discussion not only on what the claimed invention is, but also, and perhaps more importantly, on what language or parameters will be used in the claims to set out the invention to be patented. That discussion occurs in the detailed description that is an exposition of the general aspects and preferred embodiments of invention followed by specific working examples of those aspects and embodiments. The detailed description will contain a general description of the inventive amorphous sold dispersion, its characteristics, its properties, its uses, method of making the amorphous solid dispersion, pharmaceutical compositions containing the amorphous solid dispersion, methods of treatment using the amorphous solid dispersion, and the pharmaceutical compositions. The working examples show the preparation of specific amorphous solid dispersions of the invention, and their solid-state characterization as well as tests/data to establish their beneficial properties/uses. Often the characterization and testing data reported in the examples are depicted in figures.

A patent specification, either in the detailed description or in a separate glossary, will define key terms in the course of describing the technical aspects of the invention. Those terms are typically used to delineate the claimed invention. For an amorphous solid dispersion patent, perhaps the one key term to be defined is the word "amorphous." This is not as easy as it may seem. And the "ordinary and customary" meaning of "amorphous" may not be well suited for patenting an amorphous solid dispersion.

Amorphous solids, lacking the three-dimensional long-range order of a crystalline material, possess a more random arrangement of molecules, exhibit short-range order over a few molecular dimensions, and have physical properties quite different from those of their corresponding crystalline states [2]. It has been reported, however, that just what the truly "amorphous" state looks like in terms of molecular ordering is more complex than it may seem [3]. This can introduce ambiguity into the patent claims. For patenting purposes, then, the definition of "amorphous" takes a slightly different focus. Inventors and patent drafters want to define "amorphous" and "amorphous solid dispersion" in such a way that it is clear and that patent infringement can be readily proved. One does not want to overburden the claim with excessive limitations that may be easily avoided by competitors or hard to prove for infringement. Rather, the definition of amorphous (actually any defined term) should be one that is readily understandable, accessible, and provable for infringement purposes. This is where the patent law permits an inventor to define terms, to be his/her own lexicographer can be used to advantage. Thus, the definition of "amorphous" may vary somewhat from patent to patent and be designed to meet the needs of the amorphous solid dispersion.

In an amorphous solid dispersion patent, the specification will contain a definition of what it means to be "amorphous" and will also provide at least one measure of how to determine that the composition is, in fact, an amorphous solid dispersion. The particular definition of "amorphous" will often depend on the technique(s) employed by the inventor to characterize the amorphous sold dispersion. To adequately define a patentable invention, and provide proof of infringement, the definition of "amorphous" or "amorphous sold dispersion" should contain objectively measurable aspects. Those objectively measurable aspects can be used to define terms in the specification and/or as claim limitations to delineate and set out the invention to be patented. For an amorphous solid dispersion, this means adding at least one objectively measurable element to the definition by which one of ordinary skill can determine whether or not the combination of an API and matrix material is an amorphous solid dispersion of the invention. The definition of an amorphous solid dispersion typically relies on an analytical technique to provide an objectively measurable indicator of the amorphous state of the API within the solid dispersion.

A number of techniques have been used to characterize amorphous solid dispersions and determine the solid state of the API within the dispersion. The more common techniques include X-ray powder diffraction (XRPD), differential scanning calorimetry (DSC), and solid-state NMR (SSNMR) [4]. Each of these techniques can differentiate an amorphous solid from its crystalline counterpart. Perhaps the most common technique is XRPD that is used both for crystalline solids and for amorphous solids and is familiar to patent examiners. Thus, as an example, an inventive amorphous solid dispersion characterized by XRPD could be described in this way:

> The invention relates to amorphous solid dispersions of amorphous [API] in a [polymer matrix]. The amorphous nature of the solid dispersions is determined by X-ray powder diffraction (XRPD). A material is "X-ray amorphous" when there are no sharp peaks observed in its XRPD pattern, only broad halos. Sharp peaks in the XRPD indicate the presence of crystalline material.

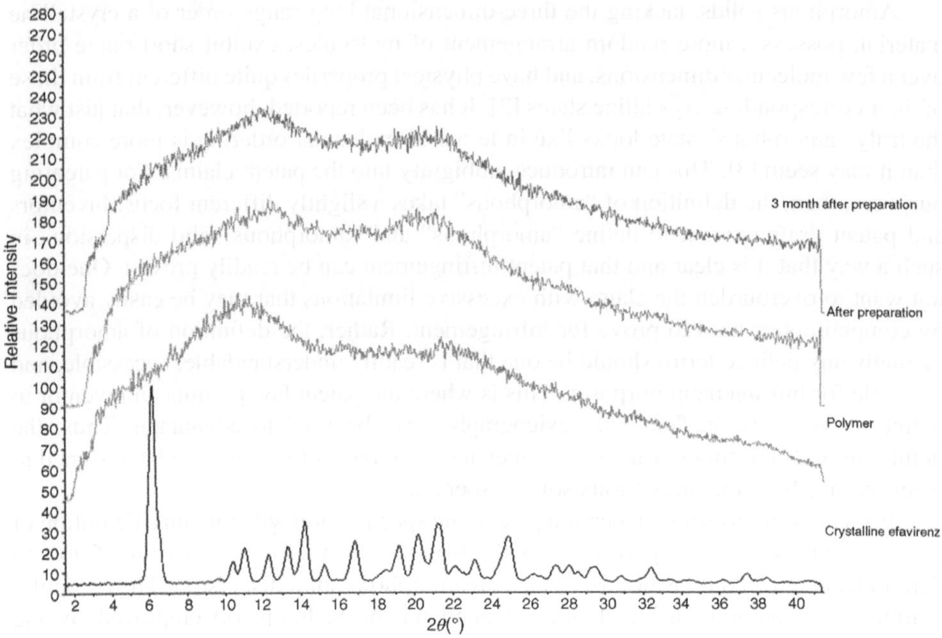

Figure 13.1. XRPD stack plot for an amorphous dispersion of efavirenz. (Courtesy of Crystallics B.V., Amsterdam, The Netherlands.)

This description relies on a definition of "X-ray amorphous" to determine whether or not the sold dispersion is amorphous. The question is whether or not sharp peaks are observed in its XRPD pattern. An illustration of this can be seen in Figure 13.1 that shows the XRPD pattern of (1) crystalline efavirenz (bottom), (2) a dispersing polymer, (3) the amorphous dispersion after preparation, and (4) the amorphous dispersion 3 months after preparation. Advantageously, this figure shows not only that an amorphous dispersion of efavirenz is formed but also that it is stable.

This description and definition of what it means to be amorphous from the patent specification is also reflected in the patent's claims. The claim does not have to contain the definition itself. Consider claim 1 at the beginning of the chapter, "A solid dispersion of amorphous drug in a polymer matrix." Construing claim 1 in view of the explicit definition of being "X-ray amorphous" in the specification, claim 1 would be understood to require the claimed solid dispersion to have no sharp peaks in the XRPD, only broad halos. It is likely that the application would have an example describing the amorphous solid dispersion and a figure depicting its XRPD pattern confirming its amorphous state.

Where other or additional techniques are used to characterize the amorphous solid dispersion, those techniques can be similarly described, defined, and used. For example, if the API contains a fluorine atom, the ^{19}F SSNMR spectra of the API may be used to define amorphous based on the broadening of the ^{19}F peak in the amorphous solid dispersion SSNMR spectrum over that of the corresponding crystalline SSNMR spectrum.

Alternatively, the glass transition temperature (T_g) may also be used to define and claim the amorphous solid dispersion. It is also possible to use these various characteristics in combination or as alternatives to describe or define an inventive amorphous sold dispersion.

Another way to describe and claim an amorphous solid dispersion is by reference to its composition rather than its analytical characterization. Although it is not a solid dispersion, an example of this is a claim to amorphous cefuroxime axetil, found to be patentable over its crystalline form.[22] The patent claim at issue read:

> Cefuroxime axetil in amorphous form essentially free of crystalline material, and having a purity of at least 95% aside from residual solvents.

Here the amorphous state was described by the lack or absence of the crystalline material and compositional purity. Taking the first property of the cefuroxime axetil as the definitional basis of an amorphous solid dispersion, the amorphous state of the drug could be defined in the specification as being "essentially free of crystalline drug." Note that it may be necessary or appropriate to put numerical limits on what is and what is not "essentially free of crystalline drug.' The possible patent claim could read this way:

> A solid dispersion of amorphous drug in a polymer matrix wherein the solid dispersion is essentially free of crystalline drug.

The point is that the definition of "amorphous" and/or "amorphous solid dispersion" can be developed from whatever means were used by the inventor to characterize the amorphous solid dispersion. The characterization method used by the inventor is often also one that points out not only features but also advantages of the new dispersion. In this way, the inventor may act as his/her own lexicographer. Doing that, however, one must understand that the definitions in the specification will control the construction of those terms in the claims, as discussed above. As a consequence, the language of the specification and of the claims should be consistent in concept and in usage in order to have clear and strong patent claims.

The written description of the inventive amorphous solid dispersion would also include a description of its properties and advantages. Amorphous forms of APIs are often used when the crystalline API has poor solubility, poor dissolution characteristics, and/or poor bioavailability. Amorphous forms tend to exhibit high levels of supersaturation and thus higher levels of apparent solubility relative to their crystalline forms [5]. Amorphous forms are, however, thermodynamically unstable relative to the crystalline form. There is the tendency for the amorphous form to revert back to the crystalline form, particularly upon storage or upon dissolution, which in both cases can negate the advantages, for example, increased solubility or dissolution, achieved with the amorphous form. One challenge, therefore, is to inhibit crystallization over the time period of product storage and to maintain a sufficient level of supersaturation upon oral administration without crystallization [5]. It is at least one of these advantages that can be

[22] See *Glaxo Group Ltd. v. Apotex, Inc.*, 268 F. Supp. 2d 1013 (N.D. Ill., 2003).

achieved with an amorphous solid dispersion or the challenges of stability that are overcome that lead to patentability of the particular inventive amorphous solid dispersion. These aspects of the invention that should be discussed in the specification show the patentability of the amorphous solid dispersion over the prior art. The key is that these advantages should be tied to the dispersion itself. Patenting the amorphous solid dispersion as a composition of matter means that the properties showing the superiority or unexpected advantages are properties of the composition as a whole.

Claims 3 and 4 at the beginning of the chapter capture these ideas. In claim 3, the solid dispersion has a single glass transition temperature (T_g). This is an indication of the miscibility of the amorphous drug in the polymer matrix. This property indicates the homogeneity and single-phase character of the solid dispersion [4]. The stability of an amorphous sold dispersion, as mentioned, is an important property of the dispersion and can be used as evidence of patentability but also to claim the dispersion. According to claim 4, the amorphous sold dispersion is stable for at least 72 h at 60 °C and 75% relative humidity. This serves to define what solid dispersions of the amorphous drug in the polymer matrix are within claim 4, those with the recited stability. But, this also serves to distinguish the claimed solid dispersions from prior art dispersions that are not stable by comparison. While such properties are described in the specification, that data supporting them are typically the subject of working examples.

This leads to the second function of the specification—enabling one of ordinary skill in the art to make and use or "practice" the invention. The specification must also "enable" the invention, that is, teach a person skilled in the art how to make and use the invention. This is a second purpose of the written description. The specification is not written to the public at large but to those skilled in the relevant art. The question of enablement looks at whether a particular claim is "supported" by the specification, that is, whether the specification contains sufficient disclosure for one of ordinary skill to practice the claimed invention "without undue experimentation." The test of enablement is not whether any experimentation is needed, but whether, if experimentation is necessary, it is undue.[20,23] Many factors may be considered when looking at enablement or asking whether "undue experimentation" is necessary. As set out in *In re Wands*, factors to be considered include, but are not limited to, (a) the breadth of the claims, (b) the nature of the invention, (c) the state of the prior art, (d) the level of one of ordinary skill, (e) the level of predictability in the art, (f) the amount of direction provided by the inventor, (g) the existence of working examples, and (h) the quantity of experimentation needed to make or use the invention based on the content of the disclosure.[23]

The specification can meet its enablement function in the expositive discussion of the invention. That discussion often walks through the various aspects of the invention explaining them first as broad concepts and then refining them in terms of specific embodiments. The expositive discussion in the specification would describe the class of polymers by their common features, for example, monomer content, molecular weight, and their ability to disperse and stabilize the amorphous API. This type of disclosure

[23] *In re Wands*, 858 F.2d 731, 737 (Fed. Cir. 1988) (reversing the PTO's determination that claims directed to methods for detection of hepatitis B surface antigens did not satisfy the enablement requirement); *see also* MPEP § 2164.01.

would set out an inventive dispersion of an amorphous API in a class of generic polymers.

In other amorphous dispersions, it may be that one matrix material or a particular combination of matrix materials could be found to disperse and stabilize a class of APIs. Here the inventive dispersion is for a genus of amorphous APIs dispersed within a specific matrix material. The specification would describe the APIs by their common features, perhaps by way of a general structural formula, and the method by which they may be dispersed within the matrix material.

Another common variant for amorphous sold dispersions is the range of relative amounts of the API and the polymer matrix that yield successful dispersions. See claim 2 of the sample claim set, "A solid dispersion of claim 1, wherein the weight ratio of drug to polymer matrix ranges from about 20:80 to about 5:95." To claim amorphous solid dispersions of a generic class of amorphous APIs and/or a generic class of matrix materials, the specification must contain sufficient supporting disclosure to satisfy the enablement requirement. This is typically done with a full description of each component of the invention and then examples showing specific embodiments with different weight ratios as part of the broader description.

While the expositive discussion of the detailed description supports the invention by describing its various aspects, the working examples of the invention come at it from the other side. The working examples are often specific examples describing the actual preparation of amorphous solid dispersion(s) according to the invention, characterization of the amorphous solid dispersion(s) by one or more analytical techniques, and tests to show the properties of the amorphous solid dispersion(s). A typical example set may be outlined as indicated below (with the normal experimental discussion omitted). The example set may contain preparative examples, characterization examples and formulation examples.

Examples

Description of analytical techniques used (e.g., XRPD and DSC methods and parameters).

Example 1: Preparation of an amorphous solid dispersion.

Example 2: Characterization of an amorphous solid dispersion.

2.1 XRPD characterization with its XRPD pattern shown in Figure X.

2.2 DSC characterization with a reported T_g and DSC pattern in Figure Y.

Example 3: Stability testing of the amorphous solid dispersion(s).

Example 4: Dissolution testing of the amorphous solid dispersion(s).

Example 5: Pharmaceutical composition containing the amorphous solid dispersion.

The example set may contain preparative and characterization examples, etc., for a representative number of amorphous solid dispersions of the invention in order to show the variation and breadth of the claimed invention. The additional examples could vary the relative amounts of API and polymer matrix in the dispersions, include additional components to the amorphous solid dispersions, report different techniques to prepare amorphous sold dispersions, show different APIs, and show that different polymers within a class or a mixture of polymers may be used. For pharmaceutical inventions, the examples may also set out pharmaceutical compositions and have tests to demonstrate pharmaceutical properties or efficacy. In this way, the working examples set out distinct embodiments of the invention that are then typically themselves the subject of specific claims while at the same time supporting and enabling broader aspects of the claimed invention.

13.5 TYPES OF PATENT APPLICATIONS AND PATENT EXAMINATION

To obtain a patent, an inventor files a patent application with a patent office, for example, the U.S. Patent and Trademark Office. There are three "common" types of patent applications that one sees in the patenting of amorphous solid dispersions of pharmaceutical. These common patent applications include a U.S. provisional patent application, a nonprovisional application, and a Patent Cooperation Treaty (PCT) application. Of the three, only a nonprovisional application leads to a granted patent.

The preceding discussion on having an adequate written description and enablement for an amorphous solid dispersion and on consideration for patent claims is most relevant to "nonprovisional" patent applications. The nonprovisional application (in contrast to a "provisional" application, discussed below) is the patent application[24] that is filed, is examined by a patent office, and eventually, if successful, becomes a patent. A nonprovisional application is more formal and formatted patent application than a provisional patent application. A typical nonprovisional application includes the following sections in order: (1) title of the invention; (2) cross-reference to related applications; (3) statement regarding federally sponsored research or grants; (4) background of the invention; (5) summary of the invention; (6) brief description of the figures; (7) detailed description of the invention; (8) claim(s); (9) abstract; and (10) the figures. Once filed, generally speaking, the nonprovisional application contains all the disclosure the patent applicant is entitled to rely upon during patent examination. Preparing a nonprovisional application then with its constituent parts is careful work. A useful, and detailed, discussion of the formal requirements and component parts of a U.S. nonprovisional application can be found at the USPTO website.[25]

Perhaps the most important impact of filing a nonprovisional application is that the 20-year patent term begins running from the application's filing date. Even if later continuation or divisional applications are filed, that first effective nonprovisional filing date determines the patent term.

Typically, patent applicants will prepare one nonprovisional application and use that application for filing in all countries. Some aspects of the application in a particular country may be revised somewhat to satisfy local patent law.

It is also the nonprovisional application that is published as a published application and later as the actual patent grant. In the United States, as in many countries, applications are published approximately 18 months from their filing date.

[24] The term "nonprovisional" is used here to distinguish this type of patent application from a U.S. provisional application. This is really only relevant for the United States that provides for "provisional" and "nonprovisional" patent applications in its law. In other countries, this is simply a "patent application." Discussing such patent applications as "nonprovisional" patent applications has become common parlance and so is used here.

[25] See http://www.uspto.gov/patents/resources/types/utility.jsp (last accessed April 2, 2014).

A patent applicant may file a U.S. provisional patent application[26] to obtain a U.S. filing date and pay a lower filing fee than with a nonprovisional application. A provisional application is often the first type of patent application to be filed. Later nonprovisional applications (in the United States and internationally) can claim benefit of the provisional application. Filing a provisional application has several advantages but also carries some important limitations that the applicant must bear in mind.

One advantage is that the provisional application may be a "less formal" patent application and has minimal filing requirements.[26] A provisional application must have a written description of the invention that meets the requirements of 35 U.S.C. § 112, paragraph (A), that is, a description of the invention such that one of ordinary skill could practice the invention. Preparing such a description is what was discussed above. A provisional application, however, does not require the format and organization of the nonprovisional application. The description may be more like a scientific paper, although its disclosure must still meet these disclosure requirements. A provisional application does not require patent claims, although many patent drafters include claims as they are helpful when preparing the patent application and provide a more complete application, especially if the provisional application is to serve as a priority document for subsequent U.S. or international patent applications.

Another advantage of filing a provisional application is obtaining a filing date. This filing date stops the "prior art" clock. The provisional application filing date establishes the date before which publications, patents, other patent applications, and so on may act as prior art for determining the patentability of the invention. A provisional application, however, is not examined, it is not published, nor does it become a patent.

A provisional application may serve as a "priority application" for a later filed patent application in the United States or internationally. The benefit here is that the later filed patent application will enjoy the benefit of the provisional application's filing date for the invention described in the provisional application.

Once filed, a provisional application expires after 1 year. This means that a nonprovisional application claiming benefit to the provisional application must be filed during that provisional application year. If a nonprovisional application is filed after the provisional application expires, one cannot claim benefit to the provisional application or enjoy its earlier filing date. While a provisional application expires after 1 year, it does not start the 20-year patent term. As mentioned above, it is the filing date of the nonprovisional application that begins the patent term. Thus, filing a provisional application first provides an additional year to the patent applicant. Having filed a provisional application, an inventor can continue to develop the invention and/or explore commercial opportunities before filing a nonprovisional application.

The ability to file a "less formal" patent application at lower cost and obtain a filing date without starting the patent term are some of the advantages of a provisional application. Choosing to file a provisional application, one must keep in mind its limitations as well, chief among those are that a provisional application expires after 1 year and all action must

[26] A helpful and more detailed discussion of U.S. provisional applications can be found at the USPTO website (www.uspto.gov). One particular discussion can be found at http://www.uspto.gov/inventors/independent/eye/201004/provisional.html (last accessed April 2, 2014).

be done within that year to enjoy the benefit of the provisional application's filing date. One word of caution—it is best to make the provisional application as complete as it can be when it is filed. There is also a benefit to preparing a provisional application using the format and meeting the formalities needed for a nonprovisional application. It is all too easy to come to the end of the provisional application year and realize that there is much more left to do to get the provisional application into shape for nonprovisional filing.

The third common patent application is a "PCT patent application." The PCT, or the Patent Cooperation Treaty, is an international patent system.[27] The PCT allows a patent applicant to file one international application and simultaneously seek patent protection in all PCT member states, approximately 148 countries throughout the world at this time. Many, if not all, countries of interest to most patent applicants are members of the PCT.[28] For example, Australia, Canada, China, European countries individually and through the European Patent Office, India, Japan, and the United States, all of which are common countries of interest to pharmaceutical patent applicants, are members of the PCT. A PCT application is a formal application in the same way as a nonprovisional application and has the same basic structure and parts. A PCT application contains patent claims that are examined as part of the PCT process.

Filing a PCT application gives the patent applicant an effective filing date in the PCT member states. The PCT, however, does not grant a patent. Instead, under the PCT, an international application proceeds through two phases—the international phase and the national phase. During the international phase, the application is published as an International Application. Published PCT applications have numbers beginning with the letters "WO" indicating that the application was published by the World Intellectual Property Organization (WIPO) that administers the PCT. PCT applications are published approximately 18 months from their earliest filing date. A PCT application also undergoes an international search that identifies published documents relevant to the patentability of the claimed invention. This search is reported in a published document known as an International Search Report (ISR). Following the ISR, a preliminary and nonbinding opinion on whether or not the invention is patentable under the PCT criteria is prepared. This opinion is first issued as a written opinion to which the patent applicant may respond and later published as an International Preliminary Report on Patentability (IPRP) The PCT patentability criteria are based on patentability criteria generally accepted and in place internationally. There is also a subsequent optional international preliminary examination available to PCT applicants. This international phase of the PCT application takes place over about 28 months after earliest priority date or the PCT application filing date. The PCT international phase thus allows a patent applicant to establish a filing date in all PCT member states, to assess the patentability of the invention in view of the ISR and IPRP, and then make a decision about entering the next phase of the PCT process, the national phase.

[27] For information on the Patent Cooperation Treaty, see http://www.wipo.int/pct/en/ (last accessed April 3, 2014).

[28] For a listing of PCT member states, see http://www.wipo.int/pct/en/pct_contracting_states.html (last accessed April 3, 2014).

In the PCT national phase, the PCT applicant selects which, if any, of the PCT member states he or she wants to file the PCT application. In the national phase, the PCT application is filed in a member state and undergoes examination according to the patent laws of that member state. It is the member state that grants a patent for its country. Depending on the member state, the national phase application is typically due at 30 or 31 months from the PCT application's earliest priority date. There are typically national fees and other formal requirements upon entering the national phase in a member state. Once those are completed, the examination process proceeds in the member state's patent office.

It is not unusual for a PCT application to enter the national phase in 5 or even 10 member states. All information from the PCT international phase is provided to the national phase member state and considered in the national phase examination. Thus, national phase filing can be an expensive undertaking. When used most effectively, the PCT international phase informs not only the national phase examination but also the selection of national phase states in which to file the PCT application.

Before a patent is granted in any country, the patent application is examined to determine whether it satisfies the particular country's legal requirements for patentability. While the criteria applied in most countries are similar to one another, each country has its own patent laws and its own patent procedure. The first step to obtaining a patent is filing a patent application such as a nonprovisional application or a PCT national phase application. A patent examiner then examines the patent application.

To be patentable, an invention must be new and useful (have a practical utility). As discussed above, the patent application must describe the invention and have one or more patent claims that set out the invention to be patented. During examination, the patent examiner focuses on the claimed invention to ensure that it meets statutory requirements of novelty and nonobviousness as well as on the patent specification that must describe the invention. During the course of examination, the patent examiner issues "office actions" to the patent applicant. In these office actions, the examiner states whether or not the claims are patentable and, if they are not, gives reasons why. The examiner's reasons are stated as "objections" that generally relate to matters of form or as "rejections" that give reasons why the claims do not meet the statutory requirements for a patent. Where there are issues of novelty or obviousness, the examiner may rely on "prior art" documents (e.g., earlier published articles or patents) that the examiner believes show that the claimed invention lacks novelty (is not new) or would have been obvious to a person skilled in the art. A patent applicant is entitled to respond by providing arguments to rebut the examiner's positions, by amending (revising or rewriting) the claims, or both. Typically, there are about two office actions and responses during examination before other procedures become available as part of patent examination. Patent examination does not necessarily end with the two office actions and responses but can continue down various avenues in each patent office. A patent is granted when the patent examiner concludes that all statutory requirements are met. Generally speaking, patent examination proceeds along similar avenues in all patent offices, although the preceding discussion only sketches the contours of patent examination.

13.6 CONCLUSIONS

As the brief overview of the types of patent applications (provisional applications, nonprovisional applications, and PCT applications) shows, patenting amorphous solid dispersion takes into account aspects of patent procedure and strategy—what to file and where to file—to say the least. The success of any patent application, however, ultimately depends on the substance of its disclosure. A full and complete description of the invention almost always yields a strong patent.

When patenting amorphous sold dispersions of pharmaceuticals, two considerations seem most important. First, an amorphous sold dispersion is a patentable composition of matter. As such, there may be patentable embodiments beyond a single composition of matter as represented by the amorphous solid dispersion and its uses. This includes variation of the composition in such a way that generic patent claims may be possible. The properties of the composition may be used to establish patentability of an amorphous solid dispersion or a generic class of dispersion. Second, the patent specification should include a definition of "amorphous" and/or "amorphous solid dispersion" that employs objectively measurable aspects of the composition. The objectively measurable aspect typically derives from the analytical characterization of the amorphous solid dispersion. These two considerations then inform the preparation of the patent specification, particularly the discussion of the invention in the detailed description and the working examples, to support the claims. Having described and exemplified the amorphous solid dispersions of the invention, the patent concludes with the claims. The claims are presented and construed in view of the specification with wording and limitations that set out the boundaries of the intellectual property embodied in the patent grant. The measure of any patent is the strength and scope of its claims.

REFERENCES

1. U.S. Patent and Trademark Office (2014) Manual of Patent Examining Procedure (MPEP), 9th edition.
2. Hancock, B. and Zografi, G. (1997) Characteristics and significance of the amorphous state in pharmaceutical systems. *J. Pharm. Sci.*, 86(1):1–12.
3. Bates, S., Zografi, G., Engers, D., Morris, K., Crowley, K., and Newman, A. (2006) Analysis of amorphous and nanocrystalline solids from their X-ray diffraction patterns. *Pharm. Res.*, 23(10):2333–2349.
4. For a discussion of these and other techniques used to characterize amorphous solid dispersions, see Newman, A. and Munson, E. (2012) Characterizing miscibility in amorphous solid dispersions. *Am. Pharm. Rev.*, April:92-08.
5. Newman, A., Knipp, G., and Zografi, G. (2012) Commentary: assessing the performance of amorphous solid dispersions. *J. Pharm. Sci.*, 101(4):1355–1377.

14

MONOGRAPHS ON POLYMERS AND SURFACTANTS

Xia Lu,[1] Robert Wenslow,[2] and Ann Newman[3]

[1]*Crystal Pharmatech, Suzhou, China*
[2]*Crystal Pharmatech, North Brunswick, NJ, USA*
[3]*Seventh Street Development Group, Lafayette, IN, USA*

14.1 PART I: POLYMERS

This section is not a comprehensive list of polymers used in amorphous solid dispersions, but it is meant to highlight common polymers that have been used in various studies.

14.1.1 Notes

14.1.1.1 Synonyms and Molecular Weights Synonyms were taken from Ref. [1]. Molecular weights were taken from Refs [1,2]. Most polymers exhibit a range of molecular weights. Where possible, the grade, molecular weight, and glass transition temperature (T_g) are given.

Pharmaceutical Amorphous Solid Dispersions, First Edition. Edited by Ann Newman.

14.1.1.2 *Solubility Parameter* Solubility parameters reported are the Hildebrand solubility parameter (δ) and were taken from Ref. [2].

14.1.1.3 *Glass Transition and Melting Temperatures* The glass transition temperature (T_g) and melting temperature (T_m) were taken from Ref. [2] and are given as a range since different molecular weights will change these properties. Some semi-crystalline polymers will exhibit both temperatures and these are presented in the table. Where possible, the grade, molecular weight, and glass transition temperature (T_g) are given.

14.1.1.4 *Solvent Solubility* Solvent solubility was obtained from Ref. [1]. Extent of solubility was listed when possible based on definitions in:[1]

Descriptive Term	Parts of Solvent Required for 1 Part of Solute
Very soluble	<1
Freely soluble	1–10
Soluble	10–30
Sparingly soluble	30–100
Slightly soluble	100–1000
Very slightly soluble	1000–10,000
Practically insoluble or insoluble	≥10,000

14.1.1.5 *pH Range Where Soluble* The values for pH solubility were taken from Ref. [2]. Many enteric polymers are only soluble near neutral pH (5–7), while others are soluble over a larger pH range.

14.1.1.6 *Amenable Methods of Preparation* Melt extrusion and spray drying were the primary production methods considered in this section. The methods are based on studies and products that employ the various polymers. It should not be considered a definitive option, as innovations may lead to more polymers being used in both production methods.

14.1.1.7 *Use in Marketed Dispersion Products* The use of the specific polymers is based on package inserts and open literature as of the publication date of this book.

[1] United States Pharmacopeia and National Formulary (USP 34-NF 29), Rockville, MD: United States Pharmacopeia Conventions; 2011, 6.

14.1.1.8 *Cellulosic Derivatives*

Cellulose Acetate		
Structure		

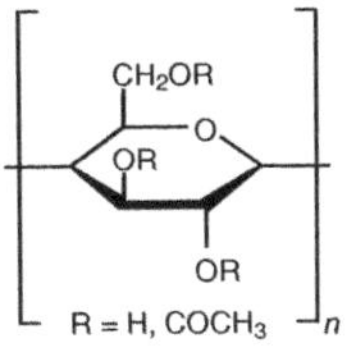

Synonyms	Acetic acid; cellulose ester; acetyl cellulose; cellulose diacetate; cellulose triacetate; cellulosi acetas	
Molecular weight range (g/mol)	30,000–60,000	
Solubility parameter, δ ($MPa^{0.5}$)	25.8–26.2	
T_g range (°C)	170–190	
T_m range (°C)	230–260	
Water uptake (% water at 100% RH)	10%	
Solvent solubility	Cellulose acetates are soluble in acetone–water blends of varying ratios, dichloromethane–ethanol blends, dimethyl formamide, and dioxane; insoluble in alcohol	
pH range where soluble	NA	
Amenable methods of dispersion manufacture	Melt extrusion, spray drying	
Use in marketed dispersion products	None	
Grade (acetyl %)	MW	T_g (or T_m) (°C)
CA-320S	38,000	180 (230–250)
CA-398-3	30,000	180 (230–250)
CA-398-6	35,000	182 (230–250)
CA-398-10	40,000	185 (230–250)
CA-398-30	50,000	189 (230–250)
CA-394-60S	60,000	186 (240–260)

Cellulose Acetate Butyrate

Structure

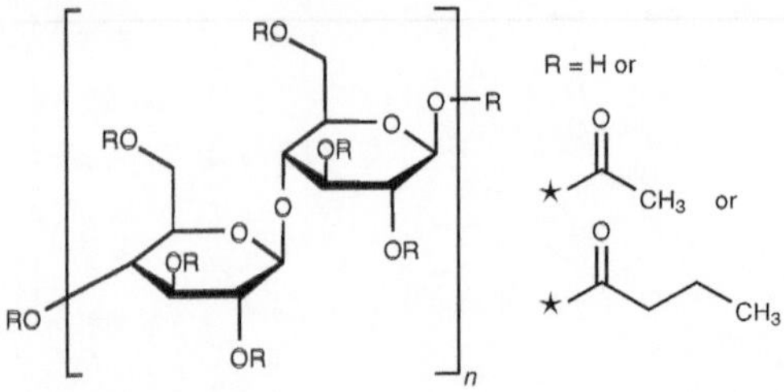

Synonyms	CAB, cellaburate
Molecular weight range (g/mol)	30,000
Solubility parameter, δ ($MPa^{0.5}$)	28.7
T_g range (°C)	130
T_m range (°C)	155–165
Water uptake (% water at 100% RH)	1–2%
Solvent solubility	Alcohol/aromatic hydrocarbon mixtures
pH range where soluble	Negligible
Amenable methods of dispersion manufacture	Melt extrusion, spray drying
Use in marketed dispersion products	None

Cellulose Acetate Phthalate

Structure

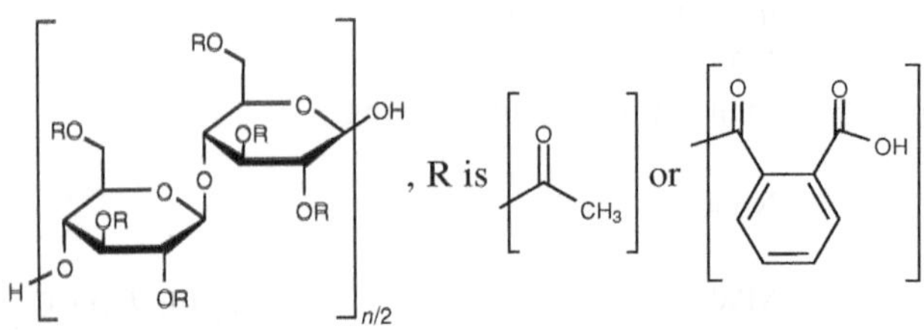

Synonyms	CAP, cellacefate, cellulosi acetas phthalas, Aquacoat cPD
Molecular weight range (g/mol)	48,000
Solubility parameter, δ ($MPa^{0.5}$)	27
T_g range (°C)	160–170
T_m range (°C)	192
Water uptake (% water at 75% RH)	7–8%
Solvent solubility	Freely soluble in acetone, soluble in diethylene glycol, and practically insoluble in ethanol and methylene chloride

(Continued)

Cellulose Acetate Phthalate	
pH range where soluble	>6.0
Amenable methods of dispersion manufacture	Melt extrusion, spray drying
Use in marketed dispersion products	None

Ethyl Cellulose	
Structure	

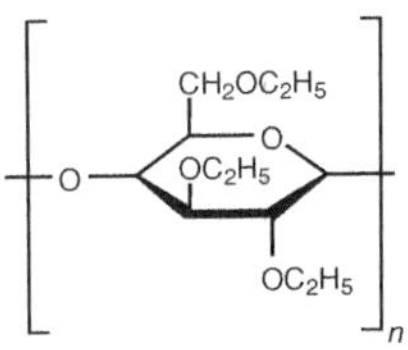

Synonyms	Aquacoat ECD; Aqualon; Ashacel; E462; Ethocel; ethylcellulosum Surelease
Molecular weight range (g/mol)	40,000–140,000
Solubility parameter, i ($MPa^{0.5}$)	18.6–18.8
T_g range (°C)	129–133
Water uptake (% water at 75% RH)	~3%
Solvent solubility	Soluble in chloroform, methyl acetate, and tetrahydrofuran, and in mixtures of aromatic hydrocarbons with ethanol (95%); chloroform, ethanol (95%), ethyl acetate, methanol, and toluene
pH range where soluble	Insoluble
Amenable methods of dispersion manufacture	Melt extrusion, spray drying
Use in marketed dispersion products	None

Hydroxypropyl Cellulose		
Structure		

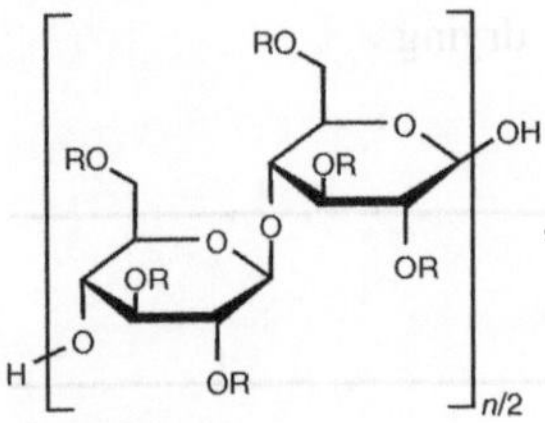

, R is H or $[CH_2CH(CH_3)O]_mH$

Synonyms	HPC, Hyprolose, oxypropylated cellulose, Klucel	
Molecular weight range (g/mol)	50,000–1,250,000	
Solubility parameter, δ ($MPa^{0.5}$)	31.5	
T_g range (°C)	100–150	
Water uptake (% water at 75% RH)	4% (@50% RH) 12% (@84% RH)	
Solvent solubility	Soluble 1 in 10 parts dichloromethane; 1 in 2.5 parts ethanol (95%); 1 in 2 parts methanol; 1 in 5 parts propan-2-ol; 1 in 5 parts propylene glycol	
pH range where soluble	1–10	
Amenable methods of dispersion manufacture	Melt extrusion, spray drying	
Use in marketed dispersion products	None	
Grade	MW	Moisture content (%)
EF	80,000	0.59
LF	95,000	2.21
JF	140,000	1.44
GF	370,000	1.67
MF	850,000	1.52
HF	1,150,000	4.27

Hydroxyethyl Cellulose

Structure

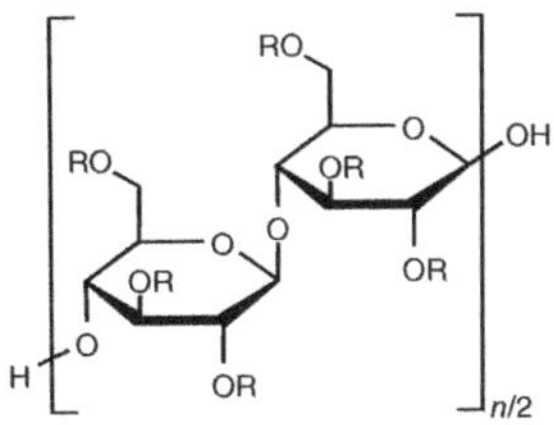

R is H or $[-CH_2CH_2O-]_mH$

Synonyms	HEC, cellulose hydroxyethylate, CELLOSIZE HEC, Natrosol
Molecular weight range (g/mol)	95,000–1,150,000
Solubility parameter, δ ($MPa^{0.5}$)	31
T_g range (°C)	100–150
Water uptake (% water at 84% RH)	29%
Solvent solubility	Partially soluble in polar organic solvents, such as glycols; practically insoluble in acetone, ethanol (95%), ether, toluene, and most other organic solvents
pH range where soluble	1–10
Amenable methods of dispersion manufacture	Spray drying
Use in marketed dispersion products	None
Grade	MW
LF	95,000
HF	1,150,000

Hydroxypropyl Methylcellulose

Structure

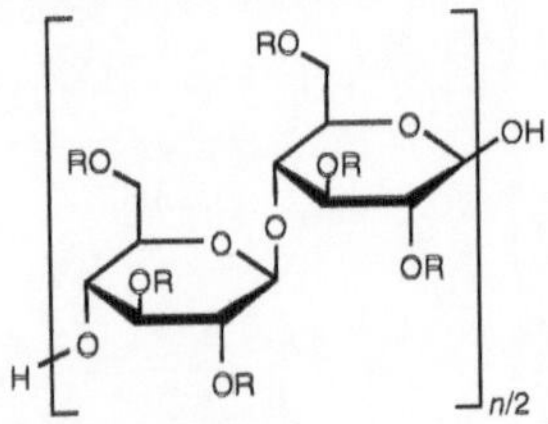

R is H, CH_3, or $CH_3CH(OH)CH_2$

Synonyms	HPMC, hypromellose, Hypromellosum, Methocel, Pharmacoat, Metolose
Molecular weight range (g/mol)	10,000–1,500,000
Solubility parameter, δ ($MPa^{0.5}$)	23.8
T_g range (°C)	170–180
Water uptake (% water at 75% RH)	~10%
Solvent solubility	Practically insoluble in chloroform, ethanol (95%), and ether, but soluble in mixtures of ethanol and dichloromethane, mixtures of methanol and dichloromethane, and mixtures of water and alcohol
pH range where soluble	1–10
Amenable methods of dispersion manufacture	Melt extrusion, spray drying
Use in marketed dispersion products	Certican®, Intelence®, Isoptin SRE-240®, LCP-Tacro®, Nivadil®, Prograf®, Sporonox®, Zortress®

Hydroxypropyl Methylcellulose Acetate Succinate

Structure

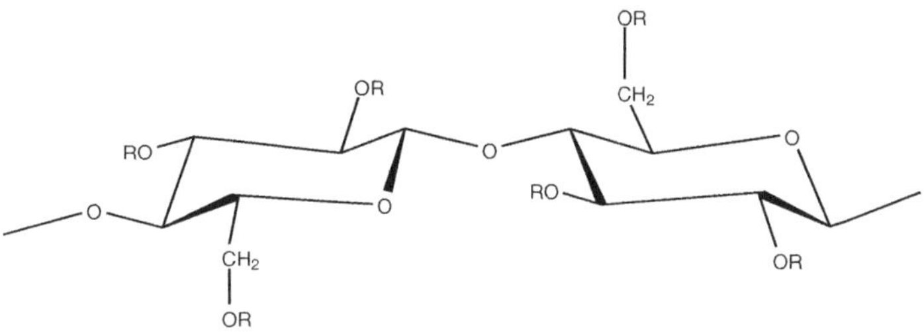

–OR represents hydroxyl, methoxyl, 2-hydroxypropoxyl, acetyl, or succinoyl

Synonyms	HPMCAS, hypromellose acetate succinate, AQOAT
Molecular weight range (g/mol)	55,000–93,000
Solubility parameter, δ ($MPa^{0.5}$)	40.5 (LF)
	31.2 (MF)
T_g range (°C)	113
Water uptake (% water at 75% RH)	7–8% (LF)
	6–7% (MF)
	5–6% (HF)
Solvent solubility	Practically insoluble in ethanol (95%), hexane, unbuffered water, and xylene; clear or turbid viscous solution obtained on the addition of acetone, or a mixture of ethanol (95%) and dichloromethane (1:1)
pH range where soluble	>5.5 (LF)
	>6.0 (MF)
	>6.5 (NF)
Amenable methods of dispersion manufacture	Melt extrusion, spray drying
Use in marketed dispersion products	Incivek®, Incivo®, Kalydeco®, Noxafil®, Zelboraf®

Hydroxypropyl Methylcellulose Phthalate

Structure

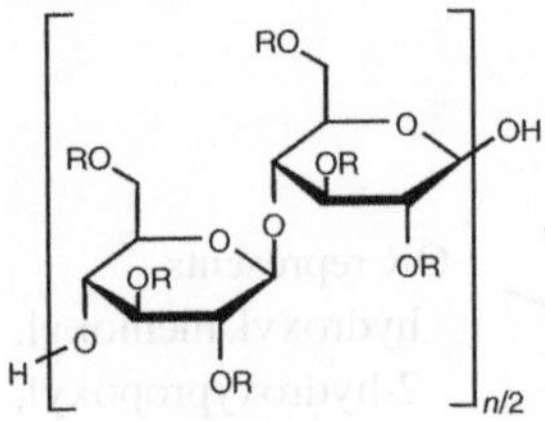

, R is H, CH_3, or $CH_3CH(OH)CH_2$

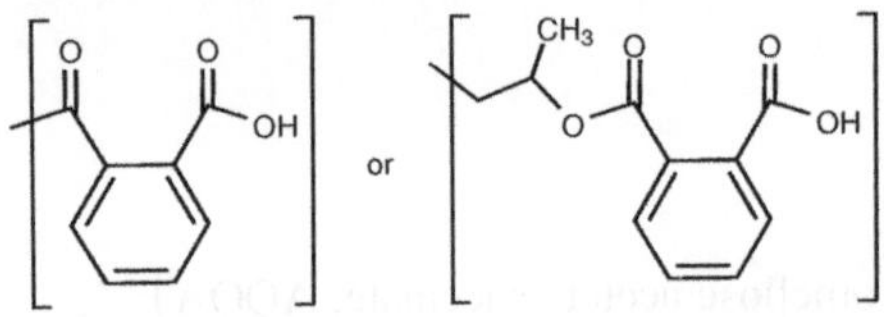

Synonyms	HPMCP, hypromellose phthalate, cellulose phthalate hydroxypropyl methyl ether, methylhydroxypropyl cellulose phthalate, Mantrocel HP-55	
Molecular weight range (g/mol)	20,000–200,000	
Solubility parameter, δ ($MPa^{0.5}$)	28	
T_g range (°C)	133–137	
T_m range (°C)	150	
Water uptake (% water at 75% RH)	7–8%	
Solvent solubility	Readily soluble in a mixture of acetone and methyl or ethyl alcohol (1:1), in a mixture of methyl alcohol and dichloromethane (1:1), practically insoluble in dehydrated alcohol and very slightly soluble in acetone	
pH range where soluble	>5.0 (HP-50), >5.5 (HP-55)	
Amenable methods of dispersion manufacture	Melt extrusion, spray drying	
Use in marketed dispersion products	None	
Grade	MW	T_g (or T_m) (°C)
HP-50	84,000	137
HP-55	78,000	133
HP-555	132,000	–

14.1.1.9 Methacrylate Polymers

Polymethacrylates	
Structure	

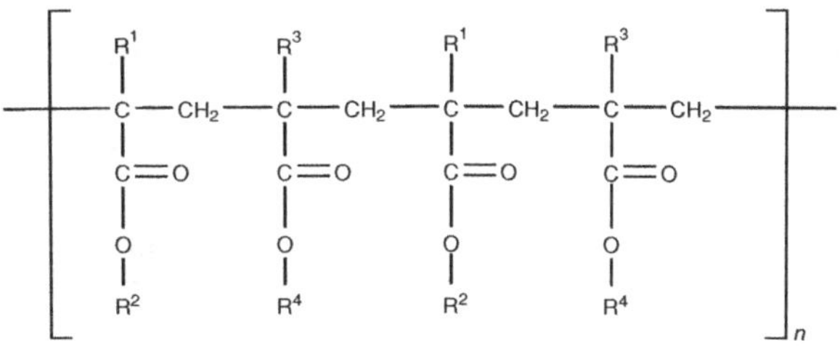

$R^1 = H$ or CH_3; $R^2 = H$, CH_3, C_2H_5, or $CH_2CH_2N(CH_3)$; $R^3 = CH_3$; $R^4 = CH_3$, C_4H_9, $CH_2CH_2N(CH_3){+}Cl^-$, C_2H_5
R^1, R^2, and R^4 depend on the grade

Synonyms	EUDRAGIT, Kollicoat, methacrylic acid copolymer, ethyl acrylate, and methyl methacrylate copolymer
Molecular weight range (g/mol)	30,000–278,000
Solubility parameter, δ ($MPa^{0.5}$)	19.3 (E100)
	23.4 (L100-55)
	23.5 (L100)
	18.9 (RS)
T_g range (°C)	48 to >150
Water uptake (% water at 75% RH)	0.3–2% (for PMMA)
Solvent solubility	E100, RL, and RS are soluble in acetone, alcohols, dichloromethane, and ethyl acetate
pH range where soluble	<5.0 (E100)
	>5.5 (L100-55)
	>6.0 (L100)
	7.0 (S100)
	Insoluble (RL)
	Insoluble (RS)
Amenable methods of dispersion manufacture	Melt extrusion, spray drying
Use in marketed dispersion products	None

(*continued*)

(Continued)

Polymethacrylates		
Grade	MW	T_g (or T_m) (°C)
E100	147,000	48
L100-55	278,000	110
L100	123,000	>150
S100	123,000	>150
RL	31,000	70
RS	**30,000**	**65**

14.1.1.10 Oxyethylene Polymers and Copolymers

Polyethylene Glycol

Structure

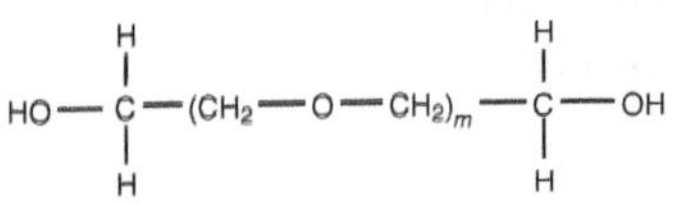

m ranges from 4 to 181

Synonyms	PEG, macrogol, polyoxyethylene glycol, Carbowax, Lipoxol, Lutrol E, Pluriol E
Molecular weight range (g/mol)	600–25,000 *Note:* Grades with MW ~190–600 are liquids, grades with MW >1000 are solids
Solubility parameter, δ ($MPa^{0.5}$)	24
T_g range (°C)	−35 to −20
T_m range (°C)	37–63
Water uptake (% water at 75% RH)	0.9%, nonhygroscopic
Solvent solubility	Soluble in acetone, dichloromethane, ethanol (95%), and methanol; they are slightly soluble in aliphatic hydrocarbons and ether, but insoluble in fats, fixed oils, and mineral oil
pH range where soluble	1–10
Amenable methods of dispersion manufacture	Melt extrusion, spray drying
Use in marketed dispersion products	Fenoglide®, Gris-PEG®, Nimotop®

(Continued)

Polyethylene Glycol		
Grade	MW	T_g (or T_m) (°C)
6000	6000	(55–63)
3500 S	35,000	(64–66)

Poloxamer		
Structure		

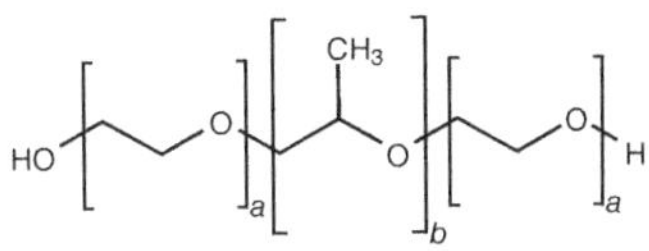

$a = 64, 80, 101, 141; b = 20, 27, 37, 44, 56$

Synonyms	Poloxalkol, poloxamera, polyethylene-propylene glycol copolymer, polyoxyethylene-polyoxypropylene copolymer, Lutrol, Monolan, Pluronic, Supronic, Synperonic	
Molecular weight range (g/mol)	7600–17,400	
Solubility parameter, δ ($MPa^{0.5}$)	23.7	
T_m range (°C)	52–57	
Water uptake (% water at 75% RH)	<0.5%	
Solvent solubility	Soluble in acetone, dichloromethane, ethanol (95%), and methanol; they are slightly soluble in aliphatic hydrocarbons and ether, but insoluble in fats, fixed oils, and mineral oil	
pH range where soluble	1–10	
Amenable methods of dispersion manufacture	Spray drying	
Use in marketed dispersion products	None	
Grade	MW	T_m (°C)
188	7680–9510	52–57
237	6840–8830	49
338	12,700–17,400	57
407	9840–14,600	52–57

Solutol	
Structure	

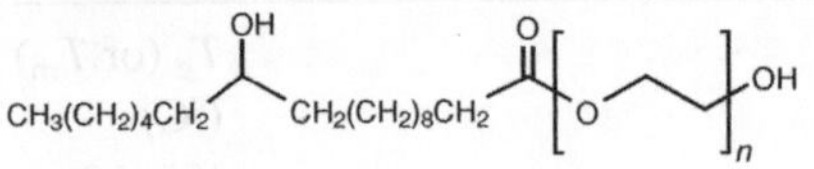

Synonyms	Macrogol 15 hydroxystearate, polyoxyl 15 hydroxystearate, macrogoli 15 hydroxystearas, polyethylene glycol-15-hydroxystearate, Kolliphor
Molecular weight range (g/mol)	344
Solubility parameter, δ ($MPa^{0.5}$)	-
T_g range (°C)	(solidification @ 25–30)
Water uptake (% water at 75% RH)	NA
Solvent solubility	Soluble in water, ethanol, and 2-propanol; insoluble in liquid paraffin
pH range where soluble	1–10
Amenable methods of dispersion manufacture	Spray drying
Use in marketed dispersion products	None

14.1.1.11 Vinyl Pyrrolidone Polymers and Copolymers

Polyvinyl Alcohol	
Structure	

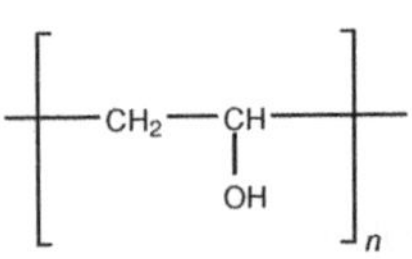

Synonyms	PVA, vinyl alcohol polymer, Polyvinol, Elvanol
Molecular weight range (g/mol)	20,000–200,000
Solubility parameter, δ ($MPa^{0.5}$)	27.6
T_m range (°C)	180–190 (partially hydrolyzed grades), 228 (fully hydrolyzed grades)
Water uptake (% water at 75% RH)	22% (for PVA fibers)
Solvent solubility	Slightly soluble in ethanol (95%); insoluble in organic solvents
pH range where soluble	1–10
Amenable methods of dispersion manufacture	Melt extrusion, spray drying
Use in marketed dispersion products	None
Grade (viscosity)	MW

(Continued)

Polyvinyl Alcohol	
High viscosity	200,000
Medium viscosity	130,000
Low viscosity	20,000

Polyvinylpyrrolidone-vinyl Acetate	
Structure	

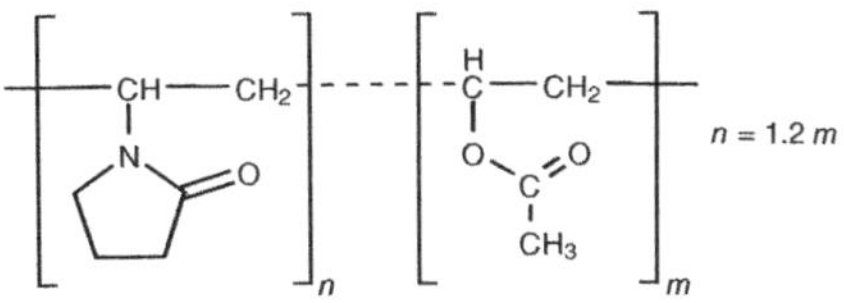

Synonyms	PVP/VA, copovidone, polyvinylpyrrolidone-*co*-vinyl acetate, copolyvidone, copovidonium, Kollidon VA 64, Plasdone S-630
Molecular weight range (g/mol)	45,000–70,000
Solubility parameter, δ ($MPa^{0.5}$)	25.6
T_g range (°C)	106
Water uptake (% water at 50% RH)	10%
Solvent solubility	Soluble in 1,4-butanediol, glycerol, butanol, chloroform, dichloromethane, ethanol (95%), glycerol, methanol, polyethylene glycol 400, propan-2-ol, propanol, propylene glycol; insoluble in cyclohexane, diethyl ether, liquid paraffin, and pentane
pH range where soluble	1–10
Amenable methods of dispersion manufacture	Melt extrusion, spray drying
Use in marketed products	Kaletra®, Norvir®, Onmel®

Polyvinylpyrrolidone		
Structure		

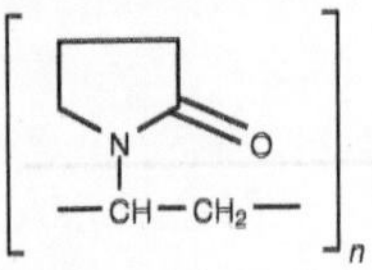

$n > 1{,}000{,}000$

Synonyms	PVP, povidone, polyvidone, 1-vinyl-2-pyrrolidinone, Kollidon, Polyplasdone, Povipharm	
Molecular weight range (g/mol)	2500–3,000,000	
Solubility parameter, δ ($MPa^{0.5}$)	27.7	
T_g range (°C)	120–174	
Water uptake (% water at 75% RH)	30–40%	
Solvent solubility	Freely soluble in chloroform, 95% ethanol, ketones, methanol, and water	
pH range where soluble	1–10	
Amenable methods of dispersion manufacture	Melt extrusion, spray drying	
Use in marketed products	Cesamet®, Rezulin®[1]	
Grade (K-value)	~MW	T_g (°C)
12	2,500	120
15	8000	–
17	10,000	126
25	30,000	160
29/32	5000	164
60	400,000	–
90	1,000,000	174
120	3,000,000	–

Soluplus

Structure

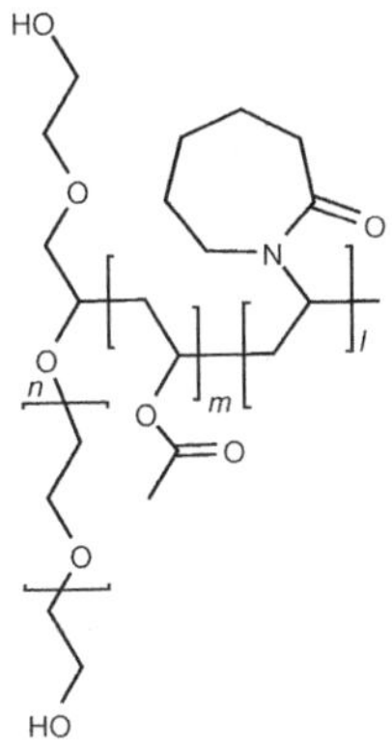

Synonyms	Polyethyleneglycol-polyvinyl caprolactam-polyvinyl acetate-grafted copolymer
Molecular weight range (g/mol)	64,000
Solubility parameter, δ ($MPa^{0.5}$)	22.1
T_g range (°C)	60–70
Water uptake (% water at 75% RH)	–
Solvent solubility	Freely soluble in water and organic solvents
pH range where soluble	1–10
Amenable methods of dispersion manufacture	Melt extrusion, spray drying
Use in marketed dispersion products	None

14.2 PART II. SURFACTANTS

This section is not a comprehensive list of surfactants used in amorphous solid dispersions, but it is meant to highlight common surfactants that have been used in various studies.

14.2.1 Notes

14.2.1.1 Synonyms, Critical Micelle Concentration, and Solubility Synonyms, critical micelle concentration, and solubility values were taken from Ref. [1]. Some surfactants exhibit a range of molecular weights; where applicable, the grade and molecular weight are given.

Polyoxyl 40 Hydrogenated Castor Oil	
Structure	

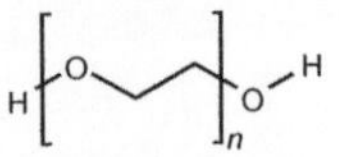

Synonyms	PEG-40 hydrogenated castor oil, Cremophor RH 40
Critical micelle concentration (CMC)	–
Solubility	Soluble in water, soluble in ethanol, 2-propanol, ethyl acetate, *n*-propanol, chloroform, carbon tetrachloride, toluene, and xylene; insoluble in mineral oil

Polysorbates	
Structure	

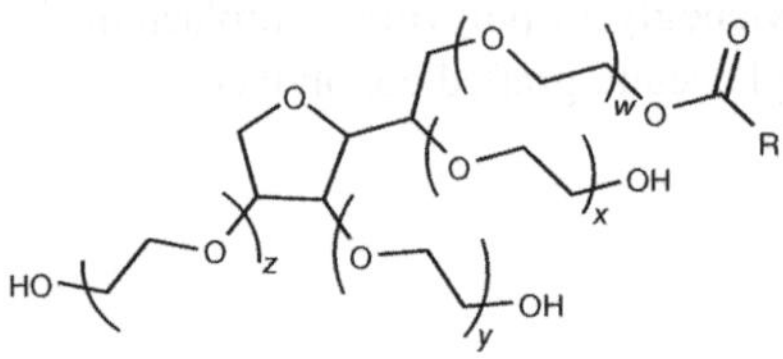

TWEEN 20	$R=(CH_2)10CH_3$
TWEEN 40	$R=(CH_2)14CH_3$
TWEEN 60	$R=(CH_2)16CH_3$
TWEEN 80	$R=(CH_2)7(CHCH)(CH_2)7CH_3$
	$w+x+y+z=20$
Synonyms	TWEENS
Critical micelle concentration (CMC)	0.0499 mM (TWEEN 20), 0.0333 mM (TWEEN 40), 0.0167 mM (TWEEN 60), 0.015 mM (TWEEN 80)
Solubility	Very soluble in water TWEEN 20: miscible with alcohol, dioxane, and ethyl acetate; practically insoluble in liquid paraffin and fixed oils TWEEN 80: soluble in ethanol, cottonseed oil, corn oil, ethyl acetate, methanol, and toluene
Grade	MW (g/mol)
TWEEN 20	1227
TWEEN 40	1283
TWEEN 60	1312
TWEEN 80	1310

Sodium Docusate	
Structure	

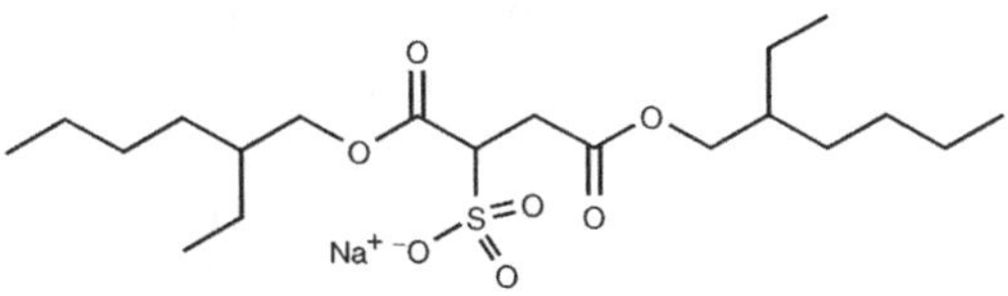

Synonyms	Dioctyl sodium sulfosuccinate
Critical micelle concentration (CMC)	0.17% at 21 °C
Solubility	Soluble in ethanol, chloroform, and diethylether Highly soluble in petroleum ether, tetrahydrofuran (despite being ionic), and glycerol
Other properties	MW: 444.56 g/mol, T_g: 153–157 °C

Sodium Lauryl Sulfate	
Structure	

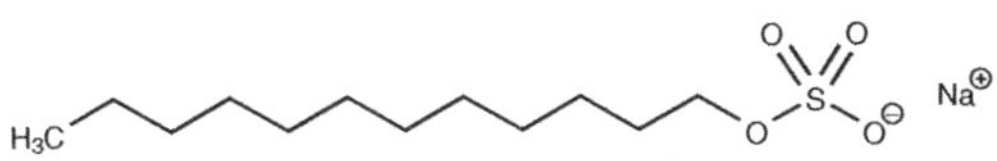

Synonyms	Sodium dodecyl sulfate, SLS, or SDS
Critical micelle concentration (CMC)	0.0082 M (pure water, 25 °C)
Solubility	0.1 M in water, slightly soluble in alcohol; insoluble in chloroform and ether
Other properties	MW: 288 g/mol, T_g: 206 °C

Sorbitans	
Structure	

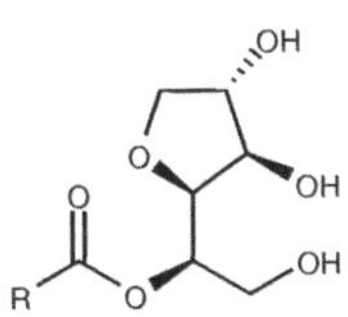

SPAN 20	R = $(CH_2)10CH_3$
SPAN 40	R = $(CH_2)14CH_3$

(continued)

(*Continued*)

Sorbitans		
	SPAN 60	R = $(CH_2)16CH_3$
	SPAN 80	R = $(CH_2)7(CHCH)(CH_2)7CH_3$
	SPAN 65	R = $(CH_2)16CH_3$
	SPAN 85	R = $(CH_2)7(CHCH)(CH_2)7CH_3$
Synonyms	SPAN	
Critical micelle concentration (CMC)	$(2.2 \pm 0.2) \times 10^{-5}$ mol/l for SPAN 20	
Solubility	Insoluble in water	
Grade	MW (g/mol)	T_g or T_m (°C)
SPAN 20	346.46	14–16
SPAN 40	402.57	46–47
SPAN 60	430.6	54–57
SPAN 65	963.54	55–57
SPAN 80	428.61	10~12
SPAN 85	957.49	10

D-α-Tocopheryl Polyethylene Glycol 1000 Succinate

Structure

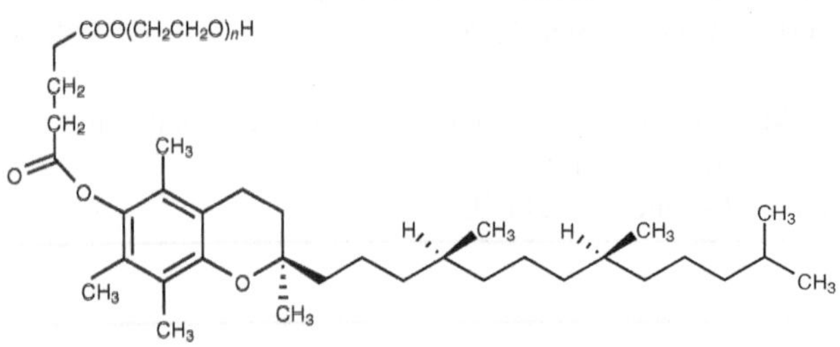

$n = 23$

Synonyms	TPGS
Critical micelle concentration (CMC)	0.02 w% at 37 °C
Solubility	Water soluble (~20% at 20 °C)
Other properties	MW: ~1513 g/mol

REFERENCES

1. Rowe, R.C., Sheskey, P.J., and Quinn, M.E. (Eds.) (2009) *Handbook of Pharmaceutical Excipients*, 6th edition, Pharmaceutical Press, London.
2. Shah, N., Sandhu, H., Choi, D.S., Kalb, O., Page, S., and Wyttenbach, N. Structured development approach for amorphous systems. In: Williams, R.O., III, Watts, A.B., and Miler, D.A. (Eds.) (2012) *Formulating Poorly Water Soluble Drugs*, Springer, New York, pp. 267–310.

APPENDIX A

APPENDIX

Xia Lu,[1] Robert Wenslow,[2] and Ann Newman[3]

[1]*Crystal Pharmatech, Suzhou, China*
[2]*Crystal Pharmatech, North Brunswick, NJ, USA*
[3]*Seventh Street Development Group, Lafayette, IN, USA*

Pharmaceutical Amorphous Solid Dispersions, First Edition. Edited by Ann Newman.

TABLE A.1. Common Polymers and Properties

	Abbreviation	Grade	MW	T_g (or T_m) (°C)	Solubility Parameter	pH Solubility	Hygroscopicity	Amenable Methods of Dispersion Manufacture	Use in Marketed Products
Cellulosic derivatives									
Cellulose acetate	CA	(acetyl %)	30,000–60,000	170–190 (230–300)	25.8–26.2	NA	10% (100% RH)	Melt extrusion, spray drying	None
		CA-320S	38,000	180					
		CA-398-3	30,000	180					
		CA-398-6	35,000	182					
		CA-398-10	40,000	185					
		CA-398-30	5,000	189					
		CA-394-60S	6,000	186					
Cellulose acetate butyrate	CAB		30,000	130 (155–165)	28.7	Negligible	1–2% (100% RH)	Melt extrusion, spray drying	None
Cellulose acetate phthalate	CAP		48,000	160–170 (192)	27	>6.0	7–8% (75% RH)	Melt extrusion, spray drying	None
Ethyl cellulose	E462	(Tensile strength)	40,000–140,000	129–133	18.6–18.8	Insoluble	~3% (75% RH)	Melt extrusion, spray drying	None
		T10	80,000						
		N10	80,000						
		N22	130,000						
Hydroxypropyl cellulose	HPC	(Moisture content)	50,000–1,250,000	100–150	31.5	1–10	4% (50% RH), 12% (84% RH)	Melt extrusion, spray drying	None
		EF	80,000						
		LF	95,000						
		JF	140,000						
		GF	370,000						
		MF	850,000						
		HF	1,150,000						

Hydroxyethyl cellulose LF	HEC	LF	95,000	100–150	31	1–10	29% (84% RH)	Spray drying	None
		HF	1,150,000						
Hydroxypropyl methylcellulose (hypromellose)	HPMC	2910	10,000–1,500,000	170–180	23.8	1–10	~10% (75% RH)	Melt extrusion, spray drying	Certican®, Intelence®, Isoptin SRE-240®, LCP-Tacro®, Nivadil®, Prograf®, Sporonox®, Zortress®
Hydroxypropyl methylcellulose acetate succinate (hypromellose acetate succinate)	HPMCAS	LF	55,000–93,000	113	40.5	>5.5	7–8% (75% RH)	Melt extrusion, spray drying	Incivek®, Incivo®, Kalydeco®, Noxafil®, Zelboraf®
		MF	55,000–93,000	113	31.2	>6.0	6–7% (75% RH)		
		HF	55,000–93,000	113		>6.5	5–6% (75% RH)		
Hydroxypropyl methylcellulose phthalate (hypromellose phthalate)	HPMCP		20,000–200,000	133–137 (150)	28		7–8% (75% RH)	Melt extrusion, spray drying	None
		HP-50	84,000	137		>5.0			
		HP-55	78,000	133		>5.5			
		HP-555	132,000						
Methacrylate polymers									
Polymethacrylates	EUDRAGIT	E100	147,000	48	19.3	<5.0	0.3–2% (75% RH for PMMA)	Melt extrusion, spray drying	None
		L100-55	278,000	110	23.4	>5.5			
		L100	123,000	>150	23.5	>6.0			
		S100	123,000	>150		7.0			
		RL	31,000	70		Insoluble			
		RS	30,000	65	18.9	Insoluble			

(continued)

TABLE A.1. (*Continued*)

	Abbreviation	Grade	MW	T_g (or T_m) (°C)	Solubility Parameter	pH Solubility	Hygroscopicity	Amenable Methods of Dispersion Manufacture	Use in Marketed Products
Oxyethylene polymers and copolymers									
Polyethylene glycol	PEG	6000	6,000	(55–63)	24	1–10	0.9% (75% RH)	Melt extrusion, spray drying	Fenoglide®, Gris-PEG®, Nimotop®
		35000S	35,000	(64–66)	24				
Poloxamer			7,600–17,400	(52–47)	23.7	1–10	<0.5% (75% RH)	Spray drying	None
		188	7,680–9,510	(52–57)					
		237	6,840–8,830						
		338	12,700–17,400						
		407	9,840–14,600	(52–57)					
Solutol		HS 15	344	(Solidification @ 25–30)		1–10	NA	Spray drying	None
Vinyl-pyrrolidone polymers and copolymers									
Polyvinyl alcohol	PVA	(Viscosity)	20,000–200,000	(228, 180–190)	27.6	1–10	22% (75% RH for PVA fibers)	Melt extrusion, spray drying	None
		High viscosity	200,000						
		Medium viscosity	130,000						
		Low viscosity	20,000						

Crospovidone	PVPP		>1,000,000			Insoluble	Max 60%		
Polyvinylpyrrolidone (povidone)	PVP	(K-value)	2,500–3,000,000		27.7	1–10	30–40% (75% RH)	Melt extrusion, spray drying	Rezulin™, Kaletra™,
		12	2,500	120					
		15	8,000						
		17	10,000	126					
		25	30,000	160					
		29/32	58,000	164					
		30	50,000	175					
		60	400,000						
		90	100,000	190					
		120	3,000,000						
Soluplus			64,000	60–70	22.1	1–10		Melt extrusion, spray drying	None

TABLE A.2. Polymers: Alphabetical Order

Polymer	Abbreviation	Grade	MW	T_g (or T_m) [°C]	Solubility parameter
Cellulose acetate	CA	CA-398-3	30,000	180	–
Cellulose acetate	CA	All	30,000–60,000	170–190 (230–300)	25.8–26.2
Cellulose acetate	CA	CA-398-6	35,000	182	–
Cellulose acetate	CA	CA-320S	38,000	180	–
Cellulose acetate	CA	CA-398-10	40,000	185	–
Cellulose acetate	CA	CA-398-30	50,000	189	–
Cellulose acetate	CA	CA-394-60S	60,000	186	–
Cellulose acetate butyrate	CAB		30,000	130 (155–165)	28.7
Cellulose acetate phthalate	CAP		48,000	160–170 (192)	27
Ethyl cellulose	EC	All	40,000–140,000	129–133	18.6–18.8
Ethyl cellulose	EC	T10	80,000	–	–
Ethyl cellulose	EC	N10	80,000	–	–
Ethyl cellulose	EC	N22	123,000	–	–
Hydroxyethyl cellulose LF	HEC	LF	95,000	100–150	31
Hydroxyethyl cellulose LF	HEC	HF	1,150,000	–	–
Hydroxypropyl methylcellulose phthalate (hypromellose phthalate)	HPMCP	HP-50	84,000	137	–
Hydroxypropyl methylcellulose (hypromellose)	HPMC	2910	10,000–1,500,000	170–180	23.8
Hydroxypropyl methylcellulose acetate succinate (hypromellose acetate succinate)	HPMCAS	LF	55,000–93,000	113	40.5
Hydroxypropyl methylcellulose acetate succinate (hypromellose acetate succinate)	HPMCAS	MF	55,000–93,000	113	31.2
Hydroxypropyl methylcellulose acetate succinate (hypromellose acetate succinate)	HPMCAS	NF	55,000–93,000	113	–

Hydroxypropyl methylcellulose phthalate (hypromellose phthalate)	HPMCP	All	20,000–200,000	133–137 (150)	28
Hydroxypropyl methylcellulose phthalate (hypromellose phthalate)	HPMCP	HP-55	78,000	133	–
Hydroxypropyl cellulose	HPC	All	50,000–1,250,000	100–150	31.5
Hydroxypropyl cellulose	HPC	EF	80,000	–	–
Hydroxypropyl cellulose	HPC	JF	132,000	–	–
Hydroxypropyl cellulose	HPC	GF	278,000	–	–
Hydroxypropyl cellulose	HPC	MF	400,000	–	–
Hydroxypropyl cellulose	HPC	LF	850,000	–	–
Hydroxypropyl cellulose	HPC	HF	1,150,000	–	–
Hypromellose phthalate	HPMCP	HP-555	130,000	–	–
Poloxamer		237	6,840–8,830	–	–
Poloxamer		188	7,680–9,510	(52–57)	–
Poloxamer		All	7,600–17,400	50–60	23.7
Poloxamer		407	9,840–14,600	(52–57)	–
Poloxamer		338	12,700–17,400	–	–
Polyethylene glycol	PEG	6000	6,000	(55–63)	24
Polyethylene glycol	PEG	35000S	25,000	(64–66)	24
Polymethacrylates	EUDRAGIT	RS	30,000	65	18.9
Polymethacrylates	EUDRAGIT	RL	31,000	70	–
Polymethacrylates	EUDRAGIT	L100	100,000	>150	23.5
Polymethacrylates	EUDRAGIT	S100	123,000	>150	–
Polymethacrylates	EUDRAGIT	E100	140,000	48	19.3
Polymethacrylates	EUDRAGIT	L100-55	200,000	110	23.4
Polyvinyl alcohol	PVA	Low viscosity	20,000	–	–
Polyvinyl alcohol	PVA	All	20,000–200,000	(228, 180–190)	27.6

(continued)

TABLE A.2. (*Continued*)

Polymer	Abbreviation	Grade	MW	T_g (or T_m) [°C]	Solubility parameter
Polyvinyl alcohol	PVA	Medium viscosity	130,000	–	–
Polyvinyl alcohol	PVA	High viscosity	147,000	–	–
Polyvinylpyrrolidone (povidone)	PVP	12	2,500	–	–
Polyvinylpyrrolidone (povidone)	PVP	All	2,500–3,000,000	–	–
Polyvinylpyrrolidone (povidone)	PVP	15	8,000	–	–
Polyvinylpyrrolidone (povidone)	PVP	17	10,000	–	–
Polyvinylpyrrolidone (povidone)	PVP	25	30,000	–	–
Polyvinylpyrrolidone (povidone)	PVP	30	50,000	175	27.7
Polyvinylpyrrolidone (povidone)	PVP	90	95,000	190	27.7
Polyvinylpyrrolidone (povidone)	PVP	60	370,000	–	–
Polyvinylpyrrolidone (povidone)	PVP	120	3,000,000	–	–
Polyvinylpyrrolidone-vinyl acetate (copovidone)	PVP VA	64	45,000–70,000	106	25.6
Soluplus			64,000	60	22.1
Solutol		HS 15	344	(Solidification @ 25–30)	–

TABLE A.3. Polymers: Sorted by Increasing Solubility Parameter

Solubility Parameter	Polymer	Abbreviation	MW	T_g (or T_m)	pH solubility	Hygroscopicity
18.6–18.8	Ethyl cellulose	E462	40,000–140,000	129–133	Insoluble	~3%
18.9	Polymethacrylates	EUDRAGIT RS	30,000	65	Insoluble	
19.3	Polymethacrylates	EUDRAGIT E100	30,000	48	<5.0	0.3–2% (for PMMA)
22.1	Soluplus		64,000	60	1–10	
23.4	Polymethacrylates	EUDRAGIT L100-55	278,000	110	>5.5	
23.5	Polymethacrylates	EUDRAGIT L100	123,000	>150	>6.0	
23.7	Poloxamer		7,600–17,400	50–60	1–10	<0.5%
23.8	Hydroxypropyl methylcellulose (hypromellose)	HPMC	10,000–1,500,000	170–180	1–10 in cold water	~10%
24	Polyethylene glycol	PEG	6,000	(55–63)	1–10	0.90%
24	Polyethylene glycol	PEG	25,000	(64–66)		Nonhygroscopic
25.6	Copovidone	PVP VA	45,000–70,000	106	1–10	10% (@50% RH)
25.8–26.2	Cellulose acetate	CA	30,000–60,000	170–190 (230–300)	Negligible	10% (@ 100% RH)
27	Cellulose acetate phthalate	CAP	48,000	160–170 (192)	>6.0	7–8%
27.6	Polyvinyl alcohol	PVA	20,000–200,000	(228, 180–190)	1–10	22% (for PVA fibers)
27.7	Povidone	PVP	2500–3,000,000	175–90	1–10	40%
28	Hydroxypropyl methylcellulose phthalate (hypromellose phthalate)	HPMCP	20,000–200,000	133–137 (150)	>5.0	7–8%

(continued)

TABLE A.3. *(Continued)*

Solubility Parameter	Polymer	Abbreviation	MW	T_g (or T_m)	pH solubility	Hygroscopicity
28.7	Cellulose acetate butyrate	CAB	30,000	130 (155–165)	Negligible	1–2% (@ 100% RH)
31	Hydroxyethyl cellulose	HEC	95,000–1,150,000	100–150	1–10	29% (@84% RH)
31.2	Hypromellose acetate succinate	HPMCAS	55,000–93,000	113	>6.0	6–7%
31.5	Hydroxypropyl cellulose	HPC	50,000–1,250,000	100–150	1–10 (T < 38)	4% (@50% RH), 12% (@84% RH)
40.5	Hydroxypropyl methylcellulose acetate succinate (hypromellose acetate succinate)	HPMCAS	55,000–93,000	113	>5.5	7–8%

TABLE A.4. Polymers: Sorted by Increasing T_g

T_g (or T_m) [°C]	Polymer	Abbreviation	MW	Solubility Parameter	pH solubility
(Solidification @ 25–30)	Solutol	–	344	–	1–10
48	Polymethacrylates	EUDRAGIT E100	30,000	19.3	<5.0
50–60	Poloxamer		7,600–17,400	23.7	1–10
(55–63)	Polyethylene glycol	PEG	6,000	24	1–10
60	Soluplus	–	64,000	22.1	1–10
(64–66)	Polyethylene glycol	PEG	25,000	24	–
65	Polymethacrylates	EUDRAGIT RS	30,000	18.9	Insoluble
70	Polymethacrylates	EUDRAGIT RL	31,000	–	Insoluble
106	Polyvinylpyrrolidone-vinyl acetate	PVP VA	45,000–70,000	25.6	1–10
110	Polymethacrylates	EUDRAGIT L100-55	278,000	23.4	>5.5
113	Hydroxypropyl methylcellulose acetate succinate	HPMCAS MF	55,000–93,000	31.2	>6.0
113	Hydroxypropyl methylcellulose acetate succinate	HPMCAS LF	55,000–93,000	40.5	>5.5
113	Hydroxypropyl methylcellulose acetate succinate	HPMCAS HF	55,000–93,000	–	>6.5
>150	Polymethacrylates	EUDRAGIT L100	123,000	23.5	>6.0
>150	Polymethacrylates	EUDRAGIT S100	123,000	–	7.0
100–150	Hydroxyethyl cellulose LF	HEC	95,000–1,150,000	31	1–10
100–150	Hydroxypropyl cellulose	HPC	50,000–1,250,000	31.5	1–10 (T < 38°C)
129–133	Ethyl cellulose	E462	40,000–140,000	18.6–18.8	Insoluble
130 (155–165)	Cellulose acetate butyrate	CAB	30,000	28.7	Negligible
133–137 (150)		HPMCP	20,000–200,000	28	>5.0
160–170 (192)	Cellulose acetate phthalate	CAP	48,000	27	>6.0

(continued)

TABLE A.4. (*Continued*)

T_g (or T_m) [°C]	Polymer	Abbreviation	MW	Solubility Parameter	pH solubility
170–180	Hydroxypropyl methylcellulose	HPMC	10,000–1,500,000	23.8	1–10 in cold water
170–190 (230–300)	Cellulose acetate	CA	30,000–60,000	25.8–26.2	Negligible
175–190	Polyvinylpyrrolidone	PVP	2,500–3,000,000	27.7	1–10
(228, 180–190)	Polyvinyl alcohol	PVA	20,000–200,000	27.6	1–10

TABLE A.5. Polymers: Sorted by Increasing Molecular Weight

MW	Polymer	Abbreviation	Grade	T_g (or T_m) [°C]	Solubility Parameter
344	Solutol	–	HS 15	(Solidification @ 25–30)	–
2,500	Povidone	PVP	12	–	–
2,500–3,000,000	Povidone	PVP	All	–	–
6,000	Polyethylene glycol	PEG	6000	(55–63)	24
6,840–8,830	Poloxamer	–	237	–	–
7,680–9,510	Poloxamer	–	188	(52–57)	–
7,600–17,400	Poloxamer	–	All	50–60	23.7
8,000	Povidone	PVP	15	–	–
9,840–14,600	Poloxamer		407	(52–57)	–
10,000	Povidone	PVP	17		–
10,000–1,500,000	Hypromellose	HPMC	2910	170–180	23.8
12,700–17,400	Poloxamer	–	338	–	–
20,000	Polyvinyl alcohol	PVA	Low viscosity	–	–
20,000–200,000	Hypromellose phthalate	HPMCP	All	133–137 (150)	28
20,000–200,000	Polyvinyl alcohol	PVA	All	(228, 180–190)	27.6
25,000	Polyethylene glycol	PEG	35000S	(64–66)	24
30,000	Cellulose acetate butyrate	CAB	–	130 (155–165)	28.7
30,000	Polymethacrylates	EUDRAGIT	RS	65	18.9
30,000	Cellulose acetate	CA	CA-398-3	180	–
30,000	Povidone	PVP	25	–	–
30,000–60,000	Cellulose acetate	CA	All	170–190 (230–300)	25.8–26.2
31,000	Polymethacrylates	EUDRAGIT	RL	70	–
35,000	Cellulose acetate	CA	CA-398-6	182	–
38,000	Cellulose acetate	CA	CA-320S	180	–
40,000	Cellulose acetate	CA	CA-398-10	185	–

(continued)

TABLE A.5. *(Continued)*

MW	Polymer	Abbreviation	Grade	T_g (or T_m) [°C]	Solubility Parameter
40,000–140,000	Ethyl cellulose	EC	All	129–133	18.6–18.8
45,000–70,000	Copovidone	PVP VA	64	106	25.6
48,000	Cellulose acetate phthalate	CAP		160–170 (192)	27
50,000	Cellulose acetate	CA	CA-398-30	189	
50,000	Povidone	PVP	30	175	27.7
50,000–1,250,000	Hydroxypropyl cellulose	HPC	All	100–150	31.5
55,000–93,000	Hypromellose acetate succinate	HPMCAS	LF	113	40.5
55,000–93,000	Hypromellose acetate succinate	HPMCAS	MF	113	31.2
55,000–93,000	Hypromellose acetate succinate	HPMCAS	NF	113	–
60,000	Cellulose acetate	CA	CA-394-60S	186	–
64,000	Soluplus			60	22.1
78,000	Hypromellose phthalate	HPMCP	HP-55	133	–
80,000	Ethyl cellulose	EC	T10	–	–
80,000	Ethyl cellulose	EC	N10	–	–
80,000	Hydroxypropyl cellulose	HPC	EF	–	–
84,000	Hypromellose phthalate	HPMCP	HP-50	137	–
95,000	Hydroxyethyl cellulose LF	HEC	LF	100–150	31
95,000	Povidone	PVP	90	190	27.7
100,000	Polymethacrylates	EUDRAGIT	L100	>150	23.5
123,000	Polymethacrylates	EUDRAGIT	S100	>150	–
123,000	Ethyl cellulose	EC	N22	–	–
130,000	Polyvinyl alcohol	PVA	Medium viscosity	–	–
130,000	Hypromellose phthalate	HPMCP	HP-555	–	–
132,000	Hydroxypropyl cellulose	HPC	JF	–	–
140,000	Polymethacrylates	EUDRAGIT	E100	48	19.3
147,000	Polyvinyl alcohol	PVA	High viscosity	–	–

200,000	Polymethacrylates	EUDRAGIT	L100-55	110	23.4
278,000	Hydroxypropyl cellulose	HPC	GF	–	–
370,000	Povidone	PVP	60	–	–
400,000	Hydroxypropyl cellulose	HPC	MF	–	–
850,000	Hydroxypropyl cellulose	HPC	LF	–	–
1,150,000	Hydroxyethyl cellulose LF	HEC	HF	–	–
1,150,000	Hydroxypropyl cellulose	HPC	HF	–	–
3,000,000	Povidone	PVP	120	–	–

APPENDIX B

MARKETED PRODUCTS

Ann Newman

Seventh Street Development Group, Lafayette, IN, USA

Marketed Dispersions

Product	API	Company	BCS Class	Polymer	Dispersion Process
Afeditab®	Nifedipine	Elan/Watson	2	Poloxamer or PVP	Melt/absorb on carrier
Certican® (non-US)	Everolimus	Novartis	3	HPMC	Melt or spray drying
Cesamet®	Nabilone	Valeant Pharmaceuticals	2 or 4	PVP	Melt extrusion
Fenoglide®	Fenofibrate	LifeCycle Pharma	2	PEG	Spray melt
Gris-PEG®	Griseofulvin	Novartis/Pedinol	2	PEG	Melt extrusion
Ibuprofen	Ibuprofen	Soliqs	2	Various	Melt extrusion
Incivek® (US)	Telaprevir	Vertex Pharmaceuticals	2 or 4	HPMCAS	Spray drying
Incivo® (Europe)	Telaprevir	Janssen Pharmaceuticals	2 or 4	HPMCAS	Spray drying

Pharmaceutical Amorphous Solid Dispersions, First Edition. Edited by Ann Newman.

(Continued)

Product	API	Company	BCS Class	Polymer	Dispersion Process
Intelence®	Etravirine	Janssen Pharmaceuticals	4	HPMC	Spray drying
Isoptin SRE-240®	Verapamil	AbbVie Inc.	2	HPC/ HPMC	Melt extrusion
Kalydeco®	Ivacaftor	Vertex	2 or 4	HPMCAS	Spray drying
Kaletra®	Lopinavir and ritonavir	AbbVie	2 and 4	PVP VA	Melt extrusion
LCP-Tacro®	Tacrolimus	LifeCycle Pharma/Veloxis	2	HPMC	Melt granulation
Nimotop®	Nimodipine	Bayer	2	PEG	Spray drying/ fluid bed
Nivadil®	Nilvadipine	Fujisawa Pharmaceutical	2	HPMC	
Norvir®	Ritonavir	AbbVie	4	PVP VA	Melt extrusion
Noxafil®	Posaconazole	Merck	2	HPMCAS	Melt extrusion
Onmel®	Itraconazole	GlaxoSmithKline/ Stiefel	2	PVP VA	Melt extrusion
Prograf®	Tacrolimus	Astellas Pharma Inc.	2	HPMC	Spray drying/ fluid bed
Rezulin®[a]	*Troglitazone*	Pfizer (Parke–Davis)	2	PVP	Melt extrusion
Sporonox®	Itraconazole	Janssen Pharmaceuticals	2	HPMC	Spray layering
Torcetrapib[b]	*Torcetrapib*	Pfizer	2	HPMCAS	Spray drying
Zelboraf®	Vemurafenib	Roche	4	HPMCAS	Antisolvent precipitation
Zortress® (US)	Everolimus	Novartis Pharmaceuticals	3	HPMC	Melt or spray drying

[a] Recalled in 2000 due to toxicity issues.
[b] Program killed in phase III.

INDEX

Pharmaceutical Amorphous Solid Dispersions, First Edition. Edited by Ann Newman.

E

I

Printed and bound by CPI Group (UK) Ltd, Croydon, CR0 4YY
16/09/2026
14963427-0002